T0211069

# Lecture Notes in Computer Science    14556

Founding Editors

Gerhard Goos
Juris Hartmanis

## Editorial Board Members

Elisa Bertino, *Purdue University, West Lafayette, IN, USA*
Wen Gao, *Peking University, Beijing, China*
Bernhard Steffen **(D)**, *TU Dortmund University, Dortmund, Germany*
Moti Yung **(D)**, *Columbia University, New York, NY, USA*

The series Lecture Notes in Computer Science (LNCS), including its subseries Lecture Notes in Artificial Intelligence (LNAI) and Lecture Notes in Bioinformatics (LNBI), has established itself as a medium for the publication of new developments in computer science and information technology research, teaching, and education.

LNCS enjoys close cooperation with the computer science R & D community, the series counts many renowned academics among its volume editors and paper authors, and collaborates with prestigious societies. Its mission is to serve this international community by providing an invaluable service, mainly focused on the publication of conference and workshop proceedings and postproceedings. LNCS commenced publication in 1973.

Stevan Rudinac · Alan Hanjalic · Cynthia Liem ·
Marcel Worring · Björn Þór Jónsson · Bei Liu ·
Yoko Yamakata
Editors

# MultiMedia Modeling

30th International Conference, MMM 2024
Amsterdam, The Netherlands, January 29 – February 2, 2024
Proceedings, Part III

*Editors*
Stevan Rudinac (iD)
University of Amsterdam
Amsterdam, The Netherlands

Alan Hanjalic (iD)
Delft University of Technology
Delft, The Netherlands

Cynthia Liem (iD)
Delft University of Technology
Delft, The Netherlands

Marcel Worring (iD)
University of Amsterdam
Amsterdam, The Netherlands

Björn Þór Jónsson (iD)
Reykjavik University
Reykjavik, Iceland

Bei Liu (iD)
Microsoft Research Lab – Asia
Beijing, China

Yoko Yamakata (iD)
The University of Tokyo
Tokyo, Japan

ISSN 0302-9743          ISSN 1611-3349 (electronic)
Lecture Notes in Computer Science
ISBN 978-3-031-53310-5          ISBN 978-3-031-53311-2 (eBook)
https://doi.org/10.1007/978-3-031-53311-2

© The Editor(s) (if applicable) and The Author(s), under exclusive license
to Springer Nature Switzerland AG 2024

This work is subject to copyright. All rights are reserved by the Publisher, whether the whole or part of the material is concerned, specifically the rights of translation, reprinting, reuse of illustrations, recitation, broadcasting, reproduction on microfilms or in any other physical way, and transmission or information storage and retrieval, electronic adaptation, computer software, or by similar or dissimilar methodology now known or hereafter developed.
The use of general descriptive names, registered names, trademarks, service marks, etc. in this publication does not imply, even in the absence of a specific statement, that such names are exempt from the relevant protective laws and regulations and therefore free for general use.
The publisher, the authors, and the editors are safe to assume that the advice and information in this book are believed to be true and accurate at the date of publication. Neither the publisher nor the authors or the editors give a warranty, expressed or implied, with respect to the material contained herein or for any errors or omissions that may have been made. The publisher remains neutral with regard to jurisdictional claims in published maps and institutional affiliations.

This Springer imprint is published by the registered company Springer Nature Switzerland AG
The registered company address is: Gewerbestrasse 11, 6330 Cham, Switzerland

Paper in this product is recyclable.

# Preface

These four proceedings volumes contain the papers presented at MMM 2024, the International Conference on Multimedia Modeling. This 30th anniversary edition of the conference was held in Amsterdam, The Netherlands, from 29 January to 2 February 2024. The event showcased recent research developments in a broad spectrum of topics related to multimedia modelling, particularly: audio, image, video processing, coding and compression, multimodal analysis for retrieval applications, and multimedia fusion methods.

We received 297 regular, special session, Brave New Ideas, demonstration and Video Browser Showdown paper submissions. Out of 238 submitted regular papers, 27 were selected for oral and 86 for poster presentation through a double-blind review process in which, on average, each paper was judged by at least three program committee members and reviewers. In addition, the conference featured 23 special session papers, 2 Brave New Ideas and 8 demonstrations. The following four special sessions were part of the MMM 2024 program:

- FMM: Special Session on Foundation Models for Multimedia
- MDRE: Special Session on Multimedia Datasets for Repeatable Experimentation
- ICDAR: Special Session on Intelligent Cross-Data Analysis and Retrieval
- XR-MACCI: Special Session on eXtended Reality and Multimedia - Advancing Content Creation and Interaction

The program further included four inspiring keynote talks by Anna Vilanova from the Eindhoven University of Technology, Cees Snoek from the University of Amsterdam, Fleur Zeldenrust from the Radboud University and Ioannis Kompatsiaris from CERTH-ITI.

In addition, the annual MediaEval workshop was organised in conjunction with the conference. The attractive and high-quality program was completed by the Video Browser Showdown, an annual live video retrieval competition, in which 13 teams participated.

We would like to thank the members of the organizing committee, special session and VBS organisers, steering and technical program committee members, reviewers, keynote speakers and authors for making MMM 2024 a success.

December 2023

Stevan Rudinac
Alan Hanjalic
Cynthia Liem
Marcel Worring
Björn Þór Jónsson
Bei Liu
Yoko Yamakata

# Organization

## General Chairs

Stevan Rudinac      University of Amsterdam, The Netherlands
Alan Hanjalic      Delft University of Technology, The Netherlands
Cynthia Liem      Delft University of Technology, The Netherlands
Marcel Worring      University of Amsterdam, The Netherlands

## Technical Program Chairs

Björn Þór Jónsson      Reykjavik University, Iceland
Bei Liu      Microsoft Research, China
Yoko Yamakata      University of Tokyo, Japan

## Community Direction Chairs

Lucia Vadicamo      ISTI-CNR, Italy
Ichiro Ide      Nagoya University, Japan
Vasileios Mezaris      Information Technologies Institute, Greece

## Demo Chairs

Liting Zhou      Dublin City University, Ireland
Binh Nguyen      University of Science, Vietnam National University Ho Chi Minh City, Vietnam

## Web Chairs

Nanne van Noord      University of Amsterdam, The Netherlands
Yen-Chia Hsu      University of Amsterdam, The Netherlands

## Video Browser Showdown Organization Committee

| | |
|---|---|
| Klaus Schoeffmann | Klagenfurt University, Austria |
| Werner Bailer | Joanneum Research, Austria |
| Jakub Lokoc | Charles University in Prague, Czech Republic |
| Cathal Gurrin | Dublin City University, Ireland |
| Luca Rossetto | University of Zurich, Switzerland |

## MediaEval Liaison

| | |
|---|---|
| Martha Larson | Radboud University, The Netherlands |

## MMM Conference Liaison

| | |
|---|---|
| Cathal Gurrin | Dublin City University, Ireland |

## Local Arrangements

| | |
|---|---|
| Emily Gale | University of Amsterdam, The Netherlands |

## Steering Committee

| | |
|---|---|
| Phoebe Chen | La Trobe University, Australia |
| Tat-Seng Chua | National University of Singapore, Singapore |
| Kiyoharu Aizawa | University of Tokyo, Japan |
| Cathal Gurrin | Dublin City University, Ireland |
| Benoit Huet | Eurecom, France |
| Klaus Schoeffmann | Klagenfurt University, Austria |
| Richang Hong | Hefei University of Technology, China |
| Björn Þór Jónsson | Reykjavik University, Iceland |
| Guo-Jun Qi | University of Central Florida, USA |
| Wen-Huang Cheng | National Chiao Tung University, Taiwan |
| Peng Cui | Tsinghua University, China |
| Duc-Tien Dang-Nguyen | University of Bergen, Norway |

# Special Session Organizers

## FMM: Special Session on Foundation Models for Multimedia

| | |
|---|---|
| Xirong Li | Renmin University of China, China |
| Zhineng Chen | Fudan University, China |
| Xing Xu | University of Electronic Science and Technology of China, China |
| Symeon (Akis) Papadopoulos | Centre for Research and Technology Hellas, Greece |
| Jing Liu | Chinese Academy of Sciences, China |

## MDRE: Special Session on Multimedia Datasets for Repeatable Experimentation

| | |
|---|---|
| Klaus Schöffmann | Klagenfurt University, Austria |
| Björn Þór Jónsson | Reykjavik University, Iceland |
| Cathal Gurrin | Dublin City University, Ireland |
| Duc-Tien Dang-Nguyen | University of Bergen, Norway |
| Liting Zhou | Dublin City University, Ireland |

## ICDAR: Special Session on Intelligent Cross-Data Analysis and Retrieval

| | |
|---|---|
| Minh-Son Dao | National Institute of Information and Communications Technology, Japan |
| Michael Alexander Riegler | Simula Metropolitan Center for Digital Engineering, Norway |
| Duc-Tien Dang-Nguyen | University of Bergen, Norway |
| Binh Nguyen | University of Science, Vietnam National University Ho Chi Minh City, Vietnam |

## XR-MACCI: Special Session on eXtended Reality and Multimedia - Advancing Content Creation and Interaction

| | |
|---|---|
| Claudio Gennaro | Information Science and Technologies Institute, National Research Council, Italy |
| Sotiris Diplaris | Information Technologies Institute, Centre for Research and Technology Hellas, Greece |
| Stefanos Vrochidis | Information Technologies Institute, Centre for Research and Technology Hellas, Greece |
| Heiko Schuldt | University of Basel, Switzerland |
| Werner Bailer | Joanneum Research, Austria |

# Program Committee

| | |
|---|---|
| Alan Smeaton | Dublin City University, Ireland |
| Anh-Khoa Tran | National Institute of Information and Communications Technology, Japan |
| Chih-Wei Lin | Fujian Agriculture and Forestry University, China |
| Chutisant Kerdvibulvech | National Institute of Development Administration, Thailand |
| Cong-Thang Truong | Aizu University, Japan |
| Fan Zhang | Macau University of Science and Technology/Communication University of Zhejiang, China |
| Hilmil Pradana | Sepuluh Nopember Institute of Technology, Indonesia |
| Huy Quang Ung | KDDI Research, Inc., Japan |
| Jakub Lokoc | Charles University, Czech Republic |
| Jiyi Li | University of Yamanashi, Japan |
| Koichi Shinoda | Tokyo Institute of Technology, Japan |
| Konstantinos Ioannidis | Centre for Research & Technology Hellas/Information Technologies Institute, Greece |
| Kyoung-Sook Kim | National Institute of Advanced Industrial Science and Technology, Japan |
| Ladislav Peska | Charles University, Czech Republic |
| Li Yu | Huazhong University of Science and Technology, China |
| Linlin Shen | Shenzhen University, China |
| Luca Rossetto | University of Zurich, Switzerland |
| Maarten Michiel Sukel | University of Amsterdam, The Netherlands |
| Martin Winter | Joanneum Research, Austria |
| Naoko Nitta | Mukogawa Women's University, Japan |
| Naye Ji | Communication University of Zhejiang, China |
| Nhat-Minh Pham-Quang | Aimesoft JSC, Vietnam |
| Pierre-Etienne Martin | Max Planck Institute for Evolutionary Anthropology, Germany |
| Shaodong Li | Guangxi University, China |
| Sheng Li | National Institute of Information and Communications Technology, Japan |
| Stefanie Onsori-Wechtitsch | Joanneum Research, Austria |
| Takayuki Nakatsuka | National Institute of Advanced Industrial Science and Technology, Japan |
| Tao Peng | UT Southwestern Medical Center, USA |

| | |
|---|---|
| Thitirat Siriborvornratanakul | National Institute of Development Administration, Thailand |
| Vajira Thambawita | SimulaMet, Norway |
| Wei-Ta Chu | National Cheng Kung University, Taiwan |
| Wenbin Gan | National Institute of Information and Communications Technology, Japan |
| Xiangling Ding | Hunan University of Science and Technology, China |
| Xiao Luo | University of California, Los Angeles, USA |
| Xiaoshan Yang | Institute of Automation, Chinese Academy of Sciences, China |
| Xiaozhou Ye | AsiaInfo, China |
| Xu Wang | Shanghai Institute of Microsystem and Information Technology, China |
| Yasutomo Kawanishi | RIKEN, Japan |
| Yijia Zhang | Dalian Maritime University, China |
| Yuan Lin | Kristiania University College, Norway |
| Zhenhua Yu | Ningxia University, China |
| Weifeng Liu | China University of Petroleum, China |

## Additional Reviewers

| | |
|---|---|
| Alberto Valese | Dimitris Karageorgiou |
| Alexander Shvets | Dong Zhang |
| Ali Abdari | Duy Dong Le |
| Bei Liu | Evlampios Apostolidis |
| Ben Liang | Fahong Wang |
| Benno Weck | Fang Yang |
| Bo Wang | Fanran Sun |
| Bowen Wang | Fazhi He |
| Carlo Bretti | Feng Chen |
| Carlos Cancino-Chacón | Fengfa Li |
| Chen-Hsiu Huang | Florian Spiess |
| Chengjie Bai | Fuyang Yu |
| Chenlin Zhao | Gang Yang |
| Chenyang Lyu | Gopi Krishna Erabati |
| Chi-Yu Chen | Graham Healy |
| Chinmaya Laxmikant Kaundanya | Guangjie Yang |
| Christos Koutlis | Guangrui Liu |
| Chunyin Sheng | Guangyu Gao |
| Dennis Hoppe | Guanming Liu |
| Dexu Yao | Guohua Lv |
| Die Yu | Guowei Wang |

Gylfi Þór Guðmundsson
Hai Yang Zhang
Hannes Fassold
Hao Li
Hao-Yuan Ma
Haochen He
Haotian Wu
Haoyang Ma
Haozheng Zhang
Herng-Hua Chang
Honglei Zhang
Honglei Zheng
Hu Lu
Hua Chen
Hua Li Du
Huang Lipeng
Huanyu Mei
Huishan Yang
Ilias Koulalis
Ioannis Paraskevopoulos
Ioannis Sarridis
Javier Huertas-Tato
Jiacheng Zhang
Jiahuan Wang
Jianbo Xiong
Jiancheng Huang
Jiang Deng
Jiaqi Qiu
Jiashuang Zhou
Jiaxin Bai
Jiaxin Li
Jiayu Bao
Jie Lei
Jing Zhang
Jingjing Xie
Jixuan Hong
Jun Li
Jun Sang
Jun Wu
Jun-Cheng Chen
Juntao Huang
Junzhou Chen
Kai Wang
Kai-Uwe Barthel
Kang Yi

Kangkang Feng
Katashi Nagao
Kedi Qiu
Kha-Luan Pham
Khawla Ben Salah
Konstantin Schall
Konstantinos Apostolidis
Konstantinos Triaridis
Kun Zhang
Lantao Wang
Lei Wang
Li Yan
Liang Zhu
Ling Shengrong
Ling Xiao
Linyi Qian
Linzi Xing
Liting Zhou
Liu Junpeng
Liyun Xu
Loris Sauter
Lu Zhang
Luca Ciampi
Luca Rossetto
Luotao Zhang
Ly-Duyen Tran
Mario Taschwer
Marta Micheli
Masatoshi Hamanaka
Meiling Ning
Meng Jie Zhang
Meng Lin
Mengying Xu
Minh-Van Nguyen
Muyuan Liu
Naomi Ubina
Naushad Alam
Nicola Messina
Nima Yazdani
Omar Shahbaz Khan
Panagiotis Kasnesis
Pantid Chantangphol
Peide Zhu
Pingping Cai
Qian Cao

Qian Qiao
Qiang Chen
Qiulin Li
Qiuxian Li
Quoc-Huy Trinh
Rahel Arnold
Ralph Gasser
Ricardo Rios M. Do Carmo
Rim Afdhal
Ruichen Li
Ruilin Yao
Sahar Nasirihaghighi
Sanyi Zhang
Shahram Ghandeharizadeh
Shan Cao
Shaomin Xie
Shengbin Meng
Shengjia Zhang
Shihichi Ka
Shilong Yu
Shize Wang
Shuai Wang
Shuaiwei Wang
Shukal Liu
Shuo Wang
Shuxiang Song
Sizheng Guo
Song-Lu Chen
Songkang Dai
Songwei Pei
Stefanos Iordanis Papadopoulos
Stuart James
Su Chang Quan
Sze An Peter Tan
Takafumi Nakanishi
Tanya Koohpayeh Araghi
Tao Zhang
Theodor Clemens Wulff
Thu Nguyen
Tianxiang Zhao
Tianyou Chang
Tiaobo Ji
Ting Liu
Ting Peng
Tongwei Ma

Trung-Nghia Le
Ujjwal Sharma
Van-Tien Nguyen
Van-Tu Ninh
Vasilis Sitokonstantinou
Viet-Tham Huynh
Wang Sicheng
Wang Zhou
Wei Liu
Weilong Zhang
Wenjie Deng
Wenjie Wu
Wenjie Xing
Wenjun Gan
Wenlong Lu
Wenzhu Yang
Xi Xiao
Xiang Li
Xiangzheng Li
Xiaochen Yuan
Xiaohai Zhang
Xiaohui Liang
Xiaoming Mao
Xiaopei Hu
Xiaopeng Hu
Xiaoting Li
Xiaotong Bu
Xin Chen
Xin Dong
Xin Zhi
Xinyu Li
Xiran Zhang
Xitie Zhang
Xu Chen
Xuan-Nam Cao
Xueyang Qin
Xutong Cui
Xuyang Luo
Yan Gao
Yan Ke
Yanyan Jiao
Yao Zhang
Yaoqin Luo
Yehong Pan
Yi Jiang

Yi Rong

Yi Zhang

Yihang Zhou

Yinqing Cheng

Yinzhou Zhang

Yiru Zhang

Yizhi Luo

Yonghao Wan

Yongkang Ding

Yongliang Xu

Yosuke Tsuchiya

Youkai Wang

Yu Boan

Yuan Zhou

Yuanjian He

Yuanyuan Liu

Yuanyuan Xu

Yufeng Chen

Yuhang Yang

Yulong Wang

Yunzhou Jiang

Yuqi Li

Yuxuan Zhang

Zebin Li

Zhangziyi Zhou

Zhanjie Jin

Zhao Liu

Zhe Kong

Zhen Wang

Zheng Zhong

Zhengye Shen

Zhenlei Cui

Zhibin Zhang

Zhongjie Hu

Zhongliang Wang

Zijian Lin

Zimi Lv

Zituo Li

Zixuan Hong

# Contents – Part III

# Global-to-Local Feature Mining Network for RGB-Infrared Person Re-Identification

Qiang Chen[iD], Fuxiao He[iD], and Guoqiang Xiao[✉][iD]

College of Computer and Information Science, Southwest University, Chongqing,
China
{cq0907,kylekyle}@email.swu.edu.cn, gqxiao@swu.edu.cn

**Abstract.** RGB-Infrared person Re-Identification (RGB-IR ReID) is
a challenging matching task that retrieves a RGB/infrared pedestrian
image from the existing infrared/RGB set captured by non-overlapping
visible or infrared cameras. Existing works mainly focus on how to alleviate the intra-modality variations and inter-modality discrepancies by
data augmentation or feature alignment. Although these methods enlarge
the diversity of the training set and, to some extent, reduce the gap
between modalities, insufficient mining of discriminative and invariant
features between modalities limits the performance of RGB-IR ReID
algorithms. To remedy this, we propose a global-to-local feature mining network (**GFMNet**) to further mine discriminative and invariant
features. Specifically, GFMNet contains two feature mining modules:
Attention-aware Feature Mining Module (**AFMM**) and Local Information Mining Module (**LIMM**). AFMM aims to learn global discriminative features by attention mechanism; LIMM mines potential local invariant features between modalities by shortest path exploration. Besides,
to reduce modality discrepancies and define a unified convergence direction, we introduce distribution consistency (**DC**) loss, which encourages
RGB and infrared modalities toward intermediate modality. Extensive
experiments on the SYSU-MM01 and RegDB datasets show that GFM-
Net achieves competitive RGB-IR ReID performance. The code will be
announced at https://github.com/cq0907/GFMNet.

**Keywords:** RGB-Infrared person Re-Identification · feature mining ·
attention mechanism · shortest path · distribution consistency

## 1 Introduction

Person re-identification (ReID) aims to query the specific pedestrian from the
known gallery set captured visible cameras [15,18]. However, visible cameras will
not work well under low-light environments, e.g., at night. In these cases, infrared
cameras are utilized to capture the pedestrian due to they rarely depend on

Supported by organization Southwest University.

© The Author(s), under exclusive license to Springer Nature Switzerland AG 2024
S. Rudinac et al. (Eds.): MMM 2024, LNCS 14556, pp. 1–13, 2024.
https://doi.org/10.1007/978-3-031-53311-2_1

visible light. Accordingly, RGB-infrared person re-identification (RGB-IR ReID), that is, given a RGB/infrared image matches the infrared/RGB images with same identity, has aroused extensive research.

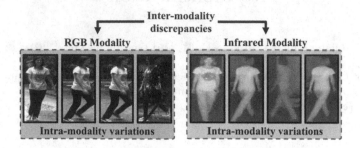

**Fig. 1.** Examples of the intra-modality variations and inter-modality discrepancies.

Two main challenges in RGB-IR ReID are intra-modality variations and inter-modality discrepancies, as shown in Fig. 1. The former is caused by different person poses, views, and occlusions in single modality. The latter stems from the data heterogeneity of different spectrum cameras. Therefore, how to overcome the above challenges becomes a thorny problem. Firstly, the images captured by non-overlapping cameras are not aligned, which distracts the learning of discriminative features. Thus, some methods based on feature alignment [12,21] are proposed to exploit correspondences between cross-modality images. In addition, to narrow the gap of data heterogeneity, some data augmentation methods [9,24,28] are introduced to generate the intermediate modality for alleviating the color discrepancy. Unfortunately, the common limitation of these methods is that they fail to fully exploit discriminative and invariant features between modalities.

To alleviate the limitation, We propose a novel global-to-local feature mining network (GFMNet) to further mine discriminative and invariant features. As shown in Fig. 3, similar to AGW [26], we adopt ResNet-50 [5] with non-local operation [16] as the backbone. The convolution block-2 and block-5 of ResNet-50 are respectively followed by the two core modules: Attention-aware Feature Mining Module (AFMM) and Local Information Mining Module (LIMM). The former is to mine global discriminative features by attention mechanism, which provides a foundation for later mining local invariant features. The latter is to further explore local invariant features between modalities by shortest path exploration. Although the ideas of shortest path exploration of local information have appeared in RGB-RGB ReID [10], we introduce it into RGB-IR ReID to mine potential invariant features, and achieve competitive performance. Additionally, motivated by the "*shortest geodesic path*" definition [4], i.e., the intermediate domains along the shortest geodesic path between the two extreme domains can play a better bridging role, we generate intermediate modality and introduce distribution consistency (DC) loss to encourage RGB and IR modalities toward intermediate modality.

Overall, our main contributions can be summarized in three-fold:

(1) We propose a global-to-local feature mining network (GFMNet) for RGB-IR ReID, in which attention-aware feature mining module (AFMM) is designed to mine global discriminative features and local information mining module (LIMM) is introduced to mine local invariant features between modalities.
(2) To narrow the gap between modalities, distribution consistency (DC) loss is introduced into the GFMNet and defines a unified convergence direction to boost our algorithm performance.
(3) On the public SYSU-MM01 and RegDB datasets, our approach achieves better performance compared to state-of-the-art methods.

## 2    Related Work

### 2.1    RGB-Infrared Person Re-Identification

RGB-infrared person re-identification (RGB-IR ReID) has aroused extensive research in recent years due to its importance in video surveillance and intelligent applications. To address the problem of modality discrepancies caused by different spectrum cameras, many typical RGB-IR ReID methods have been proposed. Wu et al. [20] first defined the task and proposed a single-stream network along with the deep zero-padding strategy to learn the shared features of different modalities. Subsequently, to better feature extraction and fusion, Ye et al. [23] changed the single-stream framework into a two-stream framework and achieved the improvement of performance. Later on, [27] further developed this framework from the viewpoint of metric learning and proposed a dual-constrained top-ranking loss for learning discriminative feature representations. Besides, Dai et al. [3] used generative adversarial networks (GANs) to learn modality-invariant representations. Recently, some works took into account part-level informations. Liu et al. [8] extracted the local features by dividing the global feature map into multiple horizontal parts. However, all of the above methods mainly focus on extracting and learning global or local features and fail to fully exploit discriminative and invariant features between modalities. Therefore, it motivates us to construct an algorithm that fully explores the discriminative and invariant features between modalities by adopting a global-to-local strategy, as shown in Fig. 2(a).

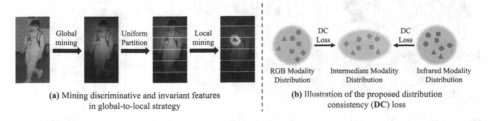

(a) Mining discriminative and invariant features in global-to-local strategy

(b) Illustration of the proposed distribution consistency (DC) loss

**Fig. 2.** The motivation of proposing GFMNet.

## 2.2   Intermediate Modality Learning

To narrow the gap between modalities, current some works focused on generating intermediate modality to assist other modalities learning. Ye et al. [24] proposed a channel augmentation method to transform visible images into grayscale images which have smaller gap with infrared images. Similarly, Liu et al. [9] reduced the modality discrepancies by employing infrared images to fully replace RGB images for feature learning by channel conversion. Although these methods generate intermediate modality by data augmentation and then reduce the discrepancies between modalities to some extent, they do not impose a unified constraint to encourage RGB and IR modalities toward intermediate modality. Thus, it inspires us to define a distribution consistency loss function to constrain the convergence direction of the three modalities, as illustrated in Fig. 2(b).

## 3   Proposed Method

In this section, we first describe the overview of the proposed network (Sect. 3.1), and then make a description of attention-aware feature mining module (AFMM, Sect. 3.2) and local information mining module (LIMM, Sect. 3.3) in detail. Finally, the objective function (Sect. 3.4) will be introduced.

### 3.1   Overview

As shown in Fig. 3, Firstly, we use the stage-1 and stage-2 of ResNet-50 to extract the modality-specific features of RGB modality, intermediate (IM) modality and infrared (IR) modality, in which the weights are not shared. Then the AFMM is used to learn global important features along the channel and spatial dimensions. Next, to better feature fusion, the features augmented by AFMM are fed into the weight-shared stages embedded non-local operation to extract the modality-shared features. Finally, two branches, global branch and local branch (LIMM), are designed to complementarily mine the invariant features. Note that, we generate IM modality by channel augmentation [24].

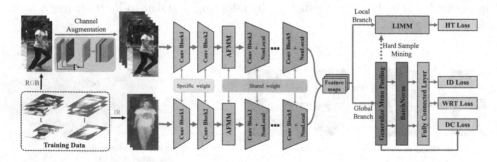

**Fig. 3.** The pipeline of the proposed global-to-local feature mining network (GFMNet) for RGB-IR ReID.

## 3.2  Attention-Aware Feature Mining Module

AFMM, shown in Fig. 4, is located behind the stage-2 of the ResNet-50 to reweight and fuse global features. We denote the input feature maps by $Z \in \mathbb{R}^{c \times h \times w}$, where $c$, $h$ and $w$ represent height, width, and the number of channels of the feature maps, respectively. Following CBAM [19], we first mine important channel informations $m_c \in \mathbb{R}^{c \times 1 \times 1}$ and spatial informations $m_s \in \mathbb{R}^{1 \times h \times w}$, they are beneficial to emphasize the distinctive object features. Both can be computed by:

$$m_c = \sigma(W_2\delta(W_1 f_{avg}(Z)) + W_2\delta(W_1 f_{max}(Z))), \tag{1}$$

$$m_s = \sigma(W_3(f_{avg}(Z) \otimes f_{max}(Z))), \tag{2}$$

where $f_{avg}(\cdot)$ and $f_{max}(\cdot)$ denote the average- and max-pooling operations. $W_1$, $W_2$ and $W_3$ are learnable parameters which are represented through the fully connected layer in our experiment. $\delta(\cdot)$ and $\sigma(\cdot)$ are ReLU and sigmoid activation functions. $\otimes$ represents elements concatenated together. Afterwards, the $m_c$ and $m_s$ are applied on $Z$ to emphasize the learned important features, respectively. To further augment features, the emphasized features are fused and fed into the non-local operation to enhance the relationship of each position of important features and then generate the attention-aware features $\hat{Z} \in \mathbb{R}^{c \times h \times w}$. In summary, the overall computation process can be summarized as:

$$\hat{Z} = h((m_c \odot Z) \oplus (m_s \odot Z)), \tag{3}$$

$$h(x_i) = \frac{1}{\mathcal{C}(x)} \sum_{\forall j} f(x_i, x_j) g(x_j), \tag{4}$$

where $\oplus$ and $\odot$ are the element-wise addition and multiplication. $h(\cdot)$ is the non-local operation for reweighting each position feature. For the input feature map $x$, here $x_i \in \mathbb{R}^{c \times 1 \times 1}$ and $x_j \in \mathbb{R}^{c \times 1 \times 1}$ represent the feature vector at the $i^{th}$ and $j^{th}$ position. $f(\cdot, \cdot)$ computes the cosine similarity of feature vectors at the $i^{th}$ position and $j^{th}$ position. $g(\cdot)$ extracts the representation from $j^{th}$ position by a convolution operation. $\mathcal{C}(\cdot)$ returns a normalized factor. Note that, different from CBAM, we found in our experiment that parallel structure gives a better, and AFMM enhances the relationships between discriminative features from a global perspective through the non-local operation.

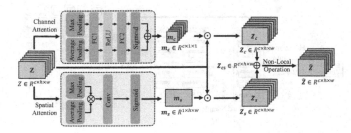

**Fig. 4.** The structure of **AFMM**. $\oplus$ and $\odot$ denote the element-wise addition and multiplication, respectively. $\otimes$ represents elements concatenated together.

### 3.3 Local Information Mining Module

LIMM, illustrated in Fig. 5, is used as a local branch to mine the local potential invariant features between modalities. Inspired by [10], we divide the feature maps horizontally into $N$ parts by horizontal convolution. Assume that backbone have extracted feature maps (anchor) of size $c_1 \times h_1 \times w_1$. Through horizontal convolution, we can get local features with a height of $\frac{h_1}{N}$, denoted by $F \in \mathbb{R}^{c_1 \times \frac{h_1}{N} \times 1}$, which is reshaped into the size of $c1 \times \frac{h_1}{N}$ later. Meanwhile, we mine the hard-positive sample (HPS) and hard-negative sample (HNS) of anchor by feature similarity calculation, denoted by $F_{HPS} \in \mathbb{R}^{c_1 \times h_1 \times w_1}$ and $F_{HNS} \in \mathbb{R}^{c_1 \times h_1 \times w_1}$, both perform the same partition operation. Note that, if HPS or HNS is mined from the same modality as anchor, LIMM is advantageous for feature alignment to overcome intra-modality variations. Conversely, if HPS or HNS is mined from a different modality than the anchor, LIMM contributes to exploring invariant features between different modalities and reducing inter-modality discrepancies. Accordingly, we compute similarity of anchor $F$ and $F_{HPS}/F_{HNS}$ as follows:

$$\mathcal{M}(F, F_H) = \|F - F_H\|_2^2, \tag{5}$$

where $\| \cdot \|_2^2$ computes the L2 distance. $F_H$ represents $F_{HPS}$ or $F_{HNS}$. Note that, the per element of matrix $\mathcal{M}$ represents the similarity between the $i^{th}$ part of anchor and the $j^{th}$ part of HPS/HNS.

Next, we mine the invariant features of intra- and inter-modality by exploring the shortest path from $(1, 1)$ to $(N, N)$ in the $\mathcal{M}$, which can be calculated by:

$$S_{i,j} = \begin{cases} e_{i,j} & i = 1, j = 1 \\ S_{i-t,j} + e_{i,j} & i \neq 1, j = 1 \\ S_{i,j-t} + e_{i,j} & i = 1, j \neq 1 \\ min(S_{i-t,j}, S_{i,j-t}) + e_{i,j} & i \neq 1, j \neq 1 \end{cases} \tag{6}$$

where $e_{i,j}$ is the element of $\mathcal{M}$ on row $i$ and column $j$. $S_{i,j}$ is the total distance of the shortest path when walking from $(1, 1)$ to $(i, j)$ in the $\mathcal{M}$. $t$ represents the number of steps which is greater than or equal to 1. $S_{N,N}$ is the shortest distance between anchor and HPS/HNS, which is calculated by dividing the total path by the number of steps

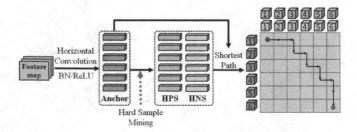

**Fig. 5.** The structure of **LIMM**. HPS and HNS denote hard-positive and hard-negative samples, respectively. Black arrows represents the shortest path

## 3.4 Objective Function

In order to guide the network to extract discriminative features, we optimize local branch with hard-mining triplet loss ($\mathcal{L}_{HT}$) and global branch with identity loss ($\mathcal{L}_{ID}$), weighted regularization triplet loss ($\mathcal{L}_{WRT}$) and distribution consistencies loss ($\mathcal{L}_{DC}$).

**HT Loss** ($\mathcal{L}_{HT}$). Hard-mining triplet (HT) loss is used for pulling the distance of anchor and hard-positive sample (HPS) and, whilst, pushing the distance of anchor and hard-negative sample (HNS), which is defined as:

$$\mathcal{L}_{HT} = max(S_{N,N}^{HPS} - S_{N,N}^{HNS} + m, 0), \tag{7}$$

where $S_{N,N}^{HPS}$ and $S_{N,N}^{HNS}$ are the final shortest paths between the anchor and HPS, and the anchor and HNS, respectively. $m$ is the margin parameter. The purpose of the HT loss is to minimize the value of $S_{N,N}^{HPS}$, that is, to mine as many invariant features as possible between the anchor and HPS to reduce the intra-class feature distance. Meanwhile, increasing the inter-class feature distance by maximize the value of $S_{N,N}^{HNS}$.

**ID Loss** ($\mathcal{L}_{ID}$). Given an arbitrary modality image $x$ with corresponding identity label $y$, ID loss aims to increase the probability that $x$ belongs to class $y$. After obtaining the prediction logits of $x$ through global branch, a *softmax* function is used to calculate the probability $p(y|x)$, formulated as:

$$p(y|x) = \frac{exp(P_x)}{\sum_{k=1}^{C} exp(F_x)}, \tag{8}$$

where $P_x$ represents the prediction logits of $x$. $C$ is the total number of identities. With the calculated probability scores, we use cross-entropy loss to optimize the prediction logits, denoted by:

$$\mathcal{L}_{ID} = -\frac{1}{n} \sum_{j=1}^{n} log(p(y|x)), \tag{9}$$

where $n$ refers to the number of samples in each mini-batch during training.

**WRT Loss** ($\mathcal{L}_{WRT}$). Following [6], we adopt weighted regularization triplet (WRT) loss to weight and optimize the feature distances between each sample and its positive and negative samples, formulated as:

$$\mathcal{L}_{WRT} = \frac{1}{n} \sum_{u=1}^{n} log(1 + exp(\sum_{v=1}^{n} w_{u,v}^{pos} d_{u,v}^{pos} - w_{u,w}^{neg} d_{u,w}^{neg})),$$
$$w_{u,v}^{pos} = \frac{exp(d_{u,v}^{pos})}{\sum_{d_{u,v}^{pos} \in \mathcal{K}_u} exp(d_{u,v}^{pos})}, \quad w_{u,v}^{neg} = \frac{exp(-d_{u,w}^{neg})}{\sum_{d_{u,w}^{neg} \in \mathcal{N}_u} exp(-d_{u,w}^{neg})}, \tag{10}$$

where $(u, v, w)$ represent a triplet. For the anchor $c_u$. $d_{u,v}^{pos}$ and $d_{u,w}^{neg}$ are the L2 distance between anchor and positive sample, and anchor and negative sample, denoted by $d_{u,v}^{pos} = \|c_u - c_v\|_2^2$ and $d_{u,w}^{neg} = \|c_u - c_w\|_2^2$. $\mathcal{P}_u$ and $\mathcal{N}_u$ denote the positive and negative sample sets for the anchor $c_u$.

**DC Loss** ($\mathcal{L}_{DC}$). Due to the data heterogeneity of different modalities, we know that it is challenging to directly fit the RGB modality to the IR modality or vice versa. Therefore, many current methods use an intermediate (IM) modality to reduce the discrepancies between modalities. However, they do not define a unified convergence direction, which may affect performance to some extent. Inspired by [4], we propose the distribution consistency (DC) loss, which aims to encourage the feature vector distribution of RGB and IR modalities toward the distribution of IM modality. The mathematical form is denoted by:

$$P_1 = \frac{P^{IR} + P^{IM}}{2}, P_2 = \frac{P^{RGB} + P^{IM}}{2}, \tag{11}$$

$$\mathcal{L}_{DC} = \frac{1}{2}(\sum_{i=1}^{L} P_i^{IR} \log \frac{P_i^{IR}}{P_1} + \sum_{i=1}^{L} P_i^{IM} \log \frac{P_i^{IM}}{P_1}) +$$
$$\frac{1}{2}(\sum_{i=1}^{L} P_i^{RGB} \log \frac{P_i^{RGB}}{P_2} + \sum_{i=1}^{L} P_i^{IM} \log \frac{P_i^{IM}}{P_2}) \tag{12}$$

where $P^{RGB}$, $P^{IM}$ and $P^{IR}$ denote the feature vector distribution of RGB modality, IM modality and IR modality. $L$ is the length of feature vector. By optimizing the DC loss, the feature vector distributions of different modalities with same identity will be as close as possible, which is beneficial for reducing the discrepancies between modalities.

---

**Algorithm 1:** The algorithm flow of the GFMNet.

---

**Input**: RGB modality sample $x_v$ and IR modality sample $x_i$ with their labels;
**Output**: The trained network;
**Initialization**: Initializing the network by ResNet-50, optimizer SGD and other hyper-parameters.
**for** i = 1 to epoch **do**
  - Generating intermediate modality sample $x_m$ based on $x_v$ through channel augmentation.
  - Using stage-1 and stage-2 of ResNet-50 to extract modality-specific features $Z$;
  - Mining global discriminative features $\hat{Z}$ by AFMM and then feeding forward the batch into remaining stages of ResNet-50 to obtain fused feature maps;
  - Feeding forward the feature maps into global branch and local branch (LIMM);
  - Calculating the overall loss by Eq.13 and Updating the network by SGD.
**end**
**Until** model convergence or the fixed epoch;

---

Therefore, combining HT loss, ID loss, WRT loss and DC loss, we design the following overall objective function:

$$\mathcal{L} = \mathcal{L}_{ID} + \mathcal{L}_{WRT} + \mathcal{L}_{DC} + \lambda_{HT}\mathcal{L}_{HT}, \tag{13}$$

where $\lambda_{HT}$ is hyper-parameter which balances the contribution of local branch.

# 4  Experiments

## 4.1  Datasets and Settings

**Dataset.** There are two available benchmark datasets for evaluating the performance of our approach: SYSU-MM01 [20] and RegDB [11]. The former is captured by six cameras. 22258 RGB and 11909 infrared images with 395 identities are used for training and 301 RGB and 3803 infrared images with 96 identities are utilized for testing. The latter is collected by dual camera systems, it contains in total 412 identities, half for training and half for testing. In the end, Rank-1, mean Average Precision (mAP), and mean Inverse Negative Penalty (mINP) are adopted as the evaluation protocol.

**Settings.** Following the classic methods of RGB-IR ReID, each person image is resized to $288 \times 144$. The total epoch is 100 and we use SGD optimizer with momentum 0.9 and weight decay 5e-4 to optimize our model. The initial learning rate is 0.1, which will decay by a factor of 10 at the 20-th and 50-th epoch. The hyper-parameters in the formula are set as: $n = 8$, $N = 6$, $\lambda_{HT} = 0.1$ and $m = 0.3$. We implement the proposed method with PyTorch and use single Geforce RTX 3090 GPU for acceleration.

**Table 1.** Comparison with state-of-the-art methods on SYSU-MM01 and RegDB datasets. Rank-1 accuracy (%), mAP (%) and mINP (%) are reported. The hyphen (-) indicates that the method has not computed the metric.

| Method | Venue | SYSU-MM01 | | | | | | RegDB | | | | | |
|---|---|---|---|---|---|---|---|---|---|---|---|---|---|
| | | All Search | | | Indoor Search | | | Visible to Infrared | | | Infrared to Visible | | |
| | | Rank-1 | mAP | mINP | Rank-1 | mAP | mINP | Rank-1 | mAP | mINP | Rank-1 | mAP | mINP |
| D$^2$RL | CVPR(2019) | 28.9 | 29.2 | – | – | – | – | 43.4 | 66.1 | – | – | – | – |
| AlignGAN | ICCV(2019) | 42.4 | 40.7 | – | 45.9 | 54.3 | – | 57.9 | 53.6 | – | 56.3 | 53.4 | – |
| Xmodal | AAAI(2020) | 49.92 | 50.73 | – | – | – | – | 62.21 | 60.18 | – | – | – | – |
| DDAG | ECCV(2020) | 54.75 | 53.02 | 39.62 | 61.02 | 67.98 | 62.61 | 69.34 | 63.46 | 49.24 | 68.06 | 61.8 | 48.62 |
| Hi-CMD | CVPR(2020) | 34.94 | 35.94 | – | – | – | – | 70.93 | 66.04 | – | – | – | – |
| AGW | TPAMI(2021) | 47.5 | 47.65 | 35.3 | 54.17 | 62.97 | 59.23 | 70.05 | 66.37 | 50.19 | 70.04 | 65.9 | 51.24 |
| LbA | ICCV(2021) | 55.41 | 54.14 | – | 58.46 | 66.33 | – | 74.17 | 67.64 | – | 72.43 | 65.46 | – |
| CAJ | ICCV(2021) | 69.88 | 66.89 | 53.61 | 76.26 | 80.37 | 76.79 | 85.03 | 79.14 | 65.33 | 84.75 | 77.82 | 61.56 |
| SPOT | TIP(2022) | 65.34 | 62.25 | 48.86 | 69.42 | 74.63 | 70.48 | 80.35 | 72.46 | 56.19 | 79.37 | 72.26 | 56.06 |
| DART | CVPR(2022) | 68.72 | 66.29 | 53.26 | 72.52 | 78.17 | 74.94 | 83.60 | 75.67 | 60.60 | 81.97 | 73.78 | 56.70 |
| GFMNet(Ours) | – | **73.04** | **69.71** | **56.62** | **79.12** | **82.70** | **79.22** | **87.14** | **80.65** | **66.83** | **85.55** | **78.96** | **62.92** |

## 4.2  Comparison with State-of-the-Art Methods

We compare our GFMNet with the state-of-the-art methods for RGB-IR ReID, including D$^2$RL [17], AlignGAN [14], Xmodal [7], DDAG [25], Hi-CMD [2], AGW [26], LbA [12], CAJ [24], SPOT [1] and DART [22].

As shown in Table 1, we conduct a quantitative analysis to compare our method with the state-of-the-art methods for RGB-IR ReID. In the all-search and indoor-search mode of SYSU-MM01 dataset, GFMNet achieves the improvement of 3.16%/2.86% Rank-1, 2.82%/2.33% mAP and 3.36%/2.43% mINP over CAJ, respectively. Meanwhile, compared with CAJ in *visible2infrared* and *infrared2visible* mode of RegDB dataset, the Rank-1, mAP and mINP are improved by 2.14%/0.80%, 1.51%/1.14% and 1.50%/1.36%. In short, quantitative comparison shows that the GFMNet can effectively mine discriminative and invariant features and outperform the state-of-the art methods. Note that, we don't use the re-ranking algorithms as post-processing.

**Table 2.** Ablation study on SYSU-MM01 dataset.

| Baseline | AFMM | LIMM | $\mathcal{L}_{DC}$ | Rank-1 | mAP | mINP |
|---|---|---|---|---|---|---|
| √ | | | | 69.88 | 66.89 | 53.61 |
| √ | √ | | | 71.72 | 67.07 | 54.33 |
| √ | √ | √ | | 72.03 | 67.81 | 55.12 |
| √ | √ | √ | √ | **73.04** | **69.71** | **56.62** |

## 4.3 Ablation Study and Visualization

To verify the effectiveness of each component in our method, we conduct ablation studies on SYSU-MM01 dataset. Here we adopt CAJ as the baseline. From the results in Table 2, we can see that three components (AFMM, LIMM and $\mathcal{L}_{DC}$) continuously boost the performance of baseline. We also visualize the class activation mapping (CAM) [13] on two datasets. As illustrated in Fig. 6, the highlighted areas show that our model focuses on discriminative features, such as face, clothing and shoes logos, package, etc., which are distinctive characters of one identity.

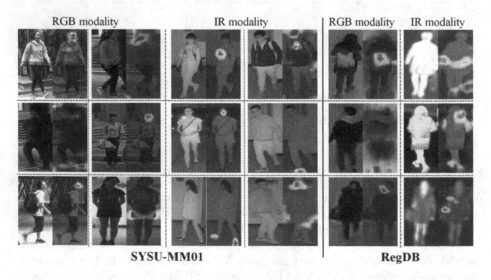

**Fig. 6.** The Class Activation Mapping visualization results.

# 5    Conclusion

We have presented a global-to-local feature mining network (GFMNet) for RGB-IR ReID, which is composed of two core modules: the attention-aware feature mining module (AFMM) and local information mining Module (LIMM). AFMM mines discriminative features from the level of global features, which encourages the network to focus on discriminative features and provides a foundation for later mining local invariant features. From the perspective of local features, LIMM mines the potential invariant features between modalities through shortest path exploration. In addition, to reduce modality discrepancies, we propose distribution consistency (DC) loss to build the bridge between RGB modality and IR modality through intermediate modality learning. Moreover, compared with the state-of-the-art methods, the experimental results on SYSU-MM01 and RegDB dataset have demonstrated the effectiveness of each component. We hope the proposed GFMNet can be a useful baseline for future researches on global and local feature mining for RGB-IR ReID.

# References

1. Chen, C., Ye, M., Qi, M., Wu, J., Jiang, J., Lin, C.W.: Structure-aware positional transformer for visible-infrared person re-identification. IEEE Trans. Image Process. **31**, 2352–2364 (2022)
2. Choi, S., Lee, S., Kim, Y., Kim, T., Kim, C.H.C.: Hierarchical cross-modality disentanglement for visible-infrared person re-identification. In: Proceedings of the IEEE Conference on Computer Vision and Pattern Recognition, Seattle, WA, USA, pp. 13–19 (2020)
3. Dai, P., Ji, R., Wang, H., Wu, Q., Huang, Y.: Cross-modality person re-identification with generative adversarial training. In: IJCAI, vol. 1, p. 6 (2018)
4. Dai, Y., Liu, J., Sun, Y., Tong, Z., Zhang, C., Duan, L.Y.: IDM: an intermediate domain module for domain adaptive person re-id. In: Proceedings of the IEEE/CVF International Conference on Computer Vision, pp. 11864–11874 (2021)
5. He, K., Zhang, X., Ren, S., Sun, J.: Deep residual learning for image recognition. In: Proceedings of the IEEE Conference on Computer Vision and Pattern Recognition, pp. 770–778 (2016)
6. Hermans, A., Beyer, L., Leibe, B.: In defense of the triplet loss for person re-identification. arXiv preprint arXiv:1703.07737 (2017)
7. Li, D., Wei, X., Hong, X., Gong, Y.: Infrared-visible cross-modal person re-identification with an x modality. In: Proceedings of the AAAI Conference on Artificial Intelligence, vol. 34, pp. 4610–4617 (2020)
8. Liu, H., Tan, X., Zhou, X.: Parameter sharing exploration and hetero-center triplet loss for visible-thermal person re-identification. IEEE Trans. Multimedia **23**, 4414–4425 (2020)
9. Liu, H., Ma, S., Xia, D., Li, S.: SFANET: a spectrum-aware feature augmentation network for visible-infrared person reidentification. IEEE Trans. Neural Netw. Learn. Syst. **34**, 1958–1971 (2021)
10. Luo, H., Jiang, W., Zhang, X., Fan, X., Qian, J., Zhang, C.: Alignedreid++: dynamically matching local information for person re-identification. Pattern Recogn. **94**, 53–61 (2019)

11. Nguyen, D.T., Hong, H.G., Kim, K.W., Park, K.R.: Person recognition system based on a combination of body images from visible light and thermal cameras. Sensors **17**, 605 (2017)
12. Park, H., Lee, S., Lee, J., Ham, B.: Learning by aligning: visible-infrared person re-identification using cross-modal correspondences. In: Proceedings of the IEEE/CVF International Conference on Computer Vision, pp. 12046–12055 (2021)
13. Selvaraju, R.R., Cogswell, M., Das, A., Vedantam, R., Parikh, D., Batra, D.: Grad-cam: visual explanations from deep networks via gradient-based localization. In: Proceedings of the IEEE International Conference on Computer Vision, pp. 618–626 (2017)
14. Wang, G., Zhang, T., Cheng, J., Liu, S., Yang, Y., Hou, Z.: Rgb-infrared cross-modality person re-identification via joint pixel and feature alignment. In: Proceedings of the IEEE/CVF International Conference on Computer Vision, pp. 3623–3632 (2019)
15. Wang, W., Shen, J., Lu, X., Hoi, S.C., Ling, H.: Paying attention to video object pattern understanding. IEEE Trans. Pattern Anal. Mach. Intell. **43**(7), 2413–2428 (2020)
16. Wang, X., Girshick, R., Gupta, A., He, K.: Non-local neural networks. In: Proceedings of the IEEE Conference on Computer Vision and Pattern Recognition, pp. 7794–7803 (2018)
17. Wang, Z., Wang, Z., Zheng, Y., Chuang, Y.Y., Satoh, S.: Learning to reduce dual-level discrepancy for infrared-visible person re-identification. In: Proceedings of the IEEE/CVF Conference on Computer Vision and Pattern Recognition, pp. 618–626 (2019)
18. Watson, G., Bhalerao, A.: Person re-identification combining deep features and attribute detection. Multimedia Tools Appl. **79**(9–10), 6463–6481 (2020)
19. Woo, S., Park, J., Lee, J.Y., Kweon, I.S.: CBAM: convolutional block attention module. In: ECCV, pp. 3–19 (2018)
20. Wu, A., Zheng, W.S., Yu, H.X., Gong, S., Lai, J.: Rgb-infrared cross-modality person re-identification. In: Proceedings of the IEEE International Conference on Computer Vision, pp. 5380–5389 (2017)
21. Wu, Q., et al.: Discover cross-modality nuances for visible-infrared person re-identification. In: Proceedings of the IEEE/CVF Conference on Computer Vision and Pattern Recognition, pp. 4330–4339 (2021)
22. Yang, M., Huang, Z., Hu, P., Li, T., Lv, J., Peng, X.: Learning with twin noisy labels for visible-infrared person re-identification. In: Proceedings of the IEEE/CVF Conference on Computer Vision and Pattern Recognition, pp. 14308–14317 (2022)
23. Ye, M., Lan, X., Li, J., Yuen, P.: Hierarchical discriminative learning for visible thermal person re-identification. In: AAAI, vol. 32 (2018)
24. Ye, M., Ruan, W., Du, B., Shou, M.Z.: Channel augmented joint learning for visible-infrared recognition. In: Proceedings of the IEEE/CVF International Conference on Computer Vision, pp. 13567–13576 (2021)
25. Ye, M., Shen, J., J. Crandall, D., Shao, L., Luo, J.: Dynamic dual-attentive aggregation learning for visible-infrared person re-identification. In: Vedaldi, A., Bischof, H., Brox, T., Frahm, J.-M. (eds.) ECCV 2020. LNCS, vol. 12362, pp. 229–247. Springer, Cham (2020). https://doi.org/10.1007/978-3-030-58520-4_14
26. Ye, M., Shen, J., Lin, G., Xiang, T., Shao, L., Hoi, S.C.: Deep learning for person re-identification: a survey and outlook. IEEE Trans. Pattern Anal. Mach. Intell. **44**(6), 2872–2893 (2021)

27. Ye, M., Wang, Z., Lan, X., Yuen, P.C.: Visible thermal person re-identification via dual-constrained top-ranking. In: IJCAI, vol. 1, p. 2 (2018)
28. Zhong, Z., Zheng, L., Kang, G., Li, S., Yang, Y.: Random erasing data augmentation. In: Proceedings of the AAAI Conference on Artificial Intelligence, vol. 34, pp. 13001–13008 (2020)

# Semantic Transition Detection for Self-supervised Video Scene Segmentation

Lu Chen[1,2], Jiawei Tan[1,2], Pingan Yang[1,2], and Hongxing Wang[1,2](✉)

[1] Key Laboratory of Dependable Service Computing in Cyber Physical Society
(Chongqing University), Ministry of Education, Chongqing, China
{alchen,jwtan,pnyang,ihxwang}@cqu.edu.cn
[2] School of Big Data and Software Engineering, Chongqing University,
Chongqing, China

**Abstract.** Video scene segmentation is a crucial task in temporally parsing long-form videos into basic story units. Most advanced self-supervised methods of video scene segmentation focus heavily on learning video shot features in the pre-training stage. However, these methods ignore to encode shot relations, which are essential to video scene segmentation, resulting in over-segmentation of video scenes. A straightforward solution to the above problem is to use sufficient scene boundaries to model the shot relations. In this paper, we introduce a high-quality pseudo-scene boundary generation method, Semantic Transition Detection (STD), by discovering semantic inconsistencies in temporal video chunks. Taking scene boundary prediction as the pretext task, we propose a self-supervised method for video scene segmentation, by which we can fine-tune a STD pre-trained model with limited scene boundary ground truths. In addition, considering the impact of shot duration on the segmentation results, we integrate the shot duration information into the fine-tuning stage. Experiments on widely used benchmark datasets demonstrate that our approach effectively mitigates over-segmentation and achieves remarkable results in comparison with the state of the arts.

**Keywords:** Video Scene Segmentation · Self-Supervised Learning · Pseudo-Scene Boundary · Semantic Transition

## 1 Introduction

A long video, such as a movie or TV show, generally consists of various scenes, each of which is a collection of semantically related and temporally adjacent shots [30]. Once obtained video scene segments, one can easily parse a long video, which facilitates a wide range of long video understanding tasks such as movie preview retrieval [9], movie story understanding [21] and movie summarization [12].

One of the challenges in video scene segmentation is shot relation modeling. To tackle this challenge, previous methods [1,18,25] extract features to comprehend shot relations through the supervision from scene annotations, yielding promising results. However, annotating video scenes is time-consuming and

© The Author(s), under exclusive license to Springer Nature Switzerland AG 2024
S. Rudinac et al. (Eds.): MMM 2024, LNCS 14556, pp. 14–27, 2024.
https://doi.org/10.1007/978-3-031-53311-2_2

expensive. To alleviate annotation burdens, recent self-supervised learning methods [6,20,38] pre-train models by designed pretext tasks over unlabeled data to learn useful representations for video scene segmentation. Most pretext tasks suffer from over-segmentation due to inaccessible scene boundaries. Although pseudo boundary prediction has shown great potential [20], little research has been conducted on this aspect. Pseudo-boundary prediction can help pre-model discriminative shot relations with guidance of pseudo labels, which facilitates the transfer of pre-trained models. In such cases, obtaining high-quality pseudo labels becomes particularly important.

In this paper, we adopt pseudo boundary prediction as the pretext task to pre-encode shot contexts from unlabeled videos. We aim to improve the affinity between the pseudo labels and ground-truth labels to catch more accurate supervising signals which enable us to learn high-quality context representations. Motivated by the fact that scene boundaries arise from inter-scene inconsistencies, we propose the Semantic Transition Detection algorithm (STD). It leverages semantic inconsistencies between each shot and its context within temporal video chunks to detect scene boundaries more accurately. Therefore, it can improve the modeling capability of relation shots and reduce over-segmentation.

On the other hand, most existing methods for video scene segmentation, such as [6,11,20,22,25,33], fail to consider the impact of shot duration on the segmentation results. However, in movie production, shots with longer duration, known as long takes, are typically used for single-shot scenes or at the beginning of scenes. Inspired by this observation, we incorporate shot duration information in the fine-tuning stage. Experiments demonstrate that incorporating shot duration information leads to improvements in segmentation performance.

Our main contributions are summarized as follows:

- We introduce the Semantic Transition Detection algorithm (STD), which utilizes semantic inconsistencies between shots to extract high-quality pseudo-scene boundaries from unlabeled video data.
- We propose to incorporate shot duration information into the training process of video scene segmentation and design a duration encoder for this purpose.
- We apply STD to train a self-supervised video scene segmentation model that effectively mitigates over-segmentation and achieves remarkable performance on public benchmarks.

## 2   Related Work

### 2.1   Long Video Scene Segmentation

Scenes in videos such as movies or TV shows usually contain rich semantic relations and complex storylines. Early unsupervised methods [17,26,27,29,31,35, 41] detect scenes using low-level features. They can hardly capture the contextual relations among shots to parse long-form videos into scenes. With the development of deep learning, many supervised methods [18,23,25,33,34] capture the relations between semantic features extracted by neural network models for scene segmentation. Supervised methods have achieved better results than early methods, but they rely heavily on a large amount of high-quality scene boundary annotations. Since annotating scene boundaries for long videos is time-consuming and labor-intensive, the demand for annotations is difficult to satisfy.

To resolve the issue of lacking annotations, Chen *et al.* [6] and Wu *et al.* [38] contrast sampled image pairs for learning scene consistency. However, for video scene segmentation, it is essential to comprehend the contextual relationships among shots. To this end, Mun *et al.* [20] introduce a pseudo-boundary prediction task that generates pseudo-boundaries as supervising signals, showing promising results. They employ a modified Dynamic Time Wrapping (DTW) [3] to generate pseudo-boundaries, which relies on intra-scene consistency to segment a sequence of consecutive video shots. However, the fact is overlooked that scene boundaries arise due to inter-scene inconsistencies. Considering this oversight, we propose Semantic Transition Detection (STD) which depends on the semantic inconsistencies between each shot and its context for pseudo-boundary generation.

### 2.2   Self-supervised Leaning in Videos

Video-based self-supervised learning aims to learn video representations without relying on human annotation. To achieve this, researchers exploit the inherent spatio-temporal structure of videos and design pretext tasks tailored for learning video representations. Some methods focus on learning spatio-temporal correlations between frames by contrasting pairs of videos [24] or reconstructing missing parts of videos [40]. Meanwhile, other studies employ temporal-related pretext tasks, including frame order prediction [15], speed detection [2], and boundary classification [39]. However, these aforementioned methods primarily address frame-level relations in short videos, which may not be suitable for long-video scene segmentation.

## 3   Method

We work on exploring shot relations and propose a self-supervised method for video scene segmentation. As shown in Fig. 1, our method has two training stages. During pre-training, we design a pretext task, which leverages the pseudo-boundaries extracted by Semantic Transition Detection (STD) to pre-train a context encoder. When fine-tuning, we attach shot duration information and train the entire model by predicting ground-truth scene boundaries.

### 3.1   Pseudo-Boundary Extraction

For the pseudo-boundary task, the more similar the pseudo-label is to the ground-truth label, the better context representation can be learned. To this end, Mun *et al.* [20] modifies the Dynamic Time Warping (DTW) algorithm [36] to extract pseudo-boundaries based on intra-scene consistency. However, it ignores

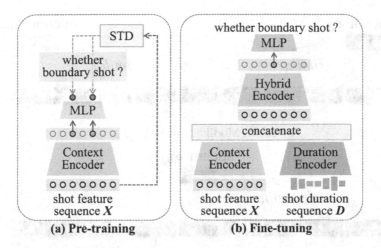

**Fig. 1.** Overview of the proposed method. (a) We rely on STD to extract pseudo-boundaries and pre-train a context encoder for pseudo-boundary prediction. (b) Aided by shot duration embedding, the entire model is fine-tuned by predicting ground-truth scene boundaries.

the fact that scene boundaries exist due to inter-scene inconsistencies. With this in consideration, we propose STD which focuses on the semantic inconsistencies between each shot and its context. Before presenting the proposed STD, we revisit modified-DTW proposed by [20].

**Modified-DTW in [20].** Given a sequence $S = \{s_0, \cdots, s_{n-1}\}$ of shots, DTW in [20] finds the optimal matching for two pairs, *i.e.*, $s_0$ *v.s.* $S_{left} = \{s_0, \cdots, s_{b-1}\}$ and $s_n$ *v.s.* $S_{right} = \{s_b, \cdots, s_{n-1}\}$ by traversing the candidate boundary offset $b$. As shown in Fig. 2(a), the DTW matching function is formulated by the intra-scene similarity in feature space, with $s_0$ and $s_{n-1}$ being two pseudo-scene anchors:

$$f_{\mathrm{DTW}}(b) = \frac{\sum_{i=1}^{b-1} \mathrm{sim}(\boldsymbol{x}_0, \boldsymbol{x}_i)}{b-1} + \frac{\sum_{j=b}^{n-2} \mathrm{sim}(\boldsymbol{x}_j, \boldsymbol{x}_{n-1})}{n-b-1}, \tag{1}$$

where $\boldsymbol{x}$ is the feature vector of the shot $\boldsymbol{s}$. Cosine similarity $\mathrm{sim}(\boldsymbol{x}_i, \boldsymbol{x}_j) = \frac{\boldsymbol{x}_i^\top \boldsymbol{x}_j}{\|\boldsymbol{x}_i\| \cdot \|\boldsymbol{x}_j\|}$ gives the matching score between shots $\boldsymbol{s}_i$ and $\boldsymbol{s}_j$. As a result, pseudo-boundary offset $b^*$ can be obtained by intra-scene similarity maximization:

$$b^* = \underset{b=2,\cdots,n-2}{\arg\max} \ f_{\mathrm{DTW}}(b). \tag{2}$$

**Semantic Transition Detection.** Different from the modified-DTW, our proposed STD extracts pseudo-boundaries by detecting the semantic inconsistencies

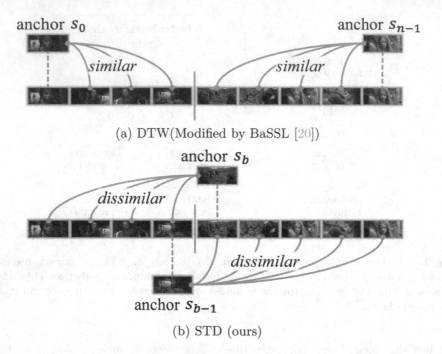

(a) DTW(Modified by BaSSL [20])

(b) STD (ours)

**Fig. 2.** Visualization of the matching processes. (a) DTW takes the shots at both ends of the sequence as anchors to match similar shots. (b) STD takes two shots adjacent to the candidate boundary as anchors to detect semantic transitions.

in given sequences. As shown in Fig. 2(b), STD can be described as minimizing similarity scores between pseudo scenes and anchors:

$$b^* = \operatorname*{arg\,min}_{b=1,\cdots,n-1} f_{\text{STD}}(b), \tag{3}$$

where

$$f_{\text{STD}}(b) = \sum_{i=0}^{b-1} \text{sim}(\boldsymbol{x}_b, \boldsymbol{x}_i) + \sum_{j=b}^{n-1} \text{sim}(\boldsymbol{x}_{b-1}, \boldsymbol{x}_j). \tag{4}$$

As indicated in Eq. 4, we employ candidate boundary shots $\boldsymbol{s}_b$ and $\boldsymbol{s}_{b-1}$ as scene anchors. Where, a boundary shot is the shot next to the pseudo scene boundary.

**Anchor-Based Normalization.** Due to the fact that videos often contain a large number of abnormal shots with unclear semantics, and STD based on Eq. 4 is easily disturbed by abnormal shots. When an anchor shot itself is an abnormal shot, Eq. 4 can obtain a higher response value, and thus misjudge the abnormal shot as the boundary shot. To make STD robust to abnormal shots, we design an Anchor-Based Normalization (ABN) for similarity scores:

$$w_{a,j} = \frac{\text{sim}(\boldsymbol{x}_a, \boldsymbol{x}_j)}{\sqrt{\sum_{k=0}^{n-1} \text{sim}(\boldsymbol{x}_a, \boldsymbol{x}_k)}}. \tag{5}$$

Based on Eq. 5, our STD in Eq. 3 will be reformulated into

$$b^* = \underset{b=1,\cdots,n-1}{\arg\min} \left( \sum_{i=0}^{b-1} w_{b,i} + \sum_{j=b}^{n-1} w_{b-1,j} \right). \tag{6}$$

## 3.2 Shot Duration

In movies, a shot with a long time duration, known as a long take, tends to be a single-shot scene or the beginning of a scene. Motivated by this fact, we introduce a duration encoder to embed shot duration to help fine-tuning our model. Specifically, we leverage a linear layer as the duration encoder to map each duration $\delta$ into a vector $\boldsymbol{d}$, and compose all the duration representations of shots from the given sequence $\boldsymbol{S}$ into $\boldsymbol{D}$:

$$\boldsymbol{D} = \text{Linear}([\delta_0, \cdots, \delta_{n-1}]), \tag{7}$$

where Linear($\cdot$) denotes the duration encoder.

## 3.3 Pre-training

Since Transformer [37] is suitable for processing sequence data, we built a two-layer Transformer as a context encoder to model shot contexts. To pre-train the context encoder, we inject the concatenated sequence of learnable position embedding [10] $\boldsymbol{P} = [\boldsymbol{p}_0, \cdots, \boldsymbol{p}_{n-1}]$ and feature sequence $\boldsymbol{X} = [\boldsymbol{x}_0, \cdots, \boldsymbol{x}_{n-1}]$ into the context encoder to output a feature sequence $\boldsymbol{R} = [\boldsymbol{r}_0, \cdots, \boldsymbol{r}_{n-1}]$, which carries inter-shot relation information:

$$\boldsymbol{R} = \text{Context}(\text{concat}(\boldsymbol{P}, \boldsymbol{X})), \tag{8}$$

where Context($\cdot$) denotes the context encoder and concat($\cdot$) stands for shot-by-shot concatenation.

For each sequence $\boldsymbol{X} = [\boldsymbol{x}_0, \cdots, \boldsymbol{x}_{n-1}]$, we obtain the boundary shot $s_{b^*}$ by performing STD. We take $\boldsymbol{x}_{b^*}$ as a positive sample and randomly select an element from $\boldsymbol{X} \setminus \boldsymbol{x}_{b^*}$ as a negative sample. Through feeding $\boldsymbol{X}$ into the context encoder, we obtain the contextual embedding sequence $\boldsymbol{R}$. We use binary cross entropy loss to train the context encoder:

$$\mathcal{L}_p = -\log(h_p(\boldsymbol{r}_{b^*})) - \log(1 - h_p(\boldsymbol{r}_o)), \tag{9}$$

where $\boldsymbol{r}_{b^*}$ and $\boldsymbol{r}_o$ are from the features sequence $\boldsymbol{R} = [\boldsymbol{r}_0, \cdots, \boldsymbol{r}_{n-1}]$, and $h_p(\cdot)$ denotes a two-layer MLP head.

### 3.4  Fine-Tuning

After pre-training, we use videos with ground-truth labels to fine-tune the whole model. For a shot sequence $S = \{s_0, \cdots, s_{n-1}\}$, we embed shot duration into features $D = [d_0, \cdots, d_{n-1}]$, which are coupled with $R = [r_0, \cdots, r_{n-1}]$ and position embedding $P = [p_0, \cdots, p_{n-1}]$ for shot-by-shot concatenation and then inject them into a hybrid encoder to obtain the contextual embedding $E = [e_0, \cdots, e_{n-1}]$:

$$E = \text{Hybrid}(\text{concat}(R, P, D)), \tag{10}$$

where Hybrid$(\cdot)$ denotes the hybrid encoder, a single-layer Transformer [37]. Then we examine only the contextual embedding $e_c$ of the center shot $s_c$ in $S$ to fine-tune the whole model by minimizing the cross-entropy loss:

$$\mathcal{L}_f = - y_c \log\left(h_f\left(e_c\right)\right) \\ + (1 - y_c) \log\left(1 - h_f\left(e_c\right)\right), \tag{11}$$

where $h_f(\cdot)$ is a two-layer MLP head different from $h_p(\cdot)$, and $y_c = 1$ if shot $s_c$ is a boundary shot, 0 otherwise.

## 4  Experiments

### 4.1  Experimental Setup

**Metrics.** For evaluation, we follow [6,25,38] to use the mean of Average Precision (AP) on each video. Also, we report mIoU, AUC, and F1 as has been done in [20]. For each metric, a higher value indicates a better result.

**Datasets.** We evaluate our method on the MovieNet [14], OSVD [28], and BBC [1] datasets. MovieNet [14] consists of 1,100 movies, where 190, 64, and 64 movies are labeled with scene boundaries in the training, validation, and test sets. Similar to recent self-supervised work [6,20], we pre-train our video scene segmentation model using all videos in MovieNet [14] without any ground-truth labels, fine-tune the model on the training set, and report results on the test set. Besides the MovieNet [14] test set, we also test on two small datasets, $i.e.$, OSVD [28] and BBC [1], which have 21 and 11 videos, respectively.

Considering knowledge leakage caused by test set in pre-training, we also report results on test set of MovieNet [14] with only train set used in pre-training.

**Implementation Details.** Following the preprocessing in [14], we parse every long video into shots using PySceneDetect [4] and sample each shot evenly to three frames. For shot representation, we utilize a cascade structure of ResNet-50 [13] pre-trained on ImageNet [8] and maximum pooling on sampled frames, having shot feature vectors initialized. By sliding a window on each parsed video, sequences containing $n = 13$ shots are produced. We apply the dropout technique [32] with a big probability of 40% in the context encoder and the hybrid

**Table 1.** Performance comparison between our proposed method and other baselines on MovieNet [14]. Bold numbers indicate the best results, and numbers with underlines indicate the second-best results.

| Method | Venue | AP | mIoU | ROC | F1 |
|---|---|---|---|---|---|
| *Unsupervised Learning* | | | | | |
| GraphCut [27] | TMM | 14.10 | 29.70 | - | - |
| SCSA [5] | TMM | 14.70 | 30.50 | - | - |
| DP [11] | ICME | 15.50 | 32.00 | - | - |
| StoryGraph [35] | CVPR | 25.10 | 35.70 | - | - |
| Grouping [28] | IJSC | <u>33.60</u> | 37.20 | - | - |
| BaSSL [20] w/o fine-tuning | ACCV | 31.55 | <u>39.36</u> | <u>71.67</u> | <u>32.55</u> |
| Ours w/o fine-tuning | - | **37.22** | **39.58** | **78.24** | **34.35** |
| *Supervised Learning* | | | | | |
| Siamese [1] | MM | 28.10 | 36.00 | - | - |
| MS-LSTM [14] | ECCV | 46.50 | 46.20 | - | - |
| LGSS [25] | CVPR | <u>47.10</u> | <u>48.80</u> | - | - |
| Temporal Perceiver [33] | TPAMI | **51.20** | **52.50** | - | - |
| *Self-supervised Pre-trained* | | | | | |
| ShotCoL [6] | CVPR | 53.37 | - | - | - |
| Movie2Scenes [7] | CVPR | 54.20 | - | - | - |
| SCRL [38] | CVPR | 54.82 | - | - | 51.43 |
| BaSSL [20] | ACCV | 57.40 | 50.69 | 90.54 | 47.02 |
| Ours w/o duration | - | <u>60.31</u> | <u>51.17</u> | <u>90.92</u> | 46.95 |
| Ours | - | **60.61** | **51.72** | **90.97** | <u>48.26</u> |

encoder against overfitting. In optimization, we use SGD with an initial learning rate of 0.3 during pre-training and use Adam [16] with an initial learning rate of $10^{-5}$ during fine-tuning. We train our model for 21 epochs at a mini-batch size of 256 in both pre-training and fine-tuning. Then we use the boundary prediction results of the middle shot in shot sequence to perform the test.

## 4.2    Comparison with State-of-the-Art Methods

Table 1 summarizes the comparison between our approach and various ones, including (i) unsupervised methods: GraphCut [27], SCSA [5], DP [11], StoryGraph [35], Grouping [28], and BaSSL [20] w/o fine-tuning, (ii) supervised methods: Siamese [1], MS-LSTM [14], LGSS [25] and Temporal Perceiver [33], and (iii) self-supervised pre-trained methods: ShotCoL [6], Movie2Scenes [7], SCRL [38], and BaSSL [20]. It is worth noting that, without fine-tuning, our method can be viewed as unsupervised, relying solely on visual features. The results demonstrate that ours w/o fine-tuning outperforms other unsupervised

**Table 2.** Comparison of self-supervised pre-trained methods. MovieNet [14] is split into training, validation and testing set with 660, 220 and 220 movies respectively. Train. only refers to training set excluding validation and test set. We exclude Movie2Scenes [7] for no code provided in their paper. Bold numbers indicate the best results, and numbers with underlines indicate the second-best results.

| Method | Venue | Pre-Training Data | AP | mIoU | AUC | F1 |
|--------|-------|-------------------|-----|------|-----|-----|
| ShotCol [6] | CVPR | Train. only | 48.21 | - | - | 46.52 |
| SCRL [38] | CVPR | Train. only | 54.55 | - | - | **51.39** |
| BaSSL [20] | ACCV | Train. only | 53.36 | 48.32 | 89.26 | 43.64 |
| Ours w/o duration | – | Train. only | <u>57.49</u> | <u>50.93</u> | <u>90.29</u> | 46.90 |
| Ours | – | Train. only | **58.98** | **51.61** | **90.71** | <u>48.30</u> |

**Table 3.** Cross dataset transfer results (AP) on OVSD [28] and BBC [1] without further fine-tuning. Bold numbers indicate the best results, and numbers with underlines indicate the second-best results.

| Method | Venue | OVSD | BBC |
|--------|-------|------|-----|
| ShotCoL [6] | CVPR | 25.53 | 27.98 |
| SCRL [38] | CVPR | <u>38.80</u> | 30.18 |
| BaSSL [20] | ACCV | 28.68 | **39.98** |
| Ours | – | **42.76** | <u>37.02</u> |

methods by 2.77% in AP, suggesting STD provides high-quality pseudo labels for pre-training the context encoder. As a result, our model effectively mitigates overfitting. By fine-tuning the entire model, shot duration information helps improve all metrics. Our model shows superior performance compared with the state-of-the-art method (*i.e.*, BaSSL [20]) by a large margin (3.21%↑ in AP).

Table 2 summarizes segmentation results of our proposed method and Shot-Col [6], SCRL [38], BaSSL [20]. It is noted that all of these methods employ the training set of MovieNet [14] excluding the test and validation data. The results demonstrate that our method achieves a notable improvement of 4.43% in AP, surpassing the improvement seen in Table 1.

To evaluate the generalization ability of our model, we compared it with previous self-supervised methods. We trained our model on MovieNet [14] and transferred it to OVSD [28] and BBC [1] for testing without any additional training. The BBC [1] dataset includes annotation results from five different annotators, and we report the average performance following [38]. As shown in Table 3, our method outperforms recent methods [6,20,38] on OVSD [28] and achieves competitive performance on BBC [1], demonstrating its considerable generalization.

**Table 4.** Comparison of different pseudo-boundary extraction algorithms, where we use the test set of MovieNet [14] to evaluate the accuracy of each algorithm. Bold number indicate the best result, and number with underline indicate the second-best result.

| Method | Acc. |
| --- | --- |
| DTW(Implemented by tslearn [36]) | 14.68 |
| DTW(Modified by BaSSL [20]) | 17.34 |
| DTW(Modified by BaSSL [20]) w/ ABN | 14.10 |
| Our STD w/o ABN | 15.51 |
| Our STD w/ ABN | **31.97** |

## 4.3 Ablation Studies

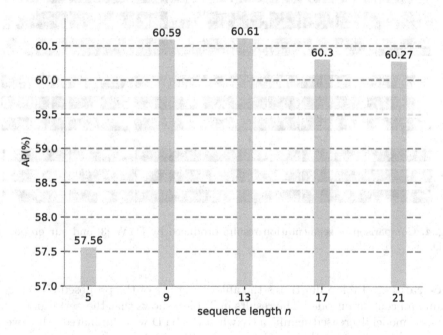

**Fig. 3.** Performance changes *wrt.* the sequence length $n$.

**Pseudo-Boundary.** To evaluate pseudo-boundary extraction algorithms, we collect shot sequences from the MovieNet [14] test set and execute algorithms to extract pseudo boundaries. By counting the proportion of the pseudo-boundaries matching the ground-truth scene boundaries, we calculate the Accuracy (Acc.) of each algorithm. As reported in Table 4, our STD with Anchor-Based Normalization (ABN) performs significantly better than other methods, almost twice as much.

**Table 5.** Ablation study of STD and Context Encoder(CE). + signifies "included" while − "excluded".

| STD | CE | AP | mIOU | ROC | F1 |
|---|---|---|---|---|---|
| − | − | 49.96 | 46.51 | 87.12 | 40.99 |
| − | + | 54.01 | 47.44 | 88.20 | 41.41 |
| + | + | 60.61 | 51.72 | 90.97 | 48.26 |

**Sequence Length.** We study the effect of the sequence length $n$ on the performance of our method in Fig. 3. The results show that increasing the values of $n$ steadily improves the performance and the result reaches the peak at $n = 13$.

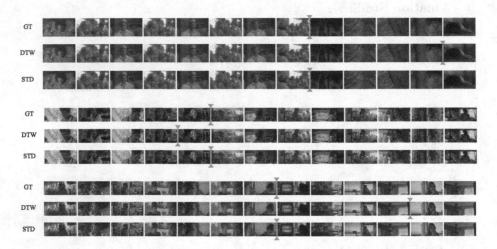

**Fig. 4.** Comparison of segmentation results produced by DTW [3] and our proposed STD.

**Pre-training.** Table 5 illustrates the ablation study of the proposed STD algorithm and context encoder. The results in Table 5 shows that the performance of our best model decreased significantly when the STD was eliminated. Moreover, when we also removed the context encoder, the performance of the proposed method drop further.

## 4.4    Visualization of Pseudo-Boundary Results

Figure 4 shows three segmentation instances, where the orange line indicates the segmentation position. Showing ground-truth annotations, we also select the DTW [3] algorithm as a comparative study. Compared to DTW, our proposed STD can handle more complex semantic transition environment, to produce more precise segmentation results.

## 4.5   Visualization of Context Embedding Distribution

To examine the impact of the proposed method on shot features, we employ t-SNE [19] for visualizing the distribution of context embedding features in Fig. 5. These features are extracted from the movie "The Ninth Gate" using BaSSL [20] and our method. In sub-figure (a), the feature space of BaSSL exhibits confusion between non-boundary and boundary shot. However, in sub-figure (b), our context embeddings shows better compactness in the feature distribution of boundary shots, which enhances the ability of the model to discriminate boundary shots.

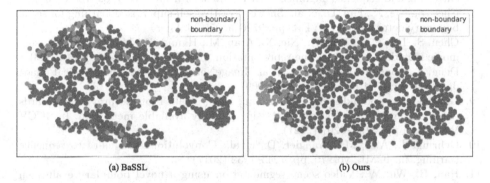

(a) BaSSL                                         (b) Ours

**Fig. 5.** Comparison on context embeddings $R$ defined in $Eq$. 8 for BaSSL [20] and our method. Blue points denote context embeddings of non-boundary shots, and orange points denote those of boundary shots.

## 5   Conclusion

This paper studies shot relations modeling in self-supervised video scene segmentation. By focusing on semantic inconsistencies between shots at scene boundaries, we design STD to extract pseudo-boundaries from unlabeled video data, which enables us to pre-train a video scene segmentation model of high quality. In addition, we incorporate shot duration information to help fine-tune our STD pre-trained model. Experimental results demonstrate that our presented self-supervised video scene segmentation method effectively alleviates over-segmentation.

**Acknowledgements.** This work is supported in part by the National Natural Science Foundation of China under Grant 61976029, the Key Project of Chongqing Technology Innovation and Application Development under Grant cstc2021jscx-gksbX0033, and the Fundamental Research Funds for the Central Universities Under Grant 2023CDJXY-035.

# References

1. Baraldi, L., Grana, C., Cucchiara, R.: A deep siamese network for scene detection in broadcast videos. In: ACM Multimedia, pp. 1199–1202 (2015)
2. Benaim, S., et al.: SpeedNet: learning the speediness in videos. In: CVPR, pp. 9919–9928 (2020)
3. Berndt, D.J., Clifford, J.: Using dynamic time warping to find patterns in time series. In: KDD Workshop, pp. 359–370 (1994)
4. Castellano, B.: Pyscenedetect: intelligent scene cut detection and video splitting tool. https://pyscenedetect.readthedocs.io/en/latest/ (2018)
5. Chasanis, V., Likas, A., Galatsanos, N.P.: Scene detection in videos using shot clustering and sequence alignment. IEEE Trans. Multim. **11**(1), 89–100 (2009)
6. Chen, S., Nie, X., Fan, D., et al.: Shot contrastive self-supervised learning for scene boundary detection. In: CVPR, pp. 9796–9805 (2021)
7. Chen, S., Liu, C.H., Hao, X., Nie, X., Arap, M., Hamid, R.: Movies2Scenes: using movie metadata to learn scene representation. In: CVPR, pp. 6535–6544 (2023)
8. Deng, J., Dong, W., Socher, R., et al.: ImageNet: a large-scale hierarchical image database. In: CVPR, pp. 248–255 (2009)
9. Gaikwad, B., Sontakke, A., Patwardhan, M.S., et al.: Plots to previews: towards automatic movie preview retrieval using publicly available meta-data. In: ICCV Workshop, pp. 3198–3207 (2021)
10. Gehring, J., Auli, M., Grangier, D., et al.: Convolutional sequence to sequence learning. In: ICML, vol. 70, pp. 1243–1252 (2017)
11. Han, B., Wu, W.: Video scene segmentation using a novel boundary evaluation criterion and dynamic programming. In: ICME, pp. 1–6 (2011)
12. Haq, I.U., Muhammad, K., Hussain, T., et al.: Quicklook: movie summarization using scene-based leading characters with psychological cues fusion. Inf. Fusion **76**, 24–35 (2021)
13. He, K., Zhang, X., Ren, S., et al.: Deep residual learning for image recognition. In: CVPR, pp. 770–778 (2016)
14. Huang, Q., Xiong, Y., Rao, A., et al.: MovieNet: a holistic dataset for movie understanding. In: ECCV, vol. 12349, pp. 709–727 (2020)
15. Kim, D., Cho, D., Kweon, I.S.: Self-supervised video representation learning with space-time cubic puzzles. In: AAAI, pp. 8545–8552 (2019)
16. Kingma, D.P., Ba, J.: Adam: a method for stochastic optimization. In: ICLR (2015)
17. Liang, C., Zhang, Y., Cheng, J., et al.: A novel role-based movie scene segmentation method. In: PCM, vol. 5879, pp. 917–922 (2009)
18. Liu, D., Kamath, N., Bhattacharya, S., et al.: Adaptive context reading network for movie scene detection. IEEE Trans. Circuits Syst. Video Technol. **31**(9), 3559–3574 (2021)
19. Van der Maaten, L., Hinton, G.: Visualizing data using t-SNE. J. Mach. Learn. Res. **9**(11), 2579–2605 (2008)
20. Mun, J., Shin, M., Han, G., et al.: BaSSL: boundary-aware self-supervised learning for video scene segmentation. In: ACCV, pp. 4027–4043 (2022)
21. Na, S., Lee, S., Kim, J., et al.: A read-write memory network for movie story understanding. In: ICCV, pp. 677–685 (2017)
22. Nicolas, H., Manoury, A., Benois-Pineau, J., et al.: Grouping video shots into scenes based on 1d mosaic descriptors. In: ICIP, pp. 637–640 (2004)
23. Protasov, S., Khan, A.M., Sozykin, K., et al.: Using deep features for video scene detection and annotation. Sig. Image Video Process. **12**(5), 991–999 (2018)

24. Qian, R., et al.: Spatiotemporal contrastive video representation learning. In: CVPR, pp. 6964–6974 (2021)
25. Rao, A., Xu, L., Xiong, Y., et al.: A local-to-global approach to multi-modal movie scene segmentation. In: CVPR, pp. 10143–10152 (2020)
26. Rasheed, Z., Shah, M.: Scene detection in hollywood movies and TV shows. In: CVPR, pp. 343–350 (2003)
27. Rasheed, Z., Shah, M.: Detection and representation of scenes in videos. IEEE Trans. Multim. 7(6), 1097–1105 (2005)
28. Rotman, D., Porat, D., Ashour, G.: Optimal sequential grouping for robust video scene detection using multiple modalities. Int. J. Semantic Comput. 11(2), 193–208 (2017)
29. Rui, Y., Huang, T.S., Mehrotra, S.: Exploring video structure beyond the shots. In: ICMCS, pp. 237–240 (1998)
30. Rui, Y., Huang, T.S., Mehrotra, S.: Constructing table-of-content for videos. Multim. Syst. 7(5), 359–368 (1999)
31. Sidiropoulos, P., Mezaris, V., Kompatsiaris, I., et al.: Temporal video segmentation to scenes using high-level audiovisual features. IEEE Trans. Circuits Syst. Video Technol. 21(8), 1163–1177 (2011)
32. Srivastava, N., Hinton, G.E., Krizhevsky, A., et al.: Dropout: a simple way to prevent neural networks from overfitting. J. Mach. Learn. Res. 15(1), 1929–1958 (2014)
33. Tan, J., Wang, Y., Wu, G., et al.: Temporal perceiver: a general architecture for arbitrary boundary detection. IEEE Trans. Pattern Anal. Mach. Intell. 45, 12506–12520 (2023)
34. Tan, J., Wang, H., Yuan, J.: Characters link shots: character attention network for movie scene segmentation. ACM Trans. Multim. Comput Commun, Appl. 20(4), 1–23 (2023)
35. Tapaswi, M., Bäuml, M., Stiefelhagen, R.: StoryGraphs: visualizing character interactions as a timeline. In: CVPR, pp. 827–834 (2014)
36. Tavenard, R., Faouzi, J., Vandewiele, G., et al.: Tslearn, a machine learning toolkit for time series data. J. Mach. Learn. Res. 21, 118:1–118:6 (2020)
37. Vaswani, A., Shazeer, N., Parmar, N., et al.: Attention is all you need. In: NIPS, pp. 5998–6008 (2017)
38. Wu, H., Chen, K., Luo, Y., Qiao, R., Ren, B., Liu, H., Xie, W., Shen, L.: Scene consistency representation learning for video scene segmentation. In: CVPR, pp. 14001–14010 (2022)
39. Xu, M., Pérez-Rúa, J.M., Escorcia, V., et al.: Boundary-sensitive pre-training for temporal localization in videos. In: ICCV, pp. 7200–7210 (2021)
40. Yang, H., et al.: Self-supervised video representation learning with motion-aware masked autoencoders. CoRR abs/2210.04154 (2022)
41. Yeung, M.M., Yeo, B., Liu, B.: Segmentation of video by clustering and graph analysis. Comput. Vis. Image Underst. 71(1), 94–109 (1998)

# Multi-task Collaborative Network for Image-Text Retrieval

Xueyang Qin[1], Lishuang Li[1(✉)], Jing Hao[1], Meiling Ge[2], Jiayi Huang[1], and Guangyao Pang[3]

[1] School of Computer Science and Technology, Dalian University of Technology, Dalian 116024, China
qinxueyang@snnu.edu.cn, lils@dlut.edu.cn, jinghao6936@163.com, huangjiayi@mail.dlut.edu.cn

[2] School of Computer Engineering, Weifang University, Weifang 261061, China
gemeiling@wfu.edu.cn

[3] Guangxi Key Laboratory of Machine Vision and Intelligent Control, Wuzhou University, Wuzhou 543002, China
pangguangyao@gmail.com

**Abstract.** Image-text retrieval aims to capture semantic relevance between images and texts. Most existing approaches rely solely on the image-text pairs to learn visual-semantic representation through fine-grained alignments while neglecting the potential beneficial impact of unimodal tasks on cross-modal retrieval. To this end, we present a Multi-Task Collaborative Network (MTCN) that leverages the synergy between multiple tasks to enhance the performance of image-text retrieval. Specifically, we introduce three unimodal tasks, including text-text matching, image multi-label classification, and text multi-label classification, and train together with target task image-text retrieval from the perspective of semantic constraints. Additionally, we employ a modality interaction module for image-text retrieval to discover interrelationships between these two modalities. Subsequently, a cascaded graph convolutional network combined with a multi-layer perceptron is used to infer the correlation scores between images and texts. We conduct comprehensive experiments on two benchmark datasets, Flickr30K and MSCOCO, and the quantitative and qualitative experimental results demonstrate the effectiveness of the proposed method. The source code is available at https://github.com/FlyCuteBird/MTCN.

**Keywords:** Image-text retrieval · Cross-modal retrieval · Multi-task learning · Graph convolutional network

## 1 Introduction

Vision and language are the two most prevalent modalities presenting information in people's daily lives. Exploring their relationship has attracted increasing

© The Author(s), under exclusive license to Springer Nature Switzerland AG 2024
S. Rudinac et al. (Eds.): MMM 2024, LNCS 14556, pp. 28–42, 2024.
https://doi.org/10.1007/978-3-031-53311-2_3

attention and spawned some basic cross-modal applications, ranging from cross-modal image-text retrieval [3,21] to multimodal recommendation [14,20]. This paper focuses on image-text retrieval, *i.e.*, retrieving relevant images given a query text and vice versa. Due to the heterogeneity of information presentation and the difference in data distribution, image-text retrieval remains a challenge in accurately measuring the semantic relevance of image-text pair.

To address the challenge, existing methods for learning semantic correlation of image-text pair primarily rely on global or local alignments. Global-based alignment methods [7,11,15] aim to project the entire visual and textual information into a unified subspace, where the similarity of image-text pair is easily measured. Although global-based approaches have high retrieval efficiency, they are limited to considering the global semantic information and ignore fine-grained correspondences between visual regions and textual words. As a response to this limitation, some researchers have proposed local-based alignment approaches [10,13,22,23,29,30] that prioritize learning local fragment alignment to infer the semantic relevance between images and texts. Since global-based or local-based alignment methods only consider the semantic correlations between images and texts from one aspect, some studies [6,24,28] have sought to exploit the complementarity of global and local information to model multi-level alignment jointly.

Although the aforementioned approaches are impressive, they ignore the beneficial effects of incorporating unimodal tasks on the performance of image-text retrieval. Intuitively, when optimizing the loss of the target task, the network's parameters are updated in the direction of loss reduction. If the network becomes trapped in a local minimum during the process, single-task training may struggle to further optimize it, resulting in suboptimal performance. To mitigate this issue, introducing auxiliary tasks during the target task training can help the network escape local optima, providing a possibility for seeking the global optimum. Theoretically, different tasks may have distinct optimization directions. By jointly training multiple related tasks, the network can learn from a broader spatial perspective that may not exist in single-task learning, ultimately improving the performance of the target task.

Based on the analysis presented above, we propose a Multi-Task Collaborative Network (MTCN) that utilizes some related unimodal auxiliary tasks to optimize the network's training process to improve the performance of image-text retrieval in this paper. Specifically, we introduce two multi-label classification tasks, including image and text classification, along with a text-text matching task in cross-modal retrieval, all three of which can be considered semantic constraints. The multi-label classification tasks ensure the global semantic consistency of images and texts, and text-text matching task encourages texts that are similar to images to attain high similarity scores. Furthermore, we present a cross-modal interaction approach to explore the relationship between these two modalities to obtain weighted visual and textual features. These weighted features are utilized in both image and text multi-label classification tasks, as well as in computing the similarity matrix vector representation of image-text pairs. Subsequently, we employ a cascaded graph convolutional network coupled with a multi-layer perceptron to infer the relevance of images and texts. Overall, the main contributions of this paper are as follows.

- We propose a Multi-Task Collaborative Network (MTCN) to achieve cross-modal image-text retrieval. To the best of our knowledge, this is the first framework that integrates several unimodal tasks into cross-modal retrieval to optimize visual-semantic embedding.
- We design three unimodal auxiliary tasks to constrain the global semantics and adjust the optimization direction of the target task from the perspective of training, which is beneficial for enhancing the performance of image-text retrieval.
- Extensive experiments on two benchmark datasets, Flickr30K and MSCOCO, show the advantages of the proposed MTCN on integrating multi-task collaborative learning, and experimental results outperform many recent methods by a clear margin.

## 2  Related Work

### 2.1  Image-Text Retrieval

Existing image-text retrieval methods can be categorized into three groups, including global-based methods, local-based methods, and multi-level based methods. Global-based methods typically employ two independent encoders to map images and texts into the same subspace to calculate their correlations. For instance, Peng et al. [15] adopted convolutional neural networks to encode these two modalities and jointly learned their representations through generative adversarial networks. Li et al. [11] utilized graph convolutional networks to enhance visual feature representation and implemented global semantic reasoning through GRUs. Similar methods such as [4,18] also employ global semantic information to explore the correlations of image-text pairs. Compared with global-based methods, local-based methods [3,10,21,29,30] focus on the interaction between fragments in different modalities. Lee et al. [10] developed a stacked attention network to discover all potential aligned pairs between visual regions and textual words. Wei et al. [22] presented a multi-modality attention network that jointly models the intra-modality and inter-modality relationships. Wu et al. [23] proposed a region reinforcement network to pay more attention to important visual regions. Subsequently, to explore the correspondences between these two modalities from different perspectives, some researchers proposed multi-level based approaches [8,12,17,28] for image-text retrieval. For example, Yuan et al. [28] jointly modeled global and local alignment into a unified framework. Based on this, Ji et al. [8] further incorporated global-local alignment.

### 2.2  Multi-task Learning

Multi-task learning aims to simultaneously train multiple related tasks from an optimization perspective, which has been applied successfully in a wide

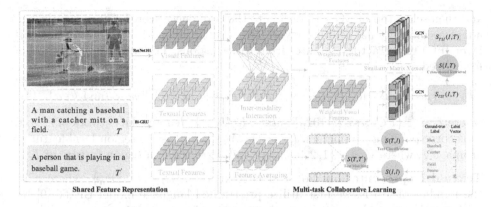

**Fig. 1.** The overall framework of MTCN.

range of domains, such as multimodal sentiment analysis [25,27], recommendation system [5,19], facial expression recognition [2,26], and so on. For example, Yu et al. [27] utilized supervised learning to integrate three unimodal tasks into sentiment analysis to train and learn the consistency and difference jointly. Tang et al. [19] proposed a progressive layered extraction mechanism to extract and separate more profound semantic knowledge, improving multi-task representation learning. Chen et al. [2] adopted the auxiliary task head pose estimation to improve the performance of multi view facial expression recognition. Additionally, some methods such as [11,24] also employ a multi-task approach to model image-text retrieval. Xu et al. [24] exploited image multi-label prediction task to constrain the global semantic information. Li et al. [11] optimized the network's learning through the image-to-text generation task. Although these methods also adopt multi-task learning, the proposed method differs from them as follows. First, we introduce two text-related unimodal tasks while existing approaches ignore this. Second, we utilize graph convolutional networks followed by a fully-connected layer to infer the similarity between images and texts. Differently, these methods rely more on metric representations such as cosine similarity and inner product.

## 3   Methodology

In this section, we will describe the proposed Multi-Task Collaborative Network (MTCN) in detail. As shown in Fig. 1, the model primarily consists of two main components: shared feature representation, and multi-task collaborative learning.

### 3.1   Shared Feature Representation

**Visual Representation.** Given an image $I$, we initially employ Faster-RCNN pre-trained by Anderson et al. [1] on Visual Genomes to detect salient regions, and then select top-$m$ ROIs according to the confidence scores. Subsequently, we adopt a convolutional neural network $ResNet101$ followed by a fully-connected layer (FC) to extract the features from these regions. The whole process can be formalized as

$$F_V = f_{FC}(f_{ResNet101}(f_{Faster-RCNN}(I, \theta_{Faster-RCNN}), \theta_{ResNet101}), \theta_{FC}), \quad (1)$$

where $f_r$, $r \in \{FC, ResNet101, Faster-RCNN\}$, denotes the network of $r$, and $\theta_r$ is corresponding parameters. $F_V = \{V_1, V_2, ..., V_m\}$ is the region-level features of image $I$, where $V_i$, $i \in \{1, 2, 3, ..., m\}$, indicates the $i$-th region feature of $I$.

**Textual Representation.** For a given text $T = \{w_1, w_2, ..., w_n\}$, where $n$ is the number of words, we first utilize pre-trained language model BERT [9] to initialize the embedding of each word, represented as $F_E = \{e_1, e_2, ..., e_n\}$. Then, the embedding vector $e_j$, $j \in \{1, 2, 3, ..., n\}$, is fed into a Bi-directional Gated Recurrent Unit (Bi-GRU) to generate:

$$\begin{aligned} \overrightarrow{h_j} &= \overrightarrow{GRU}(\overrightarrow{h_{j-1}}, e_j, \overrightarrow{\theta}), \\ \overleftarrow{h_j} &= \overleftarrow{GRU}(\overleftarrow{h_{j+1}}, e_j, \overleftarrow{\theta}), \end{aligned} \quad (2)$$

where $\overrightarrow{GRU}$ and $\overleftarrow{GRU}$ denote the forward and backward GRU, respectively. $\overrightarrow{\theta}$ and $\overleftarrow{\theta}$ are the corresponding parameters. The final representation $t_j$ of word $w_j$ is the average of $\overrightarrow{h_j}$ and $\overleftarrow{h_j}$, i.e., $t_j = (\overrightarrow{h_j} + \overleftarrow{h_j})/2$. Therefore, we can obtain the feature representation $F_T = \{t_1, t_2, ...., t_n\}$ of text $T$. Analogously, the feature representation of $T'$ can be written as $F_{T'} = \{t_1', t_2', ..., t_{n'}'\}$.

### 3.2   Multi-task Collaborative Learning

In this subsection, we will introduce the proposed multi-task learning framework. Since the inter-modality interaction module is used in image-text retrieval and multi-label classification tasks, we first present it. Subsequently, we introduce the target task and three auxiliary tasks. Finally, we present the optimization objective.

**Inter-modality Interaction.** In general, each fragment in the text describes a specific region of the image, and different textual fragments may have varying degrees of importance to the image. As shown in Fig. 1, for the text "*A man catching a baseball with a catcher mitt on a field*", it is evident that the word fragment "*man*" is particularly relevant to the man depicted in the image. Motivated by this observation, we intend to compute the importance of each textual fragment with respect to the image and vice versa. Specifically, given a matching image-text pair $(I, T)$ and their feature representations $F_V$ and $F_T$, we first calculate attention coefficients between all visual region features and textual fragment features, denotes as $F_V(F_T)^T$. We then apply the "*LeakyRelu*" activation function followed by a "*softmax*" function to adjust the attention coefficients. This process can be formalized as

$$Att_{t2i} = f_s(\alpha \cdot f_l(F_V(F_T)^T)), \quad (3)$$

where $\alpha$ is a hyper-parameter used to control the size of the attention coefficient. $f_s$ and $f_l$ denote the *"softmax"* and *"LeakyRelu"* function, respectively.

After obtaining the attention coefficient $Att_{t2i}$, we utilize matrix multiplication to acquire weighted textual feature $F_T'$, i.e., $F_T' = Att_{t2i} F_T$. Likewise, the weighted visual feature can be denoted as $F_V'$.

**Image-text Retrieval Task.** To learn the semantic relevance of images and texts, we first apply a vector similarity function [6] to obtain similarity matrix vector representation between visual (textual) feature and weighted textual (visual) feature. Taking visual feature $F_V$ and weighted textual feature $F_T'$ as an example, the similarity vector matrix $S_{t2i}$ is computed by

$$S_{t2i}(F_V, F_T', W) = \frac{W|F_V - F_T'|^2}{||W|F_V - F_T'|^2||_2}, \tag{4}$$

where $W$ is a learnable parameter matrix, $|\cdot|^2$ and $||\cdot||_2$ denote element-wise square and $L_2$-norm respectively.

Subsequently, we regard $S_{t2i}$ as a fully connected graph $G_{t2i}$, where each row in $S_{t2i}$ corresponds to a node in $G_{t2i}$. To facilitate the interaction between these nodes, we adopt a multi-layer graph convolutional network (GCN) in a cascaded manner to enhance their associations. The update methods for node features and edge weights can be formalized as

$$\begin{aligned} F^{(k+1)} &= f_r(\tilde{D}^{-\frac{1}{2}} \tilde{A} \tilde{D}^{-\frac{1}{2}} F^k W^k), \\ W^k &= f_s(f_{FC1}^k(F^k, \theta_{FC1}^k)(f_{FC2}^k(F^k, \theta_{FC2}^k))^T), \end{aligned} \tag{5}$$

where $F^k$ is the nodes' feature representation in the $k$-th graph layer. $f_r$ is the *ReLu* activation function. $\tilde{A} = A + \tilde{I}$ denotes the adjacency matrix, where $\tilde{I}$ is the identity matrix. $\tilde{D}$ is the diagonal node degree matrix of $\tilde{A}$. $W^k$ is the edge weight matrix of $k$-th graph layer. Subsequently, a multi-layer perceptron (MLP) is used to infer the relevance of image-text pair, i.e.,

$$S_{T2I} = f_m(F^k, \theta_m), \tag{6}$$

where $f_m$ performs MLP and $\theta_m$ is learnable parameter. Similarly, we can obtain $S_{I2T}$. For image-text retrieval, we adopt the triplet ranking loss to optimize the network and define it as:

$$L_R = max[0, \varphi_1 - S(I, T) + S(I, \hat{T})] + max[0, \varphi_1 - S(I, T) + S(\hat{I}, T)], \tag{7}$$

where $\varphi_1$ is a margin that is set to 0.2. $\hat{T}$ and $\hat{I}$ are the corresponding negative examples. For each batch-size $b$, we choose top $c/b + 1$ least relevant examples as negative examples, where $c$ is the number of $S + \varphi_1 - f_{diag}(S) > 0$, $S \in \{S_{T2I}, S_{I2T}\}$.

**Text-text matching Task.** Theoretically, if both text $T$ and $T'$ are the ground-true captions of image $I$, then the semantics of text $T$ and $T'$ should be consistent. Based on this premise, we introduce text-text matching as an auxiliary task to enhance the performance of image-text retrieval. Taking text $T$ as an example, we utilize the average of all word features as the semantic feature of $T$, then map it to a shared subspace using a fully connected layer. The entire process is expressed as

$$F_T^S = f_{FC}^T(f_{mean}(F_T), \theta_{FC}^T), \tag{8}$$

where $F_T^S$ denotes the global semantic feature of $T$. $f_{mean}$ indicates the summation function in the row direction. Likewise, we can obtain the semantic feature $F_{T'}^S$ of $T'$. Subsequently, we exploit cosine similarity to compute the semantic relevance score $S(T, T')$ of $T$ and $T'$, that is, $S(T, T') = cos(F_T^S, F_{T'}^S)$.

Similar to image-text retrieval, the triplet ranking loss is also adopted to optimize text-text matching. The equation is as follows:

$$L_S = \max[0, \varphi_2 - S(T, T') + S(T, C')] + \max[0, \varphi_2 - S(T, T') + S(C, T')], \tag{9}$$

where $\varphi_2$ is a margin that is set to 0.1. $C$ and $C'$ are text negative examples of $T$ and $T'$ respectively, and they are obtained in the same way as the negative example of $I$.

**Multi-label classification Task.** To constrain the global semantics of matching image-text pairs, we design two supervised multi-label classification tasks to ensure the semantic consistency of images and texts in image-text retrieval. For convenience, we first introduce semantic labels for matching image-text pairs. Concretely, we select the top-$p$ frequently occurred nouns in all the texts to construct the label dictionary $l_D \in R^{1 \times p}$ to ensure its diversity, where $p$ relies on the frequency distribution of a specific dataset. Based on this, each image-text pair can be assigned with one or more semantic labels. In the experiment, we utilize one-hot encoding to represent the label information of image-text pairs. If a noun in the text appears in the label dictionary, the corresponding position is assigned a value of 1, otherwise it is set to 0.

After obtaining the semantic labels of images and texts, we exploit weighted visual feature $F_V'$ for multi-label image classification. Firstly, we use fully connected layers to predict the semantic label $\hat{y}_l^I$ of images, i.e.,

$$\hat{y}_l^I = f_{FC2}^I(f_t(f_{FC1}^I(f_{mean}(F_V'), \theta_{FC1}^I)), \theta_{FC2}^I), \tag{10}$$

where $f_t$ denotes the *tanh* activation function. Then, the predicting label $\hat{y}_l^I = [\hat{y}_{l,1}^I, \hat{y}_{l,2}^I, ..., \hat{y}_{l,p}^I]$ and ground-truth label $y_l = [y_{l,1}, y_{l,2}, ..., y_{l,P}]$ can be derived as a set of the binary classification task, that is,

$$L_{I,l} = -\sum_{p=1}^{P} y_{l,p} \cdot \log \sigma(\hat{y}_{l,p}^I) + (1 - y_{l,p}) \cdot \log \sigma(1 - \hat{y}_{l,p}^I), \tag{11}$$

where $\sigma$ denotes the *Sigmoid* activation function. Likewise, when we utilize weighted textual feature for multi-label text classification, the loss $L_{T,l}$ can be written as

---

**Algorithm 1:** Model the training process of MTCN

---

**Input**: Image-text pairs $(I, T)$; Text positive sample $T'$, batch-size $b$;
   parameters $\varphi_2$, $\varphi_1$, $\alpha$; learning network parameters: $\theta$.
**Output**: $\theta$.

1  **for** $epoch = 1, 2, ..., E$ **do**
2      **for** *each batch size b* **do**
3          Initial feature representation: $F_V$, $F_T$, $F_{T'}$;
4          Obtain weighted features via Eq.(3) and $F_V$, $F_T$, $F_{T'}$;
5          Compute the loss $L_R$ of image-text retrieval via Eqs.(4)-(7);
6          Calculate the loss $L_S$ of text-text matching via Eqs.(8) and (9);
7          Acquire the losses $L_{I,l}$ and $L_{T,l}$ via Eqs.(10)-(12);
8      **end**
9      Obtain losses $L_R$,$L_S$,$L_{I,l}$ and $L_{T,l}$ ;
10     Compute $L$ via Eq.(13);
11     $\theta \leftarrow$ Backward $(L)$.
12 **end**

---

$$L_{T,l} = - \sum_{p=1}^{P} y_{l,p} \cdot \log \sigma(\hat{y}_{l,p}^T) + (1 - y_{l,p}) \cdot \log \sigma(1 - \hat{y}_{l,p}^T) \tag{12}$$

**Optimization Objective.** As mentioned earlier, the proposed multi-task learning framework MTCN includes a total of 4 tasks, and Algorithm 1 is the training process of MTCN. In our experiments, we optimize all loss functions together, and the final optimization loss $L$ can be written as

$$L = L_R + L_S + L_{I,l} + L_{T,l} \tag{13}$$

## 4 Experiments

### 4.1 Dataset and Protocols

We evaluate our proposed method MTCN and all baselines on two large-scale image-text retrieval datasets, Flickr30K [10] and MSCOCO [13]. Flickr30K consists of $31,000$ images collected from the Flickr website, where each image is annotated with five text descriptions. Following the split in [16], we utilize $29,000$ images for training, $1,000$ images for validation and the remaining images for testing. MSCOCO comprises $123,287$ images and each image is accompanied by five manually annotated texts. Following the split in [6], we adopt $113,287$ images for training, $5,000$ images for validation and $5,000$ images for testing. We report results on MSCOCO by averaging over 5-folds of 1K test images and testing all 5K images.

Following standard practice in image-text retrieval, we measure the performance by Recall at K ($R@K, K = 1, 5, 10$), which is defined as the proportion of matching examples in the top-K retrieval results. In addition, we adopt an

**Table 1.** Performance comparison between our proposed MTCN and several advanced baselines on Flickr30K test set.

| Methods | Image-to-Text | | | Text-to-Image | | | rSum |
|---------|------|------|------|------|------|------|------|
| | R@1 | R@5 | R@10 | R@1 | R@5 | R@10 | |
| SCAN (2018) [10] | 67.4 | 90.3 | 95.8 | 48.6 | 77.7 | 85.2 | 465.0 |
| VSRN (2019) [11] | 71.3 | 90.6 | 96.0 | 54.7 | 81.8 | 88.2 | 482.6 |
| CAAN (2020) [29] | 70.1 | 91.6 | 97.2 | 52.8 | 79.0 | 87.9 | 478.6 |
| MMCA (2020) [22] | 74.2 | 92.8 | 96.4 | 54.8 | 81.4 | 87.8 | 487.4 |
| VSR++ (2021) [28] | 72.6 | 92.7 | 97.2 | 56.3 | 82.7 | 89.0 | 490.5 |
| SHAN (2021) [8] | 74.6 | 93.5 | 96.9 | 55.3 | 81.3 | 88.4 | 490.0 |
| RRTC (2022) [23] | 72.7 | 93.8 | 96.8 | 54.2 | 79.4 | 86.1 | 483.0 |
| CGMN (2022) [3] | 77.9 | 93.8 | 96.8 | 59.9 | 85.1 | **90.6** | 504.1 |
| RAAN (2023) [21] | 77.1 | 93.6 | 97.3 | 56.0 | 82.4 | 89.1 | 495.5 |
| GLFN (2023) [30] | 75.1 | 93.8 | 97.2 | 54.5 | 82.8 | 89.9 | 493.3 |
| MTCN (i2t) | 73.2 | 92.5 | 96.4 | 55.6 | 81.8 | 87.9 | 487.4 |
| MTCN (t2i) | 75.3 | 94.3 | 97.4 | 57.7 | 83.6 | 89.6 | 497.9 |
| MTCN (i2t+t2i) | **78.9** | **95.5** | 97.6 | **60.8** | **85.3** | 90.4 | **508.5** |

extra evaluation metric $rSum$ to demonstrate the effectiveness of the proposed method. $rSum$ is the summation of bi-directional retrieval results of $R@K$.

### 4.2 Comparison with Existing Methods

**Baselines.** To verify the effectiveness of our proposed method MKTLON, we compare MTCN with current some advanced approaches in image-text retrieval, including (1) local-based embedding methods SCAN [10], CAAN [29], MMCA [22], RRTC [23] that learn local image-text correspondences by designing different attention networks. (2) Global-based embedding methods VSRN [11] and CGMN [3] that learn global semantic relevances between images and texts. (3) VSR++ [28] and SHAN [8] that combine global and local alignment. (4) Some recent methods RAAN [21] and GLFN [30].

**Result on Flickr30K.** Table 1 reports the quantitative results of bi-directional retrieval on Flickr30K dataset, where the proposed MTCN method surpasses all baselines on evaluation metric $rSum$, with the least improvement of 4.4%. Compared with the advanced global-based embedding method CGMN [3], our method exhibits performance gains of 1.0% on text retrieval and 0.9% on image retrieval ($R@1$). Furthermore, compared with some recent approaches, such as RAAN [21] and GLFN [30], our method outperforms them on all evaluation metrics by a large margin, which verifies the effectiveness of the proposed approach MKTLON.

**Result on MSCOCO.** Tables 2 and 3 show the experimental results on MSCOCO 5-folds 1K and 5K test sets, respectively. Clearly, our method achieves the best results on evaluation metric $rSum$ compared with all baselines, especially on the 5K test set, with a considerable improvement of 7.4%. Additionally, compared with the recent method GLFN in 5-folds 1K test set, our method obtains a boost of 0.4% on text retrieval and 1.1% on image retrieval ($R@1$), providing further evidence of the effectiveness of multi-task collaborative learning in image-text retrieval.

**Table 2.** Performance comparison between our proposed MTCN and several advanced baselines on MSCOCO 5-fold 1K test set.

| Methods | Image-to-Text | | | Text-to-Image | | | $rSum$ |
|---|---|---|---|---|---|---|---|
| | $R@1$ | $R@5$ | $R@10$ | $R@1$ | $R@5$ | $R@10$ | |
| SCAN (2018) [10] | 72.7 | 94.8 | 98.4 | 58.8 | 88.4 | 94.8 | 507.9 |
| VSRN (2019) [11] | 76.2 | 94.8 | 98.2 | 62.8 | 89.7 | 95.1 | 516.8 |
| CAAN (2020) [29] | 75.5 | 95.4 | 98.5 | 61.3 | 89.7 | 95.2 | 515.6 |
| MMCA (2020) [22] | 74.8 | 95.6 | 97.7 | 61.6 | 89.8 | 95.2 | 514.7 |
| VSR++ (2021) [28] | 76.6 | 95.2 | 98.2 | 63.4 | 90.6 | 95.7 | 519.7 |
| SHAN (2021) [8] | 76.8 | 96.3 | 98.7 | 62.6 | 89.6 | 95.8 | 519.8 |
| RRTC (2022) [23] | 76.2 | 96.3 | **98.9** | 61.6 | 89.3 | 94.6 | 516.9 |
| CGMN (2022) [3] | 76.8 | 95.4 | 98.3 | 63.8 | **90.7** | 95.7 | 520.7 |
| RAAN (2023) [21] | 76.8 | **96.4** | 98.3 | 61.8 | 89.5 | 95.8 | 518.6 |
| GLFN (2023) [30] | 78.4 | 96.0 | 98.5 | 62.6 | 89.6 | 95.4 | 520.5 |
| MTCN (i2t) | 75.0 | 95.5 | 98.5 | 60.7 | 89.5 | 95.5 | 514.7 |
| MTCN (t2i) | 75.0 | 95.2 | 98.4 | 61.5 | 89.5 | 95.6 | 515.2 |
| MTCN (i2t+t2i) | **78.8** | **96.4** | 98.7 | 63.7 | **90.7** | **96.1** | **524.4** |

**Table 3.** Performance comparison between our proposed MTCN and several advanced baselines on MSCOCO 5K test set.

| Methods | Image-to-Text | | | Text-to-Image | | | $rSum$ |
|---|---|---|---|---|---|---|---|
| | $R@1$ | $R@5$ | $R@10$ | $R@1$ | $R@5$ | $R@10$ | |
| SCAN (2018) [10] | 50.4 | 82.2 | 90.0 | 38.6 | 69.3 | 80.4 | 410.9 |
| VSRN (2019) [11] | 53.0 | 81.1 | 89.4 | 40.5 | 70.6 | 81.1 | 415.7 |
| CAAN (2020) [29] | 52.5 | 83.3 | 90.9 | 41.2 | 70.3 | **82.9** | 421.1 |
| MMCA (2020) [22] | 54.0 | 82.5 | 90.7 | 38.7 | 69.7 | 80.8 | 416.4 |
| VSR++ (2021) [28] | 53.6 | 82.1 | 90.2 | 41.1 | 71.0 | 81.8 | 419.8 |
| CGMN (2022) [3] | 53.4 | 81.3 | 89.6 | 41.2 | **71.9** | 82.4 | 419.8 |
| MTCN (i2t) | 53.1 | 82.3 | 90.1 | 38.1 | 68.6 | 79.8 | 412.0 |
| MTCN (t2i) | 53.7 | 81.5 | 89.6 | 39.1 | 69.1 | 80.1 | 413.1 |
| MTCN (i2t+t2i) | **57.8** | **84.5** | **91.4** | **41.5** | 71.3 | 82.0 | **428.5** |

**Table 4.** Ablation studies on Flickr30K test set.

| Model | Task Setting | | | | Image-to-Text | | | Text-to-Image | | | rSum |
|---|---|---|---|---|---|---|---|---|---|---|---|
| | R | S | IC | TC | R@1 | R@5 | R@10 | R@1 | R@5 | R@10 | |
| ① | ✓ | | | | 77.5 | 94.3 | 97.3 | 60.2 | 84.8 | 90.2 | 504.3 |
| ② | ✓ | ✓ | | | 78.4 | 94.7 | 97.8 | 60.7 | 85.2 | 90.7 | 507.5 |
| ③ | ✓ | | ✓ | | 78.1 | 94.6 | 97.5 | 59.9 | 85.2 | 90.7 | 506.0 |
| ④ | ✓ | | | ✓ | 78.0 | 94.2 | 97.6 | 60.0 | 84.4 | 90.3 | 504.5 |
| ⑤ | ✓ | ✓ | ✓ | | 78.8 | 94.5 | 97.4 | **60.9** | 85.0 | 90.5 | 507.1 |
| ⑥ | ✓ | ✓ | | ✓ | 77.7 | 94.6 | **98.1** | 60.5 | 84.8 | **90.9** | 506.6 |
| ⑦ | ✓ | | ✓ | ✓ | **79.7** | 94.5 | 97.9 | 59.7 | 84.8 | 90.3 | 506.9 |
| ⑧ | ✓ | ✓ | ✓ | ✓ | 78.9 | **95.5** | 97.6 | 60.8 | **85.3** | 90.4 | **508.5** |

## 4.3   Ablation Study

To explore the impact of different tasks on the experimental results, we conduct ablation studies on Flickr30K dataset. As shown in Table 4, $R$, $S$, $IC$ and $TC$ represent image-text retrieval, text-text matching, image classification and text classification tasks, respectively. Obviously, when all three auxiliary tasks are adopted, the result achieves the best 508.5% on evaluation metric $rSum$. Besides, when we employ only a single auxiliary task, such as models ②, ③ and ④, the performance on $rSum$ is improved by 3.2%, 1.7% and 0.2% compared with model ①. Furthermore, we observe that when tasks $IC$ and $TC$ are used alone with task $S$, the model's performance will be reduced, but when they are utilized together with task $S$, the performance achieves the best.

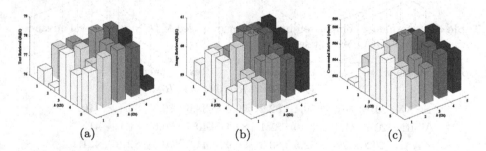

| (a) | (b) | (c) |

**Fig. 2.** Effect of the number cascaded GCNs on Flickr30k dataset.

## 4.4   Effect of Different Hyper-parameter

To assess the impact of the hyper-parameter on the model's performance, we select an important parameter $k$ to explore this as shown in Fig. 2, where $k$

denotes the number of cascaded GCNs and $t2i$ ($i2t$) indicates that $S_{t2i}$ ($S_{i2t}$) is used to compute the correlation of image-text pairs. As we can see, when $k(t2i)$ and $k(i2t)$ are combined with values ranging from 1 to 5, the performance of image-text retrieval fluctuates differently, with the highest $R@1$ values being 78.9% for text retrieval and 60.8% for image retrieval respectively. Similarly, for the evaluation metric $rSum$, the experimental result achieves the best 508.5% with $k(t2i) = k(i2t) = 3$. Also, we consider a special case, $i.e.$, $k(t2i) = k(i2t)$. It can be seen that as the value of $k(t2i)$ changes from 1 to 5, whether it is text retrieval or image retrieval on the evaluation metric $R@1$, the experimental results have a trend of first increasing and then decreasing.

### 4.5  Visualization of Retrieval Results

To further validate the effectiveness of the proposed MTCN method, we visualize the results of text retrieval and image retrieval in Figs. 3 and 4, respectively, and compare with an advanced model CGMN whose code is publicly available. From the examples in Fig. 3, we can observe that the proposed method MTCN retrieves matching texts effectively, while CGMN fails to retrieve all the matching texts in the second example. Similarly, for image retrieval in Fig. 4, MTCN consistently makes more accurate judgments than CGMN. Even for some "incorrect" retrieval images, they still exhibit a high correlation with the query text, indirectly confirming the effectiveness of our proposed method.

1. A bald man holds a fish in front of a lake while two blond young children stand near him while holding fishing poles.
2. A man and two girls show off a fish while holding fishing poles in front of a body of water.
3. A man and two young children are fishing by a pond.
4. Two young girls with fishing poles and a man displaying a fish they caught.
5. This man holds a fish while two little girls stand to either side of him.

1. A man and two young children are fishing by a pond.
2. A man and two girls show off a fish while holding fishing poles in front of a body of water.
3. A bald man holds a fish in front of a lake while two blond young children stand near him while holding fishing poles.
4. This man holds a fish while two little girls stand to either side of him.
5. An adult wearing a black shirt and tan pants is standing at the water's edge with two children.

1. Two hockey players wearing yellow and black uniforms skate on the ice.
2. A man is guiding the hockey puck across the ice while an opposing player races towards it.
3. Two ice hockey players.
4. Two people in hockey uniforms on the ice.
5. hockey player in white uniform with stick.

1. Two hockey players wearing yellow and black uniforms skate on the ice.
2. A man is guiding the hockey puck across the ice while an opposing player races towards it.
3. hockey player in white uniform with stick.
4. Two people in hockey uniforms on the ice.
5. Two ice hockey players.

MTCN                                                    CGMN

**Fig. 3.** Qualitative results of text retrieval on Flickr30K. The matched texts are in blue font. (Color figure online)

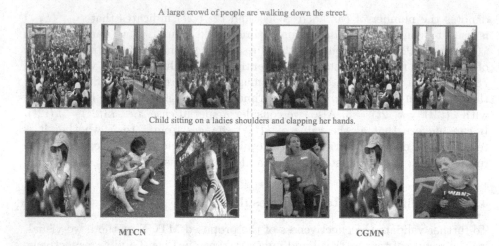

**Fig. 4.** Qualitative results of image retrieval on Flickr30K. The matched image is marked with a green box. (Color figure online)

## 5   Conclusions

This paper proposes a Multi-Task Collaborative Network for image-text retrieval. Furthermore, we introduce three unimodal tasks to enhance the performance of image-text retrieval from the perspective of model training. Quantitative and qualitative experimental results verify the effectiveness of our method. In future work, we plan to integrate the multi-task learning framework into other multi-modal application scenarios to improve their performance.

**Acknowledgement.** This work was supported by the National Natural Science Foundation of China [Grant No. 62076048].

## References

1. Anderson, P., et al.: Bottom-up and top-down attention for image captioning and visual question answering. In: Proceedings of the IEEE Conference on Computer Vision and Pattern Recognition, pp. 6077–6086 (2018)
2. Chen, J., Yang, L., Tan, L., Xu, R.: Orthogonal channel attention-based multi-task learning for multi-view facial expression recognition. Pattern Recogn. **129**, 108753 (2022)
3. Cheng, Y., Zhu, X., Qian, J., Wen, F., Liu, P.: Cross-modal graph matching network for image-text retrieval. ACM Trans. Multimedia Comput. Commun. Appl. (TOMM) **18**(4), 1–23 (2022)
4. Chi, J., Peng, Y.: Zero-shot cross-media embedding learning with dual adversarial distribution network. IEEE Trans. Circuits Syst. Video Technol. **30**(4), 1173–1187 (2019)

5. Deng, Y., Zhang, W., Xu, W., Lei, W., Chua, T.S., Lam, W.: A unified multi-task learning framework for multi-goal conversational recommender systems. ACM Trans. Inf. Syst. **41**(3), 1–25 (2023)
6. Diao, H., Zhang, Y., Ma, L., Lu, H.: Similarity reasoning and filtration for image-text matching. In: Proceedings of the AAAI Conference on Artificial Intelligence, pp. 1218–1226 (2021)
7. Gao, Q., Lian, H., Wang, Q., Sun, G.: Cross-modal subspace clustering via deep canonical correlation analysis. In: Proceedings of the AAAI Conference on Artificial Intelligence, pp. 3938–3945 (2020)
8. Ji, Z., Chen, K., Wang, H.: Step-wise hierarchical alignment network for image-text matching. In: Proceedings of the 31th International Joint Conference on Artificial Intelligence (2021)
9. Kenton, J.D.M.W.C., Toutanova, L.K.: Bert: Pre-training of deep bidirectional transformers for language understanding. In: Proceedings of NAACL-HLT, pp. 4171–4186 (2019)
10. Lee, K.H., Chen, X., Hua, G., Hu, H., He, X.: Stacked cross attention for image-text matching. In: Proceedings of the European Conference on Computer Vision, pp. 201–216 (2018)
11. Li, K., Zhang, Y., Li, K., Li, Y., Fu, Y.: Visual semantic reasoning for image-text matching. In: Proceedings of the IEEE/CVF International Conference on Computer Vision, pp. 4654–4662 (2019)
12. Li, W., Yang, S., Wang, Y., Song, D., Li, X.: Multi-level similarity learning for image-text retrieval. Inf. Process. Manage. **58**(1), 102432 (2021)
13. Liu, C., Mao, Z., Liu, A.A., Zhang, T., Wang, B., Zhang, Y.: Focus your attention: a bidirectional focal attention network for image-text matching. In: Proceedings of the 27th ACM International Conference on Multimedia, pp. 3–11 (2019)
14. Liu, K., Xue, F., Guo, D., Wu, L., Li, S., Hong, R.: MEGCF: multimodal entity graph collaborative filtering for personalized recommendation. ACM Trans. Inf. Syst. **41**(2), 1–27 (2023)
15. Peng, Y., Qi, J.: Cm-GANs: cross-modal generative adversarial networks for common representation learning. ACM Trans. Multimedia Comput. Commun. Appl. (TOMM) **15**(1), 1–24 (2019)
16. Qin, X., Li, L., Hao, F., Pang, G., Wang, Z.: Cross-modal information balance-aware reasoning network for image-text retrieval. Eng. Appl. Artif. Intell. **120**, 105923 (2023)
17. Qin, X., Li, L., Pang, G.: Multi-scale motivated neural network for image-text matching. Multimedia Tools Appl. 1–25 (2023). https://doi.org/10.1007/s11042-023-15321-0
18. Sarafianos, N., Xu, X., Kakadiaris, I.A.: Adversarial representation learning for text-to-image matching. In: Proceedings of the IEEE/CVF International Conference on Computer Vision, pp. 5814–5824 (2019)
19. Tang, H., Liu, J., Zhao, M., Gong, X.: Progressive layered extraction (PLE): a novel multi-task learning (mtl) model for personalized recommendations. In: Proceedings of the 14th ACM Conference on Recommender Systems, pp. 269–278 (2020)
20. Tao, Z., Liu, X., Xia, Y., Wang, X., Yang, L., Huang, X., Chua, T.S.: Self-supervised learning for multimedia recommendation. IEEE Trans. Multimedia **25**, 1–10 (2022)
21. Wang, Y., Su, Y., Li, W., Sun, Z., Wei, Z., Nie, J., Li, X., Liu, A.A.: Rare-aware attention network for image-text matching. Inf. Process. Manage. **60**(3), 103280 (2023)

22. Wei, X., Zhang, T., Li, Y., Zhang, Y., Wu, F.: Multi-modality cross attention network for image and sentence matching. In: Proceedings of the IEEE/CVF Conference on Computer Vision and Pattern Recognition, pp. 10941–10950 (2020)
23. Wu, J., Wu, C., Lu, J., Wang, L., Cui, X.: Region reinforcement network with topic constraint for image-text matching. IEEE Trans. Circuits Syst. Video Technol. **32**(1), 388–397 (2021)
24. Xu, X., Wang, T., Yang, Y., Zuo, L., Shen, F., Shen, H.T.: Cross-modal attention with semantic consistence for image-text matching. IEEE Trans. Neural Netw. Learn. Syst. **31**(12), 5412–5425 (2020)
25. Yang, B., Wu, L., Zhu, J., Shao, B., Lin, X., Liu, T.Y.: Multimodal sentiment analysis with two-phase multi-task learning. IEEE/ACM Trans. Audio Speech Lang. Process. **30**, 2015–2024 (2022)
26. Yu, W., Xu, H.: Co-attentive multi-task convolutional neural network for facial expression recognition. Pattern Recogn. **123**, 108401 (2022)
27. Yu, W., Xu, H., Yuan, Z., Wu, J.: Learning modality-specific representations with self-supervised multi-task learning for multimodal sentiment analysis. In: Proceedings of the AAAI Conference on Artificial Intelligence, vol. 35, pp. 10790–10797 (2021)
28. Yuan, H., Huang, Y., Zhang, D., Chen, Z., Cheng, W., Wang, L.: VSR++: improving visual semantic reasoning for fine-grained image-text matching. In: Proceedings of the 25th International Conference on Pattern Recognition, pp. 3728–3735 (2021)
29. Zhang, Q., Lei, Z., Zhang, Z., Li, S.Z.: Context-aware attention network for image-text retrieval. In: Proceedings of the IEEE/CVF Conference on Computer Vision and Pattern Recognition, pp. 3536–3545 (2020)
30. Zhao, G., Zhang, C., Shang, H., Wang, Y., Zhu, L., Qian, X.: Generative label fused network for image-text matching. Knowl.-Based Syst. **263**, 110280 (2023)

# FGENet: Fine-Grained Extraction Network for Congested Crowd Counting

Hao-Yuan Ma, Li Zhang$^{(\boxtimes)}$, and Xiang-Yi Wei

School of Computer Science and Technology, Soochow University,
Suzhou 215006, China
zhangliml@suda.edu.cn

**Abstract.** Crowd counting has gained significant popularity due to its practical applications. However, mainstream counting methods ignore precise individual localization and suffer from annotation noise because of counting from estimating density maps. Additionally, they also struggle with high-density images. To address these issues, we propose an end-to-end model called Fine-Grained Extraction Network (FGENet). Different from methods estimating density maps, FGENet directly learns the original coordinate points that represent the precise localization of individuals. This study designs a fusion module, named Fine-Grained Feature Pyramid (FGFP), that is used to fuse feature maps extracted by the backbone of FGENet. The fused features are then passed to both regression and classification heads, where the former provides predicted point coordinates for a given image, and the latter determines the confidence level for each predicted point being an individual. At the end, FGENet establishes correspondences between prediction points and ground truth points by employing the Hungarian algorithm. For training FGENet, we design a robust loss function, named Three-Task Combination (TTC), to mitigate the impact of annotation noise. Extensive experiments are conducted on four widely used crowd counting datasets. Experimental results demonstrate the effectiveness of FGENet. Notably, our method achieves a remarkable improvement of 3.14 points in Mean Absolute Error (MAE) on the ShanghaiTech Part A dataset, showcasing its superiority over the existing state-of-the-art methods. Even more impressively, FGENet surpasses previous benchmarks on the UCF_CC_50 dataset with an astounding enhancement of 30.16 points in MAE.

**Keywords:** Crowd counting · Computer vision · Convolutional neural network

## 1 Introduction

Crowd counting, a fundamental task in computer vision, aims to accurately estimate the number of individuals in an image. Crowd counting has found practical applications in various fields, such as crowd monitoring and counting customer flow [25]. At present, crowd counting methods are all based on deep learning,

© The Author(s), under exclusive license to Springer Nature Switzerland AG 2024
S. Rudinac et al. (Eds.): MMM 2024, LNCS 14556, pp. 43–56, 2024.
https://doi.org/10.1007/978-3-031-53311-2_4

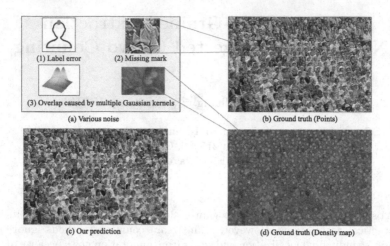

**Fig. 1.** An example from the ShangHaiTech PartA dataset, (a) various noise caused by data annotation ((1) label noise, (2) missing annotation, and (3) overlapping effect caused by Gaussian kernels), (b) the ground truth under the point framework, (c) the prediction map generated by FGENet, and (d) the ground truth under the density-map framework.

and most of them count individuals for given images from estimating density maps. This type of methods is under the density-map framework and involves converting an input image into a density map, achieved by smoothing the centroids using multiple Gaussian kernels and subsequently predicting density maps using convolutional neural network (CNN).

We can further divide these methods under the density-map framework into two groups: multi-branch [26,27,29] and single-branch [11,15]. Although these approaches have made a certain success, they come with a significant drawback of losing the precise location information pertaining to each individual. In addition, the superposition effect of Gaussian kernels exacerbates the noise during the annotation process, particularly in high-density scenarios [2], as vividly illustrated in Figs. 1(a) and 1(d).

To remedy above issues, a kind of point framework-based approaches has been proposed [12,22]. By employing the point framework, models can effectively circumvent the noise generated during the generation of density map and crucially retain the precise location information in the original data. Although methods under the point framework can solve the issues caused by the density-map framework, they cannot avoid the noise issue introduced in the annotation process. Both label noise and missing mark (Fig. 1(a)) can ruin the quality of ground truth (Fig. 1(b)) [4]. In other words, this issue can introduce inaccuracies and inconsistencies in the ground truth data, impeding the accurate estimation of crowd counting. Moreover, the preservation of fine-grained information remains a critical concern [19]. Existing methods often struggle to retain and effectively utilize intricate details in high-density crowd scenarios, limiting their accuracy and precision.

In a nutshell, crowd counting faces with three challenges at present. (1) Methods based on the density-map framework miss the precise localization of individuals during the counting process. (2) Noise, including label noise, missing mark, and superposition effect of Gaussian kernels, is inevitable when annotating data manually. (3) Existing methods are unsatisfactory in counting high-density crowd images because they may not make full use of fine-grained information.

To address these challenges, we present an end-to-end crowd counting model, named Fine-Grained Extraction Network (FGENet), that uses the point framework to learn the relationship between original images and coordinate points. For FGENet, we design a Fine-Grained Feature Pyramid (FGFP) module and an entirely novel Three-Task Combination (TTC) loss function. FGFP can adeptly capture global dependencies within top-level features and preserve crucial information pertaining to corner regions, while TTC amalgamates the outcomes of classification, regression, and counting tasks. Specially, the TTC loss function could partially mitigate the impact of noise caused by data annotation. Figure 1 gives an example from the ShangHaiTech Part_A dataset. Noise caused by data annotation would degrade the qualification of ground truth under both the point framework and the density-map framework. Figure 1(d) gives the prediction result of FGENet, which indicates that FGENet is robustness. In summary, our contributions are mainly three folds.

- We design a new module, FGFP, that not only enhances the preservation of fine-grained information in feature maps within the same layer but also facilitates effective information fusion of feature maps across different layers. By leveraging the FGFP module, we can overcome the limitations of existing methods that often struggle to capture and integrate intricate details crucial for accurate crowd counting. Through the integration of fine-grained information and the seamless combination of information across feature map layers, our proposed FGFP module serves as a key building block for achieving high counting performance in high-density crowd scenarios.
- We design the TTC loss function that combines classification, regression, and counting tasks, enhancing the generalization ability and robustness of the model. This novel loss function overcomes limitations of traditional approaches and mitigates the impact of training data noise, resulting in a more precise and resilient model that outperforms existing methods.
- On the basis of FGFP and TTC, we propose FGENet, an end-to-end counting model architecture. FGENet can preserve fine-grained information and enhance the accuracy of target counting in high-density scenarios. Extensive experiments unequivocally show that FGENet achieves state-of-the-art (SOTA) performance on diverse high-density datasets, surpassing existing benchmarks.

## 2    Related Work

In this section, we briefly review mainstream crowd counting methods and meticulously examine their respective strengths and weaknesses.

### 2.1    Methods Under Density-Map Framework

In the domain of crowd counting, Lempitsky and Zisserman [10] introduced the groundbreaking method that has revolutionized the approach to tackling high-density crowd counting tasks. Their approach constructs a linear mapping from local features of an image to a density map, thereby shifting the focus from individual targets to the collective features of target groups. As mentioned above, methods based on the density-map framework can be divided into multi-branch and single-branch ones.

**Multi-branch Models.** Early counting methods are mostly multi-branch. Some typical multi-branch models are described as follows. Multi-column Convolutional Neural Network (MCNN) is a typical multi-branch model that consists of three-column sub-networks [29]. In MCNN, three sub-networks with different convolution kernel sizes are used to extract features of the crowd image separately, and finally the features with three scales are fused by a $1 \times 1$ convolution. Multi-column Mutual Learning (McML) merges a statistical network into a multi-column network to estimate the mutual information between different columns [3]. Dilated-Attention-Deformable ConvNet (DADNet) uses dilated CNNs with different dilation rates to capture more contextual information as the front end and adaptive deformable convolution as the back end to accurately localize the location of objects [5].

Above methods are multi-branch but involve in only one task, or density-map estimation. There are multi-task multi-branch models that deal with not only density-map estimation but also other related tasks, for example, crowd count classification [21], crowd density classification [27], and crowd counting and localization [7]. However, multi-branch models have a significant drawback that they may contain more redundant parameters, leading to reduced efficiency.

**Single-Branch Models.** Recently, single-branch models become the primary counting methods, and have been proposed for making up the shortcoming of existing models. For example, Li et al. [11] proposed a Congested Scene Recognition Network (CSRNet), Liu et al. [15] proposed Context-Aware Feature Network (CAN), and Ma et al. [16] proposed the FusionCount. These methods aim at expanding the receptive field of models using different schemes.

In addition to the issue of receptive field, it is also important to eliminate the effect of backgrounds and solve the problem of extracting characters at different sizes. Thus, Miao et al. [18] proposed a Shallow Feature Based Dense Attention Network (SDANet), which is to weight fusion of multi-level features to obtain the final counting results. Moreover, researchers have noticed the problem of different distributions in images. Therefore, attention mechanisms have been introduced to solve this problem, such as Attention Scaling NetWork (ASNet) [8] and Multifaceted Attention Network (MAN) [14].

After that, it was gradually found that the noise of the data has some influence on the final model. Thus, researchers have provided different schemes to eliminate noise in data [2,4,13,17,23]. For example, Wan et al. [23] presented a Generalized Loss (GLoss) function that can smooth images and serve to eliminate data noise. Cheng et al. [2] proposed a Gaussian kernels Network (GauNet), which effectively overcomes the annotated noise in the process of generating density map by replacing the original convolution kernel with a Gaussian kernel. Dai et al. [4] proposes a Cross-head Supervision Network (CHS-Net) that solves the noise annotation problem to some extent by applying mutual supervision of convolution head and transformation head. Semi-balanced Sinkhorn with Scale consistency (S3) changes the prediction objects from density maps to the ground truth values of scatter labeling and the centroids of small Gaussian kernels separately [13].

### 2.2 Methods Under the Point Framework

To address the issue of data noise, methods under the point framework have been proposed. These methods directly predict the point locations of individuals instead of estimating density maps. Although S3 operates under the density-map framework, it essentially to estimate point objects generated from density maps.

Song et al. [22] proposed a Point to Point Network (P2PNet) that is the first crowd counting method under the point framework. P2PNet directly uses the original point coordinates as training data and retains the position information of the individuals during the counting process. After that, a Crowd Localization TRansformer (CLTR) was proposed [12].

These methods represent a notable advancement, but they neglect critical factors, such as data noise, intricate feature patterns within feature layers, and corner information-elements that play a pivotal role in accurately addressing the challenges inherent in the high-density counting task [19].

## 3    Our Method

In this section, we present a comprehensive overview of FGENet, outlining the design flow of our overall network, the innovative FGFP module, and the development of the TTC loss function.

### 3.1    Network Design

The overall architecture of our model, as depicted in Fig. 2, showcases our innovative pipeline design. FGENet mainly includes three parts: a backbone, a neck, and a head, where the neck is our designed FGFP module. First, we need to select a proper backbone. As a general backbone, FasterNet proposed in [1] has

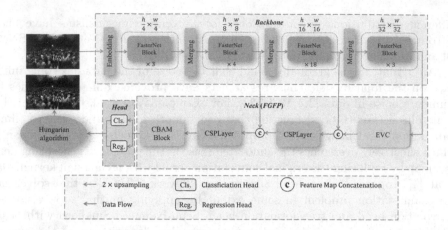

**Fig. 2.** Structure of the FGENet.

a high accuracy and operation speed because it utilizes the Partial Convolution (PConv). Therefore, we leverage the power of FasterNet-L and take it as the backbone of FGENet. Subsequently, the extracted features undergo fusion and enhancement through our meticulously designed FGFP module, a critical component that will be extensively discussed in Sect. 3.2. By leveraging the collective intelligence of these modules, our model accurately predicts the positions of targets and their corresponding confidence levels under the help of dedicated classification and regression tasks.

Following the aforementioned three parts, FGENet generates a set of precise prediction points, which serve as the foundation for the final result. Without loss of generality, let $P$ and $\hat{P}$ be the ordered sets of coordinate pairs for Ground Truth (GT) and predicted points, respectively, where $P = \{\mathbf{p}_i\}_{i=1}^N$, $\hat{P} = \{\hat{\mathbf{p}}_j\}_{j=1}^M$, $N \leq M$, $N$ is the number of GT points, and $M$ is the number of predicted points. To ensure accurate correspondence between predicted points and GT points, we employ the Hungarian matching algorithm presented by Kuhn et al. [9], leveraging a cost matrix as the basis for pairing. Let $\mathbf{C} \in \mathbb{R}^{N \times M}$ be the cost matrix. Then, the $i$th row and $j$th column of $C_{ij}$ is calculated as follows:

$$C_{ij} = \gamma \|\mathbf{p}_i - \hat{\mathbf{p}}_j\|_2 - \hat{t}_j, i = 1, \cdots, N, j = 1, \cdots, M, \tag{1}$$

where $\gamma$ is the equilibrium parameter, $\| \cdot \|_2$ denotes the $L_2$-norm Euclidean distance, and $0 \leq \hat{t}_j \leq 1$ is the confidence that the point $\hat{\mathbf{p}}_j$ is predicted as an individual. Let $\overline{P} = h(P, \hat{P}, \mathbf{C})$, where $h(\cdot)$ is the Hungarian matching algorithm and $\overline{P} = \{\overline{\mathbf{p}}_j\}_{j=1}^M$ is the matching result. In the ordered set $\overline{P}$, the first $N$ points are matched with the corresponding $N$ points in $P$, and the last $(M - N)$ points are lack of matching. Through this meticulous pairing process, our method establishes reliable associations between the predicted and GT points, culminating in a robust and accurate prediction outcome.

Finally, the obtained results undergo a critical evaluation through our meticulously designed loss function, which is comprehensively elucidated in Sect. 3.3.

This innovative TTC loss function, tailored to our specific problem domain, serves as a fundamental component in the optimization of our model parameters. Leveraging back-propagation, we iteratively refine and update the model parameters to maximize the model performance and enhance the overall accuracy and robustness of our approach.

## 3.2 FGFP Module

FGFP is a fine-grained extraction module, this module is capable of preserving the fine-grained information in the same layer of the feature map and fusing the valid information contained in different layers of the feature map.

The 1/32 feature map is initially processed by the Explicit Visual Center (EVC) block proposed in [19]. Subsequently, the information is up-sampled from the input module and fused with the 1/16 feature map. This process is repeated for the 1/8 feature map, followed by the application of the Convolutional Block Attention Module (CBAM) to enhance overall performance.

## 3.3 TTC Loss

Once the predicted points have been matched by using the Hungarian matching algorithm, we proceed to employ a meticulously designed TTC loss to guide the training process. TTC encompasses three distinct components: regression loss, classification loss, and count loss, each addressing specific aspects of our crowd counting task.

To mitigate the effect of noise in data, we design a novel regression loss $L_{reg}$, called highly smoothed $L_1$ (HSL$_1$) loss. HSL$_1$ is a variant of the smooth $L_1$ loss and can effectively minimize the influence of data noise on the final model, which has the form

$$L_{reg} = \frac{1}{N} \sum_{i=1}^{N} log(s(\mathbf{p}_i, \overline{\mathbf{p}}_i) + 1), \tag{2}$$

where the $L_1$ smoothing loss $s(\cdot, \cdot)$ can be expressed as

$$s(\mathbf{p}_i, \overline{\mathbf{p}}_i), = \begin{cases} 0.5\|\mathbf{p}_i - \overline{\mathbf{p}}_i\|_2^2, & \text{if } \|\mathbf{p}_i - \overline{\mathbf{p}}_i\|_1 < 1 \\ \|\mathbf{p}_i - \overline{\mathbf{p}}_i\|_1 - 0.5, & \text{otherwise} \end{cases} \tag{3}$$

and $\| \cdot \|_1$ denotes the $L_1$-norm.

Regarding the classification loss $L_{cls}$, we adapt the Cross Entropy (CE) loss proposed in [24] by introducing a carefully calibrated weight $\alpha$ and subsequently propose a Weighted CE (WCE) loss. Here, WCE for classification is defined as

$$L_{cls} = -\frac{1}{N} \left\{ \alpha \sum_{i=1}^{N} log\, \overline{t}_i + (1 - \alpha) \sum_{i=N+1}^{M} \left(1 - log\, \overline{t}_i\right) \right\}, \tag{4}$$

where $\bar{t}_i$ is the confidence that $\bar{p}_i$ is predicted as an individual, and the weight $\alpha \in [0,1]$.

For the counting task, we design a Highly Robust Count (HRC) loss, that is

$$L_{cou} = |M - N| \log \left( \frac{|M - N|}{N + \epsilon} + 1 \right), \tag{5}$$

where $\epsilon$ is a very small positive constant. The HRC loss effectively emphasizes errors in high-density counts and outperforms the traditional Mean Absolute Error (MAE), as demonstrated in our experiments.

Finally, the overall loss function $L$ is defined as a weighted combination of the three aforementioned components, contributing to the fine-tuning of the model performance. The comprehensive formulation of the loss function is expressed as

$$L = \lambda_1 L_{cls} + \lambda_2 L_{reg} + \lambda_3 L_{cou}, \tag{6}$$

where $\lambda_1$, $\lambda_2$, and $\lambda_3$ are the weights for combining three loss components.

## 4   Experiments

In this section, we embark on a comprehensive exploration of our proposed approach through a series of rigorous experiments. Our algorithm is coded in Python, using the Pytorch deep learning framework. Experiments are performed on an Ubuntu 18.04 system with a GeForce RTX 3090 graphics card.

### 4.1   Datasets

In our experiments, there are four public datasets, including ShangHaiTech Part_A and Part_B [29], UCF_CC_50 [6], and UCF-QNRF [7], as shown in Table 1, where SHT_A is short for ShangHaiTech Part_A, and SHT_B is for ShangHaiTech Part_B. The description about these datasets are as follows

- **SHT_A:** This dataset is from the ShangHaiTech dataset that holds immense significance in crowd counting. SHT_A is regarded as a cornerstone dataset and has earned its reputation as one of the largest and most widely used datasets presently. This dataset encompasses a comprehensive collection of 482 web images, with a division of 300 images for training and 182 images for rigorous test. In addition, SHT_A contains an extensive collection of marker points, totaling 241,677 individuals. On average, there are approximately 501.4 marker points per image, highlighting the rich and detailed annotations present in the dataset.
- **SHT_B:** This dataset is also from the ShangHaiTech dataset. SHT_B offers a unique perspective with busy street images captured across Shanghai. With a total of 716 images, there are 400 images dedicated to training and 316 images for test. This dataset presents diverse scene types, encompassing varied perspective angles and crowd density, thus providing a rich and representative collection for comprehensive evaluation.

- **UCF_CC_50:** This dataset presents a unique challenge in crowd counting research. With a total of 50 images, this dataset offers a diverse range of scenes, including concerts, protests, stadiums, and marathons. Each image encapsulates diverse crowd densities and perspectives. Crowd density in the UCF_CC_50 dataset range from 94 to 4543 individuals. This dataset serves as a valuable resource for evaluating the performance and accuracy of crowd counting methods across challenging and diverse real-world scenarios.
- **UCF-QNRF:** This dataset is a comprehensive and extensive collection that significantly contributes to the field of crowd counting research. With a staggering 1535 images, this dataset provides a rich and diverse set of data for analysis. The dataset is derived from web images and encompasses a massive 1,251,642 tags, serving as ground truth annotations for individual targets.

**Table 1.** Four crowd counting datasets used in experiments.

| Dataset | #Images | #Training/Test | Count statistics | | | | |
|---|---|---|---|---|---|---|---|
| | | | Average resolution | Total | Min | Average | Max |
| SHT_A | 482 | 300/182 | 589 × 868 | 241,677 | 33 | 501.4 | 3139 |
| SHT_B | 716 | 400/316 | 768 × 1024 | 88,488 | 9 | 123.6 | 578 |
| UCF_CC_50 | 50 | – | 2101 × 2888 | 63,974 | 94 | 1280 | 4543 |
| UCF-QNRF | 1535 | 1201/334 | 2013 × 2902 | 1,251,642 | 49 | 815 | 12,865 |

## 4.2 Model Evaluation

To benchmark our approach, we compare it against 16 methods, including MCNN [29], CSRNet [11], CAN [15], SDANet [18], ASNet [8], GLoss [23], S3 [13], Spatial Uncertainty-Aware (SUA) [17], P2PNet [22], FusionCount [16], MAN [14], GauNet [2], Characteristic Function Loss (ChfL) [20], CLTR [12], Ordinal-lEntropy [28], and CHS-Net [4]. The evaluation metrics employed for assessing the model performance are MAE and Mean Squared Error (MSE), which are widely adopted in crowd counting.

Table 2 lists the comparison of FGENet with other 16 methods for crowd counting, where the best results are in bold and the second best ones are underlined. Observation on Table 2 indicates that FGENet achieves the SOTA results on both SHT_A and UCF_CC_50. Specifically, FGENet significantly improves 5.72% MAE on SHT_A, and 17.46 % MAE on the UCF_CC_50 dataset compared to the second best method. In Fig. 3, three images from SHT_A are given, and their predictions obtained by FGENet are also provided. We can see the number of predicted points is very close to the number of GT points.

On SHT_B, FGENet ranks the third but is only 0.14 MAE less than the best model. As indicated in Table 1, there are a large amount of missing annotations in the SHT_B dataset. Although our method can capture high occlusion small density targets, it is not able to cope with this situation well.

**Table 2.** Comparison of counting performance obtained by 17 methods.

| Methods | Venue | SHT_A | | SHT_B | | UCF_CC_50 | | UCF-QNRF | |
|---|---|---|---|---|---|---|---|---|---|
| | | MAE | MSE | MAE | MSE | MAE | MSE | MAE | MSE |
| MCNN [29] | CVPR'16 | 110.2 | – | 26.4 | – | 377.6 | – | – | – |
| CSRNet [11] | CVPR'18 | 68.2 | 115 | 10.6 | 16 | 266.1 | – | 120.3 | 208.5 |
| CAN [15] | CVPR'19 | 62.8 | 101.8 | 7.7 | 12.7 | 212.2 | 243.7 | 107 | 183 |
| SDANet [18] | AAAI'20 | 63.6 | 101.8 | 7.8 | 10.2 | 227.6 | 316.4 | – | – |
| ASNet [8] | CVPR'20 | 57.78 | 90.13 | – | – | 174.84 | 251.63 | 91.59 | 159.71 |
| GLoss [23] | IJCAI'21 | 57.3 | 90.7 | 7.3 | 11.4 | – | – | 81.2 | 138.6 |
| S3 [13] | IJCAI'21 | 57 | 96 | 6.3 | 10.6 | – | – | 80.6 | 139.8 |
| SUA [17] | ICCV'21 | 68.5 | 121.9 | 14.1 | 20.6 | – | – | 130.3 | 226.3 |
| P2PNet [22] | ICCV'21 | 55.73 | 89.20 | 6.97 | 11.34 | 172.72 | 256.18 | 85.32 | 154.5 |
| FusionCount [16] | ICIP'22 | 62.2 | 101.2 | 6.9 | 11.8 | – | – | – | – |
| MAN [14] | CVPR'22 | 56.8 | 90.3 | – | – | – | – | **77.3** | **131.5** |
| GauNet [2] | CVPR'22 | 54.8 | 89.1 | **6.2** | **9.9** | 186.3 | 256.5 | 81.6 | 153.7 |
| ChfL [20] | CVPR'22 | 57.5 | 94.3 | 6.9 | 11 | – | – | 80.3 | 137.6 |
| CLTR [12] | ECCV'23 | 56.9 | 95.2 | 6.5 | 10.6 | – | – | 85.8 | 141.3 |
| OrdinalEntropy [28] | ICLR'23 | 65.6 | 105 | 9.1 | 14.5 | – | – | – | – |
| CHS-Net [4] | ICASSP'23 | 59.2 | 97.8 | 7.1 | 12.1 | – | – | 83.4 | 144.9 |
| FGENet (Ours) | – | **51.66** | **85** | 6.34 | 10.53 | **142.56** | **215.87** | 85.2 | 158.76 |

Due to images with large size in UCF-QNRF, the methods related to Transformer work better, such as MAN. In addition, our method is limited by the efficiency of the matching algorithm, so there are some issues in the process of parameter adjustment.

In summary, we find that the existing models can fit the dataset with large-scale and high-density or small-scale and low-density, but perform generally bad on small-scale and high-density datasets. Fortunately, our method can remedy this defect because experimental results demonstrate that FGENet is particularly effective on small-scale and high-density datasets.

### 4.3   Ablation Study

The goal of ablation study is to validate the efficacy and reliability of new designed FGFP module and TTC loss function. Here, ablation experiments are conducted on the SHT_A dataset.

**Experiments on FGFP.** To demonstrate the validity of our FGFP, we list the possible combinations of our model. The ablation settings and experimental results are shown in Table 3. Among these combinations, our model with the complete FGFP module (ID8 in Table 3) has the best counting performance, and our model without FGFP (ID1 in Table 3) is the worst. Observation on ID2–ID4 in Table 3 indicates that CSPLayer is the most important block among three blocks, and observation on ID5–ID7 shows that the combination of CSPLayer and CBAM is compared to that of CSPLayer and EVC.

**Fig. 3.** Three crowd images of SHT_A and their predictions obtained by FGENet.

In summary, findings demonstrate the scientific nature of our proposed FGFP module. In FGFP, the goal of CSPLayer is to fuse the information of feature maps between layers. EVC and CBAM are used to extract the fine-grained information within feature maps and enhance attention to different channels and spaces, respectively. The three blocks complement each other, and neither is dispensable.

**Table 3.** Ablation experiments on FGFP.

| ID | EVC | CSPLayer | CBAM | MAE | MSE |
|----|-----|----------|------|-----|-----|
| 1 | × | × | × | 72.19 | 112.18 |
| 2 | ✓ | × | × | 71.84 | 112.09 |
| 3 | × | ✓ | × | 57.6 | 92.27 |
| 4 | × | × | ✓ | 69.51 | 116.6 |
| 5 | ✓ | ✓ | × | 54.67 | 96.99 |
| 6 | ✓ | × | ✓ | 62.79 | 109.84 |
| 7 | × | ✓ | ✓ | 54.1 | 92.44 |
| 8 | ✓ | ✓ | ✓ | 51.66 | 85 |

**Experiments on Loss Function.** The TTC loss function (6) consists of three terms: HSL$_1$ $L_{reg}$ for regression, WCE $L_{cls}$ for classification, and HRC $L_{cou}$ for counting. To validate the performance of TTC loss, we set up 10 possible variants and show their experimental results in Table 4. Note that ID1 is common used loss, and ID2 is the baseline of our loss.

Among variants ID2–ID5, we can see that ID4 has the best result, which changes only CE to WCE for classification. The main reason is that the weight $\lambda_1$ of the classification task accounts for a higher weight in the whole loss function.

In the case of changing two terms (variants ID6–ID8), ID8 using WCE and HRC works better. We believe that the smoothness and robustness of HRC solves the problem caused by the missed mark to a certain extent. Among variants ID8–ID10, the variant ID10 performs the best. Because $HSL_1$ is less penalized for smaller position errors and overly very smooth, it can reduce the influence of labeling errors.

**Table 4.** Ablation experiments on TTC loss function.

| ID | Regression loss | Classification loss | Counting loss | MAE | MSE |
|----|-----------------|---------------------|---------------|-----|-----|
| 1  | MSE             | CE                  | –             | 59.62 | 104.12 |
| 2  | MSE             | CE                  | MAE           | 57.18 | 98.21 |
| 3  | $HSL_1$         | CE                  | MAE           | 57.115 | 97.62 |
| 4  | MSE             | WCE                 | MAE           | 53.37 | 92.72 |
| 5  | MSE             | CE                  | HRC           | 56.36 | 100.80 |
| 6  | $HSL_1$         | WCE                 | MAE           | 53.35 | 90.74 |
| 7  | $HSL_1$         | CE                  | HRC           | 56.59 | 95.02 |
| 8  | MSE             | WCE                 | HRC           | 52.14 | 87.37 |
| 9  | Smooth L1       | WCE                 | HRC           | 54.64 | 95.54 |
| 10 | $HSL_1$         | WCE                 | HRC           | **51.66** | **85.00** |

## 5    Conclusion

In our study, we have shed light on two prevalent challenges in the field of crowd counting. First, we have identified three distinct types of noise that manifest in the training data, which inevitably impact the performance of the final model. Second, given the pixel-level nature of crowd counting, the preservation and utilization of fine-grained information play a crucial role in achieving accurate results. To effectively tackle these challenges, we have introduced a novel loss function to mitigate the influence of noise, and a dedicated FGFP module to address the fine-grained information problem. Through extensive experimentation, we have demonstrated the scientific validity and efficacy of our proposed approach.

Our method has achieved state-of-the-art (SOTA) results on benchmark datasets, such as ShangHaiTech PartA and UCF_CC_50, showcasing its superior performance compared to existing methods. However, it is important to acknowledge that our method does have certain limitations. For instance, it exhibits relatively lower performance in detecting large targets, and the matching process for high-resolution images incurs high time complexity. Moving forward, we are committed to further optimizing our detection results and refining our matching algorithm to overcome these limitations.

# References

1. Chen, J., et al.: Run, Don't Walk: chasing higher FLOPS for faster neural networks. In: Proceedings of the IEEE/CVF Conference on Computer Vision and Pattern Recognition (CVPR), pp. 12021–12031 (2023)
2. Cheng, Z.Q., Dai, Q., Li, H., Song, J., Wu, X., Hauptmann, A.G.: Rethinking spatial invariance of convolutional networks for object counting. In: Proceedings of the IEEE/CVF Conference on Computer Vision and Pattern Recognition (CVPR), pp. 19638–19648 (2022)
3. Cheng, Z.Q., Li, J.X., Dai, Q., Wu, X., He, J.Y., Hauptmann, A.: Improving the learning of multi-column convolutional neural network for crowd counting (2019)
4. Dai, M., Huang, Z., Gao, J., Shan, H., Zhang, J.: Cross-head supervision for crowd counting with noisy annotations. In: Proceedings of the IEEE International Conference on Acoustics, Speech and Signal Processing (ICASSP), pp. 1–5. IEEE (2023)
5. Guo, D., Li, K., Zha, Z., Wang, M.: DADNet: dilated-attention-deformable ConvNet for crowd counting. In: Proceedings of the the 27th ACM International Conference (ACM MM) (2019)
6. Idrees, H., Saleemi, I., Seibert, C., Shah, M.: Multi-source multi-scale counting in extremely dense crowd images. In: Proceedings of the IEEE Conference on Computer Vision and Pattern Recognition (CVPR), pp. 2547–2554 (2013)
7. Idrees, H., et al.: Composition loss for counting, density map estimation and localization in dense crowds. In: Ferrari, V., Hebert, M., Sminchisescu, C., Weiss, Y. (eds.) ECCV 2018. LNCS, vol. 11206, pp. 544–559. Springer, Cham (2018). https://doi.org/10.1007/978-3-030-01216-8_33
8. Jiang, X., et al.: Attention scaling for crowd counting. In: Proceedings of the IEEE/CVF Conference on Computer Vision and Pattern Recognition (CVPR), pp. 4706–4715 (2020)
9. Kuhn, H.W.: The Hungarian method for the assignment problem. Naval Res. Logistics Q. **2**(1–2), 83–97 (1955)
10. Lempitsky, V.S., Zisserman, A.: Learning to count objects in images. In: Proceedings of the 24th Annual Conference on Neural Information Processing Systems (NeurIPS) (2010)
11. Li, Y., Zhang, X., Chen, D.: CSRNet: dilated convolutional neural networks for understanding the highly congested scenes. In: Proceedings of the IEEE Conference on Computer Vision and Pattern Recognition (CVPR), pp. 1091–1100 (2018)
12. Liang, D., Xu, W., Bai, X.: An end-to-end transformer model for crowd localization. In: Avidan, S., Brostow, G., Cissé, M., Farinella, G.M., Hassner, T. (eds.) Computer Vision – ECCV 2022. ECCV 2022. LNCS, vol. 13661, pp. 38–54. Springer, Cham (2022). https://doi.org/10.1007/978-3-031-19769-7_3
13. Lin, H., et al.: Direct measure matching for crowd counting. In: Proceedings of the Thirtieth International Joint Conference on Artificial Intelligence (IJCAI), pp. 837–844 (2021)
14. Lin, H., Ma, Z., Ji, R., Wang, Y., Hong, X.: Boosting crowd counting via multifaceted attention. In: Proceedings of the IEEE/CVF Conference on Computer Vision and Pattern Recognition (CVPR), pp. 19628–19637 (2022)
15. Liu, W., Salzmann, M., Fua, P.: Context-aware crowd counting. In: Proceedings of the IEEE/CVF Conference on Computer Vision and Pattern Recognition (CVPR), pp. 5099–5108 (2019)
16. Ma, Y., Sanchez, V., Guha, T.: FusionCount: efficient crowd counting via multiscale feature fusion. In: Proceedings of the IEEE International Conference on Image Processing (ICIP), pp. 3256–3260 (2022)

17. Meng, Y., et al.: Spatial uncertainty-aware semi-supervised crowd counting. In: Proceedings of the IEEE/CVF International Conference on Computer Vision (ICCV), pp. 15549–15559 (2021)
18. Miao, Y., Lin, Z., Ding, G., Han, J.: Shallow feature based dense attention network for crowd counting. In: Proceedings of the AAAI Conference on Artificial Intelligence (AAAI), pp. 11765–11772 (2020)
19. Quan, Y., Zhang, D., Zhang, L., Tang, J.: Centralized feature pyramid for object detection. CoRR abs/2210.02093 (2022)
20. Shu, W., Wan, J., Tan, K.C., Kwong, S., Chan, A.B.: Crowd counting in the frequency domain. In: Proceedings of the IEEE/CVF Conference on Computer Vision and Pattern Recognition (CVPR), pp. 19618–19627 (2022)
21. Sindagi, V.A., Patel, V.M.: CNN-based cascaded multi-task learning of high-level prior and density estimation for crowd counting. In: Proceedings of the 14th IEEE International Conference on Advanced Video and Signal Based Surveillance (AVSS), pp. 1–6. IEEE (2017)
22. Song, Q., et al.: Rethinking counting and localization in crowds: a purely point-based framework. In: Proceedings of the IEEE/CVF International Conference on Computer Vision (ICCV), pp. 3365–3374 (2021)
23. Wan, J., Liu, Z., Chan, A.B.: A generalized loss function for crowd counting and localization. In: Proceedings of the IEEE/CVF Conference on Computer Vision and Pattern Recognition (CVPR), pp. 1974–1983 (2021)
24. Wang, Y., Ma, X., Chen, Z., Luo, Y., Yi, J., Bailey, J.: Symmetric cross entropy for robust learning with noisy labels. In: Proceedings of the IEEE/CVF International Conference on Computer Vision (CVPR), pp. 322–330 (2019)
25. Wen, L., et al.: Detection, tracking, and counting meets drones in crowds: a benchmark. In: Proceedings of the IEEE/CVF Conference on Computer Vision and Pattern Recognition (CVPR), pp. 7812–7821 (2021)
26. Yan, L., Zhang, L., Zheng, X., Li, F.: Deeper multi-column dilated convolutional network for congested crowd understanding. Neural Comput. Appl. **34**(2), 1407–1422 (2022)
27. Zhang, L., Yan, L., Zhang, M., Lu, J.: $T^2$CNN: a novel method for crowd counting via two-task convolutional neural network. Visual Comput. **39**(1), 73–85 (2023)
28. Zhang, S., Yang, L., Mi, M.B., Zheng, X., Yao, A.: Improving deep regression with ordinal entropy (2023). https://doi.org/10.48550/arXiv.2301.08915
29. Zhang, Y., Zhou, D., Chen, S., Gao, S., Ma, Y.: Single-image crowd counting via multi-column convolutional neural network. In: Proceedings of the IEEE Conference on Computer Vision and Pattern Recognition (CVPR), pp. 589–597 (2016)

# MSMV-UNet: A 2.5D Stroke Lesion Segmentation Method Based on Multi-slice Feature Fusion

Jingjing Xie[1], JiXuan Hong[1], Manjin Sheng[1], and Chenhui Yang[2]([⊠])

[1] Department of Computer Science and Technology, Xiamen University, Xiamen, Fujian, China
[2] Information College, Xiamen University, Xiamen, Fujian, China
chyang@xmu.edu.cn

**Abstract.** Stroke is a severe and life-threatening disease. MRI plays a crucial role in the diagnosis and treatment of stroke, enabling comprehensive analysis of MRI images for accurate localization and qualitative assessment of stroke lesions. A 2.5D stroke lesion segmentation method based on the fusion of multi-slice features is proposed in this study. This method introduces the time information feature between slices in three-dimensional (3D) image data on the basis of a two-dimensional (2D) segmentation model. Multiple encoding paths are densely connected between adjacent slices to utilize the time information feature. To address the problem of difficult segmentation of stroke lesions edges, a Slice-Context Attention Module is proposed to reinforce the differences between adjacent slices in the feature maps. Additionally, considering the multi-perspective features in stroke lesions imaging data, this method proposes to train the segmentation model from three different perspectives: cross section, sagittal plane and coronal plane, and uses soft voting strategy to fuse the results of the three models to form a 2.5D method. Qualitative and quantitative experimental results demonstrate that this method has certain superiority compared to existing methods.

**Keywords:** Stroke Lesion Segmentation · MRI · Deep Learning · Computer-aided Diagnosis

## 1 Introduction

Stroke is one of the leading causes of disability and mortality worldwide [5]. Early detection, prompt action, and immediate treatment are crucial for reducing the risk of stroke-related disability and death [19]. Imaging examinations play a vital role in the diagnosis, treatment, post-treatment evaluation, and prognosis assessment of stroke. Medical imaging can help to visualise the distribution of internal brain tissues, facilitating comprehensive localization, quantification, and qualitative understanding of stroke lesions. Among various imaging modalities, magnetic resonance imaging (MRI) has emerged as the most widely used non-invasive technique for assessing structural brain changes following stroke, thanks

© The Author(s), under exclusive license to Springer Nature Switzerland AG 2024
S. Rudinac et al. (Eds.): MMM 2024, LNCS 14556, pp. 57–69, 2024.
https://doi.org/10.1007/978-3-031-53311-2_5

to its ability to accurately identify alterations in brain tissues and cells [2]. In this paper, T1-weighted magnetic resonance images of stroke patients are used to study the location segmentation of lesions in the images.

Lesion segmentation is to separate lesion tissues from normal tissues using boundary delineation. The segmentation process needs to be fast, and the results should be accurate. As an important research branch in the field of medical image segmentation, brain lesion segmentation has seen a significant number of machine learning-based solutions proposed in recent years. Lucas et al. [11] proposed an ultra-densely connected segmentation network and a hybrid loss function for ischaemic stroke lesion segmentation. This densely connected network effectively utilizes multi-modal contextual information and addresses the issues of vanishing and exploding gradients. Liu et al. [9] proposed Res-FCN, a fully convolutional network with residual structures, to improve the accuracy of stroke lesion boundary segmentation. Dolz et al. [3] extended the convolutional blocks of InceptionNet [16] to capture more contextual information and enhance performance. However, these frameworks have limitations in terms of information diversity. Qi et al. [13] proposed X-Net, a network based on depthwise separable convolutions, which introduced a Feature Similarity Module to capture long-range dependencies and improve segmentation results. Hui et al. [4] combined attention mechanisms with U-Net and improved segmentation accuracy by suppressing irrelevant regions of lesions through attention generated from multi-scale features.

Most of the aforementioned methods are based on 2D image slices, which leads to the loss of important contextual information. Zhou et al. [21] proposed D-UNet, a neural network that combines 2D and 3D convolutions to leverage the correlation between consecutive slices. Basak et al. [1] further improved this approach with DFENet, incorporating selective feature fusion and an edge-guided module for refined lesion boundaries. However, these methods do not consider feature information perpendicular to the slice plane. Additionally, some networks solely based on 3D convolutions, such as 3D MI-UNet proposed by Zhang et al. [20], achieved better segmentation results but require significant computational resources and higher data requirements.

To leverage the 3D spatial information of MRI images and conserve computational resources, we propose a 2.5D stroke lesion segmentation method and employ a soft voting fusion approach for result integration. In summary, this paper makes the following main contributions:

- We propose a stroke lesion segmentation algorithm based on multi-slice dense feature fusion. By fully utilizing the information within and between slices, the network achieves improved segmentation performance.
- We propose a stroke lesion segmentation method based on a multi-perspective soft voting strategy. This model utilizes the information differences between different viewpoints to improve the segmentation results.

## 2 Method

We propose a 2.5D stroke lesion segmentation method based on multi-slice feature fusion (MSMV-UNet). The overall framework of the method is depicted in Fig. 1 and comprises three main parts. Figure 1(a) is the data preprocessing stage, where multi-slice sequences are generated. The MRI data is sliced along the three axes to form two-dimensional slice sequences. Figure 1(b) is the training of a stroke lesion segmentation model based on multi-slice dense feature fusion. The model takes consecutive 2D slices as input. It incorporates the Inter Slice Attention Module (ISAM) to weight the differences between slices and uses dense connections across multiple encoding paths to facilitate information exchange between slices. Figure 1(c) is the stroke lesion segmentation strategy based on the multi-view soft voting approach. The method reconstructs the 2D segmentation results obtained from the three models in part (b) into 3D segmentation results based on the original slice order. These 3D results are then fused using a soft voting strategy to obtain the final segmentation result.

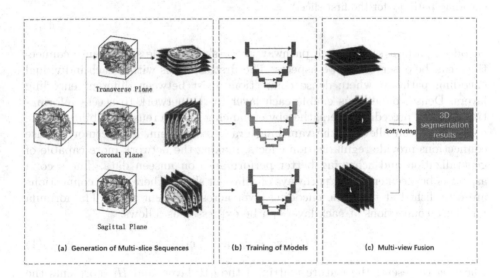

**Fig. 1.** MSMV-UNet architecture

### 2.1 Multi-slice Dense Feature Fusion

There is temporal information between the slices of MRI images. Using only 2D slices as input results in the loss of this important information. Therefore, we convert the input into a sequence of consecutive 2D slices to provide more informative cues for lesion segmentation.

To fully utilize the information between slices, we propose a multi-slice dense feature fusion framework, as shown in Fig. 2. This structure utilizes multiple

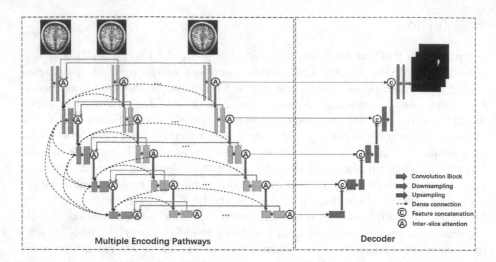

**Fig. 2.** Multi-slice Dense Feature Fusion Framework (Only illustrating the complete encoding pathway for the first slice.)

encoding pathways, with each pathway processing a single slice. Dense connections can be observed in two aspects. The first aspect is within each individual encoding pathway, where dense connections exist between different encoding layers. Dense connections enable each layer in the network to access information from all preceding layers, thereby enhancing feature reuse capability. These connections can alleviate the vanishing gradient problem. Furthermore, dense connections provide regularization effects, making the network more capable of generalization and achieving better performance on unseen data. The second aspect is between encoding pathways of different slices, where dense connections are established at the same encoding layer across different slices. The formula for dense connections at each layer can be expressed as follows:

$$x_l = H_l \left( [x_{l-1}, x_{l-2}, \ldots, x_1] \right) \tag{1}$$

where $x_l$ represents the feature matrix of the $l$-th layer, and $H_l$ represents the mapping function corresponding to the proposed convolutional block. [...] denote matrix concatenation. The formula indicates that, unlike the conventional feature output formula $x_l = H_l (x_{l-1})$, the dense connection approach of stacking previous layers features is an effective method to prevent feature loss.

To maintain the uniqueness of each path, we connect the feature maps of each branch in different paths in different orders. The feature map of the current slice branch is placed in the first position, followed by the feature maps of other slice branches in sequential order. The feature of the $p$-th slice in the $l$-th layer, after shuffling, can be represented as follows:

$$f_l^p = F_l^p \left( x_{l-1}^1, \ldots, x_{l-1}^p, \ldots, x_{l-1}^n \right) = \left[ x_{l-1}^p, x_{l-1}^1, \ldots, x_{l-1}^{p-1}, x_{l-1}^{p+1}, \ldots, x_{l-1}^n \right] \tag{2}$$

Assuming there are a total of $n$ consecutive slices numbered from 1 to $n$, the shuffling function $F_l^p$ is responsible for placing the features of the current slice at the beginning, while concatenating the features of the other slices in their original order. Combining this with the aforementioned dense connection strategy, the final formulation for the features of each layer can be expressed as follows:

$$x_l^p = H_l^p(F_l^p((x_{l-1}^1, ..., x_1^1), ..., (x_{l-1}^p, ..., x_1^p), ..., (x_{l-1}^n, ..., x_1^n))) \qquad (3)$$

## 2.2  Inter-slice Attention Module

We proposes an Inter-slice Attention Module (ISAM) for the fusion of consecutive slice features. Due to the strong similarity between consecutive slices, the differences in features between slices mainly concentrate on different positions, such as lesion edges and tissue boundaries. These regions have a significant impact on the localization of tissue lesions and can effectively assist in the segmentation of stroke lesions. The objective of this attention module is to utilize the feature differences between adjacent slices, focusing the network's attention on challenging regions that are difficult to predict and refining the decoding results. This module is applied after each convolutional block to perform feature value weighting.

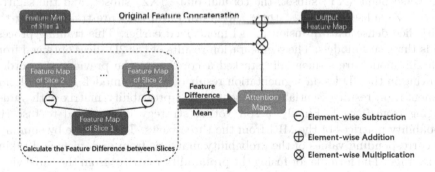

**Fig. 3.** Structure of ISAM

The module structure of ISAM is shown in Fig. 3. For the $d$-th encoding layer, there are $n$ different slice features $\{x_t\}_{t=1}^n$. The formula for the inter-slice attention map of the $d$-th layer feature map of Slice $i$ is as follows:

$$\mathcal{M}_i^d = \frac{1}{n-1} \sum_{t \neq i} (|x_t - x_i|) \qquad (4)$$

In this formula, $x_i$ represents the current slice feature map, $x_t$ represents the feature maps of the other slices except the current Slice $i$. The formula calculates the absolute difference between the feature map of slice $i$ and the feature maps of other slices, then takes the average to obtain the attention map for the current slice. In this attention map, pixels close to 0 indicate that the corresponding pixels in the current slice are more similar to other slices, while pixels closer to 1 indicate greater differences between the current slice and other slices. By multiplying this attention map with the original feature map and adding it to the original feature map, the final output feature map is obtained. This operation prevents information loss and corrects the information in the feature map. Additionally, introducing the feature information from other slices achieves the purpose of inter-slice information interaction.

## 2.3   Multi-view Soft Voting Strategy

According to our proposed multi-slice dense feature fusion method, the goal of lesion segmentation can be achieved. However, this method can only capture the 2D slice and its contextual slice information along a single view direction, leading to the loss of slice information in the direction perpendicular to the plane. In order to fully utilize the 3D spatial features of MRI, we propose a multi-view 2.5D segmentation method, which uses a soft voting strategy for result combination. The main purpose of this method is to further enhance network performance through the extraction of information from multiple views.

First, the 3D MRI model (XYZ) is sliced into n consecutive slices along the transverse, coronal, and sagittal axes, resulting in three subsets of data: the transverse plane (XY) subset, the coronal plane (XZ) subset, and the sagittal plane (YZ) subset. For these three subsets of data, they are trained using the multi-slice dense feature fusion model mentioned earlier. This training process yields three sub-models. The segmentation results of the 2D slices obtained from each sub-model are sequentially stacked according to the previous slice order, resulting in the 3D lesion segmentation result of a single model. Each 3D lesion segmentation result essentially represents a 3D probability matrix with values ranging from 0 to 1. Finally, a soft voting strategy is employed to fuse the probability matrices of the MRI from the three views. This is done by summing the corresponding values of the probability matrices from each view and taking the average. The formula for fusing the probability matrices from the three views using the soft voting strategy is as follows:

$$p_{x,y,z}^F = \begin{cases} 1, & \left( \frac{p_{x,y,z}^C + p_{x,y,z}^A + p_{x,y,z}^S}{3} \right) \geq 0.5 \\ 0, & otherwise \end{cases} \tag{5}$$

In the formula, $p$ represents the segmentation probability value of different pixels. The subscripts $x$, $y$, $z$ of $p$ respectively represent the coordinates of the pixels in the three plane directions. The superscripts $C$, $A$, $S$ of $p$ represent the results of the coronal plane model, transverse plane model, and sagittal plane model, respectively. The $p$ value is either 0 or 1. For each pixel, the average of the probability results from the three models is first calculated. Then, a threshold of 0.5 is used to correct the values, where 1 represents the lesion area (foreground) and 0 represents the non-lesion area (background). This process generates the final segmentation result image.

# 3   Experiments

## 3.1   Datasets

We used a subset of the publicly available dataset "Anatomical Tracings of Lesions After Stroke" (Atlas) [7] as our experimental dataset. This dataset includes manually segmented lesion masks and metadata provided by expert annotators. After excluding MRI images that failed registration, we were left with 229 data samples transformed to the standard MNI space.

The experimental dataset consists of 229 standardized 3D T1-weighted MRI images. First, the 3D MRI image data were cropped to remove irrelevant information and increase the proportion of stroke lesions in the entire image. The cropped 3D MRI had a resolution of $208 \times 176 \times 176$ ($X \times Y \times Z$). Subsequently, the 3D MRI images were sliced consecutively along the transverse, coronal, and sagittal axes. The resulting input data sizes were $208 \times 176 \times n$ ($X \times Y \times n$), $208 \times 176 \times n$ ($X \times Z \times n$), and $176 \times 176 \times n$ ($Y \times Z \times n$), respectively, where n represents the number of slices.

The dataset was divided into categories, with a random selection of 137 cases for the training set, 36 cases for the validation set, and 56 cases for the test set. The 3D MRI images were sliced and each slice was saved in .npy format. The training set after slicing consisted of 24,112 two-dimensional images, the validation set had 8,244 two-dimensional images, and the test set contained 12,824 two-dimensional images. Only the training dataset underwent a random shuffling operation.

## 3.2   Implementation Details

In the experiments, we implemented our model using the PyTorch framework and using the Adam optimizer [6] with an initial learning rate of $init_{lr} = 1e^{-3}$. A multi-learning rate strategy is applied to adjust the initial learning rate, where $lr = init_{lr} \times \left(1 - \frac{epoch}{nEpoch}\right)^{power}$, with $power = 0.9$ and $nEpoch = 100$. The batch size is uniformly set to 24.

We utilized a range of evaluation metrics to assess the performance of the proposed model. For each patient in the test set, we computed evaluation scores based on the 3D MRI data. We calculated and reported the mean Dice Similarity Coefficient (DSC), Precision, and Recall. Additionally, we evaluated the global DSC across all data. DSC is the primary metric used to measure the dissimilarity between the model's predictions and the ground truth, indicating the overlap ratio between the two masks. Precision represents the ratio of true positives to the overall predicted results, while recall represents the ratio of correctly classified positive samples by the model to the total number of true positive samples. These metrics are defined as follows:

$$DSC = \frac{2TP}{2TP + FP + FN} \tag{6}$$

$$Recall = \frac{TP}{TP + FN} \tag{7}$$

$$Precision = \frac{TP}{TP + FP} \tag{8}$$

A threshold of 0.5 is applied to binarize the probability map. Pixels above the threshold are set to 1, and pixels below the threshold are set to 0, generating the predicted segmentation map. Positive predictions occur when the predicted pixel value matches the true value, while negative predictions occur when they differ. True positive (TP) represents the number of samples where both the true class and the model's prediction are positive. False negative (FN) refers to the number of samples where the true class is positive, but the model predicts a negative value. False positive (FP) refers to the number of samples where the true class is negative, yet the model predicts a positive value.

### 3.3   Comparison of Different Methods

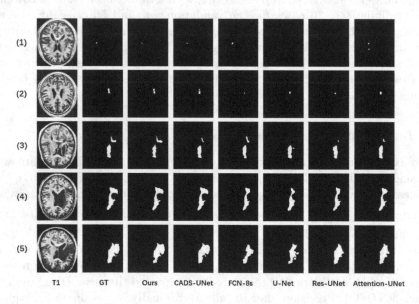

**Fig. 4.** Comparisons of our method, baseline, CADS-UNet, FCN-8s, U-Net, Res-UNet, and Attention-UNet.

We compared our proposed stroke lesion segmentation method, MSMV-UNet, with four classical models in the segmentation field: FCN-8s [10], U-Net [14], Res-UNet [17], Attention-UNet [12], as well as the stroke lesion segmentation model CADS-UNet [15]. Figure 4 displays the comparative results among these models. Since the other five methods are 2D image segmentation methods trained only on cross-sectional datasets, we obtained cross-sectional slices from our final 3D model for comparison and demonstration. The images are vertically arranged based on the size of the stroke lesions, and the second column represents the segmentation ground truth labels.

By comparing the results in the first row, we can observe that our model is capable of detecting small lesions that U-Net and Res-UNet fail to segment. In the second row, our model not only correctly detects small lesions but also achieves the highest segmentation accuracy, indicating its advantage in detecting small lesions. In the third row, it can be observed that other methods exhibit pixel-level omissions for both large and small lesions. In contrast, our model demonstrates more accurate segmentation for both regions. Comparing the results in the fourth and fifth rows, it can be seen that our model's segmentation for large lesions aligns more closely to the true labels. Through qualitative analysis of the visualized results of each model, we can observe the superiority of our model.

**Table 1.** Experimental results of different methods on the Atlas1.1 dataset

| Method | DSC | DSC (global) | Recall | Precision |
|---|---|---|---|---|
| FCN-8s [10] | 41.2% | 63.6% | 38.2% | 59.5% |
| U-Net [14] | 49.0% | 73.1% | 52.5% | 54.7% |
| Res-UNet [17] | 51.4% | 70.2% | 54.2% | 60.0% |
| Attention-UNet [12] | 52.0% | 76.2% | 56.6% | 57.5% |
| CADS-UNet [15] | 55.9% | 76.3% | 58.3% | 64.8% |
| MSMV-UNet (Ours) | 58.1% | 78.2% | 60.4% | 65.9% |

Table 1 presents the results of various models trained based on the experimental environment configuration. Analyzing the table, it can be observed that the results of Res-UNet, Attention-UNet, and CADS-UNet models, which are built upon the U-Net baseline, show improvement. This demonstrates the effectiveness of the modules incorporated into these methods on top of U-Net. Our MSMV-UNet model achieved the best Dice Similarity Coefficient (DSC) of 58.1%, which is a 9.1% improvement compared to the U-Net baseline. Furthermore, our model outperformed in terms of global DSC, Recall, and Precision, confirming the superiority of our model.

To further demonstrate the effectiveness of our model, Table 2 compares our MSMV-UNet model with other existing methods on the same dataset. Five deep learning-based methods were selected, including D-UNet [21], U-Net (9 paths) [18], DFENet [1], MSDF-Net [8], and 3D MI-UNet [20]. From Table 2, it can be observed that our model achieved the best performance compared to other methods on the same dataset. Additionally, 3D MI-UNet, which employs 3D convolution for image segmentation, achieved better results compared to other 2D models, demonstrating the effectiveness of incorporating 3D spatial information. Our module outperformed the 2D counterpart, which further confirms that we

**Table 2.** Comparison results of other methods on the Atlas1.1 dataset

| Method | DSC | Recall | Precision | Train/Test |
|---|---|---|---|---|
| D-UNet [21] | 53.5% | 52.4% | 63.3% | 183/46 |
| U-Net(9 paths) [18] | 54.0% | – | – | 0/54 |
| DFENet [1] | 54.6% | 49.7% | 63.7% | 5-fold CV |
| MSDF-Net [8] | 55.8% | – | – | 160/69 |
| 3D MI-UNet [20] | 56.7% | 59.4% | 65.5% | 5-fold CV |
| MSMV-UNet (Ours) | 58.1% | 60.4% | 65.9% | 137/56 |

introduced 3D spatial information into the model through dense feature fusion of multiple slices and a multi-view soft voting mechanism. This approach filters out redundant spatial information brought by direct 3D convolution and yields superior results.

### 3.4 Influence of Consecutive Slice Quantity

In our proposed method based on multi-slice dense feature fusion, the choice of different slice numbers n has an impact on both the overall model parameter computation and model accuracy. In this section, we have conducted experiments with varying slice numbers n(2, 3, 4, 5), using only cross-sectional data for the model dataset. The comparative results are presented in Table 3 As shown in Table 3, it can be observed that the DSC values of the model exhibit a trend of initially increasing and then decreasing. When n is set to 4, the overall model reaches its optimal performance. When the number of slices is too low, such as n = 2, although the dense feature connections between slices improve the model's performance to some extent, they do not provide sufficient temporal information between slices. On the other hand, when the number of slices is too high, such as n = 5, an excessive number of slices not only dramatically increases computational complexity but also introduces misleading information due to significant differences in inter-slice information, resulting in a decrease in model accuracy.

**Table 3.** The experimental results for different numbers of consecutive slices

| n | DSC | DSC (global) | Recall | Precision |
|---|---|---|---|---|
| 2 | 55.0% | 76.8% | 62.4% | 57.6% |
| 3 | 56.6% | 75.8% | 56.2% | 67.3% |
| 4 | 57.0% | 79.0% | 62.1% | 62.0% |
| 5 | 55.6% | 76.1% | 54.2% | 68.1% |

## 3.5    Influence of Multi-view Soft Voting Strategy

**Table 4.** Results of ablation experiments on the multi-view soft voting strategy

| Method | DSC | DSC (global) | Recall | Precision |
| --- | --- | --- | --- | --- |
| MSMV-UNet-A | 57.0% | 79.0% | 62.1% | 62.0% |
| MSMV-UNet-C | 54.0% | 79.2% | 57.8% | 61.2% |
| MSMV-UNet-S | 56.6% | 76.3% | 59.6% | 65.4% |
| MSMV-UNet | 58.1% | 78.2% | 60.4% | 65.9% |

To validate the effectiveness of the multi-view soft voting strategy, this section conducts ablation experiments on this module. MSMV-UNet-A, MSMV-UNet-C, and MSMV-UNet-S correspond to models trained on different subsets of data, namely A (axial view), C (coronal view), and S (sagittal view). MSMV-UNet is the fusion of the results from these three models, with the optimal number of consecutive slices set to n = 4. From the results in Table 4, it can be observed that among the model results from the three views, the axial model MSMV-UNet-A achieved the highest DSC value of 57.0%. This finding explains why axial views are commonly used as inputs for stroke lesion segmentation models. The fusion model MSMV-UNet, which combines the results from all three views, achieves the best DSC of 58.1%. Therefore, the multi-view soft voting strategy is proven to be effective.

## 4    Conclusion

We propose a 2.5D method based on multi-slice feature fusion for automatic segmentation of stroke lesions. The network utilizes a dense multi-slice feature fusion module to concatenate and fuse information between slices. Additionally, a slice-wise attention module is introduced to calculate the difference values between slices and enhance the features in regions with significant slice-wise differences, directing the model's attention to edge areas. By employing a multi-view soft voting strategy, a three-view segmentation model is established to improve spatial interactions and incorporate spatial positional information into the segmentation. Through a series of comparative ablation experiments, the effectiveness of the model has been demonstrated.

In the future, we plan to validate our method on a larger clinical dataset and extend the proposed modules to different applications, further verifying the practicality and generalizability of our approach.

## References

1. Basak, H., Hussain, R., Rana, A.: DFENet: a novel dimension fusion edge guided network for brain MRI segmentation. SN Comput. Sci. **2**, 1–11 (2021)

2. Burke Quinlan, E., et al.: Neural function, injury, and stroke subtype predict treatment gains after stroke. Ann. Neurol. **77**(1), 132–145 (2015)
3. Dolz, J., Ben Ayed, I., Desrosiers, C.: Dense multi-path U-Net for ischemic stroke lesion segmentation in multiple image modalities. In: Crimi, A., Bakas, S., Kuijf, H., Keyvan, F., Reyes, M., van Walsum, T. (eds.) Brainlesion: Glioma, Multiple Sclerosis, Stroke and Traumatic Brain Injuries: 4th International Workshop, BrainLes 2018, Held in Conjunction with MICCAI 2018, Granada, Spain, 16 September 2018, Revised Selected Papers, Part I 4. LNCS, vol. 11383, pp. 271–282. Springer, Cham (2019). https://doi.org/10.1007/978-3-030-11723-8_27
4. Hui, H., Zhang, X., Li, F., Mei, X., Guo, Y.: A partitioning-stacking prediction fusion network based on an improved attention U-Net for stroke lesion segmentation. IEEE Access **8**, 47419–47432 (2020)
5. Kim, J., et al.: Global stroke statistics 2019. Int. J. Stroke **15**(8), 819–838 (2020)
6. Kingma, D.P., Ba, J.: Adam: a method for stochastic optimization. arXiv preprint arXiv:1412.6980 (2014)
7. Liew, S.L., et al.: A large, curated, open-source stroke neuroimaging dataset to improve lesion segmentation algorithms. Sci. Data **9**(1), 320 (2022)
8. Liu, X., et al.: MSDF-Net: multi-scale deep fusion network for stroke lesion segmentation. IEEE Access **7**, 178486–178495 (2019)
9. Liu, Z., Cao, C., Ding, S., Liu, Z., Han, T., Liu, S.: Towards clinical diagnosis: automated stroke lesion segmentation on multi-spectral MR image using convolutional neural network. IEEE Access **6**, 57006–57016 (2018)
10. Long, J., Shelhamer, E., Darrell, T.: Fully convolutional networks for semantic segmentation. In: Proceedings of the IEEE Conference on Computer Vision and Pattern Recognition, pp. 3431–3440 (2015)
11. Lucas, C., Maier, O., Heinrich, M.P.: Shallow fully-connected neural networks for ischemic stroke-lesion segmentation in MRI. In: Bildverarbeitung für die Medizin 2017: Algorithmen-Systeme-Anwendungen. Proceedings des Workshops vom 12. bis 14. März 2017 in Heidelberg, pp. 261–266. Springer, Cham (2017). https://doi.org/10.1007/978-3-662-54345-0_59
12. Oktay, O., et al.: Attention U-Net: learning where to look for the pancreas. arXiv preprint arXiv:1804.03999 (2018)
13. Qi, K., et al.: X-Net: brain stroke lesion segmentation based on depthwise separable convolution and long-range dependencies. In: Shen, D., et al. (eds.) Medical Image Computing and Computer Assisted Intervention-MICCAI 2019: 22nd International Conference, Shenzhen, China, 13–17 October 2019, Proceedings, Part III 22. LNCS, vol. 11766, pp. 247–255. Springer, Cham (2019). https://doi.org/10.1007/978-3-030-32248-9_28
14. Ronneberger, O., Fischer, P., Brox, T.: U-Net: convolutional networks for biomedical image segmentation. In: Navab, N., Hornegger, J., Wells, W., Frangi, A. (eds.) Medical Image Computing and Computer-Assisted Intervention-MICCAI 2015: 18th International Conference, Munich, Germany, 5–9 October 2015, Proceedings, Part III 18. LNCS, vol. 9351, pp. 234–241. Springer, Cham (2015). https://doi.org/10.1007/978-3-319-24574-4_28
15. Sheng, M., Xu, W., Yang, J., Chen, Z.: Cross-attention and deep supervision UNet for lesion segmentation of chronic stroke. Front. Neurosci. **16**, 836412 (2022)
16. Szegedy, C., et al.: Going deeper with convolutions. In: Proceedings of the IEEE Conference on Computer Vision and Pattern Recognition, pp. 1–9 (2015)
17. Xiao, X., Lian, S., Luo, Z., Li, S.: Weighted Res-UNet for high-quality retina vessel segmentation. In: 2018 9th International Conference on Information Technology in Medicine and Education (ITME), pp. 327–331. IEEE (2018)

18. Xue, Y., et al.: A multi-path 2.5 dimensional convolutional neural network system for segmenting stroke lesions in brain MRI images. NeuroImage Clin. **25**, 102118 (2020)

19. Zhang, R., et al.: Automatic segmentation of acute ischemic stroke from DWI using 3-D fully convolutional DenseNets. IEEE Trans. Med. Imaging **37**, 2149–2160 (2018)

20. Zhang, Y., Wu, J., Liu, Y., Chen, Y., Wu, E.X., Tang, X.: MI-UNet: multi-inputs UNet incorporating brain parcellation for stroke lesion segmentation from T1-weighted magnetic resonance images. IEEE J. Biomed. Health Inform. **25**(2), 526–535 (2020)

21. Zhou, Y., Huang, W., Dong, P., Xia, Y., Wang, S.: D-UNet: a dimension-fusion u shape network for chronic stroke lesion segmentation. IEEE/ACM Trans. Comput. Biol. Bioinf. **18**(3), 940–950 (2019)

# Non-Local Spatial-Wise and Global Channel-Wise Transformer for Efficient Image Super-Resolution

Xiang Gao, Sining Wu, Fan Wang, and Xiaopeng Hu[✉]

School of Computer Science and Technology, Dalian University of Technology,
Dalian 116024, China
{gaoxiangv1,isaacwu}@mail.dlut.edu.cn, {wangfan,
huxp}@dlut.edu.cn

**Abstract.** Transformer-based methods have made favorable breakthroughs in image super-resolution (SR) due to the strong ability of capturing long-range dependencies in images. However, these methods mainly concentrate on capturing spatial interaction information, often ignoring to explore the global characteristics across the channel dimension. In this paper, we propose a novel Non-local Spatial-wise and Global Channel-wise Transformer (NSGCT) for efficient image SR. To comprehensively investigate inherent similarity information in both spatial and channel dimensions, we design a hybrid of Non-local Spatial-wise Self-Attention (NSSA) and Global Channel-wise Self-Attention (GCSA) within the Transformer layer. Specifically, NSSA is shifted-window-based and concentrates on the non-local spatial similarity features, while GCSA calculates the cross-covariance across the channels to exploit the global long-range image relationships. We also design an Efficient Gated Depth-wise-conv Feed-forward Network (EGDFN) as the feed-forward network to enhance and control the information flow in Transformer with an efficient implementation for further restoring the accurate texture information. Extensive quantitative and qualitative evaluations on benchmark datasets demonstrate that the proposed NSGCT performs favorably against other state-of-the-art efficient image SR methods in terms of computation costs and image reconstruction quality.

**Keywords:** Image Super-Resolution · Transformer · Efficient · Self-Attention

## 1 Introduction

Image super-resolution (SR) is a considerable research interest to reconstruct a view-pleasing high-resolution (HR) image from its low-resolution (LR) counterpart. Image SR is a challenging ill-posed problem because a single LR image can match many latent HR counterparts. Recently, deep convolutional neural network (CNN) based methods have been raised to tackle this inverse problem and achieved significant breakthroughs over conventional methods.

© The Author(s), under exclusive license to Springer Nature Switzerland AG 2024
S. Rudinac et al. (Eds.): MMM 2024, LNCS 14556, pp. 70–85, 2024.
https://doi.org/10.1007/978-3-031-53311-2_6

SRCNN [1] firstly adopts a three-layer trainable end-to-end CNN for image SR. Then, numerous SR approaches [2, 3, 5, 6] achieve better performance through deeper network and the residual learning [4]. It is a well-known cognition that deeper networks based on residual learning can obtain better reconstruction results. However, deep network layers introduce a large number of parameters and computation costs, which limits these methods in real-world applications or practice scenarios. A straightforward way to alleviate this problem is to design lightweight image SR networks for reaching a trade-off between performance and complexity. Lightweight methods reduce the complexity through specific strategies, such as recursive mechanism [3, 7], attention mechanism [8], cascading residual structure [9], information distillation mechanism [10, 11], and neural architecture search [12, 13]. Unfortunately, they may suffer from a performance drop due to the reduced network capacity and ignoring the non-local and global information in images.

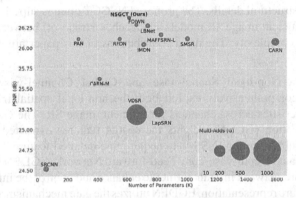

**Fig. 1.** Performance and complexity comparison on Urban100 for ×4 SR. Multi-Adds is calculated on 720p HR image. NSGCT achieves a better trade-off among PSNR, number of parameters and Multi-Adds.

Nowadays, Transformer-based image restoration methods have achieved notable progress by exploiting self-attention (SA) for the strong ability to capture long-range dependencies in images. However, the over-parametrized deep structure and high GPU memory consumption raise the training difficulty and limit them in real-world practice applications. There is limited research available on the efficient SR Transformer architecture, which attracts us to explore a lightweight Transformer structure for efficient image SR.

Since the vanilla multi-head self-attention [14] has a quadratic complexity with the spatial pixel size, this drawback limits it to applications with high-resolution images. An effective measure to reduce the introduced computation costs is conducting SA on small spatial window within local regions [20, 21]. However, this approach restricts the self-attention to focus on local spatial neighbor interactions, ignoring global characteristics across the channel dimension, which is important for the image SR reconstruction to restore fine-details. We conduct an efficient Transformer with a linear complexity by exploring non-local spatial similarity information and global channel-wise interactions in images. Meanwhile, the feed-forward network is widely employed in Transformer layers for further feature representation. To enhance and control the information flow after self-attention in Transformer, we use the gate mechanism and depth-wise convolutions to devise an Efficient Gated Depth-wise-conv Feed-forward Network (EGDFN) implemented efficiently for recovering accurate texture information.

According to the above analysis, we propose a novel Non-local Spatial-wise and Global Channel-wise Transformer (NSGCT) for efficient image SR. Specifically, shifted-window-based Non-local Spatial-wise Self-Attention (NSSA) is employed to explore intrinsic similarity characteristics in spatial dimension, while Global Channel-wise Self-Attention (GCSA) investigates the global long-range image relationships across the channel dimension. NSSA and GCSA are conducted successively within the Transformer layer, which has a linear complexity and is applicable to large input size images. NSGCT consists of continuous efficient Transformer layers in a clean structure. Additionally, an Efficient Gated Depth-wise-conv Feed-forward Network (EGDFN) is designed as the feed-forward network in the Transformer layer for further feature representation to recover accurate structural details. The proposed NSGCT obtains superior results to the state-of-the-art efficient image SR models and reaches a trade-off between performance and computation complexity. The main contributions of this paper are summarized as follows:

- We propose a Non-local Spatial-wise and Global Channel-wise Transformer (NSGCT) to comprehensively exploit both the non-local spatial interactions and global characteristics across channels for efficient image SR. The Global Channel-wise Self-Attention (GCSA) weights the spatial features extracted by Non-local Spatial-wise Self-Attention (NSSA) to acquire fine-detailed features.
- An Efficient Gated Depth-wise-conv Feed-forward Network (EGDFN) is designed as the feed-forward network in Transformer to enhance and control the information flow for further feature representation. EGDFN utilizes the gate mechanism and depth-wise convolutions with an efficient implementation to improve the feature representation ability.
- Extensive experiments demonstrate that our method outperforms other state-of-the-art image SR methods in terms of quantitative and qualitative evaluations. NSGCT reaches a better trade-off between performance and computation complexity than other state-of-the-art methods (See Fig. 1).

## 2   Related Work

Nowadays, Transformer-based methods have garnered noteworthy advancements through the utilization of self-attention (SA) [14] to capture long-range dependencies in images for image restoration tasks, including image super-resolution. SwinIR [20] is constructed upon Swin Transformer [19] for image restoration, using the shifted window mechanism to capture image interdependencies. ESRT [15] adopts a hybrid architecture of CNN and Transformer to capture deep features and extract long-range image relationships for lightweight image SR. SST [18] proposes a Spatial-Spectral Transformer to extract both spatial and spectral features for hyperspectral image denoising. LBNet [16] constitutes an additional hybrid model that integrates a recursive mechanism within its Transformer component, thereby mitigating the overall complexity of the model architecture. ELAN [17] embraces the application of shift convolution (shift-conv) in tandem with group-wise multi-scale self-attention (GMSA) and shared attention, with the primary intent of curtailing the computational overhead associated with the model. The Transformer-based image SR approaches yield quantitative performance improvement.

Since self-attention (SA) constitutes the fundamental technology in Transformer due to its strong capability to acquire extensive image dependencies over long-range. The complexity of SA experiences a quadratic escalation with the input image pixel size, consequently impeding its applicability to large resolution images. A compromise solution involves the implementation of SA exclusively within local regions of the image, such as SwinIR [20] and ELAN [17]. However, this approach confines SA to concentrate solely on interactions among spatial neighbors within local regions, ignoring global long-range characteristics across the channel dimension [22], which is important and essential for image reconstruction. Thus, we aim to investigate an efficient Transformer architecture that can acquire real long-range image dependencies through both non-local spatial-wise and global channel-wise dimensional features for image SR.

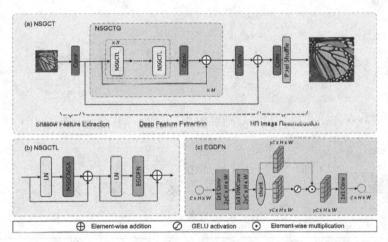

**Fig. 2.** Architecture of the proposed method. (a) Non-local Spatial-wise and Global Channel-wise Transformer (NSGCT). (b) Non-local Spatial-wise and Global Channel-wise Transformer Layer (NSGCTL). (c) Efficient Gated Depth-wise-conv Feed-forward Network (EGDFN).

## 3   Proposed Method

### 3.1   Overall Architecture

Our goal is to investigate an efficient Transformer for image SR to expansively investigate inherent similarity information in both spatial and channel dimensions. As illustrated in Fig. 2 (a), we choose a plain network architecture to avoid introducing extra parameters and complexity, which has three components: shallow feature extraction (SFE), deep feature extraction (DFE), and HR image reconstruction. NSGCT mainly consists of Non-local Spatial-wise and Global Channel-wise Transformer Groups (NSGCTGs), and each NSGCTG contains multiple consecutive Non-local Spatial-wise and Global Channel-wise Transformer Layers (NSGCTLs).

Given an input LR image $I_{LR} \in \mathbb{R}^{H \times W \times 3}$, one standard $3 \times 3$ convolution is firstly applied to extract the shallow features $F_0 \in \mathbb{R}^{H \times W \times C}$, where $H \times W$ is the spatial resolution and $C$ denotes the intermediate feature number. Then we employ the deep feature extraction processes. It conducts a serious of NSGCTGs with the total number of $M$, one $3 \times 3$ convolution, and a skip connection to extract the deep texture features $F_{DF} \in \mathbb{R}^{H \times W \times C}$ from the input shallow features $F_0$. Each NSGCTG consists of several NSGCTLs with the number of $N$. Finally, for efficient design, we utilize one $3 \times 3$ convolution with $3 \times s^2$ output feature channels and a sub-pixel convolution [23] as the pixel-shuffle operation to upsample the features back to the original HR pixel size, where $s$ is the upscale factor. A long skip residual connection is also added to fully use the shallow features for ensuring training stability. The reconstructed HR image $I_{HR} \in \mathbb{R}^{sH \times sW \times 3}$ is the final output of the model.

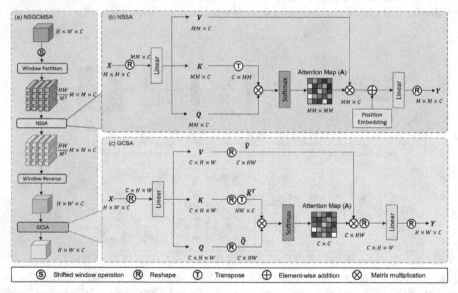

**Fig. 3.** Architecture of (a) Non-local Spatial-wise and Global Channel-wise Multi-head Self-Attention (NSGCMSA). NSGCMSA consists of (b) Non-local Spatial-wise Self-Attention (NSSA) and (c) Global Channel-wise Self-Attention (GCSA).

## 3.2 Non-local Spatial-Wise and Global Channel-Wise Transformer

The existing Transformer-based methods are not efficient enough to be applied in real-world scenarios due to the high computation costs and heavy GPU memory consumption. Meanwhile, they tend to focus on extracting the spatial features, often ignoring the global information hided in the channel dimension. To alleviate the issue, we design an efficient Non-local Spatial-wise and Global Channel-wise Transformer (NSGCT) for image SR to capture similarity characteristics in both non-local spatial dimension and global channel dimension. In addition, an Efficient Gated Depth-wise-conv Feed-forward Network (EGDFN) is introduced into Transformer as the feed-forward network (FN) for further feature representation to restore detail information. Non-local Spatial-wise and Global Channel-wise Transformer Layer (NSGCTL) is made up of layer normalization (LN), Non-local Spatial-wise and Global Channel-wise Multi-head Self-Attention (NSGCMSA) and EGDFN, as shown in Fig. 2 (b). The process in NSGCTL can be formulated as follows:

$$X = NSGCMSA(LayerNorm(X)) + X \tag{1}$$

$$X = EGDFN(LayerNorm(X)) + X \tag{2}$$

Specifically, NSGCMSA is mainly made up by the Non-local Spatial-wise Self-Attention (NSSA) and the Global Channel-wise Self-Attention (GCSA), which is shown in Fig. 3 (a). After the layer normalization operation, the regularized input features $F \in \mathbb{R}^{H \times W \times C}$ is gained. We execute a window partition operation with the window size of $M$ to segment $F$ into small patches across the spatial dimension. The input features are segmented into non-overlapping small patches $\{F_1, \cdots, F_i, \cdots, F_n\}$ with the size of $\frac{HW}{M^2}$, where $F_i \in \mathbb{R}^{M \times M \times C}$. Then each small patch $F_i$ are fed into the NSSA to exploit non-local similarity characteristics in spatial dimension. Window reverse operation is conducted to gather the output patches from NSSA. After this operation, the gathered features $F_{NSSA} \in \mathbb{R}^{H \times W \times C}$ are passed through the GCSA to obtain global image dependencies across the channel dimension. The whole process is formulated as follows:

$$\{F_i\} = WindowPartition(F), i = 1, \cdots, n \tag{3}$$

$$F_i^{NSSA} = NSSA(F_i), i = 1, \cdots, n \tag{4}$$

$$F_{NSSA} = WindowReverse\left(F_i^{NSSA}\right), i = 1, \cdots, n \tag{5}$$

$$F_{GCSA} = GCSA(F_{NSSA}) \tag{6}$$

**Non-local Spatial-Wise Self-attention (NSSA).** The vanilla self-attention [14] in Transformer causes the computation cost to grow quadratically with the spatial input pixel size. This shortcoming limits the application in image SR with large input image resolution. To overcome this problem, we conduct NSSA based on the shifted window to compute the cross-covariance in small patches to obtain non-local spatial similarity information, as is shown in Fig. 3 (b).

For each NSSA after the window partition operation, we get the non-overlapping patch $F_i \in \mathbb{R}^{M \times M \times C}$ as the input. We conduct the reshape and linear projection operation to obtain the *query* matrix $Q$, *key* matrix $K$, *value* matrix $V$. Here $Q, K, V \in \mathbb{R}^{M^2 \times C}$. The non-local spatial-wise attention matrix within a local window can be calculated by the self-attention mechanism as follows:

$$Attention(Q, K, V) = Softmax\left(QK^T/\sqrt{d} + B\right)V \qquad (7)$$

where $B$ is the relative position encoding, similar to the preview work [20]. Following [14], we execute the attention into $N$ heads to learn separate attention matrixes in parallel. $d$ is specified as $C/N$. Then the linear projection and reshape operation are acted to obtain the output features of NSSA $F_i^{NSSA} \in \mathbb{R}^{M \times M \times C}$ in Eq. (4). To achieve more comprehensive spatial information and interactions between local windows, we conduct regular and spatial shifted window operation alternately. The shifted window operation means that shifting the input features by $\lfloor\lfloor\frac{M}{2}\rfloor, \frac{M}{2}\rfloor$ pixels before partitioning. NSSA effectively reduce the complexity from quadratic to linear with the input image size. The computation complexity of NSSA is $\mathcal{O}(M^2 HWC)$.

**Global Channel-Wise Self-attention (GCSA).** The non-local spatial features are already extracted after NSSA. To further exploit long-range image dependency information, we introduce GCSA after NSSA to weight the extracted spatial features for investigating more global characteristics across the channel dimension, as shown in Fig. 3 (c). Both non-local spatial similarity information and global characteristics across channels are taken into account for restoring fine-detailed SR images.

Given the input features $F_{NSSA} \in \mathbb{R}^{H \times W \times C}$ from NSSA. We firstly conduct reshape and liner projection operation by using $1 \times 1$ convolutions to the *query* matrix $Q$, *key* matrix $K$, *value* matrix $V$. Here $Q, K, V \in \mathbb{R}^{C \times H \times W}$. Then the reshape operation is acted on $Q$ and $V$ to obtain $\hat{Q}, \hat{V} \in \mathbb{R}^{C \times HW}$. Both the reshape and transpose operations are conducted on $K$ to get $\hat{K}^T \in \mathbb{R}^{HW \times C}$. The global channel-wise attention matrix can be calculated as follows:

$$Attention\left(\hat{Q}, \hat{K}, \hat{V}\right) = Softmax\left(\hat{Q}\hat{K}^T/\alpha\right)\hat{V} \qquad (8)$$

where $\alpha$ is a learnable scaling parameter to control the magnitude of the dot product between $\hat{Q}$ and $\hat{K}^T$. Then the linear projection and reshape operation are conducted to obtain the output features of GCSA $F_{GCSA} \in \mathbb{R}^{H \times W \times C}$ in Eq. (6). GCSA also employs the multi-head self-attention mechanism and calculates the cross-covariance across the channel dimension with a linear complexity of $\mathcal{O}(C^2 HW)$.

**Efficient Gated Depth-Wise-Conv Feed-Forward Network (EGDFN).** Feed-forward network (FN) is a fundamental component in Transformer for further recovering the accurate texture information. We introduce the EGDFN in the Transformer layer for feature representation with an efficient implementation, as illustrated in Fig. 2 (c).

EGDFN utilizes gate mechanism by the element-wise multiplication of two parallel feature branches derived from the chunking technique, and one of them is passed through the GELU non-linearity activation. GELU has gradually become the major choice in recent work [19, 20]. It can be viewed as a smoother variant version of ReLU. Since the chunk operation in PyTorch automatically splits a tensor into a specified number of equally-sized small tensors, this process can be implemented efficiently. Two $1 \times 1$ convolutions are also employed in EGDFN to expand ($\gamma$ is the expansion factor) and reduce the feature dimension. Given the regularized input $X \in \mathbb{R}^{H \times W \times C}$, the formulations to obtain the output $Y \in \mathbb{R}^{H \times W \times C}$ are as follows:

$$X_1, X_2 = Chunk(H_{DWConv}(H_{1 \times 1Conv}(X))) \tag{9}$$

$$Y = H_{1 \times 1Conv}(GELU(X_1) \odot X_2) \tag{10}$$

where $Chunk(\cdot)$ denotes the chunk operation, $GELU(\cdot)$ is the GELU activation function, $H_{DWConv}(\cdot)$ and $H_{1 \times 1Conv}(\cdot)$ are the depth-wise convolution and $1 \times 1$ convolution, and $\odot$ represents the element-wise multiplication operation. EGDFN operates respective hierarchical levels to control the information flow efficiently, concentrating on the adequate details of each level as a compliment to the others.

### 3.3 Loss Function

NSGCT is optimized with $L1$ pixel loss [24] function for a fair comparison with other state-of-the-art image SR methods. Given the training set $\{I_{LR}^i, I_{HR}^i\}_{i=1}^N$, it contains $N$ pairs of the input LR images and their HR counterparts. The goal of training NSGCT is to minimize the $L1$ loss function between the restored image $I_{SR}$ and its corresponding ground truth image $I_{HR}$:

$$L(\Theta) = \frac{1}{N} \sum_{i=1}^N \left\| I_{SR}^i - I_{HR}^i \right\|_1 = \frac{1}{N} \sum_{i=1}^N \left\| H_{NSGCT}(I_{LR}^i) - I_{HR}^i \right\|_1 \tag{11}$$

where $\Theta$ denotes the parameter set of the proposed NSGCT.

## 4 Experiments

### 4.1 Experimental Setup

**Datasets and Metrics.** NSGCT is trained by the widely used public dataset DIV2K [25], which contains 800 high-quality RGB images. Following previous works [10, 11, 15, 20], five widely used standard benchmark datasets are evaluated as the testing datasets, including Set5, Set14, B100, Urban100, and Manga109. Peak signal-to-noise ratio (PSNR) and structural similarity (SSIM) [24] are conducted as the quantitative performance evaluation metrics on the Y channel of YCbCr space transformed from RGB space. We also provide the number of parameters and operations (Multi-Adds) to evaluate the model computation complexity. Multi-Adds refers to the number of composite multiply-accumulate operations, calculated on the query image size of 1280 × 720 (720p). Higher PSNR and SSIM values signify better model performance, while fewer parameters and Multi-Adds indicate lower computational complexity.

**Implementation Details.** Following [10, 11, 15, 20], the LR images are obtained by conducting the MATLAB bicubic kernel function on the training set DIV2K [25]. Data augmentation is performed by randomly rotated by 90°, 180°, 270° and flipped horizontally on the training dataset. For each training mini-batch, 32 LR image patches are randomly selected with the input size of $64 \times 64$ for the model. To train the model, we adopt $L1$ loss function [24] with Adam optimizer ($\beta_1 = 0.9$, $\beta_2 = 0.99$) for 500k iterations from scratch. The initialized maximum learning rate is $2 \times 10^{-4}$, and half it at milestones: [250k, 400k, 450k, 475k]. The weight of the exponential moving average (EMA) is set to 0.999. NSGCT contains 4 NSGCTGs, each NSGCTG contains 4 NSGCTLs, the window size is set to 8 in NSSA, and the number of input feature channels is set to 48 for NSGCTG. The model is performed by using PyTorch with a single NVIDIA GeForce RTX 3090 GPU.

## 4.2 Comparisons with State-of-the-Art Methods

NSGCT is compared with other state-of-the-art efficient image SR methods on five commonly used benchmark datasets to demonstrate its effectiveness. The compared methods include VDSR [2], DRCN [3], CARN [9], IMDN [10], PAN [29], RFDN [11], SMSR [30], LatticeNet [26], FDIWN [27], and HPUN-L [28]. Table 1 shows the quantitative comparison results on five benchmark datasets with scale factor 2 and 4 for image SR. Compared with other methods, NSGCT achieves the best reconstruction performance in terms of PSNR and SSIM with fewer parameters and Multi-Adds.

Nowadays, the Transformer-based methods have gradually been applied in image SR. As shown in Table 2, a quantitative comparison is presented on the five benchmark datasets for ×4 SR with other Transformer-based methods, including ESRT [15], LBNet [16], and SwinIR [20]. We can observe that the proposed NSGCT achieves comparable or superior performance with the fewest parameters and Multi-Adds among all the compared methods.

Additionally, a quantitative trade-off comparison between model performance and complexity is provided in Table 3. For a comprehensive evaluation, three extra metrics are introduced. "FLOPs" denotes the number of floating-point operations, calculated on a $256 \times 256$ image for ×4 SR. "Memory" is derived by using the Pytorch function *torch.cuda.max_memory_allocated()*, which indicates the maximum consumption of GPU memory. "Latency" denotes the average inference time per image on a dataset. It shows that NSGCT has the fewest GPU consumption among the Transformer-based methods. Compared with other Transformer-based models, NSGCT reduces complexity but maintains comparable or superior performance.

To further demonstrate the effectiveness and superiority of NSGCT, we provide a visual qualitative comparison for ×4 image SR in Fig. 4. While the results restored by the compared methods suffer from over-smoothing and heavy artifacts, the proposed NSGCT can recover accurate structures and textures which are the most similar to the ground-truth HR images. For instance, in image093 from Urban100, the majority of the compared methods struggle to accurately reconstruct the correct lines, leading to the creation of unwanted artifacts. On the contrary, NSGCT can reconstruct the precise line shape and fine texture details. We also have similar observations in the cases of 148026 from B100 and img078 from Urban100. The visual comparison demonstrates that our NSGCT has a strong feature representation ability to restore complex texture information.

**Table 1.** Quantitative comparison with other state-of-the-art methods on the five benchmark datasets. The best and the second-best results are **bold** and *italic*, respectively.

| Method | Scale | Params | Multi-Adds | Set5 PSNR/SSIM | Set14 PSNR/SSIM | B100 PSNR/SSIM | Urban100 PSNR/SSIM | Manga109 PSNR/SSIM |
|---|---|---|---|---|---|---|---|---|
| VDSR [2] | ×2 | 665K | 612.6G | 37.53/0.9590 | 33.05/0.9130 | 31.90/0.8960 | 30.77/0.9140 | 37.22/0.9750 |
| DRCN [3] | | 1,774K | 17,974.3G | 37.63/0.9588 | 33.04/0.9118 | 31.85/0.8942 | 30.75/0.9133 | 37.55/0.9732 |
| CARN [9] | | 1,592K | 222.8G | 37.76/0.9590 | 33.52/0.9166 | 32.09/0.8978 | 31.92/0.9256 | 38.36/0.9765 |
| IMDN [10] | | 694K | 158.8G | 38.00/0.9605 | 33.63/0.9177 | 32.19/0.8996 | 32.17/0.9283 | 38.88/0.9774 |
| PAN [29] | | 261K | 70.5G | 38.00/0.9605 | 33.59/0.9181 | 32.18/0.8997 | 32.01/0.9273 | 38.70/0.9773 |
| RFDN [11] | | 534K | 123.0G | 38.05/0.9606 | 33.68/0.9184 | 32.16/0.8994 | 32.12/0.9278 | 38.88/0.9773 |
| SMSR [30] | | 985K | 131.6G | 38.00/0.9601 | 33.64/0.9179 | 32.17/0.8990 | 32.19/0.9284 | 38.76/0.9771 |
| LatticeNet [26] | | 756K | 169.5G | 38.06/0.9607 | 33.70/0.9187 | 32.20/0.8999 | 32.25/0.9288 | -/- |
| FDIWN [27] | | 629K | 112.0G | 38.07/0.9608 | 33.75/*0.9201* | 32.23/0.9003 | 32.40/0.9305 | 38.85/0.9774 |
| HPUN-L [28] | | 714K | 151.1G | *38.09/0.9608* | **33.79**/0.9198 | *32.25/0.9006* | 32.37/*0.9307* | *39.07/**0.9779*** |
| NSGCT | | 598K | 102.4G | **38.15/0.9612** | **33.79**/*0.9200* | **32.26/0.9007** | **32.46/0.9312** | **39.09**/*0.9778* |
| VDSR [2] | ×4 | 665K | 612.6G | 31.35/0.8830 | 28.02/0.7680 | 27.29/0.7260 | 25.18/0.7540 | 28.83/0.8809 |
| DRCN [3] | | 1,774K | 17,974.3G | 31.53/0.8854 | 28.02/0.7570 | 27.23/0.7233 | 25.14/0.7510 | 28.98/0.8816 |
| CARN [9] | | 1,592K | 90.9G | 32.13/0.8937 | 28.60/0.7806 | 27.58/0.7349 | 26.07/0.7837 | 30.47/0.9084 |
| IMDN [10] | | 715K | 40.9G | 32.21/0.8948 | 28.58/0.7811 | 27.56/0.7353 | 26.04/0.7838 | 30.45/0.9075 |
| PAN [29] | | 272K | 28.2G | 32.13/0.8948 | 28.61/0.7822 | 27.59/0.7363 | 26.11/0.7854 | 30.51/0.9095 |
| RFDN [11] | | 550K | 31.6G | 32.24/0.8952 | 28.61/0.7819 | 27.57/0.7360 | 26.11/0.7858 | 30.58/0.9089 |
| SMSR [30] | | 1006K | 41.6G | 32.12/0.8932 | 28.55/0.7808 | 27.55/0.7351 | 26.11/0.7868 | 30.54/0.9085 |
| LatticeNet [26] | | 777K | 43.6G | 32.18/0.8943 | 28.61/0.7812 | 27.57/0.7355 | 26.14/0.7844 | -/- |
| FDIWN [27] | | 664K | 28.4G | 32.23/0.8955 | 28.66/0.7829 | 27.62/0.7380 | 26.28/*0.7919* | 30.63/0.9098 |
| HPUN-L [28] | | 734K | 39.7G | *32.31/0.8962* | *28.73/0.7842* | **27.66**/*0.7386* | *26.27/0.7918* | *30.77/0.9109* |
| NSGCT | | 614K | 27.1G | **32.47/0.8978** | **28.75/0.7853** | **27.66/0.7397** | **26.37/0.7938** | **30.85/0.9129** |

**Table 2.** Quantitative comparison on the five benchmark datasets with other Transformer-based methods for ×4 SR. **Bold** indicates the best results.

| Method | Params | Multi-Adds | Set5 PSNR/SSIM | Set14 PSNR/SSIM | B100 PSNR/SSIM | Urban100 PSNR/SSIM | Manga109 PSNR/SSIM | Average PSNR/SSIM |
|---|---|---|---|---|---|---|---|---|
| ESRT [15] | 751K | 67.7G | 32.19/0.8947 | 28.69/0.7833 | 27.69/0.7379 | 26.39/0.7962 | 30.75/0.9100 | 29.14/0.8244 |
| LBNet [16] | 742K | 38.9G | 32.29/0.8960 | 28.68/0.7832 | 27.62/0.7382 | 26.27/0.7906 | 30.76/0.9111 | 29.12/0.8238 |
| SwinIR [20] | 897K | 49.6G | 32.44/0.8976 | 28.77/0.7858 | 27.69/0.7406 | 26.47/0.7980 | 30.92/0.9151 | 29.26/0.8274 |
| **NSGCT** | **614K** | **27.1G** | 32.47/0.8978 | 28.75/0.7853 | 27.66/0.7397 | 26.37/0.7938 | 30.85/0.9129 | 29.22/0.8259 |

**Table 3.** Quantitative trade-off comparison between model performance and complexity on B100 for ×4 SR. **Bold** indicates the best results.

| Method | Architecture | PSNR/SSIM | Params (K) | FLOPs (G) | Memory (M) | Latency (ms) |
|---|---|---|---|---|---|---|
| RFDN-L [11] | CNN | 27.58/0.7363 | 643 | 41.54 | 38.93 | **3.85** |
| LatticeNet [26] | | 27.57/0.7355 | 777 | 49.68 | 38.25 | 6.52 |
| FDIWN [27] | | 27.62/0.7380 | 664 | 81.64 | **38.11** | 85.91 |
| ESRT [15] | CNN+Transformer | 27.69/0.7379 | 752 | 75.83 | 584.80 | 20.16 |
| LBNet [16] | | 27.62/0.7382 | 742 | 173.26 | 144.04 | 25.41 |
| SwinIR [20] | Transformer | 27.69/0.7406 | 897 | 218.78 | 71.90 | 34.25 |
| **NSGCT** | | 27.66/0.7397 | **614** | **30.27** | 69.74 | 32.77 |

### 4.3 Ablation Studies

**Effects of NSSA and GCSA.** An ablation study is provided to validate the effects of NSSA and GCSA in NSGCT. We conduct the experiments on the Transformer layer to compare the quantitative results with or without NSSA and GCSA. As shown in Table 4, we can observe that when NSSA and GCSA are both introduced into Transformer, the model performance increase without introducing excessive computation costs. It demonstrates that both NSSA and GCSA in NSGCT effectively improves the final reconstruction results.

**Effects of EGDFN.** Experiments are conducted on different feed-forward networks in NSGCT to evaluate the effects of EGDFN. As shown in Table 5, we present the quantitative comparison results between FN [14], RCAB [6], and the proposed EGDFN. It can be observed that introducing EGDFN in NSGCT obtains the best performance but with the fewest parameters and Multi-Adds. The ablation study indicates that EGDFN is effective and efficient as the feed-forward network in the Transformer layer for feature representation.

**Fig. 4.** Visual qualitative comparison in challenging cases for ×4 SR. The results reconstructed by the proposed NSGCT have accurate structures and textures similar to the ground-truth HR images. Zoom in for the best view.

**Table 4.** Ablation study of NSSA and GCSA in NSGCT on Urban100 for ×4 SR. **Bold** indicates the best results. ✗means nonexistence and ✓means existence.

| Method | NSSA | GCSA | Params | Multi-Adds | PSNR/SSIM |
|--------|------|------|--------|------------|-----------|
| w/o NSSA | ✗ | ✓ | 463K | 27.1G | 26.14/0.7861 |
| w/o GCSA | ✓ | ✗ | 466K | 18.4G | 26.26/0.7907 |
| **GCSA+NSSA** | ✓ | ✓ | 614K | 27.1G | **26.37/0.7938** |

**Table 5.** Ablation study of different feed-forward networks in NSGCT on the five benchmark datasets for ×4 SR. **Bold** indicates the best results.

| Method | Params | Multi-Adds | Set5 PSNR/SSIM | Set14 PSNR/SSIM | B100 PSNR/SSIM | Urban100 PSNR/SSIM | Manga109 PSNR/SSIM |
|---|---|---|---|---|---|---|---|
| FN [14] | 651K | 29.1G | 32.41/0.8959 | **28.76**/0.7839 | 27.64/0.7384 | 26.32/0.7913 | 30.81/0.9112 |
| RCAB [6] | 1098K | 55.2G | 32.33/0.8960 | 28.71/0.7834 | 27.63/0.7380 | 26.31/0.7915 | 30.76/0.9110 |
| **EGDFN** | 614K | 27.1G | **32.47/0.8978** | 28.75/**0.7853** | **27.66/0.7397** | **26.37/0.7938** | **30.85/0.9129** |

# 5 Conclusions

This paper designs a novel Non-local Spatial-wise and Global Channel-wise Transformer (NSGCT) for efficient image SR. NSGCT comprehensively exploit both the non-local spatial interactions and global characteristics across channels by utilizing the Non-local Spatial-wise Self-Attention (NSSA) and Global Channel-wise Self-Attention (GCSA) within the Transformer layer. GCSA weights the spatial features extracted by NSSA for acquiring fine-detailed features. Additionally, an Efficient Gated Depth-wise conv Feed-forward Network (EGDFN) is designed as the feed-forward network to enhance and control the information flow for further feature representation in Transformer. Extensive experiments demonstrate that NSGCT achieves comparable or superior results with low computation costs and GPU memory consumption. In the future, we will extend this Transformer architecture to other low-level image restoration tasks, such as image inpainting.

**Acknowledgment.** This work was supported by the National Major Special Funding Project (grant number 2018YFA0704605).

# References

1. Dong, C., Loy, C.C., He, K., Tang, X.: Image super-resolution using deep convolutional networks. IEEE Trans. Pattern Anal. Mach. Intell. **38**(2), 295–307 (2015)
2. Kim, J., Lee, J.K., Lee, K.M.: Accurate image super-resolution using very deep convolutional networks. In: Proceedings of the IEEE Conference on Computer Vision and Pattern Recognition, pp. 1646–1654 (2016)
3. Kim, J., Lee, J.K., Lee, K.M.: Deeply-recursive convolutional network for image super-resolution. In: Proceedings of the IEEE Conference on Computer Vision and Pattern Recognition, pp. 1637–1645 (2016)
4. He, K., Zhang, X., Ren, S., Sun, J.: Deep residual learning for image recognition. In: Proceedings of the IEEE Conference on Computer Vision and Pattern Recognition, pp. 770–778 (2016)
5. Lai, W.S., Huang, J.B., Ahuja, N., Yang, M.H.: Deep Laplacian pyramid networks for fast and accurate super-resolution. In: Proceedings of the IEEE Conference on Computer Vision and Pattern Recognition, pp. 624–632 (2017)
6. Zhang, Y., Li, K., Li, K., Wang, L., Zhong, B., Fu, Y.: Image super-resolution using very deep residual channel attention networks. In: Ferrari, V., Hebert, M., Sminchisescu, C., Weiss, Y.

(eds.) ECCV 2018. LNCS, vol. 11211, pp. 294–310. Springer, Cham (2018). https://doi.org/10.1007/978-3-030-01234-2_18

7. Tai, Y., Yang, J., Liu, X.: Image super-resolution via deep recursive residual network. In: Proceedings of the IEEE Conference on Computer Vision and Pattern Recognition, pp. 3147–3155 (2017)

8. Muqeet, A., Hwang, J., Yang, S., Kang, J., Kim, Y., Bae, S.-H.: Multi-attention based ultra lightweight image super-resolution. In: Bartoli, A., Fusiello, A. (eds.) ECCV 2020. LNCS, vol. 12537, pp. 103–118. Springer, Cham (2020). https://doi.org/10.1007/978-3-030-67070-2_6

9. Ahn, N., Kang, B., Sohn, K.-A.: Fast, accurate, and lightweight super-resolution with cascading residual network. In: Ferrari, V., Hebert, M., Sminchisescu, C., Weiss, Y. (eds.) ECCV 2018. LNCS, vol. 11214, pp. 256–272. Springer, Cham (2018). https://doi.org/10.1007/978-3-030-01249-6_16

10. Hui, Z., Gao, X., Yang, Y., Wang, X.: Lightweight image super-resolution with information multi-distillation network. In: Proceedings of the 27th ACM International Conference on Multimedia, pp. 2024–2032 (2019)

11. Liu, J., Tang, J., Wu, G.: Residual feature distillation network for lightweight image super-resolution. In: Bartoli, A., Fusiello, A. (eds.) ECCV 2020. LNCS, vol. 12537, pp. 41–55. Springer, Cham (2020). https://doi.org/10.1007/978-3-030-67070-2_2

12. Chu, X., Zhang, B., Xu, R.: Multi-objective reinforced evolution in mobile neural architecture search. In: Bartoli, A., Fusiello, A. (eds.) ECCV 2020. LNCS, vol. 12538, pp. 99–113. Springer, Cham (2020). https://doi.org/10.1007/978-3-030-66823-5_6

13. Chu, X., Zhang, B., Ma, H., Xu, R., Li, Q.: Fast, accurate and lightweight super-resolution with neural architecture search. In: 2020 25th International Conference on Pattern Recognition (ICPR), pp. 59–64. IEEE (2021)

14. Vaswani, A., et al.: Attention is all you need. In: Advances in Neural Information Processing Systems, vol. 30 (2017)

15. Lu, Z., Li, J., Liu, H., Huang, C., Zhang, L., Zeng, T.: Transformer for single image super-resolution. In: Proceedings of the IEEE/CVF Conference on Computer Vision and Pattern Recognition, pp. 457–466 (2022)

16. Gao, G., Wang, Z., Li, J., Li, W., Yu, Y., Zeng, T.: Lightweight bimodal network for single-image super-resolution via symmetric CNN and recursive transformer. In: International Joint Conference on Artificial Intelligence (IJCAI), pp. 913–919 (2022)

17. Zhang, X., Zeng, H., Guo, S., Zhang, L.: Efficient long-range attention network for image super-resolution. In: Avidan, S., Brostow, G., Cissé, M., Farinella, G.M., Hassner, T. (eds.) European Conference on Computer Vision, vol. 13677, pp. 649–667. Springer, Cham (2022). https://doi.org/10.1007/978-3-031-19790-1_39

18. Li, M., Fu, Y., Zhang, Y.: Spatial-spectral transformer for hyperspectral image denoising. In: Proceedings of the AAAI Conference on Artificial Intelligence, vol. 37, no. 1, pp. 1368–1376, June 2023

19. Liu, Z., et al.: Swin transformer: hierarchical vision transformer using shifted windows. In: Proceedings of the IEEE/CVF International Conference on Computer Vision, pp. 10012–10022 (2021)

20. Liang, J., Cao, J., Sun, G., Zhang, K., Van Gool, L., Timofte, R.: SwinIR: image restoration using Swin transformer. In: Proceedings of the IEEE/CVF International Conference on Computer Vision, pp. 1833–1844 (2021)

21. Wang, Z., Cun, X., Bao, J., Zhou, W., Liu, J., Li, H.: Uformer: a general U-Shaped transformer for image restoration. In: Proceedings of the IEEE/CVF Conference on Computer Vision and Pattern Recognition, pp. 17683–17693 (2022)

22. Zamir, S.W., Arora, A., Khan, S., Hayat, M., Khan, F.S., Yang, M.H.: Restormer: efficient transformer for high-resolution image restoration. In: Proceedings of the IEEE/CVF Conference on Computer Vision and Pattern Recognition, pp. 5728–5739 (2022)

23. Shi, W., et al.: Real-time single image and video super-resolution using an efficient sub-pixel convolutional neural network. In: Proceedings of the IEEE Conference on Computer Vision and Pattern Recognition, pp. 1874–1883 (2016)
24. Wang, Z., Bovik, A.C., Sheikh, H.R., Simoncelli, E.P.: Image quality assessment: from error visibility to structural similarity. IEEE Trans. Image Process. **13**(4), 600–612 (2004)
25. Agustsson, E., Timofte, R.: NTIRE 2017 challenge on single image super-resolution: dataset and study. In: Proceedings of the IEEE Conference on Computer Vision and Pattern Recognition Workshops, pp. 126–135 (2017)
26. Luo, X., Qu, Y., Xie, Y., Zhang, Y., Li, C., Fu, Y.: Lattice network for lightweight image restoration. IEEE Trans. Pattern Anal. Mach. Intell. **45**(4), 4826–4842 (2022)
27. Gao, G., Li, W., Li, J., Wu, F., Lu, H., Yu, Y.: Feature distillation interaction weighting network for lightweight image super-resolution. In: Proceedings of the AAAI Conference on Artificial Intelligence, vol. 36, no. 1, pp. 661–669, June 2022
28. Sun, B., Zhang, Y., Jiang, S., Fu, Y.: Hybrid pixel-unshuffled network for lightweight image super-resolution. In: Proceedings of the AAAI Conference on Artificial Intelligence, vol. 37, no. 2, pp. 2375–2383, June 2023
29. Zhao, H., Kong, X., He, J., Qiao, Y., Dong, C.: Efficient image super-resolution using pixel attention. In: Bartoli, A., Fusiello, A. (eds.) ECCV 2020. LNCS, vol. 12537, pp. 56–72. Springer, Cham (2020). https://doi.org/10.1007/978-3-030-67070-2_3
30. Wang, L., et al.: Exploring sparsity in image super-resolution for efficient inference. In: Proceedings of the IEEE/CVF Conference on Computer Vision and Pattern Recognition, pp. 4917–4926 (2021)

# MobileViT-FocR: MobileViT with Fixed-One-Centre Loss and Gradient Reversal for Generalised Fake Face Detection

Ting Peng[1]([✉]), Yihang Zhou[2], Rong Sun[3], Yizhi Luo[4], and Yuqi Li[4]

[1] Ningbo University of Finance and Economic, Zhejiang, China
pengting@nbufe.edu.cn
[2] The University of Queensland, St Lucia, Australia
[3] South China Normal University, Guangzhou, China
[4] Institute of Computing Technology, Chinese Academy of Sciences, Beijing, China

**Abstract.** Fake face detection is one of the most important face detection technologies, which plays an important role in preventing malicious actors from using generated fake faces for malicious purposes. However, current fake face detection techniques have poor generalisation detection ability to recognise different types of fake face images, which makes it difficult to apply this technology to real-life scenarios. Therefore, it is important to construct a fake face detection model with stronger cross-domain generalisation capabilities. In order to enhance the generalisation detection capability of the model on different face datasets, we propose a MobileViT-FocR model, which uses MobileViT to extract local and global features, and proposes the Fixed-One-Centre(FOC) loss, that is, to select a fixed centre point and focus only on similar features of real face images. The model is tuned with some of the parameters of focal loss to enhance its ability to detect more difficult fake face images. The GRL(Gradient Reversal Layer) is added based on the network to make the model better focus on the generic category differences between fake and real faces rather than the domain differences. Through experimental verification, our model has good detection capability for fake face images of different styles generated by various algorithms. Compared to the original MobileViT model, our model improved by 9.79%, 8.58%, 7.70%, and 8.23% on Internet Celebrity, Celeb-DF, DFDC, and ForgeryNet datasets respectively.

**Keywords:** Fixed One Centre Loss · Fake Face Detection · ViT model · Generalisation Detection Capability

## 1 Introduction

The rise of the Deep learning [30,32,36,37] has made it easier and more cost-effective to communicate with people. With the development of computer vision

© The Author(s), under exclusive license to Springer Nature Switzerland AG 2024
S. Rudinac et al. (Eds.): MMM 2024, LNCS 14556, pp. 86–100, 2024.
https://doi.org/10.1007/978-3-031-53311-2_7

[6,12,21,31], deep learning is applied in various fields [1,16,33], making people's lives more intelligent. However, this has also posed a new challenge: how can people effectively identify and protect against malicious actors who use false identities to commit fraud. Many applications in fields such as finance and security rely on face recognition technology for functions like phone unlocking, face payment, and video surveillance. With the emergence of fake face generation models based on generative adversarial networks and convolutional neural networks [19,27] in recent years, it is no longer difficult to generate a highly realistic fake face [25]. Existing face recognition technology is not sufficient to detect increasingly realistic fake faces, which poses a risk to applications that rely on face recognition technology of having private data or property stolen. As a result, researching how to efficiently and accurately perform face detection to distinguish between real faces and fake faces has become an important topic.

The detection of fake faces generated by algorithms has been a hot research topic in recent years, and various methods have been proposed to address this problem. Some methods decide whether a face image is fake by analysing its quality [17], such as looking for abnormal noise and compression artifacts. Some by analysing its content, such as looking for abnormal facial features [34]. Some by using facial image content and facial detail quality as a multi-modal task [35]. These methods have all made great contributions to fake face detection, but they also share a common problem. Most methods train models on image datasets generated by one or several specific algorithms. However, models trained on a particular dataset have poor generalisation ability and are usually only applicable to the data types used during training. Moreover, as fake face synthesis algorithms improve, some of the features they produce become more and more hidden, and these models also face the risk of failure. Therefore, it is necessary to propose fake face detection models with cross-domain generalisation ability.

By analysing the literature, we find the reason for the poor cross-domain generalisation of current fake face detection models. It will make it easier for the model to focus more on the similar features of the fake images by training a model using a dataset with a relatively homogeneous generation algorithm and image style. Eventually, the model is affected by the image generation algorithm and does not generalise well. Therefore, the generalisation ability of the model can be improved by intensifying its focus on similar features among real images. A Fixed-One-Centre (FOC) loss, which is an improvement on the centre loss, is used to make the model focus on intra-class features on real faces and ignore intra-class features on fake face images. This loss enables real face samples to be aggregated intra-class features and separated from fake face samples without affecting the generalisation ability of the model. Then, the model is trained by different kinds of mixed datasets to further improve the generalisation ability. Subsequently, the focal loss is added to the model to further enhance its detection accuracy, enabling the model to better learn images that are difficult to discern. Ultimately, a Gradient Reversal Layer(GRL) [8] is added to the model to enable feature extractors focused on Generic Features. The MobileVit-FocR model trained on mixed images has good performance when generalising to detect fake face images generated by StyleGAN2 [26] and Deepfake [18].

The key contributions of this article are summarised below.

- A generalisable fake face image detection model based on an improve MobileVit-FocR network is proposed. The model can generalise to detect fake face images generated by StyleGAN2 and Deepfake algorithms.
- The Gradient Reversal Layer is added to the MobileVit [20] and the model's loss function is modified with fixed one-centre loss and focal loss in this article. These modifications give the model a more effective generalisation detection ability.
- Finally, Experimental results show that the model proposed has good generalisation detection ability and performs well in recognising fake face images generated by StyleGAN2 and Deepfake.

## 2    Related Work

By analysing these researches of face authentication, the challenges that currently exist in addressing generalised detection of face authenticity are understood, and possible solutions may exist.

Some researchers have achieved the detection of fake images by analysing the content features of images. Li et al. [11] designed a single-centre loss and proposed a new learning framework for fake face detection, which uses metric learning and adaptive frequency feature learning. The entire framework can learn more discriminative features of images in an end-to-end manner. Dang et al. [2] proposed a multi-task learning model in a paper that can detect and locate fake regions in images at the same time. The model uses an attention mechanism to locate fake regions through the attention map obtained by training and further improves the classification results through the reinforcement of fake regions.

Some researchers use image quality analysis to detect whether a facial image is fake. Zhao et al. [34] see the problem as a fine-grained classification task and propose a multi-attentional deepfake detection network. By encouraging the network to attend to different local parts, amplify subtle artifacts in shallow features, and aggregate low-level texture features and high-level semantic features, they achieve the detection of image manipulation. Some researchers use a multi-modal approach, incorporating facial image content and quality of facial details as data to detect whether a facial image is fake. Zhu et al. [35] view facial images as products of underlying 3D geometry and lighting interventions, decomposing them into views and using multiple types of data as a multi-modal task for facial manipulation detection. Although the above studies are able to obtain good detection results on a specific fake face dataset, the generalised recognition capability of its model is low and difficult to be transferred to other datasets.

Some researchers have worked on the generalisability of fake face detection models. Sheng et al. [28] explore whether a more universal detector can be created to distinguish between real images and those generated by CNNs, regardless of the architecture or dataset used. Their model performs well in detecting current mainstream models that generate fake images, such as ProGAN [9], StyleGAN2 and more. Their research shows that current CNN-generated images

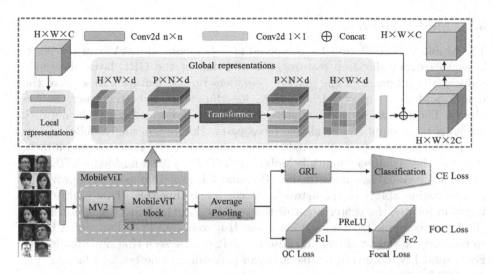

**Fig. 1.** MobileViT-FocR model. The backbone network uses the ModelViT model to extract features and improves the loss function in downstream tasks. In the ModelViT, the MobileNetV2 (MV2) module and MobileViT block module are embedded to explore the global and local features of the image, which enhances the learning ability of the network. In a branch of the downstream task, the Gradient Reversal Layer(GRL) is first designed to extract the common features of the source domain and the target domain, and finally, CE loss is used to supervise the learning. In another branch, we mainly present a Fixed One Centre (FOC) loss function to detect the central point, so that the network pays more attention to the intra-class features of real faces and features of fake faces.

have some common systematic defects, and that images generated by CNNs retain detectable fingerprints that can be distinguished from real photos. This allows forensic classifiers to generalise from one model to another without extensive modification. Sheng's research achieves generalised recognition of images produced by different kinds of CNNs, but the types of datasets used for training include animals, objects, etc., which makes the model relatively inaccurate in recognising different styles of face images.

## 3   The Proposed Method

In this section, a MobileVit model with the addition of a GRL layer is proposed, with Fixed One Centre (FOC) loss. The GRL layer enables confrontation between the model feature extractors and the classifiers so that the model's feature extractors learn generic category information rather than information about a particular domain. Meanwhile, by supervising the model focused on intra-class features of the real human dataset and giving higher learning weights to positive samples of fake faces at the same time, the loss function enhanced the generalisation ability of the model. In Sect. 3.1, the model architecture is outlined. In Sect. 3.2, the process of FOC loss function proposed is described.

### 3.1   Model Overview

The MobileViT-FocR model is composed of lightweight MobileViT as the backbone to extract the deep features, the addition of the GRL layer effectively reduces the influence of the domain classifiers on the feature extractor of the model, and it is supervised by a FOC loss function for better generalisation ability. The specific network architecture is shown in Fig. 1. Firstly, the network inputs a large number of real-fake faces pairs. Before entering MobileViT, we used Conv2d $n \times n$ (n = 3) to reduce the image to appropriate size and dimension and then combined multiple MobileNetV2(MV2) [22] and MobileViT block to learn the global features. Finally, GRL and CE loss were used to improve the generalization ability of the network. At the same time, OC (One Centre) loss function locates the center point of the real face in the Full Connection Layer (FC1), we combine OC loss and Focal loss [14] to obtain FOC loss after PReLU to balance the weight of real and false faces. It can be seen that our MobileViT-FocR model is more transferable and can generalise to detect fake face images in several datasets.

**MobileViT Block:** The core idea of the MobileViT model is to use Transformer as a convolution layer to learn global representations. During the training process, the image data passes through the MobileViT model, firstly, the feature $F \in R^{H \times W \times C}$ learns local spatial information through the $n \times n$ convolution, $1 \times 1$ convolution projects the features to a high-dimensional space to get $F' \in R^{H \times W \times d}$. When starting the global feature modeling, the $F'$ is expanded into $N$ non-overlapping patches. Then the data is unfolded into $X_f \in R^{P \times N \times d}$, where $P = w \times h$, $N = \frac{W \times H}{P}$ is the number of patches, $h$ and $w$ ($\leq N$) respectively are the width and height of patches, $p \in \{1, 2, ..., P\}$, At the same time, transformer codes the relationship between patches to get $X_f \in R^{P \times N \times d}$, which is defined as Eq. 1. Each pixel of $X_f$ encodes information about all the pixels in $F$, $X_u$ uses $n \times n$ convolution to encode local information, as shown in the Fig. 1

$$X_f(p) = Transformer(X_u(p)), 1 \leq p \leq P \tag{1}$$

In order not to lose the spatial information in each patch, we fold the expanded patch into $F_{fold} \in R^{H \times W \times d}$. Inspired by the residual network, projecting $F_{fold}$ into the low-dimensional space through the $1 \times 1$ convolution, concatenate it with $F$ to obtain $\hat{F} \in R^{H \times W \times 2C}$. Finally, using $n \times n$ convolution to learn the local and global features in the concatenation operation.

**MV2:** The module adopts an inverted residual structure, as shown in Fig. 3. In order to extract more feature information, the number of channels is reduced through $1 \times 1$ Conv2d firstly, this is because the more feature information will be learned as the number of channels increases. Then the Depthwise Conv2d $3 \times 3$ extracts spatial features, and the Conv2d $1 \times 1$ reduces dimension to filter out noise information. This not only reduces the parameter quantity and calculation amount but also can learn more feature information (Fig. 2).

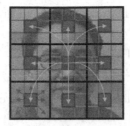

**Fig. 2.** Pixel relation. Red pixels process blue pixels (pixels in corresponding positions in other patches) through Transformer. Blue pixels use convolution to encode information about neighboring pixels, and red pixels encode information about all pixels in the image. Here, each cell in the black and gray grid represents a patch and a pixel respectively (Color figure online)

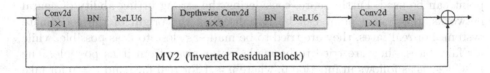

**Fig. 3.** MV2 model.

## 3.2 Loss Function Improvement

The Centre loss [29] is used to adjust the original Softmax loss of the MobileViT network so that the model can reduce the intra-class distance and focus on the commonality between samples of the same class. The original Centre loss is as follows:

$$\mathcal{L}_C = \frac{1}{2} \sum_{i=1}^{m} \|x_i - c_{y_i}\|_2^2 \tag{2}$$

where $m$ is the number of samples, $x_i$ represents the feature of the $i_{th}$ sample, $c_{y_i}$ represents the centre point. The gradients of $L_C$ with respect to $x_i$ and the update equation of $c_{y_i}$ are computed as Eq. 3, $\triangle c_j$ represents the update of the different category centres in each iteration, where $\delta(condition) = 1$ if the condition is satisfied, and $\delta(condition) = 0$ if not.

$$\frac{\partial \mathcal{L}_C}{\partial x_i} = x_i - c_{y_i}$$
$$\triangle c_j = \frac{\sum_{i=1}^{m} \delta(y_i=j)(c_j-x_i)}{1+\sum_{i=1}^{m} \delta(y_i=j)} \tag{3}$$

In the scenario of fake face detection, it is difficult to find common intra-class features between fake face images generated by different algorithms. However, real face images have some similar commonalities. Therefore it is not necessary to use multiple class-centred centre losses, but to use a one-centre loss with the real face features as a single class centre. Through one centre loss, the model can better focus on the intra-class features of the real human dataset. The OC(One Centre loss) loss is as follows.

$$\mathcal{L}_{OC} = \frac{1}{2} \sum_{i=1}^{m} \|x_i - c_{y_i}\|_2^2 \cdot \delta(y_i = 1) \tag{4}$$

where only the part of centre loss from real face is calculated, which $y_i = 1$. In the code, it is implemented by multiplying it with a mask.

$$\frac{\partial \mathcal{L}_{OC}}{\partial x_i} = x_i - c_{y_i}$$

$$\Delta c_j = \frac{\sum_{i=1}^{m} \delta(y_i=1)(c_j - x_i)}{1 + \sum_{i=1}^{m} \delta(y_i=1)} \tag{5}$$

Because only one centre is needed for the real face class, a global centre point can be used. Furthermore, considering the strong fitting ability of neural networks, the centre point can be initialised as the origin of the coordinate system. For real faces, they are tried to be made as close to it as possible, while for fake faces, they are tried to be made as far away from it as possible. The FOC loss is as follows in function 6, where $y = 1$ for real face and $y = 0$ for fake face.

$$\mathcal{L}_{FOC} = \begin{cases} \frac{1}{m-1} \sum_{i=1}^{m} \|x_i\|_2^2 & if \ y = 1 \\ 1 - \frac{1}{m-1} \sum_{i=1}^{m} \|x_i\|_2^2 & if \ y = 0 \end{cases} \tag{6}$$

At the same time, the fake faces generated by the algorithm have less obvious features, which makes it more difficult to distinguish them from real face images. Therefore, in the end, the loss function is improved by adding focal loss [14], giving higher learning weights to positive samples of fake faces. The $\gamma$ parameter is added to the loss function to increase the weight of the model in learning fake face images (positive samples), so that the model pays more attention to the false faces. The FOC loss improved by focal loss is demonstrated in function 7. The final loss function of the model is shown in function 8, where $\lambda$ represents the factor that regulates whether to focus more on difficult-to-distinguish samples.

$$\mathcal{L}_{FOC} = \begin{cases} \|x_i\|_2^{2\gamma} * \frac{1}{m-1} \sum_{i=1}^{m} \|x_i\|_2^2 & if \ y = 1 \\ \left| \|x_i\|_2^2 - 1 \right|^{\gamma} * (1 - \frac{1}{m-1} \sum_{i=1}^{m} \|x_i\|_2^2) & if \ y = 0 \end{cases} \tag{7}$$

$$\mathcal{L} = \mathcal{L}_C + \lambda \mathcal{L}_{FOC} \tag{8}$$

## 4 Experiment

### 4.1 Experiment Settings

The algorithm is implemented on the open-source PyTorch 1.8.0 framework using an NVIDIA 3090 GPU for training and the operating system is Linux. The StyleGAN2-generated Asian dataset is divided into five folds and trained on a MobileViT model for 100 epochs at a resolution of $224 \times 224$, an initial learning rate of 0.001, and a batch size of 16 as a baseline model. Accuracy (Acc) is used to measure the accuracy of fake face detection.

**Fig. 4.** To verify the generalisation detection capability of the model, part of the image data are shown collected from the Yellow Face Generator, Internet Celebrity Generator [23], DeepFakeMnist+ [7], Celeb-DF [13], DFDC [3], ForgeryNet [5], FFHQ dataset [10], and Real Internet Celebrity Face dataset [23]. (Color figure online)

### 4.2   DataSet

The dataset for this paper is built based on existing open-source datasets [5, 7, 10, 13, 23]. To verify the generalised detection capability of the model, the fake face data generated by two different algorithms (StyleGAN2 and Deepfake) with different styles is chosen as positive samples. At the same time, some publicly available real face datasets are selected as negative samples. The specific datasets and data volumes selected are shown in Table 1.

**Table 1.** Image Dataset

| Generating algorithms | Positive Sources | Positive sample size | Negative Sources | Negative Sample Size |
|---|---|---|---|---|
| StyleGAN2 | Yellow Face Generator [23] | 10000 | FFHQ dataset [10] | 10000 |
| StyleGAN2 | Internet Celebrity Generator [23] | 10000 | Real Internet Celebrity dataset [23] | 10000 |
| Deepfake | DeepFakeMnist+ [7] | 10000 | DeepFakeMnist+ | 10000 |
| Deepfake | Celeb-DF [13] | 10000 | Celeb-DF | 10000 |
| Deepfake | DFDC [3] | 10000 | DFDC | 10000 |
| Various Methods | ForgeryNet [5] | 10000 | ForgeryNet | 10000 |

In summary, all fake face images from six datasets are selected as positive samples, while the corresponding six real face datasets are selected as negative samples. The whole dataset consists of two datasets generated by StyleGAN2 [23], three datasets generated by Deepfake [3, 7, 13] and one dataset generated by multiple algorithms [5]. To expand the number and style of negative samples, the FFHQ dataset [10] is added. The images in the fake face dataset are shown in Fig. 4.

### 4.3   Choose a Good Base Model

A MobileViT-FocR model is constructed for the generalised detection of fake face images and real face images. This model is an improvement on the MobileViT network. To select a suitable base model, Asian data generated by StyleGAN2 and real data from FFHQ are used as the training set and train unmodified

MobileViT, Resnet18 [4], EfficientNet [24] and ConvNeXt [15] models respectively. These models are then tested on fake face data from internet celebrities, DFMNIST(DeepFakeMnist), Celeb-DF, DFDC, and ForgeryNet. As shown in Fig. 5, the accuracy of the MobileViT model on different data is superior to other models, it has better-generalised detection performance than other models, so the MobileViT model is chose as the base model for modification.

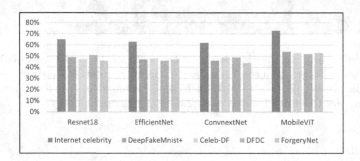

**Fig. 5.** Accuracy of different models trained with yellow face [23], FFHQ [10] data in generalising to recognise other fake face datasets (Color figure online)

**Centre, OC and FOC Loss Function Comparison.** After selecting the base model, the loss function of MobileViT is modified by adding Centre loss to the original Softmax loss. It is worth noting that directly adding unmodified Centre loss did not significantly improve the model. The reason is that the fake face images generated by different algorithms do not have some commonalities like real face image sets. Therefore, the original centre loss is modified to Fixed One centre loss, which only focuses on negative samples (i.e., real images), trying to make the intra-class distance of real faces more compact. The original Centre loss needs to maintain class centres for multiple categories and set parameters for different categories. However, our Fixed One centre (FOC) loss only needs to learn from real face images, so the zero coordinate is fixed as the class centre. The generalisation ability of the model is compared when adding the original Centre loss, adding One centre loss, and adding Fixed One Centre loss to the MobileViT network when loss weight($\lambda$) = 1. The results of the comparison are shown in Table 2. The highlighted parts of the table show the best accuracy of the model to detect each dataset.

**Table 2.** Model performance(Acc) using different loss when loss weight $\lambda = 1$

|                    | Original model | Centre loss | One Centre loss | Fixed One Centre loss |
|--------------------|----------------|-------------|-----------------|-----------------------|
| Internet celebrity | 65.51%         | 63.22%      | 73.19%          | **75.19%**            |
| DeepFakeMnist+     | **50.03%**     | 50.00%      | 49.87%          | 49.18%                |
| Celeb-DF           | 48.78%         | 49.15%      | **49.42%**      | 49.35%                |
| DFDC               | 49.04%         | 48.79%      | 48.39%          | **49.19%**            |
| ForgeryNet         | 47.04%         | 47.64%      | 48.03%          | **48.57%**            |

**Training with Mixed Data Sets.** After adding the Fixed One-centre loss, our trained model has shown significant improvement in distinguishing the celebrity face dataset generated by StyleGAN2. However, its performance in recognising dfmnst+(DeepFakeMnist+) data generated by DeepFake is still poor. The results show that the model generalises well to fake face images generated by the same algorithm, but performs poorly when generalising to fake face images generated by different algorithms. This indicates that there is a significant difference between images generated by two different algorithms. Therefore, we choose to extract fake faces generated by two different algorithms and their corresponding negative samples to generate a new training set for the model. The yellow man [23], dfmnst+ [7] datasets and their corresponding negative sample sets are chosen as the training set and validation set. The remaining positive sample sets and their corresponding negative sample sets are chosen as test sets. After generating the new dataset, the performance(Acc) of the model on the test set is tested with different weights of FOC loss in Table 3. The highlighted parts of the table show the best accuracy of the model to recognise each dataset with the addition of the Fixed One Centre loss($\lambda$).

**Table 3.** Model performance(Acc) VS weight of Fixed One Centre loss($\lambda$)

| $\lambda$ | 0.1 | 0.3 | 0.5 | 1 | 2 |
|---|---|---|---|---|---|
| Val | 99.39% | 99.40% | **99.40%** | 99.26% | 00.22% |
| Internet Celebrity | 62.46% | 64.31% | 65.75% | **66.89%** | 60.68% |
| Celeb-DF | 50.19% | 51.88% | **52.24%** | 52.19% | 51.74% |
| DFDC | 50.21% | 50.97% | 52.03% | **52.04%** | 50.34% |
| ForgeryNet | 49.11% | 50.15% | 51.23% | **51.98%** | 50.21% |

**Parameter Tuning Experiments After Adding Focal Loss.** As mentioned earlier, it is harder for the model to detect generated fake face images(positive samples) compared to real face images(negative samples). Therefore, the $\gamma$ parameter of focal loss is added to the model, allowing it to focus more on learning the more difficult-to-judge fake face images. The impact of different $\gamma$ parameters on the model's ability(Acc) to recognise the training set accuracy is compared under the condition of setting the weight of FOC loss = 1, as shown in Table 4. The case where the $\gamma$ coefficient is set to 0 means that the model is not optimised using focal loss. The highlighted parts of the table show the best accuracy of the model to recognise each dataset with the addition of the focal loss($\gamma$).

**Table 4.** Model performance(Acc) VS different $\gamma$ value

| $\gamma$ | 0 | 0.5 | 1 | 2 | 5 |
|---|---|---|---|---|---|
| Val | 99.26% | 98.97% | **99.20%** | 98.67% | 98.61% |
| Internet Celebrity | 66.89% | 67.41% | 68.14% | **70.59%** | 68.64% |
| Celeb-DF | 53.19% | 55.85% | 54.09% | 54.46% | **56.03%** |
| DFDC | 53.27% | 54.82% | 54.24% | **54.63%** | 53.82% |
| ForgeryNet | 52.81% | 53.78% | **54.01%** | 53.97% | 52.96% |

**Ablation Experiments after Adding $\gamma$.** To validate the significance of incorporating the $\gamma$ parameter, the ablation experiments are constructed where the optimised performance of the MobileViT network's Softmax layer with the addition of the $\gamma$ parameter is tested, using the mixed dataset as the training set. The results, presented in Table 5, indicate that incorporating focal loss alone resulted in an improvement of the model's generalised detection ability. The highlighted parts of the table show the best accuracy of the model to recognise each dataset in the ablation experiment for adding $\gamma$ to the loss function.

**Table 5.** Ablation experiment for adding $\gamma$ to loss function

| $\gamma$ | 0 | 0.5 | 1 | 2 | 5 |
|---|---|---|---|---|---|
| Val | 99.34% | 99.21% | **99.37%** | 98.96% | 98.81% |
| Internet Celebrity | 60.21% | 62.34% | 63.16% | **64.55%** | 63.19% |
| Celeb-DF | 52.09% | 52.85% | 53.09% | **53.40%** | 52.23% |
| DFDC | 51.37% | 51.62% | **52.24%** | 52.13% | 51.92% |
| ForgeryNet | 50.11% | 51.30% | 52.01% | **52.07%** | 51.96% |

**Parameter Tuning Experiments after Adding GRL Layer.** To enhance the model's ability to learn and detect fake face images, the GRL layer is added before the classifier to allow the model's feature extractor to be influenced as little as possible by the classifier. This allows the model to focus on any abnormal information that may exist in fake face images, rather than the domain differences between the real and fake faces. After adding the GRL layer, the $\lambda$ parameter of FOC loss with the $\gamma$ parameter of focal loss is tuned. The tuned results of the model accuracy are shown in Table 6. The highlighted parts of the table show the optimal results of the model tuning with the addition of the GRL layer.

**Table 6.** Tuning table after adding the GRL layer

| GRL | $\gamma$ | $\lambda$ | Val | Internet Celebrity | Celeb-DF | DFDC | ForgeryNet |
|---|---|---|---|---|---|---|---|
| ✗ | 0.5 | 1 | 98.97% | 67.41% | 55.85% | 54.82% | 53.78% |
| ✗ | 1 | 1 | 99.20% | 68.14% | 54.09% | 54.24% | 54.01% |
| ✗ | 2 | 1 | 98.67% | 70.59% | 54.46% | 54.63% | 53.97% |
| ✓ | 0.5 | 0.5 | 99.26% | 66.23% | 53.24% | 51.59% | 52.85% |
| ✓ | 0.5 | 1 | 98.97% | 67.53% | 53.36% | 52.69% | 53.61% |
| ✓ | 0.5 | 1.5 | 98.64% | 68.02% | 52.44% | 53.18% | 52.91% |
| ✓ | 1 | 0.5 | 98.88% | 70.13% | 57.40% | 56.39% | 54.27% |
| ✓ | 1 | 1 | **98.74%** | **75.30%** | **57.36%** | **56.74%** | **55.27%** |
| ✓ | 1 | 1.5 | 98.76% | 76.22% | 56.44% | 55.23% | 54.10% |
| ✓ | 2 | 0.5 | 99.12% | 67.85% | 55.04% | 55.23% | 53.64% |
| ✓ | 2 | 1 | 98.34% | 68.59% | 55.96% | 55.15% | 54.21% |
| ✓ | 2 | 1.5 | 98.24% | 68.52% | 56.04% | 54.38% | 53.81% |

As shown in Table 6, by adding the GRL layer, the model's generalisation recognition ability is effectively improved for different fake face datasets. When $\gamma = 1$, the accuracy improvement of the current model is compared to the accuracy improvement of using the original MobileViT model in Fig. 6. When $\gamma -1$ and $\lambda = 1$, we have effectively improved the model's generalisation ability to recognise different fake face datasets. With $\gamma = 1$ and $\lambda = 1$, the accuracy of the model in recognising the Internet Celebrity, Celeb-DF, DFDC, and ForgeryNet datasets was 75.30%, 57.36%, 56.74%, and 55.27% respectively. Compared to the baseline of the original MobileViT model, the improvement is 9.79%, 8.58%, 7.70% and 8.23%.

Through experiments, the model has a certain cross-domain generalisation ability while being able to well detect fake face images in the test set. It has a good effect when distinguishing the same internet celebrity face images, with the highest accuracy of 75.30%. However, the effect is relatively poor on the Celeb-DF, DFDC, and ForgeryNet datasets, with the highest accuracy of 57.36%, 56.74%, 55.27%. The reason is that although the styles are different, the algorithm for generating internet celebrity face images is similar to the algorithm in the training set, which is relatively easy to generalise and distinguish. However, the difference between the Celeb-DF, DFDC, and ForgeryNet data and the training set data is too large, making it difficult to do cross-domain generalisation and distinction. Nevertheless, a significant improvement has been made compared to the distinction results of the original MobileViT model.

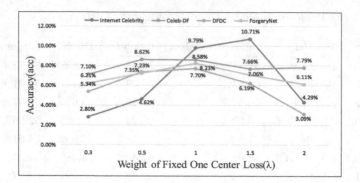

**Fig. 6.** Improvement in detection accuracy of the model compared to baseline on different datasets under $\lambda$ variations when adding the GRL layer with $\gamma = 1$

## 5 Conclusion

A detection model MobileViT-FocR is proposed for detecting fake face images generated by different algorithms. To make the model focus more on the commonality of intra-class features of real face images, a new loss function FOC loss is constructed based on centre loss. At the same time, in order to increase the learning of difficult-to-distinguish samples and achieve the best results, a $\lambda$ coefficient based on focal loss is added to FOC loss. Afterward, the GRL layer is added to the model in order to allow the model to better focus on the generic category differences between fake and real faces rather than the domain differences between real and fake faces (colour difference, illumination, clarity, and other domain information). Finally, through experimental analysis, our proposed model can distinguish between fake face images generated by different algorithms and real face images. Compared to the original MobileViT model, our model improved by 9.79%, 8.58%, 7.70%, and 8.23% on Internet Celebrity, Celeb-DF, DFDC and ForgeryNet datasets respectively. Our code has been submitted to Anonymous GitHub: https://anonymous.4open.science/r/DEEPFAKE-035F/.

## References

1. Chen, Y., et al.: Pursuing knowledge consistency: supervised hierarchical contrastive learning for facial action unit recognition. In: Proceedings of the 30th ACM International Conference on Multimedia, pp. 111–119 (2022)
2. Dang, H., Liu, F., Stehouwer, J., Liu, X., Jain, A.K.: On the detection of digital face manipulation. In: Proceedings of the IEEE/CVF Conference on Computer Vision and Pattern Recognition, pp. 5781–5790 (2020)
3. Dolhansky, B., et al.: The deepfake detection challenge (dfdc) dataset. arXiv preprint arXiv:2006.07397 (2020)
4. He, K., Zhang, X., Ren, S., Sun, J.: Deep residual learning for image recognition. In: Proceedings of the IEEE Conference on Computer Vision and Pattern Recognition, pp. 770–778 (2016)

5. He, Y., et al.: Forgerynet: a versatile benchmark for comprehensive forgery analysis. arXiv preprint arXiv:2103.05630 (2021)
6. Hu, J., Ren, Y., Yuan, Y., Li, Y., Chen, L.: Pathosisgan: sick face image synthesis with generative adversarial network. In: 2021 2nd International Conference on Artificial Intelligence and Information Systems, pp. 1–6 (2021)
7. Huang, J., Wang, X., Du, B., Du, P., Xu, C.: Deepfake mnist+: a deepfake facial animation dataset. In: Proceedings of the IEEE/CVF International Conference on Computer Vision, pp. 1973–1982 (2021)
8. Jia, Y., Zhang, J., Shan, S., Chen, X.: Single-side domain generalization for face anti-spoofing. In: Proceedings of the IEEE/CVF Conference on Computer Vision and Pattern Recognition, pp. 8484–8493 (2020)
9. Karras, T., Aila, T., Laine, S., Lehtinen, J.: Progressive growing of gans for improved quality, stability, and variation. arXiv preprint arXiv:1710.10196 (2017)
10. Karras, T., Laine, S., Aila, T.: A style-based generator architecture for generative adversarial networks. In: Proceedings of the IEEE/CVF Conference on Computer Vision and Pattern Recognition, pp. 4401–4410 (2019)
11. Li, J., Xie, H., Li, J., Wang, Z., Zhang, Y.: Frequency-aware discriminative feature learning supervised by single-center loss for face forgery detection. In: Proceedings of the IEEE/CVF Conference on Computer Vision and Pattern Recognition, pp. 6458–6467 (2021)
12. Li, W., Ma, Z., Deng, L.J., Fan, X., Tian, Y.: Neuron-based spiking transmission and reasoning network for robust image-text retrieval. IEEE Trans. Circ. Syst. Video Technol. (2022)
13. Li, Y., Yang, X., Sun, P., Qi, H., Lyu, S.: Celeb-df: a large-scale challenging dataset for deepfake forensics. In: Proceedings of the IEEE/CVF Conference on Computer Vision and Pattern Recognition, pp. 3207–3216 (2020)
14. Lin, T.Y., Goyal, P., Girshick, R., He, K., Dollár, P.: Focal loss for dense object detection. In: Proceedings of the IEEE International Conference on Computer Vision, pp. 2980–2988 (2017)
15. Liu, Z., Mao, H., Wu, C.Y., Feichtenhofer, C., Darrell, T., Xie, S.: A convnet for the 2020s. In: Proceedings of the IEEE/CVF Conference on Computer Vision and Pattern Recognition (CVPR), pp. 11976–11986 (June 2022)
16. Long, Q., Xu, L., Fang, Z., Song, G.: Hgk-gnn: heterogeneous graph kernel based graph neural networks. In: Proceedings of the 27th ACM SIGKDD Conference on Knowledge Discovery & Data Mining, pp. 1129–1138 (2021)
17. Luo, Y., Zhang, Y., Yan, J., Liu, W.: Generalizing face forgery detection with high-frequency features. In: Proceedings of the IEEE/CVF Conference on Computer Vision and Pattern Recognition, pp. 16317–16326 (2021)
18. Lyu, S.: Deepfake detection: current challenges and next steps. In: 2020 IEEE International Conference on Multimedia & Expo Workshops (ICMEW), pp. 1–6. IEEE (2020)
19. Ma, Z., Ju, W., Luo, X., Chen, C., Hua, X.S., Lu, G.: Improved deep unsupervised hashing via prototypical learning. In: Proceedings of the 30th ACM International Conference on Multimedia, pp. 659–667 (2022)
20. Mehta, S., Rastegari, M.: Mobilevit: light-weight, general-purpose, and mobile-friendly vision transformer. arXiv preprint arXiv:2110.02178 (2021)
21. Ren, Y., et al.: Crossing the gap: Domain generalization for image captioning. In: Proceedings of the IEEE/CVF Conference on Computer Vision and Pattern Recognition, pp. 2871–2880 (2023)

22. Sandler, M., Howard, A., Zhu, M., Zhmoginov, A., Chen, L.C.: Mobilenetv 2: inverted residuals and linear bottlenecks. In: Proceedings of the IEEE Conference on Computer Vision and Pattern Recognition, pp. 4510–4520 (2018)
23. Seeprettyface.com: New version of face generators based on StyleGAN2. https://github.com/a312863063/generators-with-stylegan2/, (April 26 2023)
24. Tan, M., Le, Q.: Efficientnet: Rethinking model scaling for convolutional neural networks. In: International Conference on Machine Learning, pp. 6105–6114. PMLR (2019)
25. Toshpulatov, M., Lee, W., Lee, S.: Talking human face generation: a survey. Expert Syst. Appli. 119678 (2023)
26. Viazovetskyi, Y., Ivashkin, V., Kashin, E.: StyleGAN2 distillation for feed-forward image manipulation. In: Vedaldi, A., Bischof, H., Brox, T., Frahm, J.-M. (eds.) ECCV 2020. LNCS, vol. 12367, pp. 170–186. Springer, Cham (2020). https://doi.org/10.1007/978-3-030-58542-6_11
27. Wang, H., et al.: Toward effective domain adaptive retrieval. IEEE Trans. Image Process. **32**, 1285–1299 (2023)
28. Wang, S.Y., Wang, O., Zhang, R., Owens, A., Efros, A.A.: Cnn-generated images are surprisingly easy to spot... for now. In: Proceedings of the IEEE/CVF Conference on Computer Vision and Pattern Recognition, pp. 8695–8704 (2020)
29. Wen, Y., Zhang, K., Li, Z., Qiao, Yu.: A discriminative feature learning approach for deep face recognition. In: Leibe, B., Matas, J., Sebe, N., Welling, M. (eds.) ECCV 2016. LNCS, vol. 9911, pp. 499–515. Springer, Cham (2016). https://doi.org/10.1007/978-3-319-46478-7_31
30. Xu, K., Feng, M., Huang, W.: Seeing speech: magnetic resonance imaging-based vocal tract deformation visualization using cross-modal transformer. In: Proceedings of the 30th ACM International Conference on Multimedia, pp. 6947–6949 (2022)
31. Yang, C., An, Z., Cai, L., Xu, Y.: Hierarchical self-supervised augmented knowledge distillation. In: International Joint Conference on Artificial Intelligence, pp. 1217–1223 (2021)
32. Yang, C., An, Z., Cai, L., Xu, Y.: Mutual contrastive learning for visual representation learning. In: Proceedings of the AAAI Conference on Artificial Intelligence. vol. 36, pp. 3045–3053 (2022)
33. Yang, C., Zhou, H., An, Z., Jiang, X., Xu, Y., Zhang, Q.: Cross-image relational knowledge distillation for semantic segmentation. In: Proceedings of the IEEE/CVF Conference on Computer Vision and Pattern Recognition, pp. 12319–12328 (2022)
34. Zhao, H., Zhou, W., Chen, D., Wei, T., Zhang, W., Yu, N.: Multi-attentional deepfake detection. In: Proceedings of the IEEE/CVF Conference on Computer Vision and Pattern Recognition, pp. 2185–2194 (2021)
35. Zhu, X., Wang, H., Fei, H., Lei, Z., Li, S.Z.: Face forgery detection by 3d decomposition. In: Proceedings of the IEEE/CVF Conference on Computer Vision and Pattern Recognition, pp. 2929–2939 (2021)
36. Zhu, Z., Cheng, X., Huang, Z., Chen, D., Zou, Y.: Enhancing code-switching for cross-lingual slu: a unified view of semantic and grammatical coherence. In: Proceedings of EMNLP (2023)
37. Zhu, Z., Cheng, X., Huang, Z., Chen, D., Zou, Y.: Towards unified spoken language understanding decoding via label-aware compact linguistics representations. In: Proceedings of ACL Findings (2023)

# ASF-Conformer: Audio Scoring Conformer with FFC for Speaker Verification in Noisy Environments

Xiran Zhang, Haiyan Liu, Caixia Liu$^{(\boxtimes)}$, Haiyang Zhang, and Zhiwei Huo

School of Computer Science, Inner Mongolia University, Hohhot 010021, China
cslcx@imu.edu.cn

**Abstract.** Background noise significantly impacts speech intelligibility, reducing the accuracy and reliability of the speaker verification system. Most existing noise reduction algorithms are specific to certain types of noise and have limitations, making them ineffective on eliminating background noise. Therefore, the extraction of robust features and the development of noise-resistant models that adapt to various noisy environments remain crucial challenges in the field of speaker verification. In this paper, we propose a Conformer-based Audio Scoring Conformer with Fast Fourier Convolution (ASF-Conformer), which is a speaker verification model. Firstly, the audio scoring module is introduced to evaluate and weight the audio features, aiming to select more robust features in noisy environments. Secondly, we introduce Fast Fourier Convolution as a replacement for the Conformer's convolution module, improving the model's ability to capture global features while reducing the model parameters. Finally, this paper conducts comparative tests with the current mainstream models on public dataset VoxCeleb1, and synthesized noisy dataset Mu-VoxCeleb1. The experimental results demonstrate that the proposed ASF-Conformer model, compared to the ECAPA-TDNN model with essentially the same parameters, outperforms ECAPA-TDNN by 2% and 18% respectively when evaluated using the EER metrics on the VoxCeleb1 and Mu-VoxCeleb1 datasets. These results highlight the effectiveness of the proposed model in enhancing the accuracy of speaker verification tasks, especially in noisy environments.

**Keywords:** Speaker verification · Noisy environment · Conformer

## 1 Introduction

Speaker Verification (SV) is the task of verifying whether a given audio sample belongs to a claimed enrolled speaker. It involves using feature representations of the speech signal to determine the speaker's identity. Commonly used features for SV include spectrograms, Mel-Frequency Cepstral Coefficients (MFCC), Linear Predictive Coding (LPC), etc. These features can extract individual characteristics from the voice, such as the speaker's articulation, intonation, and prosody.

Recently, deep neural networks(DNNs) have been widely applied to learn speaker embedding through classification learning processes using x-vectors [2].

© The Author(s), under exclusive license to Springer Nature Switzerland AG 2024
S. Rudinac et al. (Eds.): MMM 2024, LNCS 14556, pp. 101–111, 2024.
https://doi.org/10.1007/978-3-031-53311-2_8

The deep speaker embedding approach uses DNNs to map variable-length segment of speech into fixed-dimensional embeddings. Compared to traditional methods such as HMM-GMM [8] and i-vectors, the embedding learning methods based on DNNs [18,20] have shown superior performance. And they have been widely applied in practical scenarios such as user authentication, intelligent residential systems, law enforcement, and real-time online conferences.

Despite the speaker verification technology has made practical application, it still faces challenges such as background noise and channel distortions that greatly affect performance. To achieve practicality in speaker verification research, it is crucial to address the impact of various background noises in system. Nowadays, the noise reduction algorithms are not yet highly effective on eliminating noise, and most noise-resistant algorithms have limitations and are designed for specific scenarios. As a result, the extraction of robust features and the development of noise-resistant models that can adapt to various noisy environments are key and challenging in the field of speaker verification.

With the development of deep learning, several studies [10,24] have used deep neural networks to extract robust audio features and perform pattern matching, effectively improving the performance of speaker verification in noisy environments. Meanwhile, previous research [12] has demonstrated that local features more susceptible to noise in challenging environments, while global features can mitigate such impact and enhance the reliability and stability of speaker verification.

Inspired by these recent progresses, we have proposed ASF-Conformer based on the Convolution-augmented Transformer (Conformer). Firstly, the input acoustic features undergo downsampling. Subsequently, these audio features are scored in the channel dimension, with the objective of identifying and selecting robust audio features that can withstand noisy environments effectively. Secondly, we adopt the F-Conformer block, which integrates Fast Fourier Convolution (FFC) into the Conformer instead of the original convolution module. This modification preserves the model's original capability to extract local features while enhancing its ability to capture global features. Additionally, this approach helps reduce the number of model parameters. Finally, we concatenate the output features from all Conformer blocks to aggregate the multi-scale representations before final pooling. Our contributions can be summarized as follow:

- Before the Conformer block, we score and weight the features in the channel dimension to highlight the audio features that are more crucial and robust in noisy environments. The purpose of this step is to improve the accuracy and robustness of the model by assigning higher weights to more robust features.
- FFC was introduced to replace the original convolution module in Conformer, performing convolution operations in the frequency domain. This modification significantly enhances the model's capability to extract global features, thereby improves its overall performance in capturing long-range dependencies and contextual information.

- Experiments are conducted on publicly available dataset VoxCeleb1 and synthetic noise-corrupted dataset Mu-VoxCeleb1. The results indicate significant improvement in accuracy with ASF-Conformer compared to popular CNN-based ECAPA-TDNN [6] and MFA-Conformer [24] baseline systems.

## 2 Related Work

In recent years, x-vector [18] and its subsequent improvements have consistently provided state-of-the-art results in the field of speaker verification [6]. X-vector firstly employs the Time Delay Neural Networks (TDNNs) to map variable-length utterances into fixed-dimensional embeddings [24]. Later, the x-vector system and its variants [4,9,25] addressed the gradient vanishing problem by incorporating the ResNet architecture, which enabled the training of deeper neural networks. Recently, a significant breakthrough has been achieved by combining Time Delay Neural Networks (TDNNs) with Adaptive Context-Aware Pooling Attention (ECAPA). ECAPA-TDNN [6] and its subsequent efforts [13,19] have demonstrated notable improvements in performance.

Despite the great success of CNN-based models, they do have limitations in terms of global context fusion and handling long-range dependencies. Transformer [21] and its variants [5,23] are good at modeling long-range global context and facilitate efficient parallel training. However, Transformers struggle to achieve satisfactory performance in speaker verification without complex pre-training processes and large parameter [22]. To overcome these issues, some studies [3,11] proposed the use of Conformer, which inserts a convolution module into Transformer to increase the local information modeling. It first achieved excellent results in end-to-end speech recognition, and is later adopted in speech enhancement [11], speech separation [3], and speaker recognition [24] with remarkable performance. Zhang et al. [24] proposed to combine multi-scale feature aggregation with Conformer to further improving the performance of speaker recognition.

Most speaker embedding extractors perform reliably in clean environments, but their performance degrades in noisy environments due to background noise that interferes with speech intelligibility and quality [1,10]. Therefore, accurately extracting speaker information from noisy environments is a challenging task. The SV studies for noisy environments have been conducted from various perspectives. To build noise-robust SV systems, several researchers utilize data augmentation [1] and feature normalization techniques [16]. And some studies solve this problem by increasing the signal-to-noise ratio of the speech signal in the audio through speech enhancement, these methods often result in the loss of speaker-specific audio features and linguistic information, leading to no significant improvement in recognition accuracy. Kim et al. [10] proposed a joint-trainable framework to alleviate speaker information distortion that may occur during the noise compensation process. In addition, Cai et al. [1] proposed specialized loss functions to induce noise-mitigated speaker embeddings.

# 3  Method

## 3.1  Network Architecture

The architecture of ASF-Conformer is shown in Fig. 1. In this study, we adopt the MFA-Conformer model as our baseline model. MFA-Conformer [24] contains a downsampling module. Additionally, it adopts Conformer blocks which combine Transformers and CNNs to effectively capture both global and local features. Finally, it concatenates the output features from all Conformer blocks to aggregate the multi-scale representations before final pooling.

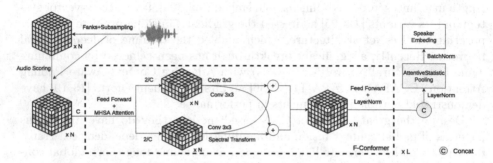

**Fig. 1.** The architecture of Audio Scoring Conformer with FFC (ASF-Conformer)

In the experiments, in order to enhance the adaptability of the speaker verification task in noisy environments, we improved the downsampling module and introduced an audio scoring (AS) block. The AS block scores the downsampled features, allowing speaker features with less noise interference to play a dominant role in the model's recognition process, effectively combating background noise interference. Additionally, we introduced Fast Fourier Convolution (FFC) in the Conformer block instead of the original convolution module. FFC transforms the convolutional computation into multiplication operations in the frequency domain and uses the Fast Fourier Transform (FFT) algorithm for fast computation of convolution results. This helps to significantly reduce computational errors on convolution calculations while accelerating the overall computation speed.

## 3.2  Downsampling Module with Audio Scoring (D-AS)

As shown in Fig. 2, the D-AS module consists of two sub-modules: downsampling and audio scoring. Firstly, the features of the input are processed through a convolutional downsampling layer to reduce computational complexity. Inspired by the VGG network, this experiment incorporates multiple convolution blocks designed to capture the deep features $X_i$ from the input audio. To retain more detailed information and avoid affecting the feature representation, pooling is not employed during the downsampling process. The input audio A undergoes pre-processed and downsampled using a $3 \times 3$ convolution with a stride of 2. To enhance the model's ability to capture nonlinear relationships and introduce a richer representation of nonlinear features, we choose the Swish activation function as the activation function for the downsampling module. The output feature $X$ is represented as Eq. (1):

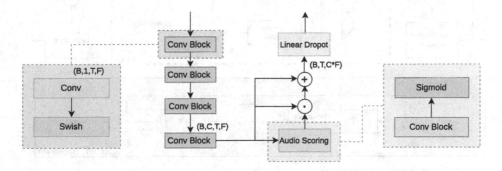

**Fig. 2.** The architecture of F-Conformer

$$X = s\left(W \times m\left(A\right) + b\right) \qquad (1)$$

where $s$ denotes swish activation function, $W$ denotes weight, $m$ denotes audio preprocessing, and $b$ denotes bias.

### 3.3   F-Conformer Block

As shown in Fig. 3, we introduce the F-Conformer block. In comparison to the Transformer block, the Conformer block consists of two Macaron-like feed forward modules (FNN) with half residual connections sandwiching the multi-head self attention module (MHSA) and convolution modules. In noisy environments, local features are prone to interference, whereas global features have the potential to mitigate this issue [12] and enhance the reliability and stability of speaker verification. Based on this experience, we use FFC after the MHSA to further enhance the model's ability to capture global features. This integration aims to improve the model's robustness in handling noisy acoustic environments. Mathematically, for each F-Conformer block the input feature $F \in R^{q*t}$, $q$ is the Conformer encoder size, $t$ is the frame length, and the output feature $f'$ is represented by Eqs. (2)–(5):

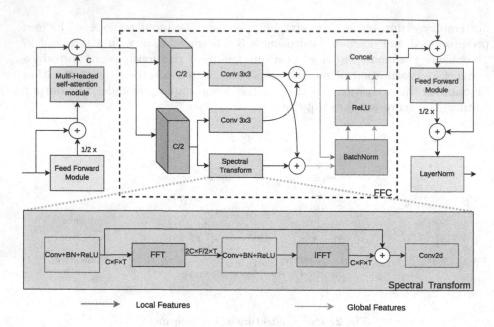

**Fig. 3.** The architecture of D-AS

$$f = f + \frac{1}{2}FFN(f) \qquad (2)$$

$$f_{MHSA} = f_{FFN1} + MHSA(f_{FFN1}) \qquad (3)$$

$$f_{FFC} = f_{MHSA} + FFC(f_{MHSA}) \qquad (4)$$

$$f' = f_{FFC} + \frac{1}{2}FFN(f_{FFC}) \qquad (5)$$

The FFC module divides the channels into two branches: local and global. The local branch uses traditional convolutional operations to extract local features from the feature maps. For the global branch, it initially applies a one-dimensional FFT along the frequency dimension of the input feature map. Then, the real and imaginary parts of the spectrum are concatenated across the channel dimension. This operation enables the model to capture global frequency patterns and dependencies in the input data. Following that, convolutional operations are applied in the frequency domain. Finally, an inverse Fourier transformation is applied to transform the processed frequency-domain feature map back into the spatial domain. Afterwards, to capture the low-level feature mappings in the speaker's audio, the experiments concatenate the output feature maps $f'_i$ obtained from each Conformer block, as shown in Eq. (6). Furthermore, the concatenated feature maps are fed into a LayerNorm layer for further normalization. Additionally, to capture the importance of each frame and extract more robust representations, we adopt the attentive statistics pooling [24]. Lastly, the speaker embeddings are derived by transforming the high-dimensional vectors into low-dimensional vectors. This transformation is accomplished using a fully-connected layer combined with BatchNorm.

$$F = Concat(f'_1, f'_2, ..., f'_3) \tag{6}$$

where $F \in R^{Q*t}$, L denotes the number of Conformer blocks and $Q = q \times L$.

## 4 Experiments

### 4.1 Datasets

Experiments were performed on VoxCeleb1 [15] dataset. The dataset comprises a training set containing 148,642 utterances from 1,211 speakers and a test set containing 4,715 utterances from 40 speakers. Although the VoxCeleb dataset collected from youtube videos is moderately noisy, we consider the original speech in the dataset to be clean. However, to further evaluate the model's performance under noisy conditions, additional noisy data was generated. For this purpose, the MUSAN corpus [17] was used as the noise source. During the training process, two different datasets were used: the original VoxCeleb1 training dataset and a synthesized noise dataset called Mu-VoxCeleb1. In the Mu-VoxCeleb1 dataset, as referenced in [7], background noise from the MUSAN corpus was incorporated into the clean VoxCeleb1 utterances. In addition, a random ratio $h$ between 0.3 and 0.5 was selected. Random noise with signal-to-noise ratios ranging (SNR) from $-5$ to 20 was then applied to the frames within this specific ratio range.

### 4.2 Implementation Details

We employed a 80-dimensional mel-spectrogram as input extracted with a 512-point FFT and a Hamming window width of 25 ms, hopped at 10 ms. No voice activity detection or augmentation was performed. In the feed-forward module, we set the linear hidden units as 2048. For the multi-head self-attention module, we set the encoder dimension as 256 and the number of attention heads as 4. The FFC module employed a convolutional kernel size of 15. The sampling rate was set to 16,000, and the time duration was set to 3 s. To ensure fair comparison, all models were trained within the same framework. We adopted the AM-Softmax loss function [24] for training. The Adam optimizer was employed, with a initial learning rate of 0.0001. To avoid overfitting, the model was warmed up in the first 10 steps. And the batch size was set to 8. Our models were implemented using PyTorch.

### 4.3 Evaluation Metrics

For scoring in our experiments, we used the cosine distance with adaptive s-norm [14]. To evaluate the performance of the model, we reported the Equal Error Rate (EER) and the minimum Detection Cost Function (minDCF) with $P_{target} = 0.01$, $C_{FA} = C_{Miss} = 1$.

### 4.4  Ablation Study

We conducted ablation experiments on our Mu-VoxCeleb1 dataset to validate the impact of the proposed D-AS and F-Conformer modules on recognition accuracy (as shown in Table 1) and parameters (as shown in Fig. 4). We used MFA-Conformer as the baseline model.

**Table 1.** Ablation study on Mu-VoxCeleb1

| Model | Mu-VoxCeleb1 | |
|---|---|---|
| | EER (%) | minDCF |
| Baseline | 11.16 | 0.78 |
| Baseline+D-ASM | 9.63 | 0.73 |
| Baseline+F-Conformer | 11.01 | 0.76 |
| Baseline+D-ASM+F-Conformer | **9.53** | **0.70** |

The results of the experiments demonstrate that D-AS and F-Conformer can effectively improve speaker verification performance in noisy environments. In

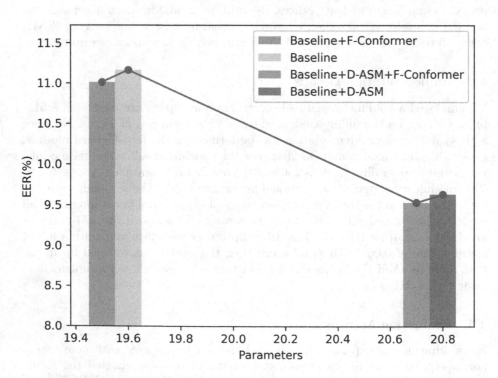

**Fig. 4.** Effect of different modules on parameters and EER of the model

particular, the D-AS module significantly improves recognition accuracy. Compared to the baseline, it achieved 14% improvement on EER, proving its ability to select robust features.

The ablation experiments have been demonstrated that the D-AS module can effectively improve model recognition accuracy but at the cost of a significant increase in model parameters. The F-Conformer can reduce the EER to a certain extent while also effectively reducing the number of model parameters.

### 4.5  Qualitative Results

As shown in Table 2, we compared our model to other state-of-the-art models on VoxCeleb1 and Mu-VoxCeleb1 datasets, including well-known ResNet34 [24], ECAPA-TDNN [24], and MFA-Conformer [24]. The results show that the ASF-Conformer model achieved an EER that is 2% better than the mainstream ECAPA-TDNN model on the publicly available VoxCeleb1 dataset, while on Mu-VoxCeleb1 dataset with noise, the ASF-Conformer model achieved an EER that is 18% better than ECAPA-TDNN.

**Table 2.** Performance overview of all systems on VoxCeleb1 and Mu-VoxCeleb1

| Model | VoxCeleb1 | | Mu-VoxCeleb1 | | Parameters(M) |
|---|---|---|---|---|---|
| | EER(%) | minDCF | EER(%) | minDCF | |
| ResNet34 | 4.61 | 0.53 | 11.62 | 0.73 | 23.2 |
| ECAPA-TDNN | 3.92 | 0.48 | 11.63 | 0.76 | 20.8 |
| MFA-Conformer | 3.90 | 0.36 | 11.16 | 0.78 | **19.6** |
| ASF-Conformerr | **3.83** | **0.32** | **9.53** | **0.70** | 20.7 |

## 5  Conclusions

In order to tackle the challenge of low accuracy in noisy environments, this paper presents ASF-Conformer, a novel model for speaker verification. During the evaluations, conducted on both publicly available VoxCeleb1 dataset and noise-synthesized dataset Mu-VoxCeleb1, ASF-Conformer achieves state-of-the-art performance. It is worth noting that ASF-Conformer demonstrates significant advantages when evaluated using the EER, especially when tested on noisy synthetic datasets. This highlights the model's robustness and ability to handle challenging acoustic conditions. Ablation experiments reveal that the incorporation of the D-AS block enables score weighting on the channel dimension of audio features, resulting in enhanced robustness in noisy environments and significant improvements in model recognition accuracy. Additionally, the F-Conformer module effectively reduces the model parameters without compromising the model's ability to extract both local and global features for modeling, thereby enhancing computational efficiency. In the future, we will explore further applications of ASF-Conformer and evaluate the proposed model in various real-world noisy environments.

# References

1. Cai, D., Cai, W., Li, M.: Within-sample variability-invariant loss for robust speaker recognition under noisy environments. In: ICASSP 2020–2020 IEEE International Conference on Acoustics, Speech and Signal Processing (ICASSP), pp. 6469–6473. IEEE (2020). https://doi.org/10.1109/ICASSP40776.2020.9053407
2. Chen, L., Liang, Y., Shi, X., Zhou, Y., Wu, C.: Crossed-time delay neural network for speaker recognition. In: Lokoč, J., et al. (eds.) MMM 2021. LNCS, vol. 12572, pp. 1–10. Springer, Cham (2021). https://doi.org/10.1007/978-3-030-67832-6_1
3. Chen, S., et al.: Continuous speech separation with conformer. In: ICASSP 2021–2021 IEEE International Conference on Acoustics, Speech and Signal Processing (ICASSP), pp. 5749–5753. IEEE (2021). https://doi.org/10.1109/ICASSP39728.2021.9413423
4. Chung, J.S., et al.: In defence of metric learning for speaker recognition. arXiv preprint arXiv:2003.11982 (2020)
5. Dai, Z., Yang, Z., Yang, Y., Carbonell, J., Le, Q.V., Salakhutdinov, R.: Transformer-xl: Attentive language models beyond a fixed-length context. arXiv preprint arXiv:1901.02860 (2019)
6. Desplanques, B., Thienpondt, J., Demuynck, K.: Ecapa-tdnn: emphasized channel attention, propagation and aggregation in tdnn based speaker verification. arXiv preprint arXiv:2005.07143 (2020)
7. Hong, J., Kim, M., Choi, J., Ro, Y.M.: Watch or listen: robust audio-visual speech recognition with visual corruption modeling and reliability scoring. In: Proceedings of the IEEE/CVF Conference on Computer Vision and Pattern Recognition, pp. 18783–18794 (2023). https://doi.org/10.1109/CVPR52729.2023.01801
8. Jin, M., Yoo, C.D.: Speaker verification and identification. In: Behavioral Biometrics for Human Identification: Intelligent Applications, pp. 264–289. IGI Global (2010)
9. Jung, J.w., Heo, H.S., Kim, J.h., Shim, H.J., Yu, H.J.: Rawnet: advanced end-to-end deep neural network using raw waveforms for text-independent speaker verification. arXiv preprint arXiv:1904.08104 (2019)
10. Kim, J.h., Heo, J., Shim, H.j., Yu, H.J.: Extended u-net for speaker verification in noisy environments. arXiv preprint arXiv:2206.13044 (2022)
11. Koizumi, Y., et al.: Df-conformer: integrated architecture of conv-tasnet and conformer using linear complexity self-attention for speech enhancement. In: 2021 IEEE Workshop on Applications of Signal Processing to Audio and Acoustics (WASPAA), pp. 161–165. IEEE (2021). https://doi.org/10.1109/WASPAA52581.2021.9632794
12. Li, Y., Lin, X.: Dual-stream time-delay neural network with dynamic global filter for speaker verification. arXiv preprint arXiv:2303.11020 (2023)
13. Liu, T., Das, R.K., Lee, K.A., Li, H.: Mfa: Tdnn with multi-scale frequency-channel attention for text-independent speaker verification with short utterances. In: ICASSP 2022–2022 IEEE International Conference on Acoustics, Speech and Signal Processing (ICASSP), pp. 7517–7521. IEEE (2022). https://doi.org/10.1109/ICASSP43922.2022.9747021

14. Matejka, P., Novotný, O., Plchot, O., Burget, L., Sánchez, M.D., Cernocký, J.: Analysis of score normalization in multilingual speaker recognition. In: Interspeech, pp. 1567–1571 (2017). https://doi.org/10.21437/Interspeech. 2017–803
15. Nagrani, A., Chung, J.S., Zisserman, A.: Voxceleb: a large-scale speaker identification dataset. arXiv preprint arXiv:1706.08612 (2017)
16. Pelecanos, J., Sridharan, S.: Feature warping for robust speaker verification. In: Proceedings of 2001 A Speaker Odyssey: The Speaker Recognition Workshop, pp. 213–218. European Speech Communication Association (2001)
17. Snyder, D., Chen, G., Povey, D.: Musan: A music, speech, and noise corpus. arXiv preprint arXiv:1510.08484 (2015)
18. Snyder, D., Garcia-Romero, D., Sell, G., Povey, D., Khudanpur, S.: X-vectors: robust dnn embeddings for speaker recognition. In: 2018 IEEE International Conference on Acoustics, Speech and Signal Processing (ICASSP), pp. 5329–5333. IEEE (2018). https://doi.org/10.1109/ICASSP.2018.8461375
19. Thienpondt, J., Desplanques, B., Demuynck, K.: Integrating frequency translational invariance in tdnns and frequency positional information in 2d resnets to enhance speaker verification. arXiv preprint arXiv:2104.02370 (2021)
20. Variani, E., Lei, X., McDermott, E., Moreno, I.L., Gonzalez-Dominguez, J.: Deep neural networks for small footprint text-dependent speaker verification. In: 2014 IEEE International Conference on Acoustics, Speech and Signal Processing (ICASSP), pp. 4052–4056. IEEE (2014). https://doi.org/10.1109/ICASSP.2014. 6854363
21. Vaswani, A., et al.: Attention is all you need. In: Advances in Neural Information Processing Systems 30 (2017)
22. Wang, C., et al.: Unispeech: unified speech representation learning with labeled and unlabeled data. In: International Conference on Machine Learning, pp. 10937–10947. PMLR (2021)
23. Yang, Z., Dai, Z., Yang, Y., Carbonell, J., Salakhutdinov, R.R., Le, Q.V.: Xlnet: generalized autoregressive pretraining for language understanding. In: Advances in Neural Information Processing Systems 32 (2019)
24. Zhang, Y., et al.: Mfa-conformer: multi-scale feature aggregation conformer for automatic speaker verification. arXiv preprint arXiv:2203.15249 (2022)
25. Zhou, T., Zhao, Y., Wu, J.: Resnext and res2net structures for speaker verification. In: 2021 IEEE Spoken Language Technology Workshop (SLT), pp. 301–307. IEEE (2021). https://doi.org/10.1109/SLT48900.2021.9383531

# Prior-Knowledge-Free Video Frame Interpolation with Bidirectional Regularized Implicit Neural Representations

Yuanjian He, Weile Zhang, Junyuan Deng, and Yulai Cong[✉]

Sun Yat-sen University, Shenzhen, China
congylai@mail.sysu.edu.cn

**Abstract.** Prevalent deep-learning-based video frame interpolation (VFI) methods are mostly pre-trained and require an optical-flow model to obtain prior knowledge. However, pre-training is often time-consuming, and may introduce unexpected artifacts when applied to a test domain that differs significantly from the training one. Alternatively, implicit neural representations have shown the ability to synthesize novel views from sparse images without pre-training. In this paper, we consider VFI as a special case of novel view synthesis and leverage implicit neural representations to perform VFI without pre-training or an optical-flow model. We propose Bidirectional Regularization Framework (BiRF), a novel VFI method that is trained per scene requiring only two input frames, which is fundamentally different from existing methods that utilize pre-trained weights containing extensive prior knowledge. We demonstrate that our BiRF, even without using prior knowledge, can generate comparable or even superior interpolated frames to prevalent pre-trained models.

**Keywords:** Video frame interpolation · Implicit neural representations · Neural fields

## 1 Introduction

Video frame interpolation (VFI) is the process of generating intermediate views from a given set of neighbouring frames. It is commonly employed to increase the frame rate of a video or image sequence, a crucial requirement for various applications. Recently, many deep-learning-based VFI methods are proposed [10,12,23,26], most of which rely heavily on a powerful optical-flow (OF) module to generate compelling results and, therefore, are sensitive to the accuracy of the underlying flow estimation [23,26]. Although some follow-up methods tried to avoid using an OF model [6], most of them still require pre-training on large datasets. Such pre-training process involves training the model on an extensive library of data to learn the correlation between frames, which often takes days or even weeks to complete; besides, hyperparameter tuning may further prolong this process multiple times. While pre-training has proven effective in many

© The Author(s), under exclusive license to Springer Nature Switzerland AG 2024
S. Rudinac et al. (Eds.): MMM 2024, LNCS 14556, pp. 112–126, 2024.
https://doi.org/10.1007/978-3-031-53311-2_9

cases, it can result in unexpected artifacts if the pre-training and testing data distributions differ significantly. The reason for this is that the model acquires all its knowledge exclusively from the pre-training data. To address this problem, some methods pre-train a dedicated model for a specific dataset, but this is not always feasible since pre-training samples may be unavailable or insufficient, especially in areas such as medical imaging and remote sensing.

Concurrent with the advancement of VFI, novel view synthesis is emerging as another highly anticipated research field in computer vision. Given only testing samples that contain views of an object from different angles, novel view synthesis aims to generate novel unseen views. Under this setup, implicit neural representation (INR) methods and their follow-up works [21,27,31] deliver state-of-the-art synthesis performance, generating almost indistinguishable novel views even in complex scenes. INR refers to fully-connected networks that take spatial coordinates as input and produce the corresponding physical quantity as output, thus representing the entire signal implicitly. Notably, these approaches are primarily trained per scene without the need for pre-training.

Noticing the resemblance of VFI and novel view synthesis, one may intuitively try to utilize the well-established INR approaches for VFI tasks. However, in reality, the VFI setup presents unpredictable variations in camera pose, target pose, and other properties, making it challenging to straightforwardly exploit novel view synthesis methods that primarily involve static scenes. Prior to our work, VIINTER [8] proposed latent code interpolation for novel view synthesis that can generate in-between views without additional context. However, directly applying latent code interpolation to VFI tasks leads to subpar results with apparent inconsistencies in fast-moving scenes. Moreover, this approach is time-consuming, requiring hours to train, making it impractical for a wide range of VFI applications.

In this paper, we propose a novel VFI method that is prior-knowledge-free, requiring only two input frames. We generalize latent code interpolation for VFI tasks and propose Bidirectional Regularization Framework (BiRF). BiRF consists of two key components: a newly introduced bidirectional regularization and an enhanced multi-MLP network architecture. The integration of these elements results in a remarkable improvement in motion consistency, generation quality, and training speed, making latent code interpolation suitable for VFI tasks.

The contribution of our work can be summarized as follows:

We propose BiRF, a novel VFI method that can generate intermediate frames without any prior knowledge based on implicit neural representations.

We show that INRs can effectively regularize themselves without additional prior knowledge using our proposed bidirectional regularization, substantially improving interpolation quality. Furthermore, BiRF utilizes a multi-MLP architecture, significantly enhancing training speed without quality loss.

We demonstrate that our method can generate comparable or superior results to prevalent pre-trained methods, even without prior knowledge. Additionally, we provide a comparison with state-of-the-art methods on custom datasets consisting of non-natural images and tiny images with extreme motion.

## 2    Related Work

### 2.1    Video Frame Interpolation

Current video frame interpolation (VFI) techniques can be broadly categorized into three groups: kernel-based [5,15,22], flow-based [1,10,12,16,23,30] and others [6], depending on the underlying technique employed.

Kernel-based methods can be traced back to an early attempt at utilizing Convolutional Neural Networks (CNN) [9,11] for VFI introduced by Long et al. [17]. However, the generated results were blurry. Significant improvements were made to this approach by Niklaus et al. [22] by generating pixel values of the interpolated frame through local convolutions of the input frames. Since then, multiple kernel-based methods have been proposed [5,15], achieving remarkable interpolation quality.

On the other hand, Flow-based methods rely on their optical flow module to find the corresponding movement of pixels. The intermediate frames are then generated by warping the pixels according to the predicted optical flow. However, these methods heavily rely on the quality of the optical flow prediction to generate high-quality interpolated frames, which are often vulnerable to occlusions. Earlier methods propose several intuitive solutions to occlusion, including learning unsupervised optical flow [30] and adding depth information [1]. More recently, Park et al. [23] proposed a VFI algorithm based on bilateral motion estimation, while Huang et al. [10] utilized a privileged distillation scheme to improve overall performance. To alleviate the Degrees of Freedom (DoF) limitations, Lee et al. [12] proposes a hybrid method that estimates both kernel weights and offset vectors for each target pixel to synthesize the output frame.

Importantly, all the methods outlined above necessitate pre-training on a vast dataset. Interpolating frames using these methods would be impossible in the absence of a pre-trained model weight comprising extensive prior knowledge.

### 2.2    Implicit Neural Representations

In recent years, Implicit Neural Representations (INR) have emerged as a powerful tool for vision tasks such as novel view synthesis [2,14], video compression [3] and image compression [7]. Despite the ability of fully connected networks to fit almost any signal, fitting an image with an INR can take hours or days to complete. Several techniques have been proposed to tackle this issue [4,20,25]. For instance, MINER [25] partitions the input image into several blocks and assigns a small MLP to each block, resulting in faster and better fitting. Prior to our work, CURE [26] claimed to be the first INR-based VFI method. It generates a local feature with the collaboration of an optical flow module and a CNN encoder. The generated feature containing motion information is then used for the MLP to decode the intermediate frame. However, the majority of motion prediction still heavily relies on its optical flow module. Recently, VIINTER [8] proposes a novel view synthesis method that generates novel views simply by interpolating the latent code conditioned on the INR. However, in our experiments, it only generates good results when the motion between frames is limited. In VFI datasets with slightly larger motion (e.g., Vimeo90K [30]), it can generate results with obvious artifacts as shown in Fig. 3 and Fig. 4.

## 3    Method

### 3.1    Implicit Neural Representation

An image can be interpreted as a mapping function that associates the coordinates of each pixel $\mathbf{x} = (x, y) \in \mathbb{R}^2$ with its corresponding RGB values $\mathbf{c} = (r, g, b) \in \mathbb{R}^3$. By training a neural network $\mathcal{M}$ to replicate this input-output relationship, we can obtain a continuous implicit representation of the image,

$$\mathcal{M}(\mathbf{x}) \to \mathbf{c}. \tag{1}$$

Here, we utilize a multi-layer perceptron (MLP) [19] to parameterize $\mathcal{M}$. Recent developments in image fitting with MLP have predominantly utilized either the sine activation function [27] or positional encoding with ReLU activation [21]. For our study, we opted for the former and utilized sine as our activation function. For all pixel coordinates $\mathbf{x}$, our MLP is trained to minimize the error between the predicted pixel color, denoted as $\mathbf{c}$, and the corresponding ground truth pixel color, denoted as $\mathbf{c}^{gt}$. We define the reconstruction loss as follows

$$L_{recon} = \sum_{\mathbf{x}} \|\mathbf{c} - \mathbf{c}^{gt}\|^2. \tag{2}$$

Multiple images can be represented with a single MLP. To achieve this, we can simply include a frame index $n$ to the above equation, namely,

$$\mathcal{M}(\mathbf{x}, n) \to \mathbf{c}. \tag{3}$$

The loss function can be modified as

$$L_{recon} = \sum_{n} \sum_{\mathbf{x}} \|\mathbf{c} - \mathbf{c}^{gt}\|^2, \tag{4}$$

where $n$ is the number of target images.

### 3.2    Latent Code Interpolation

In the context of representing multiple images, it is possible to reconstruct images that existed in the ground truth image sequence by inputting the corresponding $n$ value after training the MLP. However, attempting to interpolate views by simply inputting in-between $n$ values using this training strategy results in the generation of meaningless noise. To address this issue, VIINTER [8] proposes latent code interpolation as a solution.

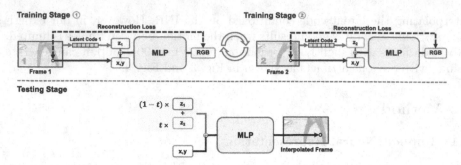

**Fig. 1.** This figure visually demonstrates latent code interpolation. Initially, a unique trainable latent code is randomly initialized for each frame. The MLP is then trained to reconstruct the corresponding frame, given a latent code. The training process involves alternating training stages 1 and 2 until convergence. During testing, a linear combination of the latent codes with $0 \leq t \leq 1$ is used to generate intermediate frames.

Instead of including $n$ to the input, a learnable vector $z \in \mathbb{R}^M$ with latent dimension $M$ is randomly initialized for each image to serve as its latent code. This vector is then concatenated with the input coordinate $\mathbf{x} = (x, y)$ to create a new input for the network. Formally, for $N$ images and $n = 1, ..., N$, the mapping function is modified in the following manner

$$\mathcal{M}(\mathbf{x}|z_n) \rightarrow \mathbf{c}_n, \tag{5}$$

note that the individual frame index $n$ is no longer necessary, as the relevant information can be encoded with latent code $z$ after training.

We train the network with the objective below

$$L_{recon} = \sum_n \sum_{\mathbf{x}} \|\mathbf{c}_n - \mathbf{c}_n^{gt}\|^2. \tag{6}$$

Directly training the network with the above procedure would result in latent codes with different scales. Interpolating them would produce a noise vector and fail to produce meaningful results. To address this issue, we regularize all latent codes with 1-norm

$$z = \frac{z}{\|z\|_1}, \forall z \in Z_N. \tag{7}$$

For two frames with latent code $z_i, z_j \in Z_N$, We can simply generate intermediate view $I^{inter}$ at desired time $t_0$ $(0 \leq t_0 \leq 1)$ by interpolating the latent code with weighted average $z_{inter} = (1 - t_0)z_i + t_0 z_j$ and concatenating it with the coordinate as input on all pixels

$$\sum_{\mathbf{x}} \mathcal{M}(\mathbf{x}|z_{inter}) \rightarrow I^{inter}. \tag{8}$$

A visual demonstration of latent code interpolation in shown in Fig. 1.

## 3.3   Bidirectional Regularization Framework (BiRF)

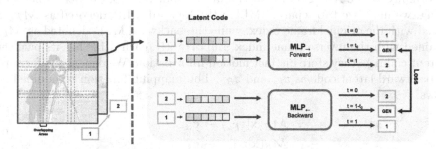

**Fig. 2.** This figure provides an overview of our proposed Bidirectional Regularization Framework (BiRF). Firstly, we partition the input frames into smaller patches with overlapping areas. For each patch, we initialize two multi-layer perceptrons (MLPs), one trained with the initial frame index and the other trained in reverse. The two MLPs are designed to regularize each other through the minimization of errors at randomly sampled timesteps $t_0$, where $0 \leq t_0 \leq 1$.

Though latent code interpolation offers a means of interpolating novel views without prior knowledge, we have observed that using this technique directly in VFI tasks is inadequate due to the larger motion present in VFI datasets. For larger motion, intermediate frames generated by latent code interpolation are inconsistent on edges and show obvious artifacts, as shown in Fig. 3. To address this issue, we propose Bidirectional Regularization Framework (BiRF) based on the proposed latent code interpolation method. BiRF consists of two key components, namely bidirectional regularization and multi-MLP architecture. A visual overview is presented in Fig. 2.

**Bidirection Regularization.** During our experiment, we observed that the presence of inconsistent edges and artifacts in fast-moving scenes can be attributed to the lack of regularization for intermediate frames. The original loss function only imposes constraints on the first and last frames, neglecting the intermediate ones. While it may be tempting to assume that consecutive frames adhere to certain conditions, such as geometry consistency, this assumption does not always hold true, especially considering the characteristics of VFI datasets.

In our observation, when we train an inverse MLP by inverting the input frame index, the inverse MLP is able to learn a distinct perspective of pixel motion in the backward direction. While the inverse MLP is similarly not consistent, it preserves some formerly scattered edges predicted by the original MLP.

Leveraging this finding, we propose bidirectional regularization as the first component of our proposed BiRF. We regularize the motion consistency of the generated frames by incorporating both the forward and backward MLPs. The forward and backward MLP learns the forward and backward motion, respectively. By imposing consistency constraints on the motion flow generated by both

MLPs, we can significantly improve the quality and consistency of the generated frames, even in the presence of occlusion and large motion.

Formally, where $N = 2$ frames are input and index $n = 0, 1$ under triplet VFI setting, we initialize two separate MLPs: the forward MLP, denoted as $\mathcal{M_F}$, is trained with the original frame index, while the backward MLP, denoted as $\mathcal{M_B}$, is trained with the inverse frame index. Importantly, each MLP has its own set of latent codes as they store distinct motion information. We denote the forward and backward latent code as $z_{F_n}$ and $z_{B_n}$. The mapping function is modified as follows:

$$
\begin{aligned}
\mathcal{M_F}(\mathbf{x}|z_{F_n}) &\rightarrow \mathbf{c}_n, \\
\mathcal{M_B}(\mathbf{x}|z_{B_n}) &\rightarrow \mathbf{c}_{1-n}.
\end{aligned}
\tag{9}
$$

During training, we randomly sample time $t_s$ with $0 \le t_s \le 1$ to generate linear combination of latent codes for both networks

$$
\begin{aligned}
z_F &= (1 - t_s)z_{F_0} + t_s z_{F_1}, \\
z_B &= t_s z_{B_0} + (1 - t_s)z_{B_1}.
\end{aligned}
\tag{10}
$$

With the interpolated latent code above, we can generate prediction $I^F$ from the forward MLP at time $t_s$ and prediction $I^B$ from the backward MLP at time $1 - t_s$

$$
\begin{aligned}
\sum_{\mathbf{x}} \mathcal{M}_F(\mathbf{x}|z_F) &\rightarrow I^F, \\
\sum_{\mathbf{x}} \mathcal{M}_B(\mathbf{x}|z_B) &\rightarrow I^B.
\end{aligned}
\tag{11}
$$

We match the two predictions with L2 loss as

$$
L_{bd} = \|I^F - I^B\|^2.
\tag{12}
$$

The gradient of this loss function is separately back-propagated by the two MLPs.

This loss is jointly trained with the reconstruction loss

$$
L = L_{bd} + L_{recon}.
\tag{13}
$$

The incorporation of bidirectional regularization enhances motion consistency considerably as the two MLPs collaboratively learn a more precise flow of motion. This is a fundamental measure in implementing latent code interpolation to VFI tasks, wherein a smooth motion prediction is the initial objective. Furthermore, this approach improves image stability and reduces artifacts without any reliance on pre-trained models. Further details on image stability are available in our ablation study.

**Multi-MLP Architecture.** The bidirectional regularization method discussed in the previous section contributes to a significant improvement in the quality of the interpolation. However, this approach poses a computational burden to the already time-consuming training of MLPs as it requires training two MLPs instead of one. To address this issue, we propose splitting the input images into smaller patches and using a set of smaller MLPs for each patch. Previous research [25] has demonstrated the effectiveness of this method in image-fitting tasks. In our experiments, we found that this approach performed exceptionally well in conjunction with latent code interpolation. Notably, the training time was reduced from several hours to just hundreds of seconds, with a minor impact on quality.

Another advantage of this approach is that not all areas of an image need to be updated in VFI tasks. For instance, for two input frames with a static background, we can blend the corresponding region from the two input frames without computing it with an MLP.

Formally, considering a case with bidirectional regularization applied to all patches, we split the input frames $I_n$ $(n = 0, 1)$ into $M \times N$ patches, which is a hyper-parameter that affects interpolation quality. Smaller patch sizes lead to faster convergence but may cause a slight decline in interpolation quality. For each patch, we initialize two MLPs with two sets of latent codes, with a total of $2 \times M \times N$ MLPs. The mapping function in (9) is modified as

$$\sum_{p=1}^{2 \times M \times N} \mathcal{M}_{F_p}(\mathbf{x}|z_{F_{n,p}}) \to I_n,$$

$$\sum_{p=1}^{2 \times M \times N} \mathcal{M}_{B_p}(\mathbf{x}|z_{B_{n,p}}) \to I_{1-n}, \tag{14}$$

where $\mathcal{M}_p$ denotes the $p$ th MLP, $z_{n,p}$ denotes the corresponding $p$ th set of latent codes for frame index $n$, $\mathbf{x}$ denotes all pixels in a single patch.

The issue with this approach is the discontinuity at the borders between patches. When pixels cross the borders, they switch from being predicted by one MLP to another, resulting in an obvious discontinuity. To address this problem, we increase the coverage area of each MLP slightly, creating overlapping areas between patches. For pixels in these overlapping areas, we output a blended value from the two corresponding MLPs that relate to this area. This technique helps to alleviate the discontinuity effect in most situations.

## 4    Experiments

### 4.1    Datasets

We perform experiments on 5 datasets.

**UCF-101** [28]. The UCF101 [28] dataset comprises action videos obtained from YouTube. For the purpose of assessment, we utilized a subset obtained by [16], which consists of 379 triplets with a resolution of $256 \times 256$.

**Vimeo-90K** [30]. Vimeo-90K is a widely used VFI dataset with a resolution of $448 \times 256$. Due to limited experiment time, we only evaluated on the first 100

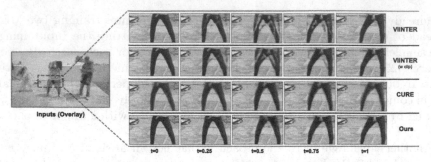

**Fig. 3.** The presented figure exhibits a demonstration extracted from Vimeo-90k [30] featuring a scenario involving large motion. Our proposed method is able to generate significantly sharper motion in comparison to other tested alternatives.

triplets from all 3782 triplets in its testing set. This is applied to all methods that are compared.

**Vimeo-90K [30] Cropped.** To further showcase the efficacy of our proposed method, we generate a sub-dataset from Vimeo-90K containing tiny images with extreme motion. We partitioned the original Vimeo-90K test set into 16 patches, each with a resolution of $112 \times 64$, and computed the LPIPS [32] score for the first and third frame of each patch. We then identified those patches for which the LPIPS score was greater than 0.3 and created a new triplet dataset with 100 samples from these patches.

**Colonoscopy Video [18].** We have also evaluated its performance on datasets involving non-natural images, such as medical imaging. Since it is often difficult to gather sufficient training data in medical imaging, fine-tuning is typically not a viable option. We used "video 32" from video with polyps, center-cropped and resized each frame to $256 \times 256$ to avoid the black circle. Our evaluation demonstrates the versatility of our method in handling a range of different datasets and applications.

**SAR Video [24].** Synthetic-aperture radar (SAR) is a form of radar used to create two-dimensional images or three-dimensional reconstructions of objects, such as landscapes. We collect "Eubank Gate and Traffic VideoSAR" from the Sandia National Laboratories website. We sample at every 5 frames and resize to $256 \times 256$ to form a dataset with 30 testing triplets.

### 4.2    Implementation Details

For our experiments on Vimeo-90k [30], SAR Video, UCF-101 [28], Colonoscopy Video [18] and Vimeo-90k Cropped, we set the number of patches to 16, 16, 16, 4, 1 respectively according to their image sizes. For each patch, we used a 5-layer MLP with a hidden dimension of 256 and a latent code dimension of 128, $\omega_0$ is set according to motion range. We employed BiRF at every step. Regarding VIINTER [8], we followed the recommendation in its paper and used a single

8-layer MLP with a hidden dimension of 512 and a latent code dimension of 128. To ensure a fair comparison between the two methods, we trained the two methods with an equal amount of time on all datasets. approximately 400 s on a RTX3090. For both methods, we used a learning rate of 1e-5 and the same batch size while employing a cosine annealing scheduler. For other pre-trained methods, we use official codes and pre-trained models for evaluation.

**Table 1.** Quantitative comparison on four datasets.

| | Vimeo-90K [30] | | | [30] Cropped | | Colonoscopy [18] | | SAR Video [24] | |
|---|---|---|---|---|---|---|---|---|---|
| | PSNR↑ | SSIM↑ | LPIPS↓ | PSNR↑ | SSIM↑ | PSNR↑ | SSIM↑ | PSNR↑ | SSIM↑ |
| Pretrained Methods | | | | | | | | | |
| SepConv [22] | 34.04 | 0.9594 | 0.011 | 22.79 | 0.8105 | 31.08 | 0.8845 | 30.73 | 0.7095 |
| BMBC [23] | 34.84 | 0.9649 | 0.013 | 24.51 | 0.8761 | 29.26 | 0.9055 | 31.15 | 0.7683 |
| AdaCoF [12] | 34.71 | 0.9641 | 0.017 | 23.68 | 0.8087 | 31.37 | 0.8964 | 31.86 | 0.7659 |
| CURE [26] | **35.69** | **0.9711** | 0.013 | 23.05 | 0.8686 | 31.52 | 0.9004 | 32.31 | 0.7719 |
| RIFE [10] | 34.20 | 0.9605 | **0.011** | 26.58 | 0.8546 | 31.15 | 0.8866 | 31.78 | 0.7489 |
| RRIN [13] | 32.68 | 0.9627 | 0.017 | 23.49 | 0.8475 | 30.19 | 0.8972 | 31.53 | 0.7666 |
| Pretrain-Free Methods | | | | | | | | | |
| VIINTER [8] | 31.28 | 0.8934 | 0.114 | 26.41 | 0.7993 | 30.37 | 0.8807 | 27.47 | 0.5946 |
| Ours | 32.89 | 0.9543 | 0.022 | **26.98** | **0.8803** | **31.54** | **0.9106** | **32.68** | **0.8074** |

**Fig. 4.** This figure shows that our method outperforms other methods in terms of LPIPS while requiring a fraction of the training time compared to VIINTER [8]. The blue box area was identified as still background and was blended without further processing (Color figure online)

## 4.3 Results

To conduct a quantitative comparison, we have included several methods as presented in Table 1 and 2. It should be noted that all of the above methods, with the exception of [8], were trained with at least 50K+ triplets for a duration ranging from days to weeks, mostly on the Vimeo-90K [30] training set. Conversely, our approach was solely provided with individual testing samples. This is an unfair comparison, so our objective in this section is not to surpass all existing methods but rather to demonstrate that our proposed method is capable of generating comparable results with prevalent pre-trained VFI methods.

**Table 2.** Quantitative comparison on UCF-101 dataset.

| UCF-101 [28] | PSNR↑ | SSIM↑ | LPIPS↓ |
|---|---|---|---|
| Pretrained Methods | | | |
| SepConv [22] | 34.84 | 0.9463 | 0.018 |
| BMBC [23] | 35.01 | 0.9501 | 0.022 |
| AdaCof [12] | 35.15 | 0.9504 | 0.022 |
| RRIN [13] | 32.51 | 0.9480 | 0.029 |
| CURE [26] | **35.21** | **0.9513** | 0.026 |
| RIFE [10] | 34.85 | 0.9475 | **0.018** |
| Pretrain-Free Methods | | | |
| VIINTER [8] | 28.02 | 0.8299 | 0.219 |
| Ours | 34.23 | 0.9503 | 0.022 |

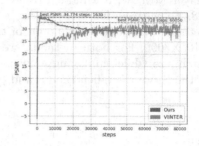

**Fig. 5.** The figure demonstrates significantly faster convergence with higher quantitative results compare to VIINTER [8]. Additionally, our training curve is more stable.

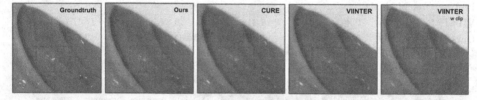

**Fig. 6.** In the highlighted area, our approach yields much sharper results with the light reflection whilst preserving a majority of the blood vessel detail.

The results on the Vimeo-90k [30] and UCF-101 [28] indicate that our method is still slightly behind some of the pre-trained methods in terms of PSNR and SSIM [29], as shown in Table 1 and 2. This could be attributed to the inherent challenges in fitting high-frequency regions with INRs. However, it is worth noting that a decrease in PSNR or SSIM results does not necessarily impact visual quality. For instance, in Fig. 4, our generated result achieved the highest LPIPS score among all other methods.

In comparison with VIINTER [8], our method achieve significantly higher quantitative results with a fraction of the required training time. In fact, VIINTER takes more than 5 h to match our quantitative results, as shown in Fig. 5. Notably, VIINTER fails to generate consistent motion on fast-moving edges, even with multiple times longer training time, as depicted in Fig. 4.

The results on custom datasets are presented in Table 1, where our method surpasses state-of-the-art methods on either PSNR or SSIM results. An example from the Colonoscopy dataset is shown in Fig. 6. Experiments on these datasets show that our method is not severely affected by the size or appearance of the

**Table 3.** Ablation study on Vimeo-90k dataset. BD-Loss is referred to Bidirectional Regularization.

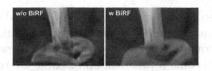

**Fig. 7.** Without regularization, the MLP generates the interpolation results with obvious artifacts, especially on moving edges.

Vimeo-90k [30]

|  | PSNR↑ | SSIM↑ | LPIPS↓ |
|---|---|---|---|
| Baseline | 31.28 | 0.8934 | 0.114 |
| +Multi-MLP | 30.87 | 0.9153 | 0.029 |
| +BD-Loss | **31.96** | **0.9360** | **0.026** |

testing samples, which is crucial in cases where other pre-trained methods may generate various artifacts under such settings.

### 4.4 Ablation Study

In this section, we discuss the individual effect of both multi-MLP architecture and bidirectional regularization. A quantitative result is shown in Table 3.

For bidirectional regularization, we have observed a significant increase in motion consistency in areas with large motion, as discussed in previous sections. Additionally, BiRF notably reduces artifacts around edges, as demonstrated in Fig. 7. However, the use of bidirectional regularization results in slower convergence times, therefore, we recommend applying bidirectional regularization in conjunction with multi-MLP architecture.

For multi-MLP architecture, one of its most prominent benefits is the acceleration of training times, albeit at the expense of a certain degree of sacrifice in quantitative results. Additionally, this approach offers the benefit of stabilizing the quality of generated images. This can be observed in the training curve presented in Fig. 5, which displays a noticeably smoother trajectory. This feature proves particularly advantageous when selecting the optimal iteration for image generation.

## 5    Conclusion

In this paper, we present Bidirectional Regularization Framework (BiRF), a novel video frame interpolation (VFI) method based on bidirectional regularized implicit neural representation. We demonstrate the feasibility of performing VFI without any prior knowledge by enabling two MLPs to mutually regulate each other. Additionally, we accelerate training by utilizing a multi-MLP architecture. Our proposed method is trained per scene without requiring pre-training or pre-trained models. This property is especially advantageous when dealing with test data that differs significantly from the pre-training data, and when data access is limited. Our method has yielded results that are comparable, if not superior, to state-of-the-art pre-trained VFI techniques. In future work, we will focus on enhancing interpolation quality and further accelerating training speed. While our proposed method is not yet suitable for real-time or production use, we are confident that with further implementation, the INR-based VFI method will become a practical and flexible alternative VFI method.

# References

1. Bao, W., Lai, W.S., Ma, C., Zhang, X., Gao, Z., Yang, M.H.: Depth-aware video frame interpolation. In: Proceedings of the IEEE/CVF Conference on Computer Vision and Pattern Recognition, pp. 3703–3712 (2019)
2. Chen, A., et al.: Mvsnerf: fast generalizable radiance field reconstruction from multi-view stereo. In: Proceedings of the IEEE/CVF International Conference on Computer Vision, pp. 14124–14133 (2021)
3. Chen, H., He, B., Wang, H., Ren, Y., Lim, S.N., Shrivastava, A.: Nerv: Neural representations for videos. Adv. Neural. Inf. Process. Syst. **34**, 21557–21568 (2021)
4. Chen, Y., Liu, S., Wang, X.: Learning continuous image representation with local implicit image function. In: Proceedings of the IEEE/CVF Conference on Computer Vision and Pattern Recognition, pp. 8628–8638 (2021)
5. Cheng, X., Chen, Z.: Video frame interpolation via deformable separable convolution. In: Proceedings of the AAAI Conference on Artificial Intelligence, vol. 34, pp. 10607–10614 (2020)
6. Choi, M., Kim, H., Han, B., Xu, N., Lee, K.M.: Channel attention is all you need for video frame interpolation. In: Proceedings of the AAAI Conference on Artificial Intelligence, vol. 34, pp. 10663–10671 (2020)
7. Dupont, E., Goliński, A., Alizadeh, M., Teh, Y.W., Doucet, A.: Coin: compression with implicit neural representations. arXiv preprint arXiv:2103.03123 (2021)
8. Feng, B.Y., Jabbireddy, S., Varshney, A.: Viinter: view interpolation with implicit neural representations of images. In: SIGGRAPH Asia 2022 Conference Papers, pp. 1–9 (2022)
9. He, K., Zhang, X., Ren, S., Sun, J.: Deep residual learning for image recognition. In: Proceedings of the IEEE Conference on Computer Vision and Pattern Recognition, pp. 770–778 (2016)
10. Huang, Z., Zhang, T., Heng, W., Shi, B., Zhou, S.: Real-time intermediate flow estimation for video frame interpolation. In: Computer Vision-ECCV 2022: 17th European Conference, Tel Aviv, Israel, 23–27 October 2022, Proceedings, Part XIV, pp. 624–642. Springer (2022). doi: https://doi.org/10.1007/978-3-031-19781-9_36
11. LeCun, Y., Bottou, L., Bengio, Y., Haffner, P.: Gradient-based learning applied to document recognition. Proc. IEEE **86**(11), 2278–2324 (1998)
12. Lee, H., Kim, T., Chung, T.Y., Pak, D., Ban, Y., Lee, S.: Adacof: adaptive collaboration of flows for video frame interpolation. In: Proceedings of the IEEE/CVF Conference on Computer Vision and Pattern Recognition, pp. 5316–5325 (2020)
13. Li, H., Yuan, Y., Wang, Q.: Video frame interpolation via residue refinement. In: ICASSP 2020–2020 IEEE International Conference on Acoustics, Speech and Signal Processing (ICASSP), pp. 2613–2617. IEEE (2020)
14. Li, Z., Niklaus, S., Snavely, N., Wang, O.: Neural scene flow fields for space-time view synthesis of dynamic scenes. In: Proceedings of the IEEE/CVF Conference on Computer Vision and Pattern Recognition, pp. 6498–6508 (2021)

15. Liu, Y.L., Liao, Y.T., Lin, Y.Y., Chuang, Y.Y.: Deep video frame interpolation using cyclic frame generation. In: Proceedings of the AAAI Conference on Artificial Intelligence, vol. 33, pp. 8794–8802 (2019)

16. Liu, Z., Yeh, R.A., Tang, X., Liu, Y., Agarwala, A.: Video frame synthesis using deep voxel flow. In: Proceedings of the IEEE International Conference on Computer Vision, pp. 4463–4471 (2017)

17. Long, G., Kneip, L., Alvarez, J.M., Li, H., Zhang, X., Yu, Q.: Learning image matching by simply watching video. In: Leibe, B., Matas, J., Sebe, N., Welling, M. (eds.) ECCV 2016. LNCS, vol. 9910, pp. 434–450. Springer, Cham (2016). https://doi.org/10.1007/978-3-319-46466-4_26

18. Ma, Y., Chen, X., Cheng, K., Li, Y., Sun, B.: LDPolypVideo benchmark: a large-scale colonoscopy video dataset of diverse polyps. In: de Bruijne, M., Cattin, P.C., Cotin, S., Padoy, N., Speidel, S., Zheng, Y., Essert, C. (eds.) MICCAI 2021. LNCS, vol. 12905, pp. 387–396. Springer, Cham (2021). https://doi.org/10.1007/978-3-030-87240-3_37

19. McClelland, J.L., Rumelhart, D.E., Group, P.R., et al.: Parallel Distributed Processing, Volume 2: Explorations in the Microstructure of Cognition: Psychological and Biological Models, vol. 2. MIT press (1987)

20. Mehta, I., Gharbi, M., Barnes, C., Shechtman, E., Ramamoorthi, R., Chandraker, M.: Modulated periodic activations for generalizable local functional representations. In: Proceedings of the IEEE/CVF International Conference on Computer Vision, pp. 14214–14223 (2021)

21. Mildenhall, B., Srinivasan, P.P., Tancik, M., Barron, J.T., Ramamoorthi, R., Ng, R.: Nerf: representing scenes as neural radiance fields for view synthesis. Commun. ACM 65(1), 99–106 (2021)

22. Niklaus, S., Mai, L., Liu, F.: Video frame interpolation via adaptive separable convolution. In: Proceedings of the IEEE International Conference on Computer Vision, pp. 261–270 (2017)

23. Park, J., Ko, K., Lee, C., Kim, C.-S.: BMBC: bilateral motion estimation with bilateral cost volume for video interpolation. In: Vedaldi, A., Bischof, H., Brox, T., Frahm, J.-M. (eds.) ECCV 2020. LNCS, vol. 12359, pp. 109–125. Springer, Cham (2020). https://doi.org/10.1007/978-3-030-58568-6_7

24. Sandia National Laboratories: Videosar. https://www.sandia.gov/app/uploads/sites/124/2021/08/eubankgateandtrafficvideosar.mp4

25. Saragadam, V., Tan, J., Balakrishnan, G., Baraniuk, R.G., Veeraraghavan, A.: Miner: Multiscale implicit neural representation. In: Computer Vision-ECCV 2022: 17th European Conference, Tel Aviv, Israel, 23–27 October 2022, Proceedings, Part XXIII, pp. 318–333. Springer (2022). https://doi.org/10.1007/978-3-031-20050-2_19

26. Shangguan, W., Sun, Y., Gan, W., Kamilov, U.S.: Learning cross-video neural representations for high-quality frame interpolation. arXiv preprint arXiv:2203.00137 (2022)

27. Sitzmann, V., Martel, J., Bergman, A., Lindell, D., Wetzstein, G.: Implicit neural representations with periodic activation functions. Adv. Neural. Inf. Process. Syst. 33, 7462–7473 (2020)

28. Soomro, K., Zamir, A.R., Shah, M.: A dataset of 101 human action classes from videos in the wild. Center Res. Comput. Vis. 2(11) (2012)

29. Wang, Z., Bovik, A.C., Lu, L.: Why is image quality assessment so difficult? In: 2002 IEEE International Conference on Acoustics, Speech, and Signal Processing, vol. 4, pp. IV-3313. IEEE (2002)

30. Xue, T., Chen, B., Wu, J., Wei, D., Freeman, W.T.: Video enhancement with task-oriented flow. Int. J. Comput. Vision **127**, 1106–1125 (2019)
31. Yu, A., Ye, V., Tancik, M., Kanazawa, A.: pixelnerf: neural radiance fields from one or few images. In: Proceedings of the IEEE/CVF Conference on Computer Vision and Pattern Recognition, pp. 4578–4587 (2021)

# Two-Stage Reasoning Network with Modality Decomposition for Text VQA

Shengrong Ling, Sisi You$^{(\boxtimes)}$, and Bing-Kun Bao

Nanjing University of Posts and Telecommunications, Nanjing, China
{1021010624,ssyou,bingkunbao}@njupt.edu.cn

**Abstract.** Text-based Visual Question Answering (Text VQA) is a challenging task that requires a comprehensive understanding of scene texts in an image. Scene texts encompass information from both textual and visual modalities. Most existing methods typically treat information of different modalities indiscriminately. However, such approaches may restrict the detailed interaction between textual and visual modalities, leading to biased or incorrect semantic understanding. To address the limitation, we propose a two-stage reasoning network with modality decomposition for Text VQA. In the first stage, we separately handle OCR textual and visual modalities through a modality-specific attention module which is adopted to capture the crucial information of each modality. In the second stage, we aim to enhance the interaction between textual and visual modalities. To achieve this, we introduce a semantic-guided interaction module that incorporates the semantic context to facilitate the alignment of the two modalities. Extensive experiments on the TextVQA and ST-VQA datasets demonstrate that our network achieves competitive performance compared with current state-of-the-art methods.

**Keywords:** Text VQA · Semantic Reasoning · Scene Text Recognition

## 1 Introduction

Text-based Visual Question Answering (Text VQA) [21] is a task that involves answering questions based on the textual content present in an image. In order to provide accurate answers, it requires the model to read scene texts in images using an OCR system and reason over the question, visual objects, and scene texts. Unlike traditional VQA methods [2,13,19], Text VQA focuses on understanding the textual information in the image, which has broad application prospects in various domains such as intelligent dialogue systems, children's education, and assisting visually impaired individuals.

To effectively understand both textual and visual information, current approaches typically merge OCR features from different modalities to form a unified representation, as shown in Fig. 1 (a), and then feed it into multimodal reasoning modules. For instance, LoRRA [21] relies on customized pairwise attention to obtain the fused OCR-question features and image-question features.

© The Author(s), under exclusive license to Springer Nature Switzerland AG 2024
S. Rudinac et al. (Eds.): MMM 2024, LNCS 14556, pp. 127–140, 2024.
https://doi.org/10.1007/978-3-031-53311-2_10

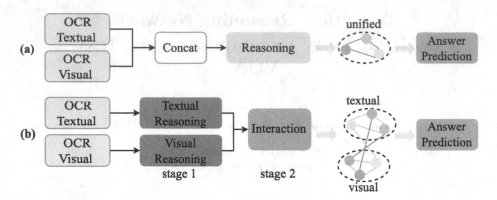

**Fig. 1.** Two types of existing Text VQA frameworks. (a) General Text VQA models [8,9,16,17,21,26]. (b) The proposed two-stage reasoning network.

M4C [9] projects all modalities into transformer layers to fuse multimodal features. Additionally, several methods focus on utilizing specific attention modules [16,17,26,29] or graph networks [8,12] to enhance the interaction between OCR tokens and visual objects. However, these approaches simply concatenate OCR textual and visual features to construct scene text information, which restricts the detailed interaction between textual and visual modalities.

To address the aforementioned issue, a feasible approach is to design a two-stage reasoning model, as depicted in Fig. 1(b), which aims to fully capture the interaction between textual and visual modalities. In the first stage, the model handles each modality independently, allowing it to focus on important textual content and visual areas. This strategy incorporates different perspectives and sources of features, enhancing the model's understanding and reasoning capabilities for scene texts. In the second stage, detailed interactions are built upon the learned representations of textual and visual modalities. Figure 1 illustrates that, compared with general methods, our approach offers greater flexibility in establishing connections not only between diverse scene texts, but also between distinct modalities. Moreover, the interaction is constructed based on the foundational understanding of each modality obtained in the first stage, reducing irrelevant information and improving efficiency. Therefore, by employing a two-stage reasoning mechanism to foster the interaction between different modalities, the model can comprehensively utilize both OCR textual and OCR visual features, contributing to a better multimodal representation for scene text understanding.

Based on above considerations, we propose a two-stage reasoning network with modality decomposition for Text VQA. To facilitate the detailed interaction between OCR textual and visual modalities, we construct the correlation between the two modalities in a staged paradigm. In the first stage, a modality-specific attention module is proposed to separately focus on the crucial information of each modality. For the textual modality, cross-modal attention layers are leveraged to construct interactions between textual features and question or visual objects, which allows the model to capture relevant OCR textual information. For the visual modality, a global image feature is introduced as auxiliary information to facilitate the attention of visual features, assisting the model to identify relevant visual areas. In the second stage, a semantic-guided interaction module is proposed to offer more flexible interactions between OCR tokens, which incorporates the semantic context to facilitate the fusion and alignment between textual and visual modalities.

In summary, the contributions of this paper are as follows:

1) We propose a two-stage reasoning network with modality decomposition for Text VQA, allowing the model to comprehensively utilize both textual and visual modalities of scene text information.
2) To capture the robust interaction between OCR textual and visual modalities, we first design a modality-specific attention module to focus on the crucial text content and visual areas in the first stage, and then employ a semantic-guided interaction module to construct the correlation in the second stage.
3) Extensive experiments on the TextVQA and ST-VQA benchmarks verify the effectiveness and superiority of our method.

## 2    Related Work

In Text VQA [21], there are typically three components: representation learning, feature interaction, and answer prediction. Concretely, the model first learns the feature representation of three input modalities, namely the question, visual objects, and scene texts, and then apply interactive reasoning to locate the answer. Our work follow the multi-step answer prediction approach along with most existing Text VQA methods. As a result, we primarily focus on introducing the representation learning and feature interaction methods, as they are most relevant to our work.

### 2.1    Representation Learning in Text VQA

The representation of multi-modality is essential for improving the performance of Text VQA. Most Text VQA models currently utilize BERT [22] to encode question words and pre-trained Faster R-CNN [20] to represent objects in the image. In terms of representing OCR tokens, existing approaches typically employ an off-the-shelf OCR model [6,23] to obtain text content and then provide a rich representation of them. For example, the early work LoRRA [21] uses FastText [11] to encode word embeddings for each OCR token. M4C [9] enhances the representation of scene texts by incorporating character-level Pyramidal Histogram of Characters (PHOC) [1], Faster-RCNN [20], and bounding box features to capture character, appearance, and spatial location information, respectively. Lately, [8,9,16,17,21,25] handle these features by concatenating them to form a unified representation, then feed it into subsequent reasoning modules. However, such approaches treat different OCR modalities indiscriminately, which may limits the understanding of each modality. In contrast, our model seeks to fully utilize the OCR textual and visual modalities, and we introduce a modality-specific reasoning module to capture the crucial information in each modality.

## 2.2  Feature Interaction in Text VQA

In the Text VQA task, another crucial aspects is establishing the interaction of multimodal features after obtaining their representations. Early work such as LoRRA [21] focuses solely on pairwise feature fusion using simple attention mechanism, which limits the interaction among multiple modalities. To overcome this limitation, M4C [9] considers projecting all modalities into transformer layers to fuse multimodal features, serving as the foundation for subsequent works [7,8,16,17,26,27]. SA-M4C [12] incorporates spatial reasoning within the transformer framework by encoding the coordinates of visual entities. CRN [17] proposes a progressive attention module to fuse multimodal information while utilizing a graph reasoning network. LaAP-Net [25] utilizes position-guided attention to merge related object features into scene texts, thereby enhancing the representation of OCR features. SC-Net [7] follows a similar mechanism but replaces the attention with two individual transformer modules. However, previous methods tend to treat OCR textual and visual features as a whole, which limits the detailed interaction between the two modalities. Thus, in our work, we aim to enhance the interaction between OCR textual and visual modalities. To achieve this, we introduce a semantic-guided interaction module that incorporates the semantic context to facilitate the alignment of the two modalities.

## 3  Methodology

In this section, we introduce our two-stage reasoning network with modality decomposition for the Text VQA task, which aims to fully exploit the detailed interaction between OCR textual and visual modalities. The overall framework is illustrated in Fig. 2. Firstly, multimodal features are extracted by pre-trained models (in Sect. 3.1). We then apply the modality-specific attention module (in Sect. 3.2) to reason over textual and visual modalities separately, which allows the model to focus on important information of each modality. Next, we utilize the semantic-guided interaction module (in Sect. 3.3) to establish interactions between the two modalities. Finally, features from different modalities are aggregated and inputted into the transformer layer for answer prediction (in Sect. 3.4).

## 3.1  Multimodal Feature Extraction

Our model takes inputs from three different modalities: question words, visual objects, and OCR tokens. We utilize pre-trained models to extract their representations and project them into a $d$-dimensional semantic space as follows:

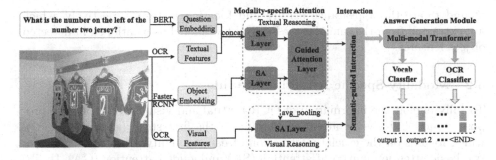

**Fig. 2.** The architecture of our method. The network consists of two main components: the modality-specific attention module and the semantic-guided interaction model.

**Question Features.** Given a question composed of $K$ words, we use a pre-trained BERT [22] model to obtain question embedding feature $Q = \{q_i\}_{i=1}^{K}$, where $q_i \in \mathbb{R}^d$ is the embedding of the $i$-th question word, and $d$ is the feature dimension. The word embeddings are fine-tuned during the training process.

**Detected Object Features.** Given an input image, we obtain a set of $M$ visual objects using the off-the-shelf object detection model Faster R-CNN [20], which extracts the appearance features $\{o_i^a\}_{i=1}^{M}$ and the corresponding bounding boxes $\{o_i^b\}_{i=1}^{M}$. To encode the position information of each visual object, we employ a 4-dimensional vector denoting the top-left position and bottom-right position of the $i$-th object, i.e., $o_i^b = [x_i^{tl}, y_i^{tl}, x_i^{br}, y_i^{br}]$. Moreover, we project both appearance and location features into a $d$-dimensional space (as $q_i \in \mathbb{R}^d$). Finally, the visual object features $O = \{o_i\}_{i=1}^{M}$ are calculated as $o_i = LN(W_a o_i^a) + LN(W_b o_i^b)$, where $W_a$ and $W_b$ are learnable projection matrices, and $LN$ denotes the layer normalization [3] function. During training, we fine-tune the last layer of the Faster R-CNN detector.

**OCR Token Features.** For the representations of scene texts, an off-the-shelf OCR model [6,23] is utilized to obtain the text contain and rich representations are provided to help answer text-related questions. In our work, we acknowledge that when dealing with text-related questions, humans tend to rely on two complementary modalities to identify the answer. They need to combine related OCR features from both visual modality and textual modality. Therefore, we split the OCR features into two modalities: textual and visual. The OCR textual features are denoted as $T^{txt} = \{t_i^{txt}\}_{i=1}^{N}$, where $t_i^{txt} = LN(W_1 t_i^{ft}) + LN(W_2 t_i^{ph}) + LN(W_3 t_i^{reg})$. Here, $t_i^{ft}$ represents the FastText [5] feature, and $t_i^{ph}$ represents the character-level Pyramidal Histogram of Characters (PHOC) [1] feature. Additionally, we add the Recog-CNN feature [18] $t_i^{reg}$, which covers both textual and visual context, to enrich the representations of both OCR textual and visual features. Similarly, we calculate the OCR visual features $T^{vis} = \{t_i^{vis}\}_{i=1}^{N}$ as $t_i^{vis} = LN(W_4 t_i^a) + LN(W_5 t_i^b) + LN(W_6 t_i^{reg})$. Here,

$t_i^a$ and $t_i^b$ are the appearance feature and location feature respectively, which are extracted from the same Faster R-CNN [20] model as mentioned in detected object features.

## 3.2  Modality-Specific Attention Module

To fully exploit the information of OCR textual and visual modalities, we propose a modality-specific attention module in the first stage (shown in Fig. 2), which focuses on the crucial information of each modality. The module utilizes two basic attention units: self-attention (SA) and guided attention (GA). Unlike previous works [7,25,29], we design two different attention reasoning modules to account for the differences between textual and visual modalities. For the textual modality, both SA and GA layers are applied to establish relationships between textual features and questions or visual objects, which allows the model to capture important information related to text. For the visual modality, we introduce a global feature as auxiliary information and feed it into the SA layer along with the OCR visual features. The details about the attention unit and the reasoning module of each modality are described below:

**Attention Unit.** The structure of our SA and GA layers follows the design of [28], where the basic attention unit can be formulated as:

$$Att(Q, K, V) = softmax(\frac{(W_q Q)(W_k K)^T}{\sqrt{d}})W_v V \qquad (1)$$

where $W_q, W_k, W_v$ are learnable parameters, and $\frac{1}{\sqrt{d}}$ is a scaling factor. We replace $Q$, $K$, $V$ with the corresponding input features in SA or GA layers. For convenience, we also denote the SA layer and the GA layer as $SA(Q, K, V)$ and $GA(Q, K, V)$ respectively.

**Textual Modality.** Considering that both question and OCR textual features contain rich semantic information, we firstly concatenate them together and feed them into the SA layer, which facilitates the interaction of semantic features. In addition, the SA layer provides the model with an initial understanding of the important information within the question and textual features. The output of the SA layer can be expressed as follows:

$$S^L = SA(S^{L-1}, S^{L-1}, S^{L-1}), S^0 = concat[Q; T^{txt}] \qquad (2)$$

where $S^0 \in \mathbb{R}^{d \times (K+N)}$, $Q$ represents question features, $T^{txt}$ represents OCR textual features, and $L$ denote the number of the SA layer. Then, we hope to establish the correlation between the output semantic feature and visual objects. Instead of directly fusing multimodal features, we also input the visual objects into the SA layer, which enables the model to initially filter out noisy information contained in detected objects [24]. The process can be formulated as:

$$O^L = SA(O^{L-1}, O^{L-1}, O^{L-1}), O^0 = O \qquad (3)$$

where $O$ is the extracted object features. After the SA layer, we construct the cross-modal relationships between the semantic feature and visual objects. Specifically, we utilize the output feature $S_L$ as guide information and feed it into the GA layer along with object features $O^L$:

$$S_e^L = GA(S^L, O^L, O^L) \tag{4}$$

In this process, we extract the related object features to augment the representation of semantic features and obtain the enhanced output $S_e^L$. We set the attention layer L=2 in our experiment.

**Visual Modality.** Here we introduce a global image feature as auxiliary information, denoted as $o^g = \frac{1}{M} \sum_{i=1}^{M} o_i^L \in \mathbb{R}^d$, which is obtained by taking the average pooling of visual object features $O^L$. By incorporating the global feature $o^g$ with OCR visual features, we aim to include global perception when establishing the inner-modal relationship among visual features. Consequently, the input of the one-layer SA is given by: $T^{gvis} = concat[T^{vis}; o^g] \in \mathbb{R}^{d \times (N+1)}$.

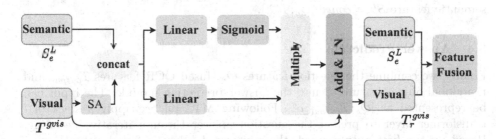

**Fig. 3.** Architecture of the Semantic-guided Modality Interaction module

### 3.3 Semantic-Guided Modality Interaction

Building upon the foundational understanding of each OCR modality in the first stage, we further establish the interaction between OCR textual and visual modalities in the second stage. Compared with general methods that merely reason over the merged OCR features, our proposed semantic-guided modality interaction module offers greater flexibility in establishing interactions not only between diverse OCR tokens, but also between distinct modalities. The overall framework is shown in Fig. 3. In this module, we incorporate the enhanced semantic feature $S_e^L$ from previous attention module as guidance, which improves the fusion and alignment between textual and visual features so that more sufficient interaction can be achieved. Specifically, an information feature ($I$) and an attention query $(Q)$ are generated through two separate linear transformations, which can be formulated as follows:

$$I = W_9 S_e^L + W_{10} SA(T^{gvis})$$
$$Q = \sigma(W_{11} S_e^L + W_{12} SA(T^{gvis})) \tag{5}$$

where $SA$ represents self-attention, $T^{gvis}$ is the global enhanced OCR visual features from the previous module, $\sigma$ denotes the sigmoid function, and $W_9 \sim W_{12}$ are learnable projection matrices. Then we obtain the result of the element-wise multiplication between information feature and attention query, which is added to the original visual-part feature $T^{gvis}$ to form the final representation: $T_r^{gvis} = Norm(T^{gvis} + I \odot Q)$. During this process, we incorporate another attention function to measure the relationship between information feature and attention query, and refines the representation of OCR visual features to correlate with semantic features. Moreover, we adaptively adjust the importance of each modality throughout the reasoning process to facilitate a more comprehensive understanding of scene text modalities. We introduce a learnable parameter $\alpha$ to dynamically adjust the contribution of textual and visual OCR features. The integration can be formulated as $T_{output} = T_e^{txt} + \alpha T_r^{vis}$, where $T_r^{vis}$ is decomposed from $T_r^{gvis} = concat[T_r^{vis}; o_r^g]$, and $T_e^{txt}$ is obtained from the enhanced semantic feature $S_e^L = concat[Q_e; T_e^{txt}]$.

### 3.4  Answer Prediction

Finally, we combine the question features $Q_e$, fused OCR features $T_{output}$ and the global object feature $o_r^g$ into the answer prediction module. The input can be represented as $[Q_e; T_{output}; o_r^g]$. Following M4C [9], we apply a stack of $l$ transformer layer to predict the question answer through iterative decoding. Based on the first output word, the next word token is found either from a predefined vocabulary or from the candidate OCR tokens extracted from the given image. And the token with the highest score will be selected as the answer of the current decoding step.

**Training Loss.** Given that the answer may come from two sources (fixed vocabulary and OCR tokens), the multi-label Binary Cross Entropy (BCE) loss as adopted. We formulated it as follows:

$$pred = \frac{1}{1+\exp(-y_{pred})}$$
$$\mathcal{L}_{bce} = -y_{gt} \log(pred) - (1 - y_{gt}) \log(1 - pred) \tag{6}$$

where $y_{pred}$ is the prediction and $y_{gt}$ is the ground truth. Besides, we adopt a new policy gradient loss $\mathcal{L}_{pg}$ [29] as the additional training loss, which can equip our model with the ability to learn fine-grained character composition alongside linguistic information. The final training loss is $\mathcal{L} = \mathcal{L}_{bec} + \beta \cdot \mathcal{L}_{pg}$, where $\beta$ is a hyper-parameter to control the trade-off of $\mathcal{L}_{pg}$.

# 4    Experiments

In this section, we conduct experiments on two benchmark datasets, including TextVQA [21]and ST-VQA [4]. The results demonstrate that our network achieves competitive performance compared with current state-of-the-art methods. Moreover, some ablation studies are performed to investigate each component of our model.

**Table 1.** Comparison on the TextVQA dataset.

| Model | OCR system | Acc on val | Acc on test |
|---|---|---|---|
| LORRA [21] | Rosetta-ml | 26.56 | 27.63 |
| M4C [9] | Rosetta-ml | 39.40 | 39.01 |
| SMA [8] | Rosetta-ml | 39.58 | 40.29 |
| CRN [17] | Rosetta-ml | 40.39 | 40.96 |
| LaAP-Net [25] | Rosetta-ml | 40.68 | 40.54 |
| SC-Net [7] | Rosetta-ml | 41.17 | 41.42 |
| RUArt-M4C [10] | Rosetta-ml | 40.61 | 40.45 |
| Ours | Rosetta-ml | **41.50** | **41.63** |
| SSbaseline [23] | SBD-Trans | 43.95 | 44.72 |
| SC-Net [7] | SBD-Trans | 44.75 | 45.65 |
| Ours | SBD-Trans | **44.86** | **45.74** |

## 4.1    Datasets and Metrics Settings

**TextVQA Dataset.** TextVQA [21] is the first proposed dataset for TextVQA task that requires understanding and reasoning over scene texts in images to answer questions. The dataset consists of 28,408 images with 45,336 questions from Open Image [15], and each question has 10 answers. The voting method is used to determine the accuracy (ACC) of question answers.

**ST-VQA Dataset.** ST-VQA [4] is another popular benchmark dataset with three tasks, which gradually increase in difficulty. The dataset contains 18,921 training-validation images and 2,971 test images from multiple datasets (ICDAR2013, ICDAR2015, ImageNet, VizWiz, III STR, Visual Genome, and COCO-Text). Following M4C, we use 17,028 images as the training set and 1,893 images as the validation set. The metric of ST-VQA dataset is Average Normalized Levenshtein Similarity (ANSL), defined as $1 - d_L(pred, gt)/ \max(|pred|, |gt|)$, where $pred$ is the prediction, $gt$ is the ground-truth answer, and $d_L$ is the edit distance.

## 4.2    Implementation Details

In this section, we describe the experiment settings of the proposed model. During training, we use Adam [14] as the optimizer with $1e-4$ learning rate. The

**Table 2.** Comparison on the ST-VQA dataset.

| Model | val Acc | ANSL on val | ANSL on test |
|---|---|---|---|
| M4C [9] | 38.05 | 0.4720 | 0.462 |
| SMA [8] | – | – | 0.466 |
| CRN [17] | – | – | 0.483 |
| LaAP-Net [25] | 39.74 | 0.4974 | 0.485 |
| SC-Net [7] | 40.41 | 0.5038 | 0.489 |
| RUArt-M4C [10] | – | – | 0.481 |
| Ours | **40.48** | **0.5113** | **0.495** |

**Table 3.** Ablation study on the TextVQA dataset.

| | SA | GA | SMI | Acc on val |
|---|---|---|---|---|
| 1 | ✗ | ✗ | ✗ | 42.29 |
| 2 | ✗ | ✗ | ✓ | 43.93 |
| 3 | ✗ | ✓ | ✓ | 44.34 |
| 4 | ✓ | ✗ | ✓ | 44.56 |
| 5 | ✓ | ✓ | ✗ | 44.58 |
| 6 | ✓ | ✓ | ✓ | **44.86** |

learning rate is multiplied with a factor of 0.1 at the 14,000 and 19,000 iteration in a total of 48,000 iterations. Dropout technique is set to 0.1 and is applied to prevent overfitting. The batch size of the model is set to 128. We follow the same input setting of M4C. The maximum number of question tokens, visual objects and OCR tokens are set to 20, 100 and 50 respectively. And we set the common dimensionality $d = 768$ and the number of transformer layers $l = 4$. We extract the question features from the first three layers of $BERT_{base}$ model, which are fine-tuned during training.

### 4.3   Comparison with State-of-the-Art Methods

**TextVQA.** We compare our model with other existing methods on TextVQA dataset. The experimental results are shown in Table 1. In the case of using the same Rosetta-en OCR system, our method achieves better performance than existing methods including M4C, SMA, CRN, LaAP-Net, SC-Net and RUArt-M4C. Especially, compared with the baseline model, our model surpasses M4C by 2.1% on the val set and 2.62% on the test set. In addition, for methods [7,23] that take similar scene text modal feature separation operations, our model still achieves better results, which demonstrates the effectiveness of establishing the correlation of scene text modalities. When upgrading the OCR system to the better SBD-Trans module, our model still owns comparable performance which achieves 44.86% validation accuracy and 45.74% test accuracy on the TextVQA dataset. The results demonstrate the effectiveness of our proposed method.

**ST-VQA.** Our model also outperforms the previous SOTA methods on the ST-VQA dataset, which validates the generalization ability of our model. We report the accuracy and ANSL in Table 2. The ANSL of the validation set and test set can reach 0.5113 and 0.495, respectively. It can be seen that our model surpasses the baseline M4C by a noticeable margin. On the validation set, we achieve a 40.48% accuracy, which outperforms M4C by 2.4%. This demonstrates that our model has a more powerful semantic understanding ability for handling scene text information. In terms of ANSL metric, the advantage of our method is relatively significant. It is evident that our method outperforms all considered methods both on the validation set and test set. Furthermore, the results also reflects the potential of our model applied to the real-word scenarios as the ST-VQA dataset has been collected from various datasets.

## 4.4  Ablation Study

To verify the validity and investigate each component of our model, we conduct elaborate ablation studies. Our experiments are built on the SBD-Trans OCR system, and the reproduced results of M4C are presented in Table 3 (row 1) under the same setting. Our model consists of two main components: the modality-specific attention module and the Semantic-guided Modality Interaction (SMI) module. We first examine the effectiveness of the two attention units in the modality-specific attention module, namely SA and GA. Ablation analyses are performed by removing either the SA or GA layer, and the corresponding results are presented in rows 2 to 4. It can be observed that both the SA and GA layers contribute positively to question answering. Next, we evaluate the performance of the SMI module by removing it from the full model (row 5). The result indicates a decrease in performance, highlighting the importance of modality interaction. Furthermore, when both the attention module and the interaction module are employed (row 6), the model's performance is significantly improved. In conclusion, these ablation experiments emphasize the significance of our proposed components and demonstrate their positive impact on enhancing the performance of our model for the Text VQA task.

## 4.5  Case Study and Visualization

In this section, we present some visualization examples from the TextVQA validation set in Fig. 4. As shown in case (a, b), the visual location of OCR tokens is crucial to correctly identify the textual word "last", and recognizing the color of "gillette" token is essential for locating the correct answer. Compared to the baseline M4C, our model effectively utilizes the visual and textual clues extracted from scene text features to provide the correct answer. For case (c, d), although M4C can recognize answer-related OCR tokens, it fails to comprehend the semantics of scene text features. In contrast, our model is capable of filtering out incorrect OCR results such as "huosod" in case c and generates the complete answer "media partner" instead of only "media" as in M4C. Additionally, for questions which are unanswerable to M4C, our model can obtain the correct answer or get the related OCR tokens as shown in case (e). These visualization examples illustrate the superiority of our model in leveraging the textual and visual cues from scene text features to accurately answer questions in the TextVQA task, surpassing the performance of the M4C baseline.

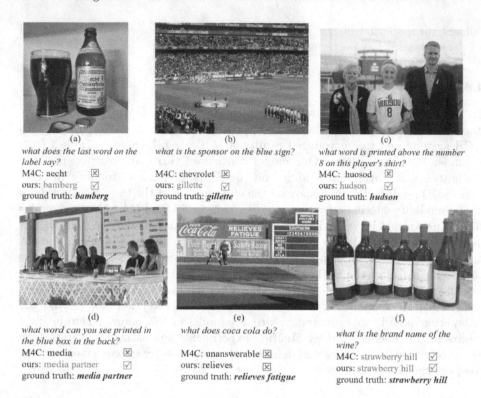

(a) what does the last word on the label say?
M4C: aecht ☒
ours: bamberg ☑
ground truth: *bamberg*

(b) what is the sponsor on the blue sign?
M4C: chevrolet ☒
ours: gillette ☑
ground truth: *gillette*

(c) what word is printed above the number 8 on this player's shirt?
M4C: huosod ☒
ours: hudson ☑
ground truth: *hudson*

(d) what word can you see printed in the blue box in the back?
M4C: media ☒
ours: media partner ☑
ground truth: *media partner*

(e) what does coca cola do?
M4C: unanswerable ☒
ours: relieves ☒
ground truth: *relieves fatigue*

(f) what is the brand name of the wine?
M4C: strawberry hill ☑
ours: strawberry hill ☑
ground truth: *strawberry hill*

**Fig. 4.** Qualitative comparison of the proposed method and MC4 on the TextVQA dataset.

## 5   Conclusion

In this paper, we propose a two-stage reasoning network with modality decomposition for the Text VQA task. Unlike existing methods that simply concatenate OCR textual and visual features to construct the scene text information, we propose to establish a more detailed interaction between the textual and visual modalities. In the first stage, our model handles each modality independently through a modality-specific attention module, allowing it to focus on important textual content and visual areas. In the second stage, a semantic-guided interaction module is introduced to offer more flexible correlations between OCR tokens. Extensive experiments on several datasets indicate that our approach achieves the state-of-the-art performance. The ablation analysis conducted on our model proves its great potential. And we believe more fine-grained interaction among OCR tokens can be investigated in the future.

**Acknowledgements.** This work was supported by the National Natural Science Foundation of China under Grants (No.62325206, 61936005, 62301276), and the Key Research and Development Program of Jiangsu Province under Grant BE2023016-4.

# References

1. Almazán, J., Gordo, A., Fornés, A., Valveny, E.: Word spotting and recognition with embedded attributes. IEEE Trans. Pattern Anal. Mach. Intell. **36**(12), 2552–2566 (2014)
2. Anderson, P., et al.: Bottom-up and top-down attention for image captioning and visual question answering. In: Proceedings of the IEEE Conference on Computer Vision and Pattern Recognition, pp. 6077–6086 (2018)
3. Ba, J.L., Kiros, J.R., Hinton, G.E.: Layer normalization. arXiv preprint arXiv:1607.06450 (2016)
4. Biten, A.F., et al.: Scene text visual question answering. In: Proceedings of the IEEE/CVF International Conference on Computer Vision, pp. 4291–4301 (2019)
5. Bojanowski, P., Grave, E., Joulin, A., Mikolov, T.: Enriching word vectors with subword information. Trans. Assoc. Comput. Linguist. **5**, 135–146 (2017)
6. Borisyuk, F., Gordo, A., Sivakumar, V.: Rosetta: large scale system for text detection and recognition in images. In: Proceedings of the 24th ACM SIGKDD International Conference on Knowledge Discovery and Data Mining, pp. 71–79 (2018)
7. Fang, C., et a.: Towards escaping from language bias and OCR error: semantics-centered text visual question answering. In: 2022 IEEE International Conference on Multimedia and Expo (ICME), pp. 01–06. IEEE (2022)
8. Gao, C., et al.: Structured multimodal attentions for textVQA. IEEE Trans. Pattern Anal. Mach. Intell. **44**(12), 9603–9614 (2021)
9. Hu, R., Singh, A., Darrell, T., Rohrbach, M.: Iterative answer prediction with pointer-augmented multimodal transformers for textVQA. In: Proceedings of the IEEE/CVF Conference on Computer Vision and Pattern Recognition, pp. 9992–10002 (2020)
10. Jin, Z.X., et al.: Ruart: a novel text-centered solution for text-based visual question answering. IEEE Trans. Multimedia **25**, 1–12 (2023). https://doi.org/10.1109/TMM.2021.3120194
11. Joulin, A., Grave, E., Bojanowski, P., Mikolov, T.: Bag of tricks for efficient text classification. arXiv preprint arXiv:1607.01759 (2016)
12. Kant, Y., et al.: Spatially aware multimodal transformers for TextVQA. In: Vedaldi, A., Bischof, H., Brox, T., Frahm, J.-M. (eds.) ECCV 2020. LNCS, vol. 12354, pp. 715–732. Springer, Cham (2020). https://doi.org/10.1007/978-3-030-58545-7_41
13. Kim, J.H., Jun, J., Zhang, B.T.: Bilinear attention networks. In: Advances in Neural Information Processing Systems, vol. 31 (2018)
14. Kingma, D.P., Ba, J.: Adam: a method for stochastic optimization. arXiv preprint arXiv:1412.6980 (2014)
15. Kuznetsova, A., et al.: The open images dataset v4: unified image classification, object detection, and visual relationship detection at scale. Int. J. Comput. Vision **128**(7), 1956–1981 (2020)
16. Li, C., Du, Q., Wang, Q., Jin, Y.: Learning hierarchical reasoning for text-based visual question answering. In: Farkaš, I., Masulli, P., Otte, S., Wermter, S. (eds.) ICANN 2021. LNCS, vol. 12893, pp. 305–316. Springer, Cham (2021). https://doi.org/10.1007/978-3-030-86365-4_25
17. Liu, F., Xu, G., Wu, Q., Du, Q., Jia, W., Tan, M.: Cascade reasoning network for text-based visual question answering. In: Proceedings of the 28th ACM International Conference on Multimedia, pp. 4060–4069 (2020)

18. Liu, Y., Zhang, S., Jin, L., Xie, L., Wu, Y., Wang, Z.: Omnidirectional scene text detection with sequential-free box discretization. arXiv preprint arXiv:1906.02371 (2019)
19. Lu, J., Batra, D., Parikh, D., Lee, S.: VilBERT: pretraining task-agnostic visiolinguistic representations for vision-and-language tasks. In: Advances in Neural Information Processing Systems, vol. 32 (2019)
20. Ren, S., He, K., Girshick, R., Sun, J.: Faster R-CNN: towards real-time object detection with region proposal networks. In: Advances in Neural Information Processing Systems, vol. 28 (2015)
21. Singh, A., et al.: Towards VQA models that can read. In: Proceedings of the IEEE/CVF Conference on Computer Vision and Pattern Recognition, pp. 8317–8326 (2019)
22. Vaswani, A., et al.: Attention is all you need. In: Advances in Neural Information Processing Systems, vol. 30 (2017)
23. Wang, P., Yang, L., Li, H., Deng, Y., Shen, C., Zhang, Y.: A simple and robust convolutional-attention network for irregular text recognition. arXiv preprint arXiv:1904.01375 6(2), 1 (2019)
24. Wang, Q., Xiao, L., Lu, Y., Jin, Y., He, H.: Towards reasoning ability in scene text visual question answering. In: Proceedings of the 29th ACM International Conference on Multimedia, pp. 2281–2289 (2021)
25. Wang, X., et al.: On the general value of evidence, and bilingual scene-text visual question answering. In: Proceedings of the IEEE/CVF Conference on Computer Vision and Pattern Recognition, pp. 10126–10135 (2020)
26. Wu, J., et al.: A multimodal attention fusion network with a dynamic vocabulary for TextVQA. Pattern Recogn. **122**, 108214 (2022)
27. Yang, Z., et al.: Tap: text-aware pre-training for text-VQA and text-caption. In: Proceedings of the IEEE/CVF Conference on Computer Vision and Pattern Recognition, pp. 8751–8761 (2021)
28. Yu, Z., Yu, J., Cui, Y., Tao, D., Tian, Q.: Deep modular co-attention networks for visual question answering. In: Proceedings of the IEEE/CVF Conference on Computer Vision and Pattern Recognition, pp. 6281–6290 (2019)
29. Zhu, Q., Gao, C., Wang, P., Wu, Q.: Simple is not easy: a simple strong baseline for TextVQA and TextCAPS. In: Proceedings of the AAAI Conference on Artificial Intelligence, vol. 35, pp. 3608–3615 (2021)

# Localization and Local Motion Magnification of Pulsatile Regions in Endoscopic Surgery Videos

Honglei Zheng[1], Wenkang Fan[1], Yinran Chen[1(✉)], and Xiongbiao Luo[1,2]

[1] Department of Computer Science and Technology,
Xiamen University, Xiamen, China
{hlzheng,23020181154270}@stu.xmu.edu.cn, yinran_chen@xmu.edu.cn,
xiongbiao.luo@gmail.com
[2] National Institute for Data Science in Health and Medicine,
Xiamen University, Xiamen, China

**Abstract.** Localization of neurovascular bundles or vessels is critical in endoscopic surgery. It still remains challenging to identify neurovascular bundles and vessels due to limited field-of-view and tactile perception loss. This paper presents a new framework for local motion magnification of the pulsatile region in endoscopic surgical videos, which can help surgeons localize these pulsatile regions from complex surgical fields. Our method consists of an autocorrelated dynamic linear model, a pulsatile region localization module, and a local motion magnification module. Specifically, the autocorrelated dynamic linear model and pulsatile region localization module can extract periodic features and position information of pulsatile regions. Then, the local motion magnification module performs local motion magnification, minimizing post-magnification artifacts. Importantly, our method also avoids distorting the surgical instruments during magnification. Experimental validations on both synthetic and clinical surgical videos demonstrate that our method outperforms the existing methods in both qualitative and quantitative assessments.

**Keywords:** Local motion magnification · Periodic features extraction · Pulsatile regions localization · Endoscopic surgery

## 1 Introduction

Minimally invasive surgery is favored by many surgeons because of its key advantages, such as small incisions and quick postoperative recovery. Procedures like nerve-preserving robotic prostatectomy [19] impose high demands on the protection of the neurovascular bundle. Damage to the neurovascular bundle can harm the nervous system and cause unpredictable bleeding, greatly affecting patients' intraoperative status and post-procedure recovery [1]. However, relying

This work was partly supported by the National Natural Science Foundation of China (Grant No. 62001403, 82272133, and 61971367).

© The Author(s), under exclusive license to Springer Nature Switzerland AG 2024
S. Rudinac et al. (Eds.): MMM 2024, LNCS 14556, pp. 141–154, 2024.
https://doi.org/10.1007/978-3-031-53311-2_11

on visual perception alone to localize neurovascular bundles is quite challenging and risky compared to open surgery because minimally invasive surgery limits the surgeon's field-of-view and deprives the surgeon of tactile perception. Therefore, accurate localization of the neurovascular bundles is of critical importance, especially in the limited visual field and varying complexities of endoscopic surgeries.

In clinical practice, preoperative imaging and interventional imaging are commonly used for localizing vessels [13,14]. However, these methods require additional equipment and many need invasive imaging, which requires a surgeon with extensive clinical experience. The advent of video motion magnification technology presents new possibilities for non-invasive neurovascular bundle localization. The pulsation of blood vessels can cause vibrations in the surrounding tissues, and magnifying this kinematic feature can help us to localize blood vessels and neurovascular bundles. McLeod et al. [15] first applied the Euler method to surgical videos, offering non-invasive support for surgeons in identifying vessels. Janatka [11] and Oh [17] improve the quality of the magnified video by improving the filtering or combining it with neural networks. However, these techniques require global video magnification, resulting in additional noise and artifacts. Therefore, local motion magnification of pulsatile regions is the best approach to reduce artifacts and prevent instrument aberrations.

In order to achieve high-quality local motion magnification of endoscopic surgical video, two important issues need to be addressed. The first issue is to determine what the pulsatile frequency is to be magnified. In this aspect, one needs to accurately estimate the frequency maps of the video. The second issue is to determine the region to be magnified. To this end, one needs to locate the pulsatile regions by using the estimated frequency map. In this paper, we propose a framework for precise local motion magnification of endoscopic video for the first time. The framework consists of three main modules, an autocorrelated dynamic linear model (ADLM), a pulsation region localization module and a local motion magnification module. The proposed method is potential to help surgeons locate the pulsatile regions visually and identify the neurovascular bundles and vessels. The technical contributions of this paper are clarified as follows:

- We propose an autocorrelated dynamic linear model (ADLM) to adaptively estimate the frequency map of pulsatile regions in the endoscopic video. The proposed ADLM can effectively extract the periodic features from the variable temporal signals of the clinical surgical videos.
- We propose a pulsatile region localization module by adopting an unsupervised cluster algorithm with maximum connectivity to discriminate the pulsatile and non-pulsatile regions in the endoscopic video by exploring the frequency map.
- We embed a local motion magnification module in the framework to magnify the pulsatile regions corresponding to the neurovascular bundles and vessels. The non-pulsatile regions and the surgical instruments remain unchanged to avoid unexpectable distortions and artifacts.

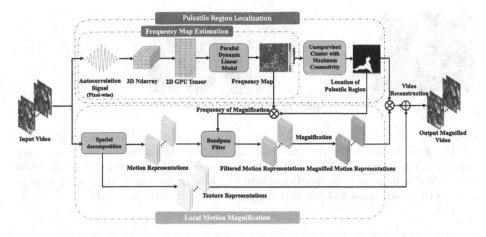

**Fig. 1.** Pipeline of our framework with (a) a frequency map estimation using auto-correlated dynamic linear model (ADLM), (b) a pulsatile region localization using unsupervised clustering, and (c) a local motion magnification using the results from the previous two modules.

## 2   Related Work

Many scholars have applied motion magnification to videos to better recognize subtle movements. Wu et al. [24] first employed the Euler method to video, proposing an approach for video motion magnification based on intensity decomposition and bandpass filter. Wadhwa et al. [21,22] improved the video decomposition method, suggesting a phase-based video magnification technique to reduce video noise. McLeod et al. [15] first combined Eulerian magnification with endoscopic surgical videos to help surgeons locate neurovascular bundles noninvasively. Janatka [11] proposed the third-order Gaussian filter on acceleration motion magnification [25], considering the characteristics of blood vessels' diastolic displacement curve. Oh [17] and Fan [8] incorporated neural networks into video decomposition and reconstruction, further minimizing noise and artifacts. However, these magnification methods can cause distortion to stationary tissues, and introduce more noise and artifacts due to global magnification. Applying local magnification to surgical videos can effectively solve this problem while retaining the advantages of intuitive and non-invasive video magnification.

Local magnification of surgical videos requires precise extraction of periodic features and localization of pulsatile regions within the videos. Methods such as fast fourier transform (FFT) [2] and quadratic peak interpolation (QPI) [23] typically extract frequencies from photoplethysmograph (PPG) signals [12,20]. Shao [18] and Doheny et al. [7] used natural videos containing hands or faces to infer heart rate or respiration rate. McLeod et al. [16] employed the dynamic linear model (DLM) to extract video periodic information. Amir-Khalili et al. [3–5] proposed auto-segmenting vessels using bandpass filtering. Although it can segment vessels, the method misses the pulsatile frequency of the vascular region,

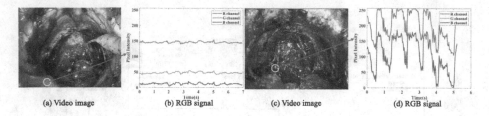

| (a) Video image | (b) RGB signal | (c) Video image | (d) RGB signal |

**Fig. 2.** The complicated RGB signals in the pulsatile regions of clinical surgical videos. (a) The region with subtle pulsatile motions exhibits similar RGB signals, showing implicit periodic features in (b). (c) The region with significant pulsatile motions have noised square-wave-like RGB signals in (d). More importantly, the R-components appears to be saturated and the peak signals are lost. (Color figure online)

which is critical in motion magnification. Additionally, most of these methods require extra filters or are not applied in endoscopic surgical scenarios.

## 3    Method

Figure 1 depicts the workflow of our framework. Specifically, the framework comprises three core components: a frequency map estimation module by using the proposed ADLM, a pulsatile region localization module, and a local motion magnification module.

### 3.1    Frequency Map Estimation

Figure 2 illustrates the raw RGB signals of the pulsatile region in two different endoscopic videos. In Fig. 2(a–b), the periodic pulsatile motion is implicit, showing that the temporal signals in the RGB channels are very close to each other. On the other hand, the region in Fig. 2(c–d) contains strong pulsatile motion, showing noised square wave-like signals. More importantly, the signals in the R channel saturate in some temporal instants since the red component dominates in most of the clinical surgical videos. To extract the periodic features from these complicated signals, we propose the ADLM with the introduction of autocorrelation processing. Specifically, let $\mathbf{V} = [\mathbf{V}(\mathbf{p},1),\ldots,\mathbf{V}(\mathbf{p},t),\ldots,\mathbf{V}(\mathbf{p},N)]_{t=1}^{N}$ be the video sequence, where N is the number of frames and $\mathbf{p} = (h,w,c)$ encapsulates the spatial and channel information of each frame. Here $h = 1,\cdots,H$ and $w = 1,\ldots,W$ specify the pixel position, and $c = 1,2,3$ denotes the RGB channel. We first conduct a temporal autocorrelation on each pixel to acquire the corresponding autocorrelated sequence $\mathbf{A_V} = [\mathbf{A_V}(\mathbf{x},1),\ldots,\mathbf{A_V}(\mathbf{x},\tau),\ldots,\mathbf{A_V}(\mathbf{x},N)]_{\tau=1}^{N}$:

$$\mathbf{A_V} = \mathcal{R}_{t,t}(\mathbf{V}) \tag{1}$$

where $\mathcal{R}_{t,t}$ denotes the temporal autocorrelation processing. $\mathbf{x} = (H,W)$ represents the spatial information of the autocorrelated sequence. $\tau$ is the time delay interval(lag) of the autocorrelation processing.

To accelerate the computation and avoid critical motion information loss due to the video frames downsampling in the original DLM [16] method, we convert the 3D autocorrelated sequence $\mathbf{A}_V$ into a 2D spatiotemporal GPU tensor $\tilde{\mathbf{A}}_V = \left[\tilde{\mathbf{A}}_V(\tilde{\mathbf{x}},1),\ldots,\tilde{\mathbf{A}}_V(\tilde{\mathbf{x}},\tau),\ldots,\tilde{\mathbf{A}}_V(\tilde{\mathbf{x}},N)\right]_{\tau=1}^{N}$ for GPU-based parallel computing, where $\tilde{x} = 1,2,\cdots,HW$. Then, we adopt a Bayesian estimator [26] to fit the periodic features of the autocorrelated sequence. Specifically, the state vector $\mathbf{S}_\tau = [\mathbf{S}_\tau(1),\mathbf{S}_\tau(2),\mathbf{S}_\tau(3)]$ that describe the periodic pulsatile motion in each pixel is modeled as a linear state prediction:

$$\mathbf{S}_\tau = \mathbf{Q}\mathbf{S}_{\tau-1} + \eta_\tau \tag{2}$$

where $\eta_\tau \sim \mathcal{N}(0,\boldsymbol{\Sigma}_\eta)$ denotes the prediction error. $\mathbf{Q}$ is a rotation matrix defined by an unknown parameter $\theta$:

$$\mathbf{Q} = \begin{bmatrix} \cos\theta & -\sin\theta & 0 \\ \sin\theta & \cos\theta & 0 \\ 0 & 0 & 1 \end{bmatrix} \tag{3}$$

The first two entries of $\mathbf{S}_\tau$ stand for the stationary cycle component of ADLM, and the last entry is used to model the random walk component. The observation of this nested ADLM is the autocorrelated sequence $\tilde{\mathbf{A}}_V$, which is described as:

$$\tilde{\mathbf{A}}_V(\tilde{\mathbf{x}},\tau) = \mathbf{H}\mathbf{S}_\tau + \varepsilon_\tau \tag{4}$$

where $\mathbf{H} = \begin{bmatrix} 1 & 0 & 1 \end{bmatrix}$ is the observation matrix. $\varepsilon_\tau \sim \mathcal{N}(0,\boldsymbol{\Sigma}_\varepsilon)$ represents the observation error. Subsequently, we perform maximum likelihood estimation (MLE) for unknown parameters in $\mathbf{Q}$, $\boldsymbol{\Sigma}_\varepsilon$ and $\boldsymbol{\Sigma}_\eta$ by numerically optimizing the log-likelihood:

$$\log L\left(\theta \mid \tilde{\mathbf{A}}_V(\mathbf{x})_t\right) = \sum_{t=1}^{N}[-\frac{1}{2}\ln 2\pi - \frac{1}{2}\ln\left[\det\left(\mathbf{H}\mathbf{D}(\tau \mid \tau-1)\mathbf{H}' + \boldsymbol{\Sigma}_\varepsilon\right)\right]$$
$$- \tilde{\mathbf{A}}_V(\tilde{\mathbf{x}},\tau) - \mathbf{Q}\mathbf{S}(\tau \mid \tau-1))' \cdot (\mathbf{H}\mathbf{D}(\tau \mid \tau-1)\mathbf{H}' + \boldsymbol{\Sigma}_\varepsilon)^{-1}$$
$$\cdot (\tilde{\mathbf{A}}_V(\tilde{\mathbf{x}},\tau) - \mathbf{Q}\mathbf{S}(\tau \mid \tau-1))] \tag{5}$$

where $\tilde{\mathbf{A}}_V(\mathbf{x})_t$ denotes the observations of ADLM up to time t. $\boldsymbol{\Sigma}_\varepsilon$ and $D$ denote the covariance matrices of the measurement noise.

In the final step of ADLM, we obtain the frequency map $\mathbf{F}_V$ by solving the log-likelihood:

$$\mathbf{F}_V(\mathbf{x}) = \mathrm{MLE}\left(\log L\left(\theta \mid \tilde{\mathbf{A}}_V(\mathbf{x})_t\right)\right) \tag{6}$$

where the frequency of the pixel is specified as $f = \theta \cdot fps/2\pi$.

## 3.2 Pulsatile Region Localization

Having obtained the frequency map $\mathbf{F}_V$, we apply unsupervised clustering and maximum connectivity to determine the pulsatile and non-pulsatile regions, in which the pulsatile region is regarded as the neurovascular bundles or vessels. In unsupervised clustering, the frequency map is roughly divided into two different regions, in which the centroid of the pulsatile one contains a frequency that is close to the heart rate of humans. Then, we employ the K-means algorithm to compare the centroid period characteristics of the two regions, and select the targeted cluster as the pulsatile region.

However, the pulsatile region obtained by unsupervised clustering alone is discontinuous due to the presence of noise. Such discontinuous region is not consistent with the anatomic morphology of neurovascular bundles or blood vessels [6]. Therefore, we perform the maximum connectivity operation on the pulsatile regions and keep the largest connected components as the mask that indicates the positional information of the pulsatile region, where $\mathcal{MC}$ denotes maximum connectivity:

$$\mathbf{PR}_{\text{mask}}(\mathbf{x}) = \mathcal{MC}\left(clustering\left(\mathbf{F_V}(\mathbf{x})\right)\right) \tag{7}$$

### 3.3   Local Motion Magnification

We proceed with local motion magnification and video reconstruction with $\mathbf{V}, \mathbf{F_V}$, and $\mathbf{PR}_{\text{mask}}$. In this step, each video frame $\mathbf{V}(\mathbf{p}, t)$ is spatially decomposed into two feature maps $\mathbf{I}_m$ and $\mathbf{I}_s$. $\mathbf{I}_m$ and $\mathbf{I}_s$ contain the motion and texture information of the corresponding video frame, respectively. Then, the pulsatile region of $\mathbf{I}_m$ specified by $\mathbf{PR}_{\text{mask}}$ is bandpass filtered using the central frequency defined in $\mathbf{A_V}$:

$$\mathbf{I}_{bp}(\mathbf{x}) = \mathcal{F}_{bp}\left(\mathbf{I}_m(\mathbf{x}) \odot \mathbf{PR}_{\text{mask}}(\mathbf{x})\right) \tag{8}$$

where $\mathcal{F}_{bp}$ denotes bandpass filtering. $\odot$ is the Hadamard product(element-wise product). We then implement the fundamental Euler method [24] on the pulsatile region of $\mathbf{I}_m$ to obtain the motion-magnified $\mathbf{I}_M$:

$$\mathbf{I}_M(\mathbf{x}) = \mathbf{I}_m(\mathbf{x}) + \alpha(\mathbf{x})\mathbf{I}_{bp}(\mathbf{x}) \tag{9}$$

where $\alpha(\mathbf{x})$ is the magnification factor defined as:

$$\alpha(\mathbf{x}) = \begin{cases} 30, & \mathbf{PR}_{\text{mask}}(\mathbf{x}) = 1 \\ 0, & \mathbf{PR}_{\text{mask}}(\mathbf{x}) = 0 \end{cases} \tag{10}$$

Finally, we reconstruct the local-motion-magnified video $\mathbf{V}_M$ by merging the magnified motion $\mathbf{I}_M$ and the original texture $\mathbf{I}_s$:

$$\mathbf{V}_M(\mathbf{p}, t) = recon\left(\mathbf{I}_M(\mathbf{x}), \mathbf{I}_s(\mathbf{x})\right) \tag{11}$$

## 4   Experiments Setting

### 4.1   Dataset

In clinical surgical videos, visually determining the regions of neurovascular bundles and blood vessels and estimating the corresponding frequencies are very

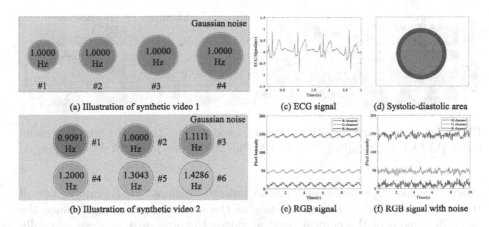

(a) Illustration of synthetic video 1    (c) ECG signal    (d) Systolic-diastolic area

(b) Illustration of synthetic video 2    (e) RGB signal    (f) RGB signal with noise

**Fig. 3.** Illustrations of the synthetic videos. (a) Video #1 contains four circles with identical 1 Hz pulsatile motions. The radius of these circles are 30 pixels, 33 pixels, 35 pixels, and 37 pixels, respectively. (b) Video #2 contains six 30-pixel-radius circles with pulsatile motions of 0.9091 Hz, 1.0000 Hz, 1.1111 Hz, 1.2000 Hz, 1.3043 Hz, and 1.4286 Hz, respectively. (c) the ECG signals selected from the MIT ITDB [9] controls the changes of RGB colors and radius of the circles. (d) Illustration of extension and extraction of the pulsatile circles. (e) noise-free RGB signals of 1 Hz pulsatile motion. (f) 1 Hz RGB-signals with 20-dB white Gaussian noise.

subjective. To quantitatively evaluate the performance of the proposed method, we generated synthetic videos to mimic the changes of morphology and colors induced by the diastolic and contractive movements of neurovascular bundles and blood vessels.Fig. 3 shows the details of the synthetic videos. As illustrated in Fig. 3(a), Video #1 contains four circles with different radii of 30 pixels, 33 pixels, 35 pixels, and 37 pixels, respectively. Identical 1 Hz pulsatile motions were added to the four circles by manipulating their radius and colors. Specifically, the four circles were rendered using a baseline color extracted from the clinical surgical video. Then, we utilized the ECG signals provided by the MIT ITDB [9] to simultaneously adjust the changes of RGB colors and radius of the circle.

The second synthetic video, Video #2, contains six circles with identical radii of 30 pixels and different frequencies. The frequency of the circular region in Video #2 is calculated by the formula: f=fps/n, where fps is the frame rate, which is 30 frames per second. The n is the number of frames in a single cycle of pulsatile motion, which are 33, 30, 27, 25, 23, and 21 for each of the circles. The ECG-induced motions are applied to the circles. Additionally, both videos have 0-Hz background. We also added 20-dB white Gaussian noise to both of the videos. The noise level was estimated from the non-pulsatile regions of the clinical surgical video.

The clinical surgical videos were collected by the da Vinci Si Surgical System in robotic laparoscopic radical prostatectomy. The data adheres to the approved protocols of Western University's Research Ethics Committee in Canada. The system captures videos automatically at a frame rate of 30 fps. We selected three

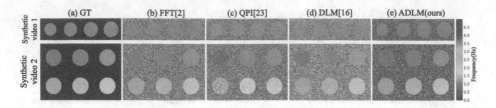

**Fig. 4.** Frequency maps of synthetic Video #1 and Video #2 obtained from (a) ground truth, (b) FFT [2], (c) QPI [23], (d) DLM [16], and (e) the proposed ADLM.

different video clips containing visible pulsatile regions. Furthermore, two of these video clips show the movements of the surgical instruments. Since the pulsatile regions of the clinical surgical videos lack ground truth, we manually depicted the regions under the guidance of experienced surgeons.

### 4.2    Comparison and Evaluation

We evaluate the proposed method by comparing it with other periodic feature extraction methods, which are FFT [2], QPI [23] and DLM [16]. On both the synthetic and clinical videos, we first performed different frequency estimations to obtain frequency maps. Then, the same pulsatile region localization strategies were employed on the respective frequency map. On the synthetic videos with ground truths, we used the mean, absolute error (Abs Err), and root-mean-square-error (RMSE) of the frequency maps to evaluate different estimators. Then we employed Dice coefficient and intersection-over-union (IoU) to evaluate the precision of pulsatile region localization. On the clinical surgical videos, we further employed both structural similarity (SSIM) and peak signal-to-noise ratio (PSNR) [10] to assess the quality of motion-magnified videos after the proposed local motion magnification or the commonly used global Euler method.

## 5    Results

### 5.1    Synthetic Video

In this section, we present the results obtained by different frequency estimators on both the synthetic and clinical surgical videos.

Figure 4 shows the frequency maps of synthetic Video #1 and Video #2 obtained by FFT [2], QPI [23] and DLM [16], and the proposed ADLM. In general, all the methods can discriminate the pulsatile circles from the surrounding backgrounds. A closer observation illustrates that QPI and DLM generate more noised frequency maps than FFT and ADLM. The quantitative measurements of each pulsatile circle in Video #1 and Video #2 are listed in Table 1 and Table 2, respectively. It can be seen that our ADLM achieves the best absolute error and RMSE in most of the pulsatile circles among all the tested methods. Particularly,

**Table 1.** Quantitative evaluations of the frequency maps in Video #1 obtained by different frequency estimators.

| Cicle | #1 | | | #2 | | | #3 | | | #4 | | |
|---|---|---|---|---|---|---|---|---|---|---|---|---|
| Metrics | Mean | Abs Err | RMSE | Mean | Abs Err | RMSE | Mean | Abs Err | RMSE | Mean | Abs Err | RMSE |
| FFT [2] | 1.0169 | 0.0169 | 0.4626 | 1.0049 | 0.0049 | 0.2448 | 1.0058 | 0.0058 | 0.2939 | 1.0189 | 0.0189 | 0.4656 |
| QPI [23] | 1.0711 | 0.0711 | 0.4796 | 1.0683 | 0.0683 | 0.4279 | 1.0586 | 0.0586 | 0.3192 | 1.0727 | 0.0727 | 0.4404 |
| DLM [16] | 1.0481 | 0.0481 | 0.1330 | 1.0513 | 0.0513 | 0.1481 | 1.0511 | 0.0511 | 0.1264 | 1.0543 | 0.0543 | 0.1524 |
| ADLM(ours) | 0.9972 | **0.0028** | **0.1151** | 0.9955 | **0.0045** | **0.1115** | 0.9958 | **0.0042** | **0.1165** | 1.0005 | **0.0005** | **0.0976** |
| GT | 1.0000 | / | / | 1.0000 | / | / | 1.0000 | / | / | 1.0000 | / | / |

**Table 2.** Quantitative evaluations of the frequency maps in Video #2 obtained by different frequency estimators.

| Cicle | #1 | | | #2 | | | #3 | | |
|---|---|---|---|---|---|---|---|---|---|
| Metrics | Mean | Abs Error | RMSE | Mean | Abs Error | RMSE | Mean | Abs Error | RMSE |
| FFT [2] | 0.9306 | 0.0215 | 0.3350 | 0.9981 | **0.0019** | 0.1059 | 1.1246 | 0.0124 | 0.3072 |
| QPI [23] | 0.7784 | 0.1307 | 0.5515 | 1.0551 | 0.0551 | 0.2428 | 1.1252 | 0.0141 | 0.5775 |
| DLM [16] | 0.9283 | 0.0192 | 0.1362 | 1.0504 | 0.0504 | 0.1115 | 1.1869 | 0.0758 | 0.1685 |
| ADLM(ours) | 0.9087 | **0.0004** | **0.1149** | 0.9922 | 0.0078 | **0.1014** | 1.1122 | **0.0011** | **0.1331** |
| GT | 0.9091 | / | / | 1.0000 | / | / | 1.1111 | / | / |
| Cicle | #4 | | | #5 | | | #6 | | |
| Metrics | Mean | Abs Error | RMSE | Mean | Abs Error | RMSE | Mean | Abs Error | RMSE |
| FFT [2] | 1.2046 | 0.0046 | 0.3026 | 1.2991 | 0.0052 | 0.3207 | 1.4325 | 0.0039 | 0.3998 |
| QPI [23] | 1.2977 | 0.0977 | 0.5558 | 1.5086 | 0.2043 | 0.4047 | 1.6590 | 0.2304 | 0.5394 |
| DLM [16] | 1.2212 | 0.0212 | 0.1408 | 1.3071 | 0.0028 | 0.1224 | 1.4273 | 0.0013 | 0.1334 |
| ADLM(ours) | 1.1072 | **0.0028** | **0.1179** | 1.3053 | **0.0010** | **0.1144** | 1.4282 | **0.0004** | **0.1057** |
| GT | 1.2000 | / | / | 1.3043 | / | / | 1.4286 | / | / |

the absolute error of ADLM is one or two order of magnitudes smaller than QPI and DLM. Additionally, while FFT achieves the second best in terms of absolute error in most of the pulsatile circles, DLM generally ranks the second in terms of RMSE.

The results in Fig. 4, Table 1, and Table 2 demonstrate that FFT creates smooth but slight biased frequency maps. DLM can provide more accurate frequencies than FFT at the expense of a higher sensitivity to noise. QPI has the worst performance among all the tested methods. In comparison, the proposed ADLM outperforms the other methods in terms of absolute error and RMSE.

Figure 5 displays the localizations of pulsatile circles obtained by different methods with the proposed unsupervised pulsatile region localization method.

**Table 3.** Quantitative evaluations of the pulsatile region localization obtained by different methods on synthetic Video #1 and Video #2.

| (Dice, IoU) | FFT [2] | QPI [23] | DLM [16] | ADLM(ours) |
|---|---|---|---|---|
| Video #1 | (0.97, 0.94) | (0.97, 0.94) | (0.97, 0.94) | **(0.98, 0.95)** |
| Video #2 | (0.95, 0.90) | (0.89, 0.82) | (0.63, 0.46) | **(0.98, 0.97)** |

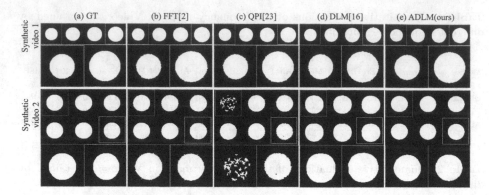

**Fig. 5.** Pulsatile region localization of synthetic Video #1 and Video #2 obtained from (a) ground truth, (b) FFT [2], (c) QPI [23], (d) DLM [16], and (e) the proposed ADLM. The first (blue) and fourth (green) circles in Video #1, and the first (purple) and sixth (yellow) circles in Video #2 are zoomed-in. (Color figure online)

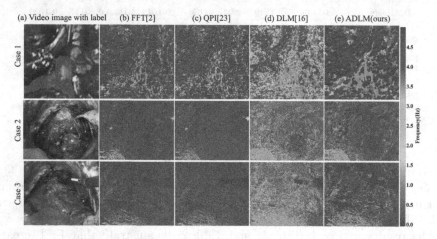

**Fig. 6.** Frequency maps of three cases in the clinical surgical videos obtained by (a) manual annotation(blue labels), (b) FFT [2], (c) QPI [23], (d) DLM [16], and (e) the proposed ADLM. (Color figure online)

The first and the fourth circles in Video #1, and the first and the sixth circles in Video #2 are zoomed-in for a better observation. Since QPI provides noised frequency maps, its localizations are also less accurate than the other methods, especially showing that the first circular region in Video #2 cannot be effectively localized. Compared to the ground truths in Fig. 5(a), ADLM produces more accurate and smoother circular pulsatile regions. Table 3 also demonstrates the outperformance of ADLM in pulsatile region localization, showing that ALDM achieves higher Dice and IoU than the other methods in both synthetic videos. These findings are consistent with the more accurate and robust frequency estimation achieved by ADLM, as illustrated in Tables 1 and 2. Table 3 presents the quantitative assessment of the location information of pulsatile regions produced by various methods. Notably, ADLM exhibits superior localization ability across different synthetic videos. This can be attributed to ADLM's ability to discern between noise points and pulsatile points, a critical aspect for creating location information in pulsatile regions and filtering for spatial decomposition.

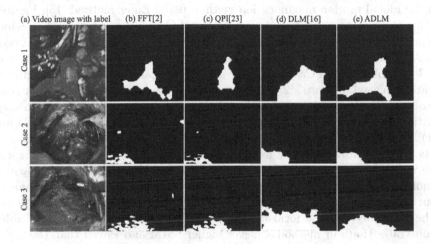

**Fig. 7.** Pulsatile region localization of the three cases in the clinical surgical datasets obtained by (a) manual annotation(blue labels), (b) FFT [2], (c) QPI [23], (d) DLM [16], and (e) the proposed ADLM. (Color figure online)

**Table 4.** SSIM and PSNR of the three cases in the clinical datasets obtained by using the global Euler method [15] and our local Euler motion magnification.

| (SSIM, PSNR) | Euler Method [15] | Ours |
|---|---|---|
| case1 | (0.62, 29.16) | **(0.91, 31.96)** |
| case2 | (0.73, 29.79) | **(0.98, 39.23)** |
| case3 | (0.62, 29.25) | **(0.95, 37.31)** |

## 5.2 Clinical Surgical Video

Figure 6 shows the frequency maps obtained by different frequency estimators on the three clinical cases. The manual-depicted pulsatile regions (blue labels) are also annotated on top of the video frames. In the frequency maps, the orange area represents the pulsatile region relating to cardiac motion, like the neurovascular bundles or blood vessels. Figure 7 presents the results of pulsatile region localization of different methods. From Fig. 6 and Fig. 7 one can see that, while FFT and ADLM create similar pulsatile regions when compared with the manual labels in Case #1, DLM and ADLM generate similar results in Case #2 and Case #3. Particularly, ADLM provides pulsatile regions that are closer to the labels than FFT [2], QPI [23], and DLM [16].

We perform local motion magnification by using the ADLM-based frequency map and the corresponding pulsatile region localizations and make a comparison with the global motion magnification method using Euler method [15]. Figure 8 presents the motion-magnified clinical videos of Case #2 and Case #3, in which the surgical instruments are also presented. The temporal cross sections of positions A–B (instrument) and C–D (pulsation) in Case #2, and E–F (instrument) and G–H (pulsation) in Case #3 before and after local/global motion magnification are displayed. In Case #2, the moment when the surgical instrument appears is also marked. We can see that both global and the proposed local motion magnification methods create very similar results in the pulsatile regions(C–D and G–H). However, in global motion magnification, the shapes of surgical instruments are inevitably distorted, as illustrated in the green dashed box. This may cause negative impacts on the surgeon's elaborate operations. In comparison, the non-pulsatile regions, including the surgical instrument, remain unchanged in our framework. Therefore, unexpected distortions, artifacts, and noise will not be introduced in the local motion-magnified videos. Furthermore, Table 4 demonstrates that our method achieves higher SSIM and PSNR than the global Euler method in all three clinical cases.

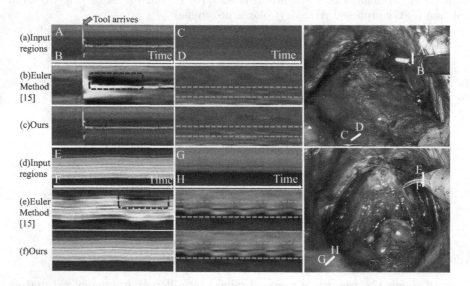

**Fig. 8.** Temporal expansions of the selected regions. (a) and (d) original videos. (b) and (e) motion-magnified videos with global Euler method [15]. (c) and (f) motion-magnified videos with our local motion magnification. A–B and E–F represent the surgical instruments. C–D and G–H denote the pulsatile regions. (Color figure online)

# 6  Conclusion

This paper proposes a framework for localization and local motion magnification of the pulsatile region in endoscopic surgical videos. Surgical videos with local motion magnification can visually and non-invasively help surgeons locate the area of the neurovascular bundles and vessels to prevent inadvertent injury. The experiments on synthetic and clinical surgical videos demonstrated the effectiveness of this framework. In future work, we plan to conduct more phantom experiments and improve the video quality after motion magnification.

# References

1. Ahlering, T.E., Skarecky, D., Borin, J.: Impact of cautery versus cautery-free preservation of neurovascular bundles on early return of potency. J. Endourol. **20**(8), 586–589 (2006)
2. Aisuwarya, R., Hendrick, H., Meitiza, M.: Analysis of cardiac frequency on photoplethysmograph (PPG) synthesis for detecting heart rate using fast Fourier transform (FFT). In: 2019 International Conference on Electrical Engineering and Computer Science (ICECOS), pp. 391–395. IEEE (2019)
3. Amir-Khalili, A., Hamarneh, G., Abugharbieh, R.: Automatic vessel segmentation from pulsatile radial distension. In: Navab, N., Hornegger, J., Wells, W.M., Frangi, A.F. (eds.) MICCAI 2015. LNCS, vol. 9351, pp. 403–410. Springer, Cham (2015). https://doi.org/10.1007/978-3-319-24574-4_48
4. Amir-Khalili, A., Hamarneh, G., Abugharbieh, R.: Modelling and extraction of pulsatile radial distension and compression motion for automatic vessel segmentation from video. Med. Image Anal. **40**, 184–198 (2017)
5. Amir-Khalili, A., et al.: Automatic segmentation of occluded vasculature via pulsatile motion analysis in endoscopic robot-assisted partial nephrectomy video. Med. Image Anal. **25**(1), 103–110 (2015)
6. Costello, A.J., Brooks, M., Cole, O.J.: Anatomical studies of the neurovascular bundle and cavernosal nerves. BJU Int. **94**(7), 1071–1076 (2004)
7. Doheny, E.P., et al.: Estimation of respiratory rate and exhale duration using audio signals recorded by smartphone microphones. Biomed. Sig. Process. Control **80**, 104318 (2023)
8. Fan, W., et al.: Robotically surgical vessel localization using robust hybrid video motion magnification. IEEE Robot. Autom. Lett. **6**(2), 1567–1573 (2021)
9. Goldberger, A.L., et al.: Physiobank, physiotoolkit, and physionet: components of a new research resource for complex physiologic signals. Circulation **101**(23), e215–e220 (2000)
10. Hore, A., Ziou, D.: Image quality metrics: Psnr vs. ssim. In: 2010 20th International Conference on Pattern Recognition, pp. 2366–2369. IEEE (2010)
11. Janatka, M., Sridhar, A., Kelly, J., Stoyanov, D.: Higher order of motion magnification for vessel localisation in surgical video. In: Frangi, A.F., Schnabel, J.A., Davatzikos, C., Alberola-López, C., Fichtinger, G. (eds.) MICCAI 2018. LNCS, vol. 11073, pp. 307–314. Springer, Cham (2018). https://doi.org/10.1007/978-3-030-00937-3_36
12. Kamshilin, A.A., Miridonov, S., Teplov, V., Saarenheimo, R., Nippolainen, E.: Photoplethysmographic imaging of high spatial resolution. Biomed. Opt. Express **2**(4), 996–1006 (2011)

13. Kothapalli, S.R., et al.: Simultaneous transrectal ultrasound and photoacoustic human prostate imaging. Sci. Transl. Med. **11**(507), eaav2169 (2019)
14. McClure, T.D., et al.: Use of MR imaging to determine preservation of the neurovascular bundles at robotic-assisted laparoscopic prostatectomy. Radiology **262**(3), 874–883 (2012)
15. McLeod, A.J., Baxter, J.S., de Ribaupierre, S., Peters, T.M.: Motion magnification for endoscopic surgery. In: Medical Imaging 2014: Image-Guided Procedures, Robotic Interventions, and Modeling, vol. 9036, pp. 81–88. SPIE (2014)
16. McLeod, A.J., Capaldi, D.P.I., Baxter, J.S.H., Parraga, G., Luo, X., Peters, T.M.: Analysis of periodicity in video sequences through dynamic linear modeling. In: Descoteaux, M., Maier-Hein, L., Franz, A., Jannin, P., Collins, D.L., Duchesne, S. (eds.) MICCAI 2017. LNCS, vol. 10434, pp. 386–393. Springer, Cham (2017). https://doi.org/10.1007/978-3-319-66185-8_44
17. Oh, T.-H., et al.: Learning-based video motion magnification. In: Ferrari, V., Hebert, M., Sminchisescu, C., Weiss, Y. (eds.) ECCV 2018. LNCS, vol. 11208, pp. 663–679. Springer, Cham (2018). https://doi.org/10.1007/978-3-030-01225-0_39
18. Shao, D., Yang, Y., Liu, C., Tsow, F., Yu, H., Tao, N.: Noncontact monitoring breathing pattern, exhalation flow rate and pulse transit time. IEEE Trans. Biomed. Eng. **61**(11), 2760–2767 (2014)
19. Tewari, A., et al.: An operative and anatomic study to help in nerve sparing during laparoscopic and robotic radical prostatectomy. Eur. Urol. **43**(5), 444–454 (2003)
20. Verkruysse, W., Svaasand, L.O., Nelson, J.S.: Remote plethysmographic imaging using ambient light. Opt. Express **16**(26), 21434–21445 (2008)
21. Wadhwa, N., Rubinstein, M., Durand, F., Freeman, W.T.: Phase-based video motion processing. ACM Trans. Graph. (TOG) **32**(4), 1–10 (2013)
22. Wadhwa, N., Rubinstein, M., Durand, F., Freeman, W.T.: Riesz pyramids for fast phase-based video magnification. In: 2014 IEEE International Conference on Computational Photography (ICCP), pp. 1–10. IEEE (2014)
23. Werner, K.J.: The xqifft: increasing the accuracy of quadratic interpolation of spectral peaks via exponential magnitude spectrum weighting. In: ICMC (2015)
24. Wu, H.Y., Rubinstein, M., Shih, E., Guttag, J., Durand, F., Freeman, W.: Eulerian video magnification for revealing subtle changes in the world. ACM Trans. Graph. (TOG) **31**(4), 1–8 (2012)
25. Zhang, Y., Pintea, S.L., Van Gemert, J.C.: Video acceleration magnification. In: Proceedings of the IEEE Conference on Computer Vision and Pattern Recognition, pp. 529–537 (2017)
26. Zyphur, M.J., Oswald, F.L.: Bayesian estimation and inference: a user's guide. J. Manag. **41**(2), 390–420 (2015)

# Co-speech Gesture Generation with Variational Auto Encoder

Shinichi Ka and Koichi Shinoda(✉)

School of Computing, Tokyo Institute of Technology, 2-12-1 Ookayama, Meguro-ku, Tokyo 152-8550, Japan
ka@ks.c.titech.ac.jp , shinoda@c.titech.ac.jp

**Abstract.** The research field of generating natural gestures from speech input is called co-speech gesture generation. Co-speech generation methods should suffice two requirements: fidelity and diversity. Several previous researches have utilized deterministic methods to establish a one-to-one mapping between speech and motion to achieve fidelity to speech, but the variety of gestures produced is limited. Other methods generate gestures probabilistically to make them various, but they often lack fidelity to the speech. To overcome these limitations, we propose Speaker-aware Audio2Gesture (SA2G) that uses a variational autoencoder (VAE) with the input of randomized speaker-aware features, an extension of the previously proposed A2G. By using ST-GCNs as encoders and controlling the variance for randomization, it can generate gestures faithful to speech content, which also have a large variety. In our evaluation on TED datasets, it improves the fidelity of the generated gestures from the baseline by 85.4, while increasing the Multimodality by $9.0 \times 10^{-3}$.

**Keywords:** Co-speech gesture generation · Human motion synthesis · Variational autoencoder · Pre-training

## 1 Introduction

In face-to-face communication, our speeches are always accompanied by body gestures. Some gestures are intentionally used to convey information complementary to speech contents [3], while other gestures are just personal habits without any intentions. In the fields of Robotics, AR/VR, and human-computer interaction, many studies have been done to artificially generate body gestures that may fit speech contents. For this purpose, there are two requirements. One is that generated gestures should not be inconsistent with speech contents. In other words, they should be faithful to speech contents. The other is that it should not be stereotypical. This is because gestures represent speakers' personalities, and we feel fixed body gestures are quite unnatural. In other words, they should be diverse enough to be felt natural.

In the past, deterministic methods have been used to estimate the correspondence between speech and gesture [2,9]. These methods produce gestures that are highly faithful to speech but are not diverse. In addition, when trained on a

© The Author(s), under exclusive license to Springer Nature Switzerland AG 2024
S. Rudinac et al. (Eds.): MMM 2024, LNCS 14556, pp. 155–168, 2024.
https://doi.org/10.1007/978-3-031-53311-2_12

large amount of data, the generated gestures become averaged motion without diversity [11].

To increase the diversity of gestures, several studies have proposed probabilistic methods [11,21] that can generate different gestures from the same speech. One example is Audio2Gesture (A2G) that uses a Variational Auto Encoder (VAE). It uses CNN to encode the time sequence of gestures and randomizes gestures based on the sample variance. While they can generate a wide variety of gestures, they are not always faithful to the speech content. Also, gestures vary from person to person. This speaker feature should be used to produce diverse gestures.

We propose a skeleton-based probabilistic method, Speaker-aware Audio2Gesture (SA2G), to generate gestures with both fidelity and diversity. As A2G, SA2G uses a VAE to generate gestures based on the input speech. We modify A2G in the following four ways. First, it employs ST-GCN as an encoder, in order to extract not only temporal features but also spatial features of human skeletons. Second, we introduced speaker-aware features to represent the speaker-specific gesture styles. Since gestures vary from person to person, considering individuality for gesture generation would contribute to generating diverse gestures. We introduce it by pre-training the ST-GCN encoder with speaker ID labels. Third, it adds Gaussian noise to the speaker-aware features, instead of replacing the features by Gaussian noise with the sample variance. This process makes the generated features less various but more faithful. We use TED Gesture Dataset [24] for evaluating our method. Finally, we use a linear mapping network instead of a VAE-based mapping network in A2G. We found a linear mapping network results in higher fidelity and diversity than a VAE-based one. We use TED [24] dataset for evaluating our method. We use FID as the measure for fidelity and Multimodality as that for diversity. Our experiments show the proposed method increases both.

The structure of this paper is as follows. Section 2 introduces the existing works on motion generation and describes their problems. Next, Sect. 3 explains the proposed method. Section 4 reports the results of evaluation experiments. Finally, Sect. 5 concludes this paper.

## 2  Related Works

### 2.1  Variational Autoencoder

Variational Autoencoder (VAE) [8] is a variational Bayesian estimator that consists of an encoder and a decoder. The objective of VAE is to estimate the true probability distribution $p(X)$ given a set of $N$ data samples $X = \{x^{(i)}\}_{i=1}^{N}$. First, a latent variable $z_i$ is generated from the prior distribution $p_\theta(z)$. Here, $x^{(i)}$ is assumed to be sampled from a posterior probability distribution $p_\theta(X|z)$. Computing the integral of the marginal likelihood $p(X) = \int p_\theta(z)p_\theta(X|z)dz$ is intractable. It is also difficult to apply the EM algorithm. To overcome this problem, a computable probability distribution $q_\phi(z|X)$ that approximates $p_\theta(z|X)$

is introduced. The parameters $\theta$ and $\phi$ are optimized by considering $q_\phi(z|X)$ as an encoder and $p_\theta(X|z)$ as a decoder.

The training process aims to estimate the probability distribution $p_\theta(X)$. Its log marginal likelihood is expressed as:

$$\log p_\theta(X) = D_{\text{KL}}(q_\phi(z|X)||p_\theta(z|X)) + L(\theta, \phi; X), \tag{1}$$

where $D_{\text{KL}}$ is the KL divergence between two distributions $q_\phi(z|X)$ and $p_\theta(z|X)$ and $L(\theta, \phi; X)$ is an evidence lower bound (ELBO)

Since the KL divergence is non-negative, the following inequality holds between the log marginal likelihood and the ELBO $L$:

$$\log p_\theta(X) \geq L(\theta, \phi; X) = -D_{\text{KL}}(q_\phi(z|X)||p_\theta(z)) + \mathbb{E}_{q_\phi(z|X)}[\log p_\theta(X|z)]. \tag{2}$$

Instead of maximizing the log marginal likelihood, the VAE parameter is estimated by maximizing $L(\theta, \phi; X)$. Since the expected value $\mathbb{E}_{q_\phi(z|X)}[\log p_\theta(X|z)]$ can be approximated by Monte Carlo estimation, $L(\theta, \phi; X)$ is estimated as:

$$L(\theta, \phi; X) \simeq -D_{\text{KL}}(q_\phi(z|X)||p_\theta(z)) + \frac{1}{M}\Sigma_{j=1}^{M}\log p_\theta(X_j|z), \tag{3}$$

where M is the number of samples.

Assuming that the prior distribution is a Gaussian distribution, then:

$$-D_{\text{KL}}(q_\phi(z|X)||p_\theta(z)) = \frac{1}{2}\Sigma_{k=1}^{K}(1 + \log((\sigma_k^{(i)})^2) - (\mu_k^{(i)})^2 \quad (\sigma_k^{(i)})^2), \tag{4}$$

where $K$ is the number of dimensions of the latent variable $z$, and $\mu$ and $\sigma$ are the mean and standard deviation obtained from the encoder, respectively.

## 2.2 Co-speech Gesture Generation

The methods to generate gestures from speech can be categorized into two: deterministic and probabilistic methods. Deterministic methods learn the one-to-one mapping between input speech and gesture. Among them, Speech2Gesture [2] employs CNN-based Generative Adversarial Networks (GANs) as the generator that is trained by the feedback from the discriminator and the reconstruction loss of gestures. Kucherenko et al. [9] proposed a generator using an autoencoder (AE) to make one-to-one mapping. While these deterministic methods can generate motions faithful to the given speech, the generated gestures tend to be similar to each other, and accordingly, they lack diversity.

In contrast, probabilistic methods estimate the probability of producing an output gesture given an input speech. Audio2Gesture(A2G) [11] randomly generates different motions from one audio sample by a Variational Auto Encoder (VAE). Trimodal-context [23] represents the speaker's motion styles in latent space during training and generates gestures by randomly sampling from the distribution of each style using the reparametrization trick in VAE. FreeMo [21] converts speech into texts and uses them as the condition in Conditional VAE. Its

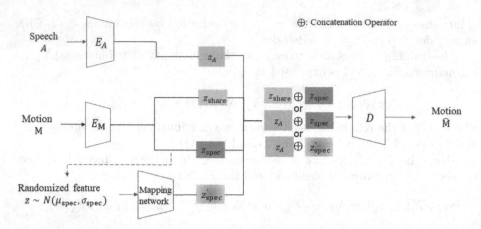

**Fig. 1.** The structure of Audio2Gesture.The mapping network is VAE-based.

generated motions represent the semantic information in speech and thus have high fidelity, but it is limited to 2D. These probabilistic methods can generate various motions, but due to the randomness of motion generation, they may not always be faithful to the input speech. In this study, we extend A2G [11] such that it is faithful. Its details will be further explained in the next section.

### 2.3  Audio2Gesture (A2G)

Figure 1 shows the structure of A2G, a VAE-based model that probabilistically infers gesture for input speech. Let $A$ be the input audio, $M = \{m_i\}_{i=1}^T$ be the continuous skeleton data with a length of $T$ frames. A2G has two encoders, one is the audio encoder $E_A$ that encodes the speech audio signals $A$ into the audio feature $z_A$ and the other is the CNN-based motion encoder that encodes the skeleton sequence signals $M$ into two different motion features, a shared feature $z_{\text{share}}$ and a motion-specific feature $z_{\text{spec}}$. Before inputting $M$ to motion encoders, each $m_i$ is converted into a joint position and flattened into one dim.

In addition, to encourage the model to generate a variety of gestures while keeping its fidelity, it utilizes a VAE mapping network to generate a randomized specific feature $z'_{\text{spec}}$. Here the mapping network takes a sampled value $z$ from a Gaussian distribution $N(\mu_{\text{spec}}, \sigma_{\text{spec}})$, where $\mu_{\text{spec}}$ is the mean of $z_{\text{spec}}$ and $\sigma_{\text{spec}}$ is the standard deviation of $z_{\text{spec}}$ per channel. It then outputs the randomized specific feature $z'_{\text{spec}}$. This randomization process can be represented as:

$$z'_{\text{spec}} = f(z), z \sim N(\mu_{\text{spec}}, \sigma_{\text{spec}}), \tag{5}$$

where $f$ is the function corresponding to the mapping network.

Let $\oplus$ be a concatenation operator of two vectors. The training of this network including the decoder $D$ is carried out using three different combinations of the four features above. First, $(z_{\text{share}} \oplus z_{\text{spec}})$ is used to reconstruct the input motion in the training phase. Second, $(z_A \oplus z_{\text{spec}})$ is used to generate gestures faithful to the input speech content. Finally, $(z_A \oplus z'_{\text{spec}})$ is used to generate different gestures even when the input audio signals are the same. This randomized feature also prevents the decoder $D$ from producing gestures only from the motion-specific features and helps the shared feature with audio control the output gesture.

A2G uses the following six loss functions as objective functions in the training. First, the VAE learning employs the reconstruction loss and KL divergence loss $L_{\text{KL}}$. The reconstruction loss is the sum of the position loss $L_{\text{pos}}$ and speed loss $L_{\text{speed}}$. $L_{\text{pos}}$ is the $L_1$ distances between each joint's coordinates of the generated motion and the correct motion. $L_{\text{speed}}$ is the $L_1$ distances of the velocity displacement per frame for each joint point of the skeleton. This reconstruction loss is calculated for the motions generated from the latent variable pairs $(z_A \oplus z_{\text{share}})$, $(z_{\text{share}} \oplus z_{\text{spec}})$, $(z_A \oplus z'_{\text{spec}})$, and the total reconstruction loss is their sum. The KL loss $L_{\text{KL}}$ is also calculated for each of the audio encoder and the motion encoder using Eq.(4).

To align the generated motion generation and audio, the code constraint is calculated as:

$$L_{\text{code}} = ||z_A - z_{\text{share}}||_1. \tag{6}$$

In order to prevent the mode collapse issue, a binding constraint $L_{\text{cyc}}$ is calculated as:

$$L_{\text{cyc}} = ||z'_{\text{spec}} - z''_{\text{spec}}||_1, \tag{7}$$

where $z''_{\text{spec}}$ are the features after passing the estimated motion through the encoders again.

Finally, a diversity loss $L_{\text{div}}$ is introduced to encourage diverse behaviors.

$$L_{\text{div}} = -||\hat{M} - \hat{M}'||_1, \tag{8}$$

where $\hat{M}$ is reconstructed gesture from $(z_A \oplus z_{\text{spec}})$, and $\hat{M}'$ is reconstructed gesture from $(z_A \oplus z'_{\text{spec}})$.

In summary, the loss functions in total is:

$$L_{\text{total}} = \lambda_{\text{pos}} L_{\text{pos}} + \lambda_{\text{speed}} L_{\text{speed}} + \lambda_{\text{KL}} L_{\text{KL}} \\ + \lambda_{\text{code}} L_{\text{code}} + \lambda_{\text{cyc}} L_{\text{cyc}} + \lambda_{\text{div}} L_{\text{div}}, \tag{9}$$

where $\lambda$ is a weight for each loss function.

The CNN motion encoder convolutes the input only in time direction and hence ignores some spatial information such as the structure of the skeleton. Also, A2G's randomization changes the specific features $z_{\text{spec}}$ randomly based on their mean and variance, but we found the variance is often too large. This may make the generated gestures too various, and accordingly, they become less faithful.

## 2.4   Human Motion Synthesis

In the field of human motion synthesis, there have been studies that use some sources other than speech such as music, and free sentences for controlling motions. The aim of these works is the same as this study, to generate diverse and faithful motions to the source contents.

Music data has been used for deterministic dance generation, where different deep learning models such as RNN, LSTM, and GAN have been employed [7,18,19]. In a recent study, Li *et al.* [12] proposed a method using attention mechanisms and Transformers to deterministically estimate dance motions, allowing for their smooth generation over a long duration of time. Huang *et al.* [6] proposed an RNN-based approach to generate various dances in accordance with the rhythms of sound sources. They addressed cumulative errors in an RNN model with curriculum learning. Dancing2Music [10], proposed by Lee *et al.*, combines VAE and GAN to generate various dance motions from the same music and the initial posture.

Some studies use free sentences and action types [4,5]. Petrovich *et al.* [17] employs a Transformer and uses action type as its condition. By utilizing Conditional VAE (CVAE), it successfully realizes a diverse range of motion representations while keeping the generated motion coherent with the action type. Tevet *et al.* [20] proposed an approach that utilizes a diffusion model to generate motion consistent with its labels and the content of the text. It utilizes random seeds to generate diverse behaviors.

## 2.5   Spatial Temporal Graph Convolutional Network (ST-GCN)

We utilize the Spatial Temporal Graph Convolutional Network (ST-GCN) [22] for the motion encoder (Fig. 2). Its main advantage is that it performs convolution in both spatial and temporal directions. It models the relationship not only between the same nodes connected by edges in the temporal direction but also the nodes in the spatial distribution of the skeleton nodes in each time frame by using a given graph structure and weighted edges. Then, the convoluted nodes contain not only their time information but also the skeleton structure information.

This point is a large advantage over CNN that convolutes temporal information and may not accurately capture spatial information.

# 3   Speaker-Aware Audio2Gesture (SA2G)

We propose a Speaker-aware Audio2Gesture (SA2G) model, an extension of the A2G model [11]. Figure 3 shows its system. To improve the performance of A2G, we have made four modifications.

First, we replaced the original motion encoder with a ST-GCN [22]. This enables the encoder to extract features with the skeleton structure information, which a CNN-based encoder cannot.

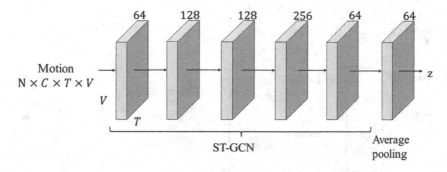

**Fig. 2.** The structure of ST-GCN that is used in our proposed method, SA2G. Where $z$ is the extracted features, and $N$, $C$, $T$, and $V$ represent the batch size, the number of channels, the motion length, and the number of graph nodes, respectively. We do not write the classification layer at the end.

Next, we pre-trained the ST-GCN encoder. We train two classifiers: one is for speakers and the other is for motions. Those encoders are trained with action and speaker ID labels. After the training, we remove its last layer, and the rest are fixed and used as encoders in our system. That is, we froze their parameters during the main training stage. We expect that the motion encoder extracts the action-related feature, which is used as $z_{\text{share}}$. Similarly, the speaker encoder extracts the speaker-related feature, which is utilized as $z_{\text{spec}}$.

Third, we add Gaussian noise for randomization. We found that the randomization process in A2G often generates gestures that lack fidelity. This is because the variance of distributions is determined by that of the specific feature $z_{\text{spec}}$. To reduce its variance, SA2G adds Gaussian noise to the speaker-aware features $z_{\text{spec}}$ instead of replacing it by Gaussian noise. More precisely, SA2G adds Gaussian noise $\epsilon \sim N(0, I)$ to the speaker-aware features $z_{\text{spec}}$ to obtain a randomized feature $z = z_{\text{spec}} + \epsilon$. This randomization process generates various randomized features while preserving more information in the speaker-aware feature than A2G. The resulting feature is then put through a linear mapping network, and the output $z'_{\text{spec}}$ is concatenated with the audio feature to create $(z_A \oplus z'_{\text{spec}})$.

Finally, we change the mapping network from a VAE-based one to a linear one. Both of them are implemented by A2G, but we experimentally found that using a linear layer for the mapping network improved both fidelity and diversity from a VAE-based mapping network.

During gesture generation training, the motion decoder $D$ reconstructs the gesture from the latent variables $(z_A \oplus z_{\text{spec}})$, $(z_{\text{share}} \oplus z_{\text{spec}})$, and $(z_A \oplus z'_{\text{spec}})$ in an epoch. We use the six loss functions defined in the A2G model [11].

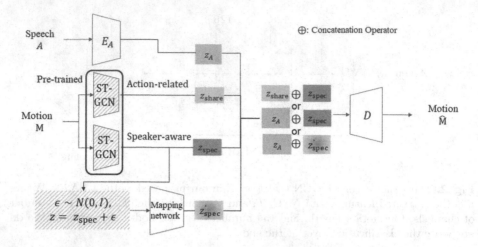

**Fig. 3.** The structure of Speaker-aware Audio2Gesture. The parts with diagonal lines are our new contributions. The ST-GCN based encoders are pre-trained to extract an action-related feature $z_{share}$ and a speaker-aware feature $z_{spec}$ with supervised training. ST-GCN is pre-trained using action type and speaker ID, and fixed during the main training phase. The mapping network is linear.

In the inference, a speaker-aware feature $z_{spec}$ is randomly selected from the training data, and Gaussian noise $\epsilon \sim N(0, I)$ is added to the feature. The resulting randomized speaker-aware features $z'_{spec}$ are then combined with the features $z_A$ obtained from the test audio, and a motion is generated from ($z_A \oplus z'_{spec}$).

## 4   Experiments

### 4.1   Experimental Setting

In our experiments, we use a 3D skeleton comprising ten upper-body joint points in TED [24] dataset. The joint points are transformed so that the length of each edge of the skeleton structure is the same overall the video frames in the dataset. In our experiments, the maximum frames of the motion are set to 34 based on [23].

We use librosa [14] to convert speech to a log-mel spectrogram. The sample rate and hop size are set to 16000 Hz and 1066, respectively. The dimension of the log-mel spectrogram is 64.

We trained the motion encoder with micro gesture labels in iMiGUE [13] during the pre-training phase. The data is converted to a 3D skeleton using VideoPose3D [15]. Additionally, we performed supervised learning of the speaker encoder using person ID labels from the TED Gesture Dataset [24] as pertaining. We set the batch size to 128, the learning rate to $10^{-3}$, and use Adam as the optimization method. The pre-training lasts for 500 epochs for both of the two encoders.

In the main training stage, Following the A2G [11], a learning rate is $10^{-4}$ and we use Adam as the optimization method. The weight of each loss is set as Eq. (10). Differing from A2G, the number of epochs of the main training is 100, and the batch size for the main training is 128.

$$\lambda_{pos} = 1, \lambda_{speed} = 5$$
$$\lambda_{KL} = 10^{-3} \qquad (10)$$
$$\lambda_{code} = 0.1, \lambda_{cyc} = 0.1, \lambda_{div} = 0.1$$

In the evaluation, we used the Frechet Inception Distance (FID) [1] as a metric to assess the fidelity of the generated motion and the Multimodality metric proposed in A2G [11] to evaluate its diversity. The FID is a commonly used evaluation metric to measure the performance of GANs in terms of how well two distributions ($X$ and $Y$) approximate each other. It is defined as :

$$FID = |\mu_X - \mu_Y|^2 + tr[\Sigma_X + \Sigma_Y - 2(\Sigma_X \Sigma_Y)^{\frac{1}{2}}], \qquad (11)$$

where their mean is $\mu_X$, $\mu_Y$ and their covariance is $\Sigma_X$, $\Sigma_Y$.

The Multimodality metric is calculated using the generated motions. Let $N$ be the number of generated motions and $\hat{M}_1, \hat{M}_2, ..., \hat{M}_N$ be the inferred motions from an audio sample $A$. Then,

$$Multimodality = \frac{1}{N \times \lceil N/2 \rceil} \Sigma_{i-1}^N \Sigma_{j=i+1}^N ||\hat{M}_i - \hat{M}_j||_1. \qquad (12)$$

We used $N = 20$. We evaluated the diversity by computing the average Multimodality scores across all test audios.

The FID and Multimodality calculations are performed 20 times, and the average values and the best values are reported below.

## 4.2    Comparison on TED Dataset

In this section, we compare our method, SA2G, with A2G [11] and the Trimodal-Context model [23]. The Trimodal-Context model does not use text data as input. Instead, we evaluate it with and without using a seed motion, which is a first 4-frame ground truth motion.

The results are shown in Table 1. Our model outperformed the A2G model by a significant margin, with a decrease of 85.4 in the FID value, indicating a significant improvement in faithful motion generation. In addition, the Multimodality metric also increased by $1.15 \times 10^{-2}$.

Second, compared with the Trimoda-Context model, SA2G achieved much better Multimodality but worse in FID. Both methods use a similar way to sample individuality during inference, but the randomizing process and mapping network in SA2G would encourage the model to generate various gestures. In terms of fidelity, the Trimodal-Context samples speaker feature probabilistically, but other parts are deterministically processed. Thus, the relationship between speech and gesture is one-to-one, and thus, it is more faithful to speech audio than SA2G.

**Table 1.** Comparison on TED dataset. ↓ means lower is better, and ↑ means higher is better. **Bold** is the best score of each metric. The mean score and the best score are reported.

| | FID ↓ | | Multimodality ↑ ($\times 10^{-2}$) | |
|---|---|---|---|---|
| | Mean | Best | Mean | Best |
| A2G [11] | 161.7 | 161.1 | 3.47 | 3.47 |
| Trimodal [23] | **3.5** | **3.4** | 1.32 | 1.32 |
| Trimodal [23] w/o seed | 52.1 | 51.8 | 3.72 | 3.73 |
| SA2G (Ours) | 76.3 | 75.7 | **4.62** | **4.62** |

### 4.3 Qualitative Comparison

We show the generated samples in Fig. 4. Compared to the A2G, the SA2G generates more diverse gestures. SA2G produces more movement than A2G, especially in the hands and elbows. There are still some noticeable stiff movements because VAE-based models tend to produce blurred images [16] and this leads to less dynamic actions.

The poses generated by SA2G are more varied than those generated by Trimodal-Context. Even when the Trimodal-Context does not utilize the seed motion, which has much effect on the generated gesture, the joint positions in Trimodal-Context are often similar between corresponding frames. On the other hand, SA2G usually starts from various poses and leads to a wide range of poses for each frame. This indicates that SA2G can provide more varied features with individuality than Trimodal-Context.

### 4.4 Ablation Study

We explore the impact of speaker-aware features and the control of variance in the randomization of the proposed method. It modifies the A2G model by (1) replacing the encoder with an ST-GCN-based encoder, (2) introducing a motion encoder and a speaker encoder, (3) adding Gaussian noise to the features to randomize the process, and (4) introducing a linear mapping network to map the randomized speaker-aware features to a more expressive space. The experimental results of the model with each modification step by step are shown in Table 2.

First, we replace the encoder with an ST-GCN-based encoder. This significantly improved FID and Multimodality values from those of the A2G model. We confirm that the ST-GCN-based encoder helped to enhance the model's expressive power by extracting the spatial features of body structures.

Next, when motion and speaker-aware features were introduced, the FID value further decreased. This may be the effect of introducing pre-training. On the other hand, the Multimodality is decreased by $1.0 \times 10^{-3}$.

Adding Gaussian noise to the speaker-aware features successfully reduced the FID value. By restricting the range of random sampling, more information

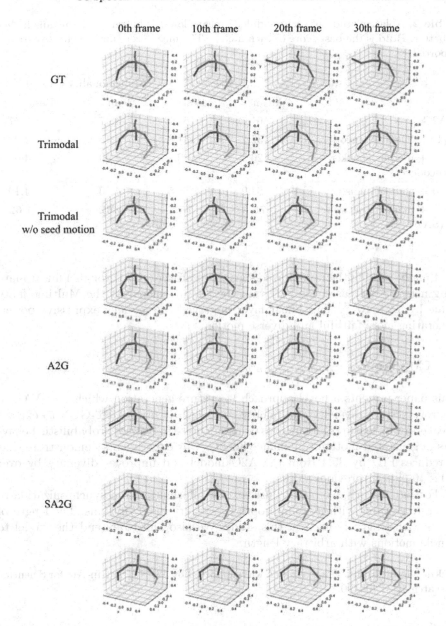

**Fig. 4.** Generated samples of Trimodal-Context, A2G, SA2G. Each snapshot corresponds to 0th, 10th, 20th, and 30th frame. The clipped input speech is "allow us to be ambitious and assertive".

in speaker-aware features is preserved than in the previous one. It prevents the model from generating too various gestures and increases its fidelity.

**Table 2.** Ablation study of each module. ↓ means lower is better, and ↑ means higher is better. **Bold** is the best score of each metric. The mean score and the best score are reported.

| | FID ↓ | | Multimodality ↑ ($\times 10^{-2}$) | |
|---|---|---|---|---|
| | Mean | Best | Mean | Best |
| A2G | 161.7 | 161.1 | 3.47 | 3.47 |
| (1) + ST-GCN encoder | 132.2 | 131.8 | 4.19 | 4.20 |
| (2) + motion, speaker encoder | 120.5 | 120.1 | 4.09 | 4.09 |
| (3) + gaussian noise | 84.9 | 84.5 | 4.10 | 4.10 |
| (4) + mapping network (SA2G) | **76.3** | **75.7** | **4.62** | **4.62** |

Finally, passing the randomized speaker-aware features through a linear mapping network decreased the FID value by 8.6 and increased the Multimodality value by $5.2 \times 10^{-3}$. This linear layer improved the model's expressive power, generating more faithful and diverse motion.

## 5    Conclusion

This paper presents a novel approach to gesture generation which uses VAE to improve gesture diversity and fidelity to speech. It utilizes ST-GCN to extract speaker-aware features and utilizes Gaussian noise to restrict probabilistic behaviors in generated motion to enhance fidelity. Our experiments demonstrate that it reduces FID by 85.4 from the A2G model and improves diversity by over $9.0 \times 10^{-3}$ from previous works.

For future work, we plan to use textual information and synchronization to generate gestures with more appropriate semantics and rhythms. The length of the motion length is fixed in this study. We also aim to extend the model to handle motions with arbitrary length.

**Acknowledgement.** This research was supported by JSPS Grant-in-Aid for Scientific Research JP23H00490.

## References

1. Dowson, D., Landau, B.: The Fréchet distance between multivariate normal distributions. J. Multivar. Anal. **12**(3), 450–455 (1982)
2. Ginosar, S., Bar, A., Kohavi, G., Chan, C., Owens, A., Malik, J.: Learning individual styles of conversational gesture. In: Proceedings of the IEEE/CVF Conference on Computer Vision and Pattern Recognition, pp. 3497–3506 (2019)
3. Graham, J.A., Argyle, M.: A cross-cultural study of the communication of extraverbal meaning by gestures (1). Int. J. Psychol. **10**(1), 57–67 (1975)

4. Guo, C., et al.: Generating diverse and natural 3D human motions from text. In: Proceedings of the IEEE/CVF Conference on Computer Vision and Pattern Recognition, pp. 5152–5161 (2022)
5. Guo, C., et al.: Action2motion: conditioned generation of 3D human motions. In: Proceedings of the 28th ACM International Conference on Multimedia, pp. 2021–2029 (2020)
6. Huang, R., Hu, H., Wu, W., Sawada, K., Zhang, M., Jiang, D.: Dance revolution: long-term dance generation with music via curriculum learning. arXiv preprint: arXiv:2006.06119 (2020)
7. Kao, H.K., Su, L.: Temporally guided music-to-body-movement generation. In: Proceedings of the 28th ACM International Conference on Multimedia, pp. 147–155 (2020)
8. Kingma, D.P., Welling, M.: Auto-Encoding variational bayes. In: 2nd International Conference on Learning Representations, ICLR 2014, Banff, AB, Canada, 14–16 April 2014, Conference Track Proceedings (2014)
9. Kucherenko, T., Hasegawa, D., Henter, G.E., Kaneko, N., Kjellström, H.: Analyzing input and output representations for speech-driven gesture generation. In: Proceedings of the 19th ACM International Conference on Intelligent Virtual Agents, pp. 97–104 (2019)
10. Lee, H.Y., et al.: Dancing to music. In: NeurIPS (2019)
11. Li, J., et al.: Audio2Gestures: generating diverse gestures from speech audio with conditional variational autoencoders. In: Proceedings of the IEEE/CVF International Conference on Computer Vision, pp. 11293–11302 (2021)
12. Li, R., Yang, S., Ross, D.A., Kanazawa, A.: AI choreographer: music conditioned 3D dance generation with AIST++. In: The IEEE International Conference on Computer Vision (ICCV) (2021)
13. Liu, X., Shi, H., Chen, H., Yu, Z., Li, X., Zhao, G.: iMiGUE: an identity-free video dataset for micro-gesture understanding and emotion analysis. In: Proceedings of the IEEE/CVF Conference on Computer Vision and Pattern Recognition (CVPR), pp. 10631–10642 (2021)
14. McFee, B., et al.: librosa/librosa: 0.9.2 (2022). https://doi.org/10.5281/zenodo.6759664
15. Pavllo, D., Feichtenhofer, C., Grangier, D., Auli, M.: 3D human pose estimation in video with temporal convolutions and semi-supervised training. In: Conference on Computer Vision and Pattern Recognition (CVPR) (2019)
16. Peng, J., Liu, D., Xu, S., Li, H.: Generating diverse structure for image inpainting with hierarchical VQ-VAE. In: Proceedings of the IEEE/CVF Conference on Computer Vision and Pattern Recognition, pp. 10775–10784 (2021)
17. Petrovich, M., Black, M.J., Varol, G.: Action-conditioned 3D human motion synthesis with transformer VAE. In: Proceedings of the IEEE/CVF International Conference on Computer Vision, pp. 10985–10995 (2021)
18. Sun, G., Wong, Y., Cheng, Z., Kankanhalli, M.S., Geng, W., Li, X.: DeepDance: music-to-dance motion choreography with adversarial learning. IEEE Trans. Multimedia 23, 497–509 (2020)
19. Tang, T., Jia, J., Mao, H.: Dance with melody: an LSTM-autoencoder approach to music-oriented dance synthesis. In: Proceedings of the 26th ACM International Conference on Multimedia, pp. 1598–1606 (2018)
20. Tevet, G., Raab, S., Gordon, B., Shafir, Y., Cohen-or, D., Bermano, A.H.: Human motion diffusion model. In: The Eleventh International Conference on Learning Representations (2023). https://openreview.net/forum?id=SJ1kSyO2jwu

21. Xu, J., Zhang, W., Bai, Y., Sun, Q., Mei, T.: Freeform body motion generation from speech. arXiv preprint: arXiv:2203.02291 (2022)
22. Yan, S., Xiong, Y., Lin, D.: Spatial temporal graph convolutional networks for skeleton-based action recognition. In: Proceedings of the AAAI Conference on Artificial Intelligence, vol. 32 (2018)
23. Yoon, Y., et al.: Speech gesture generation from the trimodal context of text, audio, and speaker identity. ACM Trans. Graph. (TOG) **39**(6), 1–16 (2020)
24. Yoon, Y., Ko, W.R., Jang, M., Lee, J., Kim, J., Lee, G.: Robots learn social skills: end-to-end learning of co-speech gesture generation for humanoid robots. In: Proceedings of The International Conference in Robotics and Automation (ICRA) (2019)

# Differentiable Neural Architecture Search Based on Efficient Architecture for Lightweight Image Super-Resolution

Chunyin Sheng, Xiang Gao, Xiaopeng Hu, and Fan Wang[✉]

School of Computer Science and Technology, Dalian University of Technology,
Dalian 116024, China
{scylovesl,gaoxiangv1}@mail.dlut.edu.cn, {huxp,wangfan}@dlut.edu.cn

**Abstract.** With the advancement of deep neural networks, image Super-Resolution (SR) has witnessed remarkable improvements in performance. However, the increasing number of parameters and computational complexity has posed challenges for the practical deployment of SR models. To address these challenges, we propose a novel approach called Differentiable Neural Architecture Search (NAS) based on Efficient Architecture for lightweight image Super-Resolution, referred to as DNAS-EASR. In DNAS-EASR, we employ the information distillation mechanism (IDM) at the cell-level space to search for key operations. Additionally, we search for attention modules at the cell-level space to determine the most suitable attention module for our architecture. Furthermore, we adopt a hierarchical architecture as our backbone network to enable multi-scale information processing and fusion. Extensive experiments conducted on benchmark datasets demonstrate that DNAS-EASR is lightweight, efficient and capable of achieving comparable performance to other lightweight methods.

**Keywords:** lightweight SR · differentiable NAS · information distillation mechanism (IDM) · attention · hierarchical architecture

## 1 Introduction

SR [1] is a well-known low-level problem in computer vision that aims to enhance a low-resolution (LR) image into its corresponding high-resolution (HR) counterpart. It holds significant value in various fields, such as surveillance videos [2], astronomical images [3] and medical images [4].

While traditional interpolation methods [5,6] for SR are fast, they suffer from low accuracy. The paradigm for image SR has shifted from traditional interpolation methods to deep learning approaches due to their superior performance, such as [7–9]. However, as deep neural networks become more complex, the number of parameters and computational complexity increase significantly. For example, RCAN [10] consists of 16M parameters, and EDSR [9] has 43M parameters.

© The Author(s), under exclusive license to Springer Nature Switzerland AG 2024
S. Rudinac et al. (Eds.): MMM 2024, LNCS 14556, pp. 169–183, 2024.
https://doi.org/10.1007/978-3-031-53311-2_13

Deploying these models on low-computing power devices becomes challenging. Therefore, there is an urgent need for lightweight models in image SR.

We have opted to utilize differentiable NAS [11] for achieving lightweight SR, as hand-crafted lightweight methods can be challenging to design. The search strategy employed in differentiable NAS is based on gradient descent. The underlying principle is to transform the architecture selection process into a super-network, where the importance of each operation is denoted by architecture parameters that are updated by gradient descent. After completing the search process, the operation corresponding to the largest architecture parameter is included as a component of the network structure.

Differentiable NAS offers the advantage of requiring minimal search time while obtaining an efficient model. For instance, HiNAS [12] firstly applied differentiable NAS to low-level vision tasks, achieving results with only 3.5 GPU hours on a GTX1080 Ti. Similarly, DLSR [13] completed the search in 2 GPU days on a Tesla V100. However, HiNAS did not perform well on SR tasks. Our analysis suggests that this can be attributed to two primary factors. Firstly, HiNAS still incorporated the Batch Normalization (BN) layer, which should be removed for SR tasks [9]. Secondly, the approach of selecting convolution operations at the cell-level space to form topology, and then stacking cells to form the entire network, is not effective for SR. Taking inspiration from DLSR, we propose a combination of differentiable NAS and efficient architecture to obtain a lightweight SR network with relatively good performance. Specifically, at the cell-level space, we search for the key operations of the distillation part based on IDM [14], as depicted in Fig. 2. However, unlike HiNAS or DLSR, we do not employ a network-level search space. We utilize a hierarchical architecture [15] as our backbone network, as shown in Fig. 1, which enables feature processing and fusion at multiple scales. Additionally, we exclude the BN layer from our network.

Here are our main contributions:

- We propose DNAS-EASR, an efficient and lightweight approach that builds upon efficient architectures, including the hierarchical architecture and IDM.
- Based on the IDM, we perform a search to identify essential operations in the information distillation part. Furthermore, we search for the most suitable attention module for our architecture.
- Our model is lightweight, efficient and capable of achieving comparable performance to other lightweight methods. Moreover, the search process for our model requires only 21 GPU hours on an Nvidia RTX3090.

## 2    Background

In this section, we focus on providing an overview of the background related to lightweight SR methods.

## 2.1  Hand-Crafted Lightweight SR Methods

In [14], IDM is proposed to extract more comprehensive features using fewer parameters, making it a relatively effective lightweight architecture. Building upon this work, IMDN [16] adopts a cascaded approach by constructing the Information Multi-Distillation Block (IMDB), which achieves a balance between lightweight design and effectiveness. Similarly, RFDN [17] reimagines the IMDB module by refining the channel split operation and the $3 \times 3$ convolution in the distillation part, resulting in a more concise and effective Residual Feature Distillation Block (RFDB). Additionally, there are other hand-crafted lightweight methods such as [18–20].

## 2.2  NAS Based Lightweight SR Methods

Two common search strategies for NAS based methods are evolutionary algorithms (EA) and reinforcement learning (RL). Both strategies generate candidate architectures iteratively and evaluate their performance, which requires a significant amount of search time. For instance, both [21] and [22] employed EA and RL search strategies, respectively, and the search process took 56 GPU days and 24 GPU days on a Tesla V100.

Another search strategy is based on gradient descent, known as differentiable NAS. Darts [11] was the first to propose differentiable NAS, while MileNAS [23] introduced a first-order method to optimize the error caused by the second-order approximation in Darts. HiNAS and DLSR are two algorithms that utilize the search frameworks of Darts and MileNAS, respectively, to search for lightweight SR networks. The search time for HiNAS was 3.5 GPU hours on a GTX1080 Ti, while DLSR took 2 GPU days on a Tesla V100. Methods based on differentiable NAS can rapidly search for relatively ideal results compared to those using EA or RL.

## 2.3  Attention Mechanism

The attention mechanism plays a crucial role in SR by enhancing specific features. Several models have incorporated attention mechanisms in their designs. For example, RCAN [10] and SAN [24] employed channel attention to rescale channel features. RFANet [25] and DLSR [13] utilized spatial attention and achieved impressive performance. Additionally, PAN [20] introduced pixel-level attention to compute weights for individual pixels.

## 2.4  Hierarchical Architecture

The method proposed by [26] highlights the importance of extracting features from multiple scales to capture diverse frequency information in images. Inspired by this idea, we have developed a hierarchical architecture for our backbone network, taking cues from the approach of [15]. This hierarchical design enables us to incorporate coarse-grained information, enhancing the overall representation capability of the network.

# 3   Methodology

In this section, we firstly introduce our **Differentiable NAS** based on **Efficient Architecture** for lightweight Image **Super-Resolution**, dubbed DNAS-EASR. Then we will discuss our search space and search strategy in detail.

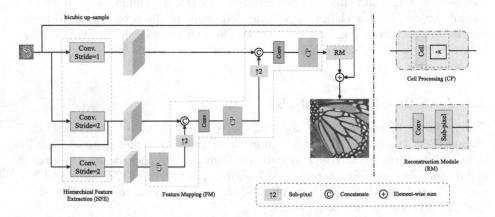

**Fig. 1.** The architecture of proposed DNAS-EASR. The network contains three main components: hierarchical feature extraction (HFE), feature mapping (FM), and reconstruction module (RM).

## 3.1   Network Architecture

As shown in Fig. 1, our DNAS-EASR consists of three main components: hierarchical feature extraction (HFE), feature mapping (FM), and reconstruction module (RM).

**HFE.** The input images are fed into the HFE module for hierarchical feature extraction. Specifically, the HFE module consists of three hierarchical convolutional layers. Following the implementation in [15], we design the high layer with k7n52s1p3, where k, n, s, and p represent kernel size, number of channels, stride, and padding, respectively. For the middle and low layers, we use k5n40s2p2 and k3n32s2p1 to extract shallow features at different scales. The whole process of HFE can be formulated in Eq. 1.

$$F_h = Conv_{k7n52s1p3}(I_{LR})$$
$$F_m = Conv_{k5n40s2p2}(I_{LR})$$
$$F_l = Conv_{k3n32s2p1}(F_m)$$

(1)

where $I_{LR}$ is the input LR image, $F_h$, $F_m$, $F_l$ represents the three shallow features which will be used in the FM module.

**FM.** In the FM process, we first fuse the low-level features with features of the current layer before processing them, especially for the top two layers. We utilize specific cells in the high, middle, and low layers as the Cell Processing (CP) module, as depicted in Fig. 1. The specific configuration of the number of cells will be thoroughly discussed in the Experiments section. It is important to note that cells within the same layer share the same architecture, while cells across different layers possess distinct architectures. As a result, our search procedure enables the discovery of three distinct types of cells. The overall FM process can be mathematically represented by Eq. 2.

$$F_l^* = CP_l(F_l)$$
$$F_m^* = CP_m(Concat(F_m, F_{l\uparrow2}^*))$$
$$F_h^* = CP_h(Concat(F_h, F_{m\uparrow2}^*))$$

(2)

where $CP_l$, $CP_m$, $CP_h$ represent the CP module at different layers respectively. We obtain $F_l^*$ from the low FM layer by $CP_l$. Then we concatenate $F_{l\uparrow2}^*$, which zooms in $F_l^*$ by a factor of 2 using sub-pixel layer, with $F_m$ as the input of the middle FM layer. We obtain $F_m^*$ using the $CP_m$ module. Similarly, we obtain $F_h^*$ in the same way we get $F_m^*$.

**RM.** The RM module is designed to enlarge the image by a certain scale. Besides, we use a global connection $f_{up}(\cdot)$ that applies bicubic interpolation to the input $I_{LR}$. This facilitates the network's training process and accelerates convergence. Hence, the output of the whole network can be formulated in Eq. 3.

$$I_{SR} = RM(F_h^*) + f_{up}(I_{LR})$$

(3)

where RM is composed of a convolution layer and a sub-pixel layer.

During the search and train process, we utilize L1 loss to optimize our network, which can be formulated as Eq. 4.

$$L(\Theta) = \frac{1}{N}\sum_{i=1}^{N}\left\|I_{SR}^i - I_{HR}^i\right\|_1 = \frac{1}{N}\sum_{i=1}^{N}\left\|H_{DNAS-EASR}(I_{LR}^i) - I_{HR}^i\right\|_1$$

(4)

where $\Theta$ represents the weight parameters and the architecture parameters during the search process, while it represents the weight parameters during the train process.

## 3.2    Search Space

Figure 2 illustrates the cell structure and two components based on NAS: the mixed residual block (MRB) and the mixed attention block (MAB). We adopt the cell structure from DLSR and utilize the MRB module for selecting optimal operations in the information distillation part. The MRB module incorporates three types of convolutions: convolution (conv), separable convolution (sep), and

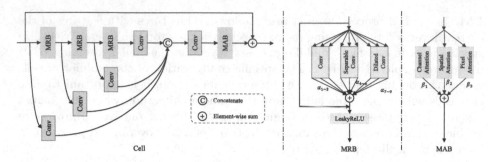

**Fig. 2.** The architecture of cell, MRB and MAB. The 'Conv' in cell represents a 1 × 1 convolution that reduces the number of channels by half. MRB is composed of 9 convolution operations weighted by parameter $\alpha$, while MAB is composed of 3 attention operations weighted by parameter $\beta$.

dilated convolution with a dilation rate of 2 (dil). These convolutions have kernel sizes of 3 × 3, 5 × 5, and 7 × 7, resulting in a total of 9 types of operators in the MRB module. The weights of these operators are determined by the parameter $\alpha$.

In MAB, we leverage NAS to determine the most suitable attention module for our architecture. We consider a total of three basic attention operations, namely channel attention (CA), spatial attention (SA), and pixel attention (PA). Unlike DLSR, we opt not to use the Enhanced Spatial Attention (ESA) module [25] in every cell. Although ESA is well-designed and has a small parameter count, incorporating it in every cell of our hierarchical architecture during the search process would lead to an excessive number of parameters. Therefore, in MAB, we simply incorporate CA, SA, and PA, weighted by the parameter $\beta$, for selection.

Each cell is composed of 3 MRB modules and 1 MAB module, resulting in a search complexity of 9 × 9 × 9 × 3 within each cell. In the FM module, cells are shared within each layer but not across layers, leading to 3 distinct types of cells. Consequently, the total complexity of the search space is $(9 \times 9 \times 9 \times 3)^3$, which reaches the magnitude of $10^{10}$. Searching for architectures within such a vast space can be time-consuming when using EA or RL methods. However, methods based on differentiable NAS can quickly search for relatively optimal results.

Here we will explain the output of MRB in the cell. The calculation process of MAB is a simplified version of the calculation process of MRB, which is omitted here. Suppose we are currently at the $k_{th}$ MRB, with input $x_k$ and operation space $O$. Each operation $o$ has a weight $\alpha_o^k$. After applying softmax, we can calculate the output of the $k_{th}$ MRB, denoted by $f(x_k)$, using Eq. 5, the residual connection and activation function have been omitted here for clarity.

$$f(x_k) = \sum_{o \in O} \frac{exp(\alpha_o^k)}{\sum_{o' \in O} exp(\alpha_{o'}^k)} o(x_k) \qquad (5)$$

## 3.3    Search Strategy

Darts firstly proposed an architecture search strategy based on gradient descent and identified it as a bilevel optimization problem. However, the use of second-order approximation to solve this problem, as seen in Darts, not only introduced errors but also consumed a lot of time. To address this, MileNAS introduced mixed-level optimization that considered training and validation errors together and employed a first-order optimization method. This method results in significant enhancements in both search time and performance. Therefore, we adopt the search framework proposed by MileNAS during our search process.

We aim to search for the target function depicted in Eq. 6.

$$\min_{A,\omega}[L_{tr}(\omega^*(A), A) + \lambda L_{val}(\omega^*(A), A)] \tag{6}$$

where $A = [\alpha, \beta]$ represent the architecture parameters, which represent the parameters of convolution and attention operations respectively, as shown in Fig. 2. $\omega$ represents the weight parameters, $\omega^*(A)$ represents the weight parameters that has converged on the training samples under the condition of $A$. $\lambda$ represents the comprehensive consideration of the influence of training errors and validation errors, which is also the core idea of the MileNAS framework. Our work follows the setting in MileNAS and sets $\lambda$ to 1. We then use a first-order optimization method to update the weight parameters and architecture parameters, as shown in the following formulas: Eq. 7 and Eq. 8.

$$\omega = \omega - \eta_\omega \nabla_\omega L_{tr}(\omega, A) \tag{7}$$

$$A = A - \eta_A(\nabla_A L_{tr}(\omega, A) + \lambda \nabla_A L_{val}(\omega, A)) \tag{8}$$

where $\eta_\omega$, $\eta_A$ represent the learning rates for weight and architecture parameters, respectively. Our search and training processes are outlined in the following Searching and Training Algorithm section.

---

**Algorithm 1.** Searching and Training Algorithm

---

**Input:** Search Dataset and Train Dataset
 1: Split Search Dataset into $\mathbb{D}_{train}$ and $\mathbb{D}_{eval}$
 2: Train the super-network on $\mathbb{D}_{train}$ for several steps to warm up
 3: **for** t = 1, 2, ..., T **do**
 4:     Sample train batch $\mathbb{B}_t = \{(x_i, y_i)\}_{i=1}^{batch}$ from $\mathbb{D}_{train}$
 5:     Optimize $\omega$ on the $\mathbb{B}_t$ by Eq. 7
 6:     Sample eval batch $\mathbb{B}_e = \{(x_i, y_i)\}_{i=1}^{batch}$ from $\mathbb{D}_{eval}$
 7:     Optimize $A = [\alpha, \beta]$ on the $\mathbb{B}_e$ by Eq. 8
 8:     Save the genotypes of the searched networks
 9: **end for**
10: Use the final genotype to obtain the sub-network
11: Use Train Dataset to train the sub-network from scratch
**Output:** A lightweight SR network

---

# 4 Experiments

## 4.1 Datasets and Implementation Details

The datasets utilized in our search and training processes comprise DIV2K [27] and Flickr2K [28]. DIV2K consists of 800 high-quality images, while Flickr2K contains 2650 images. These two datasets are combined to form DF2K, which consists of a total of 3450 images. The LR images are created by downsampling through bicubic interpolation. We test the results on four widely-used benchmark datasets: Set5 [29], Set14 [30], B100 [31], and Urban100 [32]. To assess the network's performance, we employ the peak signal-to-noise ratio (PSNR) and the structural similarity index measure (SSIM [33]). Both the search and training processes are executed on an Nvidia RTX3090 GPU.

## 4.2 Search Settings

The DF2K dataset is used as the Search Dataset and divided into two subsets: $\mathbb{D}_{train}$ and $\mathbb{D}_{eval}$. $\mathbb{D}_{train}$, which contains 2760 images (80% of the dataset), is used to update the weight and architecture parameters. $\mathbb{D}_{eval}$, which contains 690 images (20% of the dataset), is used to update the architecture parameters. To augment the Search Dataset, we randomly apply 90°, 180°, and 270° rotations, as well as horizontal and vertical flips. We conduct a total of 600 search epochs, with the initial 40 epochs serving as a warm-up process for the super-network, and the remaining 560 epochs dedicated to formal search. During the formal search, 50 iterations are performed per epoch. The size of the input LR image is 64 × 64. The batch size for both the training batch and evaluation batch is set to 32. The weight parameters are updated using L1 loss and the Adam optimizer, with the learning rate, weight decay, and betas set to 1e-3, 1e-8, and (0.9, 0.999), respectively. The architecture parameters are updated using the same loss function, optimizer, and parameter settings as the weight parameters.

## 4.3 Train Settings

We train the sub-network from scratch using the final genotype obtained from the search. During the training process, we use LR images with a size of 64 × 64 as inputs. The batch size is set to 32, and a total of 1000k iterations are performed. We employ the L1 loss function as the training objective. The Adam optimizer is utilized with a learning rate of 1e-3, weight decay of 0, and betas set to (0.9, 0.99). Furthermore, we incorporate a cosine annealing strategy to dynamically adjust the learning rate during the training process.

## 4.4 Search Results

Different search conditions can yield varying outcomes. Table 1 showcases the discovered cell structures for ×4 SR under the conditions of 5, 3, and 2 cells. The selection of 5, 3, and 2 cells is based on a consideration of performance, parameters, and the number of operations (Multi-Adds), as summarized in Table 2.

Leveraging the obtained search results, we proceed to train and obtain the ulti-mate lightweight SR network for ×4 SR. Additionally, we utilize the same search results to acquire the ultimate networks for ×2 and ×3 SR.

**Table 1.** The results of the search process for 5, 3, and 2 cells for ×4 SR. In the table, the labels "h," "m," and "l" correspond to the high, middle, and low layer of the FM module, respectively.

| FM layer | MRB1 | MRB2 | MRB3 | MAB |
|----------|------|------|------|-----|
| h | dil: $3 \times 3$ | sep: $3 \times 3$ | conv: $3 \times 3$ | PA |
| m | conv: $3 \times 3$ | conv: $3 \times 3$ | sep: $7 \times 7$ | PA |
| l | conv: $7 \times 7$ | dil: $5 \times 5$ | dil: $5 \times 5$ | PA |

## 4.5    Ablation Study

We begin by discussing the configuration of different cell numbers in each layer. Table 2 presents the results for ×4 SR obtained on four benchmark datasets under various settings. The Multi-Adds is calculated on 720p ($1280 \times 720$) HR images. PSNR and SSIM are computed on the Y channel of the transformed YCbCr color space. Analyzing the results in Table 2, we observe that the 5-3-2 configuration consistently achieves the best performance across almost all test datasets. Moreover, this configuration maintains a reasonable parameter and computational complexity. Based on these findings, we select the 5-3-2 configu-ration as the final experimental setting.

**Table 2.** Test results determined by setting different numbers of cells in different layers on benchmark datasets for ×4 SR. The best and second-best results are represented in bold and underline respectively.

| Cell number | Scale | Params(K) | Multi-Adds(G) | Set5 | Set14 | B100 | Urban100 |
|-------------|-------|-----------|---------------|------|-------|------|----------|
| | | | | PSNR/SSIM | PSNR/SSIM | PSNR/SSIM | PSNR/SSIM |
| 2-2-2 | ×4 | 367 | 12.0 | 31.91/0.8911 | 28.45/0.7776 | 27.48/0.7324 | 25.70/0.7734 |
| 2-3-4 | | 480 | 12.8 | 31.85/0.8900 | 28.47/0.7778 | 27.48/0.7327 | 25.73/0.7745 |
| 3-2-2 | | 428 | 15.7 | 32.10/0.8936 | 28.54/0.7798 | 27.52/0.7341 | 25.88/0.7794 |
| 3-3-3 | | 482 | 14.9 | 32.03/0.8926 | 28.51/0.7793 | 27.52/0.7340 | 25.87/0.7794 |
| 3-3-4 | | 509 | 16.5 | 32.09/0.8933 | 28.50/0.7794 | 27.52/0.7342 | 25.88/0.7793 |
| 4-3-2 | | 511 | 19.9 | **32.19**/<u>0.8948</u> | <u>28.60</u>/<u>0.7814</u> | <u>27.57</u>/<u>0.7358</u> | 26.00/0.7832 |
| 4-4-2 | | 539 | 19.5 | 32.16/0.8945 | 28.59/0.7811 | <u>27.57</u>/<u>0.7358</u> | <u>26.01</u>/<u>0.7838</u> |
| 5-3-2 | | 555 | 22.4 | <u>32.18</u>/**0.8952** | **28.64/0.7823** | **27.59/0.7363** | **26.07/0.7856** |

**Table 3.** Test results of different layers for the FM module for 4× SR. The best and second-best results are represented in bold and underline respectively.

| Layer number | Cell number | Scale | Params(K) | Multi-Adds(G) | Set5 | Set14 | B100 | Urban100 |
|--------------|-------------|-------|-----------|---------------|------|-------|------|----------|
| | | | | | PSNR/SSIM | PSNR/SSIM | PSNR/SSIM | PSNR/SSIM |
| 1 | 5 | ×4 | 313 | 18.0 | 32.04/0.8931 | 28.53/0.7801 | 27.54/0.7344 | 25.91/0.7797 |
| 2 | 5-3 | | 473 | 21.8 | <u>32.11</u>/<u>0.8942</u> | <u>28.57</u>/<u>0.7808</u> | <u>27.57</u>/<u>0.7356</u> | <u>26.00</u>/<u>0.7832</u> |
| 3 | 5-3-2 | | 555 | 22.4 | **32.18/0.8952** | **28.64/0.7823** | **27.59/0.7363** | **26.07/0.7856** |

We continue our assessment of the hierarchical structure by conducting experiments with 1, 2, and 3 layers. As shown in Table 3, a higher number of layers leads to an increase in Params and Multi-Adds, but it also results in improved performance. To ensure our model's comparability with other efficient lightweight models, we ultimately opt for a three-layered hierarchical structure.

### 4.6    Comparison with Hand-Crafted Lightweight Methods

We have compared our DNAS-EASR with other hand-crafted lightweight methods. The results are presented in Table 4. For ×4 SR, our results on Set14 and B100 outperform all other hand-crafted lightweight SR methods. Furthermore, our PSNR and SSIM results on Urban100 are only 0.04dB and 0.0002 lower than the best-performing methods. When compared to the high-performance RFDN, our model achieves a similar Params (555K), but our model's Multi-Adds is only 22.4G, which is less than RFDN's 31.6G. For ×2, ×3 SR, our model consistently achieves the highest performance on the B100 dataset. On other datasets, our model's performance is also competitive and does not significantly deviate from the best results.

It is noteworthy that despite our model having approximately 550K parameters, its Multi-Adds is relatively modest. This becomes particularly evident when comparing our model to PAN, which has only 270K parameters but incurs a larger Multi-Adds of 28.2G for ×4 SR, whereas our model only requires 22.4G Multi-Adds. In summary, our model exhibits performance comparable to that of other hand-crafted lightweight models. Figure 3 present visual comparisons between our model and other lightweight methods for ×4 SR.

**Table 4.** Comparisons with hand-crafted lightwieght methods for ×2, ×3, and ×4 SR are presented below. The best and second-best results are represented in bold and underline respectively.

| Method | Scale | Params(K) | Multi-Adds(G) | Set5 PSNR/SSIM | Set14 PSNR/SSIM | B100 PSNR/SSIM | Urban100 PSNR/SSIM |
|---|---|---|---|---|---|---|---|
| Bicubic | ×2 | – | – | 33.66/0.9299 | 30.24/0.8688 | 29.56/0.8431 | 26.88/0.8403 |
| LapSRN [34] | | 251 | – | 37.52/0.9591 | 32.99/0.9124 | 31.80/0.8952 | 30.41/0.9103 |
| CARN-M [18] | | 412 | 91.2 | 37.53/0.9583 | 33.26/0.9141 | 31.92/0.8960 | 31.23/0.9194 |
| IMDN [16] | | 694 | 158.8 | 38.00/0.9605 | 33.63/0.9177 | 32.19/0.8996 | 32.17/0.9283 |
| PAN [20] | | 261 | 70.5 | 38.00/0.9605 | 33.59/0.9181 | 32.18/0.8997 | 32.01/0.9273 |
| RFDN [17] | | 534 | 123.0 | **38.05/0.9606** | **33.68/0.9184** | 32.16/0.8994 | 32.12/0.9278 |
| **DNAS-EASR** | | 553 | 88.9 | 38.01/0.9605 | 33.55/0.9174 | **32.21/0.9001** | 32.11/**0.9285** |
| Bicubic | ×3 | – | – | 30.39/0.8682 | 27.55/0.7742 | 27.21/0.7385 | 24.46/0.7349 |
| LapSRN [34] | | 502 | – | 33.81/0.9220 | 29.79/0.8325 | 28.82/0.7980 | 27.07/0.8275 |
| CARN-M [18] | | 412 | 46.1 | 33.99/0.9236 | 30.08/0.8367 | 28.91/0.8000 | 27.55/0.8385 |
| IMDN [16] | | 703 | 71.5 | 34.36/0.9270 | 30.32/0.8417 | 29.09/0.8046 | 28.17/0.8519 |
| PAN [20] | | 261 | 39.0 | 34.40/0.9271 | **30.36/0.8423** | **29.11**/0.8050 | 28.11/0.8511 |
| RFDN [17] | | 541 | 55.4 | **34.41/0.9273** | 30.34/0.8420 | 29.09/0.8050 | **28.21/0.8525** |
| **DNAS-EASR** | | 554 | 39.6 | 34.36/0.9271 | 30.34/0.8418 | **29.11/0.8051** | 28.12/0.8519 |
| Bicubic | ×4 | – | – | 28.42/0.8104 | 26.00/0.7027 | 25.96/0.6675 | 23.14/0.6577 |
| LapSRN [34] | | 813 | – | 31.54/0.8852 | 28.09/0.7700 | 27.32/0.7275 | 25.21/0.7562 |
| CARN-M [18] | | 412 | 32.5 | 31.92/0.8903 | 28.42/0.7762 | 27.44/0.7304 | 25.62/0.7694 |
| IMDN [16] | | 715 | 40.9 | 32.21/0.8948 | 28.58/0.7811 | 27.56/0.7353 | 26.04/0.7838 |
| PAN [20] | | 272 | 28.2 | 32.13/0.8948 | 28.61/0.7822 | **27.59/0.7363** | **26.11**/0.7854 |
| RFDN [17] | | 550 | 31.6 | **32.24/0.8952** | 28.61/0.7819 | 27.57/0.7360 | **26.11/0.7858** |
| **DNAS-EASR** | | 555 | 22.4 | 32.18/**0.8952** | **28.64/0.7823** | **27.59/0.7363** | 26.07/0.7856 |

**Fig. 3.** Visual comparisons for ×4 SR. The test image patches are from Urban100 and Set14.

## 4.7   Comparison with NAS-Based Lightweight Methods

We conduct a comparison between our DNAS-EASR and two other NAS-based lightweight models, namely HiNAS and DLSR, in terms of search time and performance. The results are presented in Table 5 and Table 6.

While both DNAS-EASR and DLSR incur longer search costs compared to HiNAS, they exhibit significantly better performance. We attribute this enhanced performance to the absence of BN layers during the search and training processes in both DNAS-EASR and DLSR.

When comparing DNAS-EASR to DLSR, although DNAS-EASR boasts a shorter search time, it does show some disparities in terms of Params, Multi-Adds, and overall performance, falling short of DLSR's results. We hypothesize that this could be due to the attention module we discovered through the search process, which lacks the robust feature representation capabilities compared to the carefully designed, lightweight, and efficient ESA attention module used in DLSR. Therefore, in future work, we plan to conduct further optimization in this aspect.

**Table 5.** Comparisons between our DNAS-EASR and HiNAS, DLSR for ×4 SR. The best and second-best results are represented in bold and <u>underline</u> respectively.

| Method | Scale | Params(K) | Multi-Adds(G) | Set5 | Set14 | B100 | Urban100 |
| --- | --- | --- | --- | --- | --- | --- | --- |
| | | | | PSNR/SSIM | PSNR/SSIM | PSNR/SSIM | PSNR/SSIM |
| HiNAS [12] | ×4 | 290-330 | – | 31.72/0.8926 | 27.85/**0.7881** | 27.35/**0.7465** | 25.51/0.7779 |
| DLSR [13] | | 338 | 17.9 | **32.33/0.8963** | **28.68**/<u>0.7832</u> | **27.61**/<u>0.7374</u> | **26.19/0.7892** |
| **DNAS-EASR** | | 555 | 22.4 | <u>32.18/0.8952</u> | <u>28.64</u>/0.7823 | <u>27.59</u>/0.7363 | <u>26.07/0.7856</u> |

**Table 6.** Search cost of NAS based methods using the search strategy of gradient descent.

| Method | Search cost |
| --- | --- |
| HiNAS [12] | 3.5 GPU hours |
| DLSR [13] | 2 GPU days |
| **DNAS-EASR** | 21 GPU hours |

## 5   Conclusion

In this study, we propose DNAS-EASR, a differentiable neural architecture search method that incorporates the information distillation mechanism and hierarchical architecture for lightweight SR tasks. We devise a novel search space to ensure the lightweight nature of our cells, making them suitable for integration into the hierarchical architecture. Our search process is completed in only 21 GPU hours using an Nvidia RTX 3090. Experimental results demonstrate that our model is not only lightweight and efficient but also achieves comparable performance to other lightweight models.

**Acknowledgement.** This work was supported by the National Major Special Funding Project (grant number 2018YFA0704605).

# References

1. Irani, M., Peleg, S.: Improving resolution by image registration. CVGIP: Graph. Models Image Process. **53**(3), 231–239 (1991)
2. Mudunuri, S.P., Biswas, S.: Low resolution face recognition across variations in pose and illumination. IEEE Trans. Pattern Anal. Mach. Intell. **38**(5), 1034–1040 (2015)
3. Lobanov, A.P.: Resolution limits in astronomical images. arXiv preprint astro-ph/0503225 (2005)
4. Greenspan, H.: Super-resolution in medical imaging. Comput. J. **52**(1), 43–63 (2009)
5. Keys, R.: Cubic convolution interpolation for digital image processing. IEEE Trans. Acoust. Speech Signal Process. **29**(6), 1153–1160 (1981)
6. Duchon, C.E.: Lanczos filtering in one and two dimensions. J. Appl. Meteorol. Climatol. **18**(8), 1016–1022 (1979)
7. Dong, C., Loy, C.C., He, K., Tang, X.: Image super-resolution using deep convolutional networks. IEEE Trans. Pattern Anal. Mach. Intell. **38**(2), 295–307 (2015)
8. Kim, J., Lee, J.K., Lee, K.M.: Accurate image super-resolution using very deep convolutional networks. In: Proceedings of the IEEE Conference on Computer Vision and Pattern Recognition, pp. 1646–1654 (2016)
9. Lim, B., Son, S., Kim, H., Nah, S., Mu Lee, K.: Enhanced deep residual networks for single image super-resolution. In: Proceedings of the IEEE Conference on Computer Vision and Pattern Recognition Workshops, pp. 136–144 (2017)
10. Zhang, Y., Li, K., Li, K., Wang, L., Zhong, B., Fu, Y.: Image super-resolution using very deep residual channel attention networks. In: Proceedings of the European Conference on Computer Vision (ECCV), pp. 286–301 (2018)
11. Liu, H., Simonyan, K., Yang, Y.: Darts: differentiable architecture search. arXiv preprint arXiv:1806.09055 (2018)
12. Zhang, H., Li, Y., Chen, H., Gong, C., Bai, Z., Shen, C.: Memory-efficient hierarchical neural architecture search for image restoration. Int. J. Comput. Vis. 1–22 (2022)
13. Huang, H., Shen, L., He, C., Dong, W., Huang, H., Shi, G.: Lightweight image super-resolution with hierarchical and differentiable neural architecture search. arXiv preprint arXiv:2105.03939 (2021)
14. Hui, Z., Wang, X., Gao, X.: Fast and accurate single image super-resolution via information distillation network. In: Proceedings of the IEEE Conference on Computer Vision and Pattern Recognition, pp. 723–731 (2018)
15. Li, B., Li, X., Lu, Y., Liu, S., Feng, R., Chen, Z.: HST: hierarchical Swin transformer for compressed image super-resolution. In: Karlinsky, L., Michaeli, T., Nishino, K. (eds.) Computer Vision – ECCV 2022 Workshops. ECCV 2022. LNCS, vol. 13802, pp. 651–668. Springer, Cham (2023). https://doi.org/10.1007/978-3-031-25063-7_41
16. Hui, Z., Gao, X., Yang, Y., Wang, X.: Lightweight image super-resolution with information multi-distillation network. In: Proceedings of the 27th ACM International Conference on Multimedia, pp. 2024–2032 (2019)

17. Liu, J., Tang, J., Wu, G.: Residual feature distillation network for lightweight image super-resolution. In: Bartoli, A., Fusiello, A. (eds.) ECCV 2020. LNCS, vol. 12537, pp. 41–55. Springer, Cham (2020). https://doi.org/10.1007/978-3-030-67070-2_2

18. Ahn, N., Kang, B., Sohn, K.A.: Fast, accurate, and lightweight super-resolution with cascading residual network. In: Proceedings of the European Conference on Computer Vision (ECCV), pp. 252–268 (2018)

19. Tai, Y., Yang, J., Liu, X.: Image super-resolution via deep recursive residual network. In: Proceedings of the IEEE Conference on Computer Vision and Pattern Recognition, pp. 3147–3155 (2017)

20. Zhao, H., Kong, X., He, J., Qiao, Yu., Dong, C.: Efficient image super-resolution using pixel attention. In: Bartoli, A., Fusiello, A. (eds.) ECCV 2020. LNCS, vol. 12537, pp. 56–72. Springer, Cham (2020). https://doi.org/10.1007/978-3-030-67070-2_3

21. Chu, X., Zhang, B., Xu, R.: Multi-objective reinforced evolution in mobile neural architecture search. In: Bartoli, A., Fusiello, A. (eds.) ECCV 2020. LNCS, vol. 12538, pp. 99–113. Springer, Cham (2020). https://doi.org/10.1007/978-3-030-66823-5_6

22. Chu, X., Zhang, B., Ma, H., Xu, R., Li, Q.: Fast, accurate and lightweight super-resolution with neural architecture search. In: 2020 25th International Conference on Pattern Recognition (ICPR), pp. 59–64. IEEE (2021)

23. He, C., Ye, H., Shen, L., Zhang, T.: Milenas: efficient neural architecture search via mixed-level reformulation. In: Proceedings of the IEEE/CVF Conference on Computer Vision and Pattern Recognition, pp. 11993–12002 (2020)

24. Dai, T., Cai, J., Zhang, Y., Xia, S.T., Zhang, L.: Second-order attention network for single image super-resolution. In: Proceedings of the IEEE/CVF Conference on Computer Vision and Pattern Recognition, pp. 11065–11074 (2019)

25. Liu, J., Zhang, W., Tang, Y., Tang, J., Wu, G.: Residual feature aggregation network for image super-resolution. In: Proceedings of the IEEE/CVF Conference on Computer Vision and Pattern Recognition, pp. 2359–2368 (2020)

26. Chen, Y., et al.: Drop an octave: reducing spatial redundancy in convolutional neural networks with octave convolution. In: Proceedings of the IEEE/CVF International Conference on Computer Vision, pp. 3435–3444 (2019)

27. Agustsson, E., Timofte, R.: NTIRE 2017 challenge on single image super-resolution: dataset and study. In: Proceedings of the IEEE Conference on Computer Vision and Pattern Recognition Workshops, pp. 126–135 (2017)

28. Timofte, R., Agustsson, E., Van Gool, L., Yang, M.H., Zhang, L.: NTIRE 2017 challenge on single image super-resolution: methods and results. In: Proceedings of the IEEE Conference on Computer Vision and Pattern Recognition Workshops, pp. 114–125 (2017)

29. Bevilacqua, M., Roumy, A., Guillemot, C., Alberi-Morel, M.L.: Low-complexity single-image super-resolution based on nonnegative neighbor embedding (2012)

30. Yang, J., Wright, J., Huang, T.S., Ma, Y.: Image super-resolution via sparse representation. IEEE Trans. Image Process. **19**(11), 2861–2873 (2010)

31. Martin, D., Fowlkes, C., Tal, D., Malik, J.: A database of human segmented natural images and its application to evaluating segmentation algorithms and measuring ecological statistics. In: Proceedings Eighth IEEE International Conference on Computer Vision. ICCV 2001, vol. 2, pp. 416–423. IEEE (2001)

32. Huang, J.B., Singh, A., Ahuja, N.: Single image super-resolution from transformed self-exemplars. In: Proceedings of the IEEE Conference on Computer Vision and Pattern Recognition, pp. 5197–5206 (2015)

33. Wang, Z., Bovik, A.C., Sheikh, H.R., Simoncelli, E.P.: Image quality assessment: from error visibility to structural similarity. IEEE Trans. Image Process. **13**(4), 600–612 (2004)

34. Lai, W.S., Huang, J.B., Ahuja, N., Yang, M.H.: Deep Laplacian pyramid networks for fast and accurate super-resolution. In: Proceedings of the IEEE Conference on Computer Vision and Pattern Recognition, pp. 624–632 (2017)

# Learning Collaborative Reinforcement Attention for 3D Face Reconstruction and Dense Alignment

Zhengwei Yang, Yange Wang, Lei Ma, and Xiangzheng Li[(✉)]

School of Mathematics and Computer Science, Ningxia Normal University,
Xueyuan Road, Guyuan 756000, Ningxia, China
82021011@nxnu.edu.cn

**Abstract.** 3D face reconstruction from monocular outdoor images has long been a challenging problem. Traditional attention network methods that directly regress parameters may suffer from inadequate learning of discriminative features. In this paper, we propose a method called collaborative reinforcement attention module (CRAM). CRAM comprises three major modules: the perception module (PM), the channel selection module (CSM), and the multi-level feature interaction module (MFIM). CRAM leverages contextual information to simultaneously focus on multiple prominent features in facial photos. It employs multi-level and multi-angle feature extraction and fusion techniques to adaptively learn the relationship between facial regions and key feature points. This results in enhanced accuracy in 3D face reconstruction and meticulous dense alignment. Furthermore, to enhance the model's generalization performance, we introduce a regional noise injection and image composition module (RNICM) as a preprocessing step for sample data which help capture more local details and handle occluded faces, particularly under significant head rotations. Extensive experiments conducted on the AFLW2000-3D and AFLW datasets validate the effectiveness of the proposed approach.

**Keywords:** Face alignment · Face reconstruction · Collaborative reinforcement attention module

## 1 Introduction

3D face reconstruction is an important task in the fields of computer vision and computer graphics. During this process, estimating the 3D geometry and texture of a face from a single image becomes a challenging problem due to limitations in 3D scanning data and the influence of various factors such as lighting, reflectance, and geometric shape. The early 3D morphable model (3DMM) proposed by Blanz and Vetter [1] was constructed as a statistical model based on shape and texture data from hundreds of scanned faces, providing strong support for 3D face reconstruction. Subsequently, a series of methods based on 3DMM

© The Author(s), under exclusive license to Springer Nature Switzerland AG 2024
S. Rudinac et al. (Eds.): MMM 2024, LNCS 14556, pp. 184–197, 2024.
https://doi.org/10.1007/978-3-031-53311-2_14

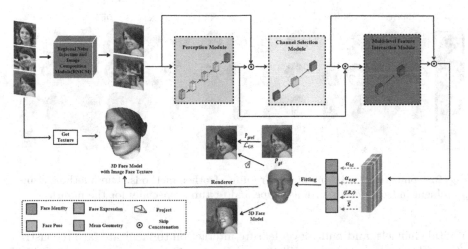

**Fig. 1.** The workflow of the collaborative reinforcement attention module network framework. The regional noise injection and image composition module generates images containing more local details and occluded faces. The preprocessed images are then sequentially input into the perception module (PM), channel selection module (CSM), and multi-level feature interaction module (MFIM) to extract salient facial features. Subsequently, we use 3DMM to fit the facial geometric contour and combine with the input image to obtain texture information. We utilize the Z-buffer to render a three-dimensional face with texture. During the network learning phase, we apply the Euclidean distance loss, denoted as $\mathcal{L}_{68}$, to constrain the shape of the facial geometry effectively. This ensures the accurate learning of the 3D facial shape.

emerged, which achieved good results in certain cases. However, these methods have limitations when it comes to handling challenges such as the large pose in the wild, extreme facial expressions, and severe local occlusions.

In recent years, with the development of deep learning, methods based on CNN network regression [2,9] have achieved remarkable results in 3D face reconstruction and dense face alignment. Many studies have introduced deep learning neural networks into 3DMM. For example, a novel hierarchical representation network (HRN) was proposed by Biwen Lei et al. [4], which can accurately reconstruct facial details from a single image. In order to achieve the separation of identity and expression and the control of expression, Fariborz Taherkhani et al. [5] proposed a new 3D face generation model. Jiang et al. [6] proposed a dual-mechanism and end-to-end 3D face alignment framework, significantly improving the face alignment effect. Using an encoder-decoder network to accurately reconstruct faces and disentangle 3D face shape features from a single 2D image is a method proposed by Feng et al. [7]. However, the above papers ignore a key problem. When dealing with large pose, occluded and unconstrained faces in the wild, these methods lack context information and the learning of local geometric features. In order to solve the above problems, in this paper, we propose CRAM, which employs a more efficient approach to capture crucial facial features. This is achieved through the perception of local features, the selection

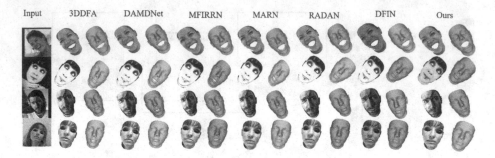

**Fig. 2.** Comparison with qualitative results of other methods. Our method is more prominent in local details such as shape and texture, especially for the mouth.

of vital channels, and multi-level feature interaction to facilitate face reconstruction and alignment. Figure 1 illustrates the workflow of our proposed CRAM. To enhance the model's generalization performance, we propose a regional noise injection and image composition module (RNICM), to assist the network in learning local details. We perform 3D face reconstruction on the AFLW2000-3D dataset, and the reconstruction results is shown in Fig. 2 to demonstrate the effectiveness of our CRAM method. From the figure, it can be observed that the facial shape reconstructed by our CRAM is smoother, the expression is more natural, and the face has finer facial details. The core contributions of our method are summarized as follows:

- In this paper, innovatively, we propose a collaborative reinforcement attention module comprising three components: the perception module (PM), the channel selection module (CSM), and the multi-level feature interaction module (MFIM), which work in coordination for 3D face reconstruction and alignment tasks. The PM focuses on extracting local facial features, while the CSM concentrates on selecting crucial channels to extract discriminative information, thereby reducing computational complexity. The MFIM extracts multi-level semantic information, such as facial texture, in a more refined manner.
- To make our proposed method more suitable for outdoor environments, we introduce an innovative regional noise injection and image composition module (RNICM). This module preprocesses input data by randomly selecting facial regions, cropping, adding Gaussian noise, and embedding noisy images into the original images. This approach enhances the model's generalization performance and its ability to capture local details.
- Extensive experimental results on the AFLW2000-3D and AFLW datasets demonstrate that our method outperforms state-of-the-art approaches.

## 2   Proposed Method

In this section, we introduce the proposed method, including the regional noise injection and image composition module and collaborative reinforcement attention module, the network architecture and loss function for the task.

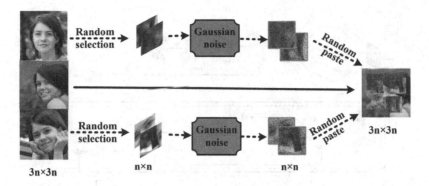

**Fig. 3.** The workflow of our proposed regional noise injection and image composition module.

## 2.1  Regional Noise Injection and Image Composition Module

In this paper, we introduce the regional noise injection and image composition module (RNICM), which innovatively combines noisy images with the original ones to enhance the robustness of 3D face reconstruction and alignment. Assuming that our RNICM takes three images, all of size $3n \times 3n$, as input, we randomly select two non-overlapping facial regions in the first two images, ensuring that these chosen regions contain sufficient facial information. Subsequently, we generate two non-overlapping sub-images, each of size $n \times n$, from these regions. To simulate real-world scenarios, we add moderate Gaussian noise to each of these sub-images. Finally, we paste the four sub-images with Gaussian noise randomly but without overlapping onto the third image. Preprocessing the original data with RNICM enables better adaptation to the captured facial semantic information under varying environmental conditions, resulting in a significant improvement in the performance of 3D face reconstruction and alignment (Fig. 3).

## 2.2  Collaborative Reinforcement Attention Module

**The Perception Module:** This module consists of multiple convolutional layers that extract various levels and angles of local facial features using different convolution operations. To enhance the expressiveness of facial features, we incorporate ReLU activation functions within the perception module (PM), facilitating non-linear transformations. Additionally, by employing the CBAM [8] attention mechanism, we proactively learn the importance of channels and spatial information. This yields the output feature, denoted as out1, which effectively combines global context information and local features. The advantage of the PM lies in its ability to progressively abstract and combine features through operations like convolution and pooling. This leads to the creation of multi-level feature representations, enabling the model to better understand different levels and semantics within the image. It aids in capturing facial details and structures while retaining crucial contextual information. Although the PM excels in

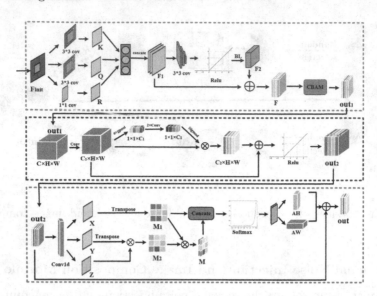

**Fig. 4.** The workflow of our proposed collaborative reinforcement attention module.

extracting local features, it may introduce some computational complexity and parameter overhead during the feature extraction process. In complex scenarios or when handling high-resolution data, it might even lead to the loss of certain details and local information. Hence, we introduce the channel selection module (CSM) to address these limitations and further enhance the performance of our model (Fig. 4).

**The Channel Selection Module:** To comprehensively assess the importance of each channel, we introduce the channel selection module (CSM) to further refine the important channels learned by the PM, thereby reducing computational complexity. Initially, the CSM evaluates the importance of each channel within the input feature out1 through global pooling and average pooling operations. It adaptively learns the significance of each channel, identifying the most critical ones to focus on key facial regions and reduce the impact of redundant information. This assists subsequent modules in utilizing discriminative features from the input image to improve the effectiveness of 3D face reconstruction. However, the CSM has certain limitations when selecting important channels. It often does not consider the interdependencies between channels, potentially overlooking latent essential feature information. Moreover, the CSM lacks the semantic information extraction of facial salient features. Therefore, we further introduce the multi-level feature interaction module (MFIM). This module excels at capturing the interrelationships between features and enhancing the network's understanding of facial semantic information.

**Multi-level Feature Interaction Module:** In order to extract facial features with rich semantic information, we have incorporated the multi-level feature interaction module (MFIM). This module utilizes an attention mechanism to enhance the expressiveness of features, helping compute correlations and weights between different features, and integrating features from different levels to improve the robustness of 3D face reconstruction. We employ the Softmax function to generate attention maps, emphasizing the importance of facial features in both horizontal and vertical directions. These attention maps are then decomposed into horizontal attention maps (AW) and vertical attention maps (AH). Subsequently, AH and AW are added to the input feature out2 to obtain the final output feature out with attention weighting. This allows the network to better understand the semantic information at different levels of the face, capturing facial details, key features, and contextual information. The output feature out combines information from different angles and levels of the face, enabling more refined facial reconstruction.

In summary, the CRAM of these three modules enhances the performance and effectiveness of the face reconstruction model. The PM extracts multi-level feature representations, the CSM optimizes feature selection and weighting, and the MFIM integrates information from different levels. These advantages collectively enable the model to better understand and reconstruct the face, resulting in more realistic and accurate facial images.

$$\bar{F}_{ij} = \frac{\exp\left(\mathbf{Q}_i \cdot \mathbf{K}_j\right)}{\sum_{n=1}^{N'} \exp\left(\mathbf{Q}_i \cdot \mathbf{K}_n\right)} \tag{1}$$

where $F_{ij}$ denotes the degree of correlation between the $i^{th}$ feature vector of Q and the $j^{th}$ feature vector of K, Q and K learned from $F_{init}$.

$$\bar{z}_c = \frac{1}{H \cdot W} \sum_{i=1}^{H} \sum_{j=1}^{W} x_{ijc} \tag{2}$$

The Eq. (2) uses global average pooling to obtain the pooling value for each channel. $z_c$ is the pooled value of channel $c$, $H$ and $W$ are the height and width of the input feature map, $x_{ijc}$ is the value of channel $c$ at position $(i, j)$ of the input feature map.

$$\bar{y}_c = \begin{cases} x_c, & \text{if } c \\ 0, & \text{otherwise} \end{cases} \tag{3}$$

where $y_c$ is the value of the output channel $c$ and $x_c$ is the value of the input channel $c$.

$$\bar{M}_{ij} = \frac{\exp\left(\mathbf{X}_i' \cdot \mathbf{Y}_j'\right)}{\sum_{n=1}^{N} \exp\left(\mathbf{X}_i' \cdot \mathbf{Y}_n'\right)} \tag{4}$$

where $M_{ij}$ denotes the degree of correlation between the $i^{th}$ feature vector of X and $j^{th}$ feature vector of Y, X and Y are both learned from out2.

$$\bar{F}_{\text{fusion}} = \sum_i w_i \cdot F_i \tag{5}$$

where $F_{\text{fusion}}$ is the fused feature, $w_i$ is the attention weight of the $ith$ feature, and $F_i$ is the $ith$ feature.

$$\bar{S}\left(\alpha_{id}, \alpha_{\text{exp}}\right) = \bar{S} + A_{id}\alpha_{id} + A_{exp}\alpha_{\text{exp}} \tag{6}$$

where $S \in R^{3n}$ is the mean geometry, $A_{id}$ and $A_{exp}$ are the basis of face identity and expression, $\alpha_{\text{id}}$ and $\alpha_{\text{exp}}$ are the corresponding face identity parameter and face expression parameter.

## 2.3   Objective Loss Function

In the learning phase, we employ three different loss functions to jointly train our model on the fully labeled data 300W-LP [9]. For the 3DMM parameters, we employ weighted parametric distance cost (WPDC) to optimize our network on the fully labeled dataset.

$$\mathcal{L}_{wpdc} = \left(P_g - \widehat{P}\right)^T W \left(P_g - \widehat{P}\right) \tag{7}$$

where $\widehat{P}$ is the predicted value and $P_g$ is the ground-truth value. The diagonal matrix $W$ contains the weights. Our ultimate goal is to accurately obtain facial information features and geometries. Therefore, we leverage the sparse 2D facial landmarks to further constrain the basic contour of the 3D face.

$$\mathcal{L}_{\text{wing}}\left(\tilde{v}\right) = \begin{cases} T\ln\left(1 + \frac{|\tilde{v}|}{\varepsilon}\right), & \text{if } |\tilde{v}| < T \\ |\tilde{v}| - \tau, & \text{otherwise} \end{cases} \tag{8}$$

where $\tilde{v} = v_g - v_p$ denote the ground-truth $v_g$ and $v_p$ the predicted 3D facial vertices obtained by using the network parameters, respectively. $T$ and $\varepsilon$ represent the log function parameters. $\tau = \mathrm{T} - \mathrm{T}\ln\left(1 + \frac{\mathrm{T}}{\varepsilon}\right)$ is a constant that smoothly links the piecewise-defined linear and nonlinear parts.

In order to better constrain the facial features of the key region, we select 68 landmarks to extract features for face reconstruction, the loss function is:

$$\mathcal{L}_{68} = \left\| \left(f * P_r * R * S_{68} + t\right) - \left(\bar{f} * P_r * \bar{R} * \bar{S}_{68} + \bar{t}\right) \right\|^2 \tag{9}$$

where $f$ is a scale factor, $P_r = \begin{pmatrix} 1 & 0 & 0 \\ 0 & 1 & 1 \end{pmatrix}$ is an orthogonal projection, $R$ is a rotation matrix conditting of 9 parameters, $t$ is a translation vector and $S_{68}$ is face geometry.

The final loss of our network is:

$$\mathcal{L}_{\text{total}} = \beta_{\text{wpdc}}\,\mathcal{L}_{\text{wpdc}} + \beta_{\text{wing}}\,\mathcal{L}_{\text{wing}} + \beta_{68}\mathcal{L}_{68} \tag{10}$$

where $\beta_{wpdc}$, $\beta_{wing}$ and $\beta_{68}$ are parameters that balance the contribution of $\mathcal{L}_{wpdc}$, $\mathcal{L}_{\text{wing}}$ and $\mathcal{L}_{68}$.

# 3    Experiments

## 3.1    Datasets

**AFLW2000-3D:** The AFLW2000-3D [12] dataset is the first 2000 face images from AFLW [10] and extended with 68 landmarks. We use this dataset to evaluate the 3D face alignment of our method.

**AFLW:** The AFLW [10] face dataset is a large-scale dataset with multi-pose and multi-view, and each face is labeled with 21 landmarks. We use part of the extreme pose face images in the AFLW [10] dataset for qualitative experiments.

**Helen:** To demonstrate the generalization ability of the model, we employ Helen [11] dataset for testing.

## 3.2    Implementation Details

During training, we use SGD to optimize our network. For network parameters, we empirically set the initial learning rate to 0.02 and the batch size to 256. Our CRAM is trained on three NVIDIA GeForce RTX 3090 with CUDA 11.4 and cuDNN 9.0.1, and runs the training for a total of 50 epochs. Our network is implemented on Pytorch deep learning framework and we empirically set the loss weights to $\beta_{wpdc} = 0.5$ and $\beta_{wing} = 1$. For the network architecture, we adopt backbone network with four ASPP convolutional layers.

## 3.3    3D Face Alignment and Reconstruction

We compare the proposed method with a variety of advanced face reconstruction and alignment methods and find that our method outperforms other advanced methods in terms of performance and effect.

**3D Face Alignment:** To evaluate the face alignment performance, we adopt the normalized mean error (NME) as the evaluation metric and restrict the box size as the normalization factor. We evaluate the dataset of face alignment performance using sparse set of face landmarks on AFLW2000-3D [12] and AFLW [10]. Figure 5 shows the qualitative results comparison between our method and other methods on AFLW2000-3D [12], and it is found that our collaborative reinforcement attention module (CRAM) significantly outperforms MARN [15], MFIRRN [16], RADAN [23] and DFIN [24] in 3D face alignment, especially large pose in the wild, occlusion, and extreme expressions. The experimental results fully show that our method overcomes the problem of 3D face alignment with inaccurate regression parameters of traditional CNN.

**Fig. 5.** Comparison of 3D facial landmark detection with 3DDFA [12], DAMDNet [13], MARN [15], MFRRN [16], RADAN [23], DFIN [24] and CRAM (Ours) on AFLW2000-3D.

Table 1 lists the NME(%) of the facial alignment results between our method and other methods, and Fig. 6 shows the CED curves of our method on AFLW2000-3D [12] and AFLW [10] datasets, respectively. In Table 1, we show quantitative results for three different yaw angles, namely $[0°, 30°]$, $[30°, 60°]$ and $[60°, 90°]$. Significance results achieved by our method in terms of mean, standard deviation or different navigation angles compared to other state-of-the-art face alignment methods. Our method on AFLW2000-3D [12] and AFLW [10] datasets achieves the best mean and standard deviation results. Experiments fully proves the generalization performance of the model.

**3D Face Reconstruction:** In this section, we evaluate our collaborative reinforcement attention module (CRAM) for 3D face reconstruction on the AFLW2000-3D [12] dataset by comparing it with MARN [15], MFIRRN [16], RADAN [23] and DFIN [24]. We perform a comparison of a 3D normalized mean error on the AFLW2000-3D [12] dataset to demonstrate the validity of our model in 3D face reconstruction. Table 2 shows the quantitative comparison results, the best result in each category is highlighted in bold. The lower it is, the better. Figure 7 shows our method against a state-of-art CED curve. This depends on our proposed CRAM that can effectively learn discriminative features and the regional noise injection and image composition module (RNICM) can improve the generalization ability of the model.

**Table 1.** The NME(%) of face alignment results on AFLW and AFLW2000-3D.

| Method | AFLW2000-3D Dataset (68 pts) | | | | | AFLW Dataset (21 pts) | | | | |
|---|---|---|---|---|---|---|---|---|---|---|
| | [0, 30] | [30, 60] | [60, 90] | Mean | Std | [0, 30] | [30, 60] | [60, 90] | Mean | Std |
| CDM [17] | – | – | – | – | – | 8.150 | 13.020 | 16.170 | 12.440 | 4.040 |
| RCPR [14] | 4.260 | 5.960 | 13.180 | 7.800 | 4.740 | 5.430 | 6.580 | 11.530 | 7.850 | 3.240 |
| ESR [18] | 4.600 | 6.700 | 12.670 | 7.990 | 4.190 | 5.660 | 7.120 | 11.940 | 8.240 | 3.290 |
| SDM [19] | 3.670 | 4.940 | 9.760 | 6.120 | 3.210 | 4.750 | 5.550 | 9.340 | 6.550 | 2.450 |
| DEFA [2] | 4.500 | 5.560 | 7.330 | 5.803 | 1.169 | – | – | – | – | – |
| 3DDFA (CVPR2016) [9] | 3.780 | 4.540 | 7.930 | 5.420 | 2.210 | 5.000 | 5.060 | 6.740 | 5.600 | 0.990 |
| Yu et al. (ICCV2017) [20] | 3.620 | 6.060 | 9.560 | – | – | 5.940 | 6.480 | 7.960 | – | – |
| Nonlinear (CVPR2018) [21] | – | – | – | 4.700 | – | – | – | – | – | – |
| DAMDNet (ICCVW2019) [13] | 2.907 | 3.830 | 4.953 | 3.897 | 0.837 | 4.359 | 5.209 | 6.028 | 5.199 | 0.682 |
| GSRN (MMM2021) [22] | 2.842 | 3.789 | 4.804 | 3.812 | 0.801 | 4.253 | 5.144 | 5.816 | 5.073 | 0.638 |
| MARN (ICPR2021) [15] | 2.989 | 3.670 | 4.613 | 3.757 | **0.666** | 4.306 | 4.965 | 5.775 | 5.015 | 0.601 |
| MFIRRN (ICASSP2021) [16] | 2.841 | 3.572 | 4.561 | 3.658 | 0.705 | 4.321 | 5.051 | 5.958 | 5.110 | 0.670 |
| RADAN (FG) [23] | 2.792 | 3.583 | 4.495 | 3.623 | 0.696 | 4.129 | 4.888 | 5.495 | 4.837 | 0.559 |
| DFIN (IEEE2023) [24] | 2.855 | 3.524 | 4.484 | 3.621 | 0.669 | 4.118 | 4.735 | 5.577 | 4.810 | 0.598 |
| **CRAM (Ours)** | **2.632** | **3.414** | **4.435** | **3.494** | 0.738 | **4.033** | **4.699** | **5.248** | **4.660** | **0.496** |

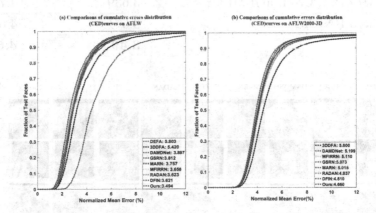

**Fig. 6.** The cumulative errors distribution (CED) curves on AFLW and AFLW2000-3D.

**Table 2.** The NME(%) of face reconstruction results on AFLW2000-3D.

| NME | 3DDFA | DAMDNet | MFIRRN | MARN | RADAN | DFIN | CRAM (Ours) |
|---|---|---|---|---|---|---|---|
| [0, 30] | 4.877 | 4.672 | 4.760 | 4.721 | 4.497 | 4.667 | **4.449** |
| [30, 60] | 6.086 | 5.619 | 5.488 | 5.535 | 5.196 | 5.197 | **5.039** |
| [60, 90] | 8.437 | 7.855 | 7.594 | 7.483 | 6.983 | **7.140** | 7.176 |
| Mean | 6.467 | 6.049 | 5.947 | 5.913 | 5.611 | 5.616 | **5.555** |
| Std | 1.478 | 1.334 | 1.196 | 1.159 | 0.991 | **1.118** | 1.172 |

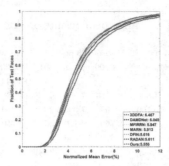

**Fig. 7.** Comparisons of cumulative errors distribution (CED) curves for 3D Face Reconstruction on AFLW2000-3D.

**Table 3.** Ablation study. We investigate the impact of our proposed different modules on the model.

| Model | $AFLW2000-3D$ | $AFLW$ |
|---|---|---|
| CRAM w/o PM CSM, MFIM, and RNICM | 3.941 | 5.303 |
| CRAM w/o PM, CSM, and MFIM | 3.639 | 4.730 |
| CRAM w/o RNICM | 3.571 | 4.691 |
| CRAM | **3.494** | **4.660** |

**Fig. 8.** 3D dense face alignment results on Helen [11]. Our proposed method outperforms the other methods.

## 3.4   Ablation Study

To demonstrate the validity of our proposed method in different modules, we evaluated normalized mean error (NME) performance on both AFLW2000-3D [12] and AFLW [10] datasets. We focus on the impact of the CRAM. Table 3 presents the comparison of using different modular methods on AFLW2000-3D [12] and AFLW [10] datasets. We find that each module is effective.

### 3.5  Evaluation of Generalization Performance

We conduct face alignment experiments on the Helen [11] dataset to demonstrate the generalization performance of our method and compare our CRAM with 3DDFA [12], DAMDNet [13], MARN [15], MFRRN [16], RADAN [23], DFIN [24] and CRAM (Ours). Figure 8 illustrates the qualitative analysis results. From the results, we can see that our CRAM outperforms the other three methods. Moreover, it has obvious advantages on the face contour and eyebrows, which further proves the effectiveness of our proposed method.

## 4  Conclusion

In this paper, we propose collaborative reinforcement attention module (CRAM) for 3D dense alignment and face reconstruction. The CRAM can capture context information and extract discriminative features from the large pose in the wild, which solves the problem of long-distance dependence. To improve the generalization performance and local details of the model, we propose the regional noise injection and image composition module (RNICM). Our method solves the problem of inaccurate 3D face alignment of traditional CNN regression parameters for large pose in the wild, extreme facial expressions, and occlusions. Extensive experiments on AFLW [10] and AFLW2000-3D [12] datasets demonstrate the effectiveness of our method.

**Acknowledgments.** This research is supported by the Ningxia Natural Science Foundation of China under Grant 2022AAC03327.

**Data Availability Statement.** The datasets generated during and or analysed during the current study are available in the https://github.com/Cristianomeko/CRAM.

**Statements and Declarations**
**Conflict of Interests.** None.

## References

1. Blanz, V., Vetter, T.: A morphable model for the synthesis of 3D faces. In: Proceedings of the 26th Annual Conference on Computer Graphics and Interactive Techniques (1999)
2. Liu, Y., et al.: Dense face alignment. In: Proceedings of the IEEE International Conference on Computer Vision Workshops (2017)
3. Jourabloo, A., Liu, X.: Large-pose face alignment via CNN-based dense 3D model fitting. In: Proceedings of the IEEE Conference on Computer Vision and Pattern Recognition (2016)

4. Lei, B., et al.: A hierarchical representation network for accurate and detailed face reconstruction from in-the-wild images. In: Proceedings of the IEEE/CVF Conference on Computer Vision and Pattern Recognition (2023)
5. Taherkhani, F., et al.: Controllable 3D generative adversarial face model via disentangling shape and appearance. In: Proceedings of the IEEE/CVF Winter Conference on Applications of Computer Vision (2023)
6. Jiang, L., Wu, X.-J., Kittler, J.: Dual attention MobDenseNet (DAMDNet) for robust 3D face alignment. In: Proceedings of the IEEE International Conference on Computer Vision Workshops (2019)
7. Liu, F., Zhu, R., Zeng, D., Zhao, Q., Liu, X.: Disentangling features in 3D face shapes for joint face reconstruction and recognition. In: Proceedings of the IEEE Conference on Computer Vision and Pattern Recognition, pp. 5216–5225 (2018)
8. Woo, S., Park, J., Lee, J.-Y., Kweon, I.S.: CBAM: convolutional block attention module. In: Ferrari, V., Hebert, M., Sminchisescu, C., Weiss, Y. (eds.) ECCV 2018. LNCS, vol. 11211, pp. 3–19. Springer, Cham (2018). https://doi.org/10.1007/978-3-030-01234-2_1
9. Zhu, X., et al.: Face alignment across large poses: a 3D solution. In: Proceedings of the IEEE Conference on Computer Vision and Pattern Recognition (2016)
10. Koestinger, M., et al.: Annotated facial landmarks in the wild: a large-scale, real-world database for facial landmark localization. In: 2011 IEEE International Conference on Computer Vision Workshops (ICCV Workshops). IEEE (2011)
11. Le, V., Brandt, J., Lin, Z., Bourdev, L., Huang, T.S.: Interactive facial feature localization. In: Fitzgibbon, A., Lazebnik, S., Perona, P., Sato, Y., Schmid, C. (eds.) ECCV 2012. LNCS, vol. 7574, pp. 679–692. Springer, Heidelberg (2012). https://doi.org/10.1007/978-3-642-33712-3_49
12. Zhu, X., et al.: Face alignment across large poses: a 3D solution. In: Proceedings of the IEEE Conference on Computer Vision and Pattern Recognition (2016)
13. Jiang, L., Wu, X.-J., Kittler, J.: Dual attention MobDenseNet (DAMDNet) for robust 3D face alignment. In: Proceedings of the IEEE International Conference on Computer Vision Workshops (2019)
14. Burgos-Artizzu, X.P., Perona, P., Dollár, P.: Robust face landmark estimation under occlusion. In: Proceedings of the IEEE International Conference on Computer Vision (2013)
15. Li, L., Li, X., Wu, K., Lin, K., Wu, S.: Multi-granularity feature interaction and relation reasoning for 3D dense alignment and face reconstruction. In: ICASSP 2021–2021 IEEE International Conference on Acoustics, Speech and Signal Processing (ICASSP), pp. 4265–4269. IEEE (2021)
16. Li, X., Wu, S.: Multi-attribute regression network for face reconstruction. In: 2020 25th International Conference on Pattern Recognition (ICPR), pp. 7226–7233. IEEE (2021)
17. Yu, X., et al.: Face landmark fitting via optimized part mixtures and cascaded deformable model. IEEE Trans. Pattern Anal. Mach. Intell. **38**(11), 2212–2226 (2015)
18. Cao, X., et al.: Face alignment by explicit shape regression. Int. J. Comput. Vis. **107**(2), 177–190 (2014). https://doi.org/10.1007/s11263-013-0667-3
19. Yan, J., et al.: Learn to combine multiple hypotheses for accurate face alignment. In: Proceedings of the IEEE International Conference on Computer Vision Workshops (2013)
20. Yu, R., et al.: Learning dense facial correspondences in unconstrained images. In: Proceedings of the IEEE International Conference on Computer Vision (2017)

21. Tran, L., Liu, X.: Nonlinear 3D face morphable model. In: Proceedings of the IEEE Conference on Computer Vision and Pattern Recognition (2018)
22. Wang, X., Li, X., Wu, S.: Graph structure reasoning network for face alignment and reconstruction. In: Lokoč, J., et al. (eds.) MMM 2021. LNCS, vol. 12572, pp. 493–505. Springer, Cham (2021). https://doi.org/10.1007/978-3-030-67832-6_40
23. Zhou, Z., Li, L., Wu, S.: Replay attention and data augmentation network for 3D dense alignment and face reconstruction. In: Proceedings of the IEEE International Conference on Automatic Face and Gesture Recognition, pp. 1–8 (2021)
24. Deng, J., et al.: Deformable feature interaction network and graph structure reasoning for 3D dense alignment and face reconstruction. In: 2023 International Joint Conference on Neural Networks (IJCNN). IEEE (2023)

# Exploring Multi-modal Fusion for Image Manipulation Detection and Localization

Konstantinos Triaridis<sup>(✉)</sup>⬤ and Vasileios Mezaris⬤

Information Technologies Institute, Centre for Research & Technology Hellas,
Thessaloniki, Greece
{triaridis,bmezaris}@iti.gr

**Abstract.** Recent image manipulation localization and detection techniques usually leverage forensic artifacts and traces that are produced by a noise-sensitive filter, such as SRM and Bayar convolution. In this paper, we showcase that different filters commonly used in such approaches excel at unveiling different types of manipulations and provide complementary forensic traces. Thus, we explore ways of merging the outputs of such filters and aim to leverage the complementary nature of the artifacts produced to perform image manipulation localization and detection (IMLD). We propose two distinct methods: one that produces independent features from each forensic filter and then fuses them (this is referred to as late fusion) and one that performs early mixing of different modal outputs and produces early combined features (this is referred to as early fusion). We demonstrate that both approaches achieve competitive performance for both image manipulation localization and detection, outperforming state-of-the-art models across several datasets (Code is publicly available at https://github.com/IDT-ITI/MMFusion-IML).

**Keywords:** Image forensics · Image manipulation localization · Image manipulation detection · Multi-modal fusion

## 1 Introduction

Editing and manipulating digital media has gotten increasingly easier and more accessible in recent years. Recent advances in image editing software, as well as deep generative models such as Generative Adversarial Networks (GANs) [15,36] and diffusion models [20,32], facilitate producing manipulations that are often imperceptible to the human eye and are widely available, even to potentially malicious users. The widespread use of smartphones and social networks also enables the spread of such manipulated media at a rapid pace. As a result, such edited images can cause social problems when used as evidence to support disinformation campaigns and stories or mislead the public by obfuscating important content from news, resulting in diminished trust. Therefore, techniques for image manipulation detection and localization, as part of complete toolboxes for media verification such as [24], are now needed more than ever.

© The Author(s), under exclusive license to Springer Nature Switzerland AG 2024
S. Rudinac et al. (Eds.): MMM 2024, LNCS 14556, pp. 198–211, 2024.
https://doi.org/10.1007/978-3-031-53311-2_15

Image forgery localization and detection are tasks the media forensics field has been working on for many years. Early works typically focused on a specific type of manipulation such as splicing [23], copy-move [3] or removal/inpainting [17]. More recently, deep-learning-based solutions of increasing robustness are proposed that are able to recognize multiple different types of manipulations [2,9,11,14,18,26,29,33]. In order to be able to perform manipulation localization in a semantic-agnostic manner these models need to suppress image contect to reveal forensic artifacts. Most approaches achieve this by applying a high-pass filter to extract noise maps [2,11,26,29,33]. The most popular high-pass filters used are the ones proposed in the Steganalysis Rich Model (SRM) [8], utilized in a wide variety of works [11,16,19,29,37], while the Bayar convolution [1] is also used in a multitude of approaches [2,11,29,33] and NoisePrint++ is used in a more recent model [9].

We hypothesize that those different forensic filters actually produce artifacts of complementary forensic capabilities. NoisePrint [4], and its successor NoisePrint++ [9] produce artifacts that relate to camera model and editing history, thus displaying limited performance for copy-move images (Sect. 4.3). On the other hand, SRM [8] filters can identify edges and boundaries without relying on camera or compression/editing artifacts, but their predetermined nature makes them vulnerable to adversarial attacks, whereas the Bayar convolution [1] learns the manipulation traces directly from data, proving more robust against malicious attacks. In this work we explore ways to expand existing state-of-the-art IMLD approaches to support multiple auxiliary forensic filters as inputs. We start with TruFor [9] as our baseline and propose utilizing NoisePrint++, SRM, and Bayar convolution as inputs auxiliary to the RGB image. We propose two different approaches: a late-fusion paradigm that extracts features from each modality separately, and an early-fusion paradigm that mixes the multi-modal features by early convolutional blocks. Our main contributions in this paper are summarized as follows:

- We compare the efficacy of different forensic filters, namely SRM, Bayar conv and NoisePrint, as inputs for deep networks performing forgery localization.
- We propose two distinct approaches for combining the outputs of different forensic filters for the purpose of image manipulation localization and detection.
- Both methods achieve state-of-the-art performance across five datasets by effectively leveraging and combining forensic cues from various input modalities.

## 2   Related Work

Image forensics methods have been based for a long time on detecting inconsistensies on low-level semantic-agnostic artifacts such as compression or internal camera filter artifacts. These artifacts are usually revealed by suppressing the image content through high-pass filtering, producing a noise-sensitive view. In recent times, various filters for noise extraction have been integrated into deep learning models to address the challenge of image manipulation localization. RGB-N [37] is a model that uses both RGB images and SRM filters together with a faster R-CNN [22] module to perform forgery detection with a bounding box, while Constrained R-CNN [33] uses a trainable noise extractor, namely Bayar convolution, to perform the same task. ManTraNet [29] integrates both SRM filters and Bayar convolution within a VGG-based architecture, while SPAN [11] enhances this approach by modeling relationships between image patches through a pyramid of local self-attention blocks. Chen et al. [2] also use the Bayar convolution together with the RGB image in a late fusion paradigm that is trained through multi-scale supervision. NoisePrint [4] is a noise extractor proposed by Cozzolino et al. that is trained in a self-supervised manner to extract camera-specific artifacts and is expanded in TruFor [9], where it is used jointly with RGB images in a dual-branch CMX [34] architecture.

Our approach innovatively explores various strategies for combining the outputs of diverse noise extractors, leveraging their complementary capabilities to develop a robust end-to-end image forgery detection localization model.

## 3   Methods

### 3.1   Encoder-Decoder Architecture

Our goal is to extend existing encoder-decoder-based architectures to be able to use multiple forensic filters (SRM [8], Bayar convolution [1], NoisePrint++ [9]) in tandem, so as to produce more robust representations for the task of Image Manipulation Localization and Detection. To this end we adopt the TruFor [9] architecture, showcased in Fig. 1, that consists of an encoder, an anomaly decoder, a confidence decoder, and a forgery detector; and we follow its two-phase training regime for anomaly localization and detection, respectively. The encoder follows the dual-branch architecture proposed in [34], comprising of 4 stages of Multi-Head Self Attention (MHSA) blocks [31] that produce feature maps $f_r^i$ of different scales: $\frac{H}{2^{i+1}} \times \frac{W}{2^{i+1}} \times C_i$, where $H$ and $W$ are the spatial dimensions of the input and $C_i$ is the channel dimension of the output at scale $i$. The two MHSA blocks' outputs in each stage are rectified through a Cross-Modal Feature Rectification Module (FRM) [34] that exploits the interactions between the two input modalities (RGB and NoisePrint in the case of TruFor). The FRM uses features from both modalities to produce weighted channel- and spatial-wise feature maps that are residually added for both modalities to perform channel- and spatial-wise rectification. The two sets of feature maps are then combined using a Feature Fusion Module (FFM) [34] whose outputs $f^i$ at each scale are combined to produce the encoder output $f$. The FFM consists of an information exchange stage, where a cross-attention mechanism exchanges information between modalities and produces two sets of mixed feature maps, and a fusion stage where the feature maps are merged into a single output through a residual MLP module that uses $1 \times 1$ convolutions. The Decoders are simple MLP decoders proposed in [31].

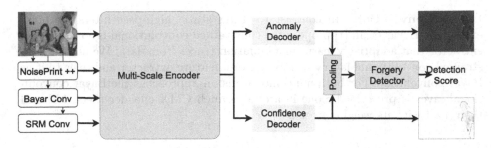

**Fig. 1.** Full encoder-decoder architecture

Utilizing this architecture one can combine RGB images with an auxiliary forensic modality to perform Image Manipulation Localization. In [9] Guillaro et al. use their own feature extractor NoisePrint++, however a multitude of other forensic filters' outputs like Bayar convolution [1] or SRM [8] can be utilized. All those filters are analyzed in Sect. 3.2. We propose two different ways of extending the encoder architecture to multiple auxiliary modal inputs: a late fusion paradigm, where each auxiliary modality is combined with RGB inputs separately using a dual-branch architecture [34] (Sect. 3.3), and an early fusion paradigm where auxiliary modalities are combined early before being utilized as input to the dual-branch encoder together with the RGB inputs (Sect. 3.4).

## 3.2  Auxiliary Modalities

For both approaches, we use the outputs of three forensic filters: NoisePrint++, SRM, and Bayar convolution as inputs together with RGB images.

**NoisePrint++.** In [4] Cozzolino et al. propose Noiseprint, a CNN-based model designed to extract camera-model-based artifacts from RGB images while suppressing image content. In [9] they expand their approach, namely NoisePrint++, to be able to recognize and extract artifacts related to the editing history of an image (e.g. compression, resizing, gamma correction). NoisePrint++ is trained in a supervised contrastive manner [12]: a batch of images is provided, from which patches are extracted from different locations. Then the patches go through different editing pipelines. Patches extracted from the same source image, the same location, and with the same editing history are considered positive samples, while others are considered negative. For our approach we use NoisePrint++ as a pretrained feature extractor.

**SRM.** Another way to suppress the image content and highlight forensic traces and noise is through static high-pass filters, the most common of which are the ones proposed for producing residual maps for the Steganalysis Rich Model (SRM) [8]. Out of the 30 high-pass filters proposed, we used the 3 most commonly used in the literature, e.g. in [11,29,37], which are displayed in Fig. 4 of [37]. They will be referred to as SRM filters.

**Bayar Convolution.** In contrast to using static high-pass filters for noise extraction Bayar et al. [1] propose the constrained convolutional layer as a noise extractor that adaptively learns manipulation traces from data. We use the constrained convolutional layer as an extra noise feature extractor and refer to it as Bayar convolution. For both multi-modal fusion approaches the Bayar convolutional layer is pretrained alone in a dual branch CMX encoder (Sect. 4.3) and then used with its weights frozen.

## 3.3    Late Fusion

For the late fusion method we extract the auxiliary representations $r_{noiseprint}$, $r_{srm}$, $r_{bayar}$ of the RGB image $x$ from the NoisePrint++, SRM and Bayar filters respectively. Then the output of each auxiliary filter is fed together with the original RGB input into a dual-branch CMX encoder to produce 4-scale feature maps $f^i_{mod} = \mathcal{E}_{mod}(x, r_{mod}), mod \in \{noiseprint, srm, bayar\}, i \in \{1, 2, 3, 4\}$ as shown in Fig. 2.

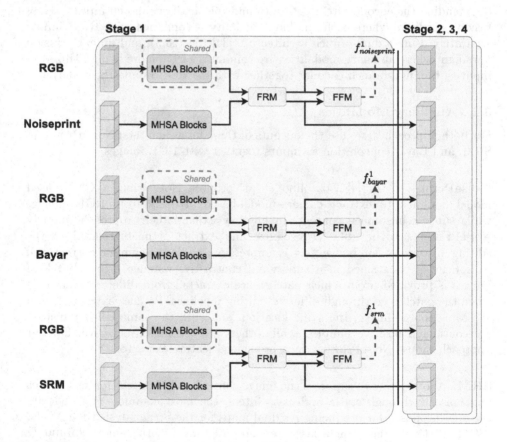

**Fig. 2.** Late Fusion with weight sharing

At each scale the outputs of the 3 encoders are concatenated to produce the final output $f$ of the encoder. We use the same decoder architecture as in TruFor for the anomaly and confidence decoders. Like other multi-modal approaches this approach is prone to overfitting and the "modality imbalance" problem [7,27], where different modalities converge and overfit at different rates, thus hindering joint optimization. To tackle this we make the weights of the modules along the RGB branch shared across all 3 encoders to increase regularization. We also employ Dropout before the anomaly decoder as the complete encoder is rather large and the simple MLP decoder is prone to overfitting.

### 3.4 Fusion by Early Convolutions

For the early fusion method (Fig. 3) we again extract the same auxiliary representations: $r_{noiseprint}, r_{srm}, r_{bayar}$ of the RGB image $x$. The inputs are then passed through our novel early fusion module $\mathcal{F}_e$ to produce the auxiliary features $f^a = \mathcal{F}_e(r_{noiseprint}, r_{srm}, r_{bayar})$. The early fusion module consists of 3 independent convolutional blocks, one for each auxiliary modality, and one final convolutional block that performs feature mixing. The convolutional blocks are good at early visual processing, resulting in a more stable optimization [30], thus aiding in mixing the features from different modalities smoothly. The mixed features $f^a$ and RGB image $x$ are used as input for a dual-branch CMX encoder [34], in the same manner as in TruFor. This is a particularly lightweight approach to expanding the TruFor architecture to handle multiple auxiliary modalities as it does not increase the number of parameters significantly (68.9M params compared to TruFor's 68.7M).

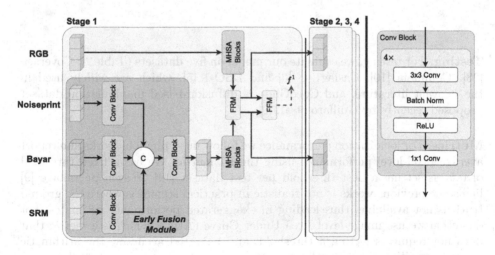

**Fig. 3.** Fusion by early convolutions

**Convolutional Block.** The convolutional block consists of four $3 \times 3$ convolutions followed by a $1 \times 1$ convolutional layer to resize the output to 3 channels. There is a batch normalization (BN) and a ReLU layer after each $3 \times 3$ convolutional layer. The output channels for the $3 \times 3$ convolutional layers are [24, 48, 96, 192].

## 4    Experiments

### 4.1    Experimental Setup

**Training.** We follow the training procedure proposed by Guillaro et al. [9]: first, we jointly train the encoder and anomaly decoder and finally, we train the confidence decoder and the forgery detector, while the encoder and anomaly decoder are kept frozen. For both training phases we use the datasets used by Kwon et al. [14], and sample an equal number of images from each one for every epoch. Training datasets are summarized in Table 1.

**Table 1.** Details for training datasets

| Dataset | Number of Images | |
|---|---|---|
| | Real | Fake |
| Casiav2 [6] | 7,491 | 5,105 |
| IMD2020 [21] | 414 | 2,010 |
| FantasticReality [13] | 16,592 | 19,423 |
| cm_coco [14] | – | 200,000 |
| bcm_coco [14] | – | 200,000 |
| bcmc_coco [14] | – | 200,000 |
| sp_coco [14] | – | 200,000 |

**Table 2.** Details for testing datasets

| Dataset | Number of Images | |
|---|---|---|
| | Real | Fake |
| Coverage [28] | 100 | 100 |
| Columbia [10] | 183 | 180 |
| Casiav1+ [6] | 800 | 921 |
| DSO-1 [5] | 100 | 100 |
| CocoGlide [9] | 512 | 512 |

**Testing.** For testing, we evaluate our model on five datasets (Table 2): Coverage [28], Columbia [10], Casiav1+[1] [6] and DSO-1 [5], which are widely used in the relevant literature, and CocoGlide, a diffusion-based manipulation dataset proposed recently by Guillaro et al. [9].

**Metrics.** For localization performance we follow most previous works and report average pixel-level performance using the F1 metric. We use a fixed threshold of 0.5, as setting a best threshold per test dataset [14] or even per image [9] like some previous works is not realistic in practical scenarios where the ground truth is not available, thus leading in exaggerated performance estimates. For detection we use image-level Area Under Curve (AUC), which is a metric that does not require selecting a threshold, and balanced accuracy, the arithmetic mean of sensitivity and specificity, with a threshold once again set to 0.5.

---

[1] Casiav1+ is a modification of the Casiav1 dataset proposed by Chen et al. [2] that replaces authentic images that also exist in Casiav2 with images from the COREL [25] dataset to avoid data contamination.

**Implementation.** All models are implemented in PyTorch and trained on an NVIDIA RTX 4090 or 3090 GPU, using an effective batch size of 24 for 100 epochs. Physical batch size ranged from 4 to 8 depending on the model and an effective batch size of 24 was reached by utilizing gradient accumulation. We use a Dropout rate of 0.3 for both multi-modal methods. The MHSA modules were initialized with ImageNet-pretrained weights as proposed in [34,35]. We utilized an SGD optimizer with an initial learning rate of 0.005, momentum of 0.9, weight decay of 0.0005 and a polynomial learning rate schedule. For training augmentations we followed Guillaro et al. [9] and resized the images in the [0.5–1.5] range, performed random cropping of size 512 × 512 and JPEG compression with a random Quality Factor QF ∈ [30,100].

## 4.2   Comparisons

We compare our methods with recent approaches for Image Manipulation Localization. Following Guillaro et al. we consider methods with open source models provided and we exclude models that use part of our testing datasets for training to avoid bias. Overall we compare with TruFor [9], CAT-Netv2 [14], ManTraNet [29], PSCC-Net [18], SPAN [11], Constrained R-CNN [33], MVSS-Net [2]. Results are presented in Table 3.

**Table 3.** Comparison for localization performance. The metric is average pixel-level F1. The best and second-best results for each dataset are presented in bold and underlined respectively. Results for all models except for the proposed ones are taken from [9].

| Model | Coverage | Columbia | Casiav1+ | CocoGlide | DSO-1 | AVG |
|---|---|---|---|---|---|---|
| TruFor | .600 | .859 | .737 | .523 | **.930** | .729 |
| CAT-Netv2 | .381 | .859 | .752 | .434 | .584 | .602 |
| ManTraNet | .317 | .508 | .180 | .516 | .412 | .387 |
| PSCC-Net | .473 | .604 | .520 | .515 | .458 | .514 |
| SPAN | .235 | .759 | .112 | .298 | .233 | .327 |
| CR-CNN | .391 | .631 | .481 | .447 | .289 | .448 |
| MVSS-Net | .514 | .729 | .528 | .486 | .358 | .523 |
| Early Fusion | **.663** | **.888** | **.784** | <u>.553</u> | .863 | <u>.750</u> |
| Late Fusion | <u>.641</u> | <u>.864</u> | <u>.775</u> | **.574** | <u>.899</u> | **.751** |

During our experiments, we replicated the training of TruFor for the purposes of our ablation study and we discovered a large variance in Localization results between training runs. For this purpose, we train our networks 4 times and report average localization performance in terms of average pixel F1 in Table 4. Both our multi-modal fusion approaches showcase state-of-the-art performance, being either the best or second-best model for every dataset. Especially for the Coverage dataset that contains only copy-move forgeries, our best approach surpasses the previous best, TruFor, by 6.3%. The only dataset where we can't achieve state-of-the-art performance is DSO-1 where our best method is 3% behind Tru-For.

**Table 4.** Comparison for localization performance for models with multiple training runs. Metric is average pixel-level F1 (± standard deviation)

| Model | Coverage | Columbia | Casiav1+ | CocoGlide | DSO-1 | AVG |
|---|---|---|---|---|---|---|
| TruFor (retrained) | .577(±.019) | .884(±.019) | .761(±.011) | .516(±.008) | .895(±.017) | .726(±.008) |
| Early Fusion | **.663**(±.011) | **.888**(±.014) | **.784**(±.001) | .553(±.015) | .863(±.025) | .750(±.005) |
| Late Fusion | .641(±.014) | .864(±.023) | .775(±.008) | **.574**(±.020) | **.899**(±.010) | **.751**(±.003) |

We also compare across models in terms of detection performance and present the results in Table 5. Notably, our early fusion method demonstrates exceptional performance, surpassing the state-of-the-art on average. Particularly noteworthy is its outstanding performance on the Coverage dataset, where it achieves a remarkable improvement of nearly 7% in terms of the Area Under the Curve (AUC) and 9% in terms of balanced accuracy (bAcc) compared to the prior leading method. Our late fusion approach also exhibits competitive AUC performance, but falls slightly behind the TruFor model in terms of bAcc. This disparity in bAcc performance could potentially be attributed to the size of our late fusion model, which may be susceptible to overfitting. Further investigation and experimentation are warranted to explore the possibility of requiring additional regularization techniques to optimize its performance for the detection task.

**Table 5.** Comparison for detection performance. Metrics are Area Under Curve (AUC) and balanced accuracy (bAcc).

| Model | Coverage | | Columbia | | Casiav1+ | | CocoGlide | | DSO-1 | | AVG | |
|---|---|---|---|---|---|---|---|---|---|---|---|---|
| | AUC | bAcc | AUC | bAcc | AUC | bAcc | AUC | bAcc | AUC | bAcc | AUC | bAcc |
| TruFor | .770 | .680 | .996 | **.984** | .916 | .813 | .752 | .639 | **.984** | .930 | .884 | .809 |
| CAT-Netv2 | .680 | .635 | .977 | .803 | **.942** | .838 | .667 | .580 | .747 | .525 | .803 | .676 |
| ManTraNet | .760 | .500 | .810 | .500 | .644 | .500 | **.778** | .500 | .874 | .500 | .773 | .500 |
| PSCC-Net | .657 | .473 | .300 | .604 | .869 | .520 | .777 | .515 | .650 | .458 | .651 | .514 |
| SPAN | .670 | .235 | **.999** | .759 | .480 | .112 | .475 | .298 | .669 | .233 | .659 | .327 |
| CR-CNN | .553 | .391 | .755 | .631 | .670 | .481 | .589 | .447 | .576 | .289 | .629 | .448 |
| MVSS-Net | .733 | .514 | .984 | .729 | .932 | .528 | .654 | .117 | .552 | .358 | .771 | .449 |
| Early Fusion | **.839** | **.770** | .996 | .962 | .929 | .845 | .755 | .660 | .966 | **.935** | .897 | **.834** |
| Late Fusion | .792 | .720 | .977 | .822 | .930 | **.860** | .760 | **.677** | .958 | .830 | .884 | .782 |

### 4.3   Ablation Study

In this section, for the purpose of contrasting various forensic filters (SRM, Bayar conv, NoisePrint++), we employ a dual-branch CMX architecture where each filter serves as an auxiliary input alongside the RGB image. The outcomes are presented in Table 6, along with the number of parameters (in millions) and runtime (for a single image on an RTX 3090 GPU) for all methods. During this training the Bayar convolutional layer is trainable, while SRM and NoisePrint are kept frozen. We can see that NoisePrint++'s editing history based training helps achieve the best performance on DSO-1, where manipulations are covered using post-processing operations, while SRM and Bayar perform better in CocoGlide and Coverage. Coverage contains only copy-move manipulations for which NoisePrint's camera model identification might not provide robust enough forensic traces, whereas CocoGlide's manipulations are diffusion-based inpaintings potentially resulting in distinct artifacts that diverge from conventional editing histories. Consequently, NoisePrint encounters difficulties in effectively handling such cases. We also compare all methods that use a single forensic filter to our multi-modal fusion approaches and we can see that both the early- and late-fusion paradigms effectively combine the forensic traces provided by the filters, resulting in increased performance. To substantiate our rationale for introducing shared weights between RGB branches in order to enhance regularization within the late fusion paradigm, we additionally evaluate a variation of our method that does not employ weight sharing, and we observe that weight-sharing does contribute to improved performance.

**Table 6.** Ablation study. Localization results in avg pixel F1. Parameter count in Millions. Runtime in milliseconds for a single image on an RTX 3090 GPU.

| Version | Coverage | Columbia | Casiav1+ | CocoGlide | DSO-1 | AVG | Params(M) | Runtime(ms) |
|---|---|---|---|---|---|---|---|---|
| CMX (RGB+NP++) | .577 | .884 | .761 | .516 | .895 | .726 | 68.3 | 73.5 |
| CMX (RGB+Bayar) | .592 | .872 | .774 | .566 | .776 | .716 | 68.1 | 60.2 |
| CMX (RGB+SRM) | .630 | .834 | **.791** | **.585** | .792 | .726 | 68.1 | 59.6 |
| Late Fusion (No weight sharing) | .611 | **.912** | .760 | .566 | .785 | .727 | 200.7 | 114.2 |
| Early Fusion | **.663** | .888 | .784 | .553 | .863 | .750 | 68.9 | 79.2 |
| Late Fusion | .641 | .864 | .775 | .574 | **.899** | **.751** | 152.3 | 110.5 |

The complementarity of the forensic filters is also apparent by the qualitative analysis in Fig. 4, where we see that the early fusion method effectively utilizes all of them. It produces accurate predictions in cases where one of the filters fail, like in the first picture where models that use Bayar or SRM can't accuarately localize the manipulation and in the second picture where the NoisePrint-based model (TruFor) fails. Even in some cases where all filters fail independently, the combined approach can produce accurate results, as shown for the fourth picture of Fig. 4.

| Image | Ground Truth | RGB+ NP++ | RGB+ Bayar | RGB+ SRM | Early Fusion |
|-------|-------------|-----------|------------|----------|--------------|

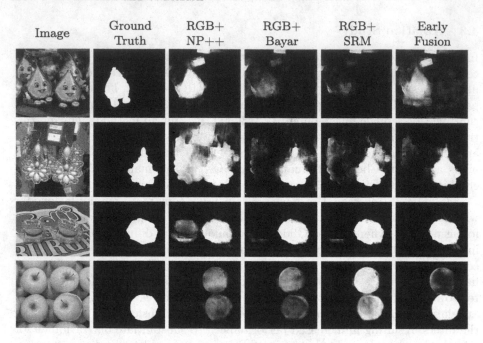

**Fig. 4.** Qualitative results.

## 4.4 Robustness Analysis

In this section, we include experiments performed on images with varying quality degradations to demonstrate the robustness of our approaches, similarly to Guillaro et al. [9]. We use the Casiav1+ dataset and perform Gaussian blurring with different kernel sizes and JPEG compression with varying quality factors and compare to our baseline, TruFor. The findings depicted in Fig. 5 demonstrate

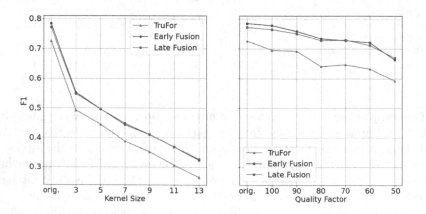

**Fig. 5.** Robustness analysis with regards to Gaussian blur (left) and JPEG compression (right)

that both of our fusion approaches exhibit good robustness across a broad spectrum of degradations, maintaining a consistent advantage over TruFor across all degradation levels employed.

## 5   Conclusion

In this work, we explore approaches toward expanding existing encoder-decoder architectures for IMLD to support multiple forensic filters as inputs. We compare the performance of approaches using Bayar conv, SRM filters, and NoisePrint++ and discover that they indeed showcase complementary forensic capabilities as was hypothesized. We propose two different modal-fusion paradigms and conduct extensive experiments to demonstrate that both approaches reach state-of-the-art across several datasets, showcasing good generalization abilities, and are effective at leveraging and combining diverse forensic artifacts from different filters. In future work, we would like to explore the performance limitations of models reliant on forensic filters against directed adversarial attacks.

**Acknowledgements.** This work was supported by the EU H2020 programme under grant agreement 101021866 (CRiTERIA). We also thank Nikolaos Gkalelis for useful discussions on our work and Myung-Joon Kwon for his help in obtaining the datasets.

## References

1. Bayar, B., Stamm, M.C.: A deep learning approach to universal image manipulation detection using a new convolutional layer. In: Proceedings of the 4th ACM Workshop on Information Hiding and Multimedia Security, pp. 5–10 (2016). https://doi.org/10.1145/2909827.2930786
2. Chen, X., Dong, C., Ji, J., Cao, J., Li, X.: Image manipulation detection by multi-view multi-scale supervision. In: Proceedings of the IEEE/CVF International Conference on Computer Vision (CVPR), pp. 14185–14193 (2021)
3. Cozzolino, D., Poggi, G., Verdoliva, L.: Copy-move forgery detection based on patchmatch. In: 2014 IEEE International Conference on Image Processing (ICIP), pp. 5312–5316. IEEE (2014)
4. Cozzolino, D., Verdoliva, L.: Noiseprint: a CNN-based camera model fingerprint. IEEE Trans. Inf. Forensics Secur. **15**, 144–159 (2019)
5. De Carvalho, T.J., Riess, C., Angelopoulou, E., Pedrini, H., de Rezende Rocha, A.: Exposing digital image forgeries by illumination color classification. IEEE Trans. Inf. Forensics Secur. **8**(7), 1182–1194 (2013)
6. Dong, J., Wang, W., Tan, T.: CASIA image tampering detection evaluation database. In: 2013 IEEE China Summit and International Conference on Signal and Information Processing, pp. 422–426. IEEE (2013)
7. Fan, Y., Xu, W., Wang, H., Wang, J., Guo, S.: PMR: prototypical modal rebalance for multimodal learning. In: Proceedings of the IEEE/CVF Conference on Computer Vision and Pattern Recognition (CVPR), pp. 20029–20038 (2023)
8. Fridrich, J., Kodovsky, J.: Rich models for steganalysis of digital images. IEEE Trans. Inf. Forensics Secur. **7**(3), 868–882 (2012). https://doi.org/10.1109/TIFS.2012.2190402

9. Guillaro, F., Cozzolino, D., Sud, A., Dufour, N., Verdoliva, L.: TruFor: leveraging all-round clues for trustworthy image forgery detection and localization. In: Proceedings of the IEEE/CVF Conference on Computer Vision and Pattern Recognition (CVPR), pp. 20606–20615, June 2023

10. Hsu, Y.F., Chang, S.F.: Detecting image splicing using geometry invariants and camera characteristics consistency. In: 2006 IEEE International Conference on Multimedia and Expo, pp. 549–552. IEEE (2006)

11. Hu, X., Zhang, Z., Jiang, Z., Chaudhuri, S., Yang, Z., Nevatia, R.: SPAN: spatial pyramid attention network for image manipulation localization. In: Vedaldi, A., Bischof, H., Brox, T., Frahm, J.-M. (eds.) ECCV 2020. LNCS, vol. 12366, pp. 312–328. Springer, Cham (2020). https://doi.org/10.1007/978-3-030-58589-1_19

12. Khosla, P., et al.: Supervised contrastive learning. Adv. Neural Inf. Process. Syst. **33**, 18661–18673 (2020)

13. Kniaz, V.V., Knyaz, V., Remondino, F.: The point where reality meets fantasy: Mixed adversarial generators for image splice detection. Adv. Neural Inf. Process. Syst. **32**, 215–226 (2019)

14. Kwon, M.J., Nam, S.H., Yu, I.J., Lee, H.K., Kim, C.: Learning jpeg compression artifacts for image manipulation detection and localization. Int. J. Comput. Vis. **130**(8), 1875–1895 (2022)

15. Lahiri, A., Jain, A.K., Agrawal, S., Mitra, P., Biswas, P.K.: Prior guided GAN based semantic inpainting. In: Proceedings of the IEEE/CVF Conference on Computer Vision and Pattern Recognition (CVPR), pp. 13696–13705 (2020)

16. Li, D., Zhu, J., Wang, M., Liu, J., Fu, X., Zha, Z.J.: Edge-aware regional message passing controller for image forgery localization. In: Proceedings of the IEEE/CVF Conference on Computer Vision and Pattern Recognition (CVPR), pp. 8222–8232 (2023)

17. Li, H., Luo, W., Huang, J.: Localization of diffusion-based inpainting in digital images. IEEE Trans. Inf. Forensics Secur. **12**(12), 3050–3064 (2017)

18. Liu, X., Liu, Y., Chen, J., Liu, X.: PSCC-Net: progressive spatio-channel correlation network for image manipulation detection and localization. IEEE Trans. Circuits Syst. Video Technol. **32**(11), 7505–7517 (2022)

19. Luo, Y., Zhang, Y., Yan, J., Liu, W.: Generalizing face forgery detection with high-frequency features. In: Proceedings of the IEEE/CVF Conference on Computer Vision and Pattern Recognition (CVPR), pp. 16317–16326 (2021)

20. Nichol, A.Q., et al.: GLIDE: towards photorealistic image generation and editing with text-guided diffusion models. In: Proceedings of the 39th International Conference on Machine Learning, pp. 16784–16804. PMLR (2022)

21. Novozamsky, A., Mahdian, B., Saic, S.: IMD 2020: a large-scale annotated dataset tailored for detecting manipulated images. In: Proceedings of the IEEE/CVF Winter Conference on Applications of Computer Vision Workshops (WACVW), pp. 71–80 (2020)

22. Ren, S., He, K., Girshick, R., Sun, J.: Faster R-CNN: towards real-time object detection with region proposal networks. Adv. Neural Inf. Process. Syst. **28**, 91–99 (2015)

23. Salloum, R., Ren, Y., Kuo, C.C.J.: Image splicing localization using a multi-task fully convolutional network (MFCN). J. Vis. Commun. Image Represent. **51**, 201–209 (2018)

24. Teyssou, D., et al.: The InVID plug-in: web video verification on the browser. In: Proceedings of the International Workshop on Multimedia Verification (MuVer) at ACM Multimedia 2017, pp. 23–30 (2017)

25. Wang, J.Z., Li, J., Wiederhold, G.: Simplicity: semantics-sensitive integrated matching for picture libraries. IEEE Trans. Pattern Anal. Mach. Intell. **23**(9), 947–963 (2001)
26. Wang, J., et al.: Objectformer for image manipulation detection and localization. In: Proceedings of the IEEE/CVF Conference on Computer Vision and Pattern Recognition (CVPR), pp. 2364–2373 (2022)
27. Wang, W., Tran, D., Feiszli, M.: What makes training multi-modal classification networks hard? In: Proceedings of the IEEE/CVF Conference on Computer Vision and Pattern Recognition (CVPR), pp. 12695–12705 (2020)
28. Wen, B., Zhu, Y., Subramanian, R., Ng, T.T., Shen, X., Winkler, S.: Coverage-a novel database for copy-move forgery detection. In: 2016 IEEE International Conference on Image Processing (ICIP), pp. 161–165. IEEE (2016)
29. Wu, Y., AbdAlmageed, W., Natarajan, P.: ManTra-Net: manipulation tracing network for detection and localization of image forgeries with anomalous features. In: Proceedings of the IEEE/CVF Conference on Computer Vision and Pattern Recognition (CVPR), pp. 9543–9552 (2019)
30. Xiao, T., Singh, M., Mintun, E., Darrell, T., Dollár, P., Girshick, R.: Early convolutions help transformers see better. Adv. Neural Inf. Process. Syst. **34**, 30392–30400 (2021)
31. Xie, E., Wang, W., Yu, Z., Anandkumar, A., Alvarez, J.M., Luo, P.: Segformer: simple and efficient design for semantic segmentation with transformers. Adv. Neural Inf. Process. Syst. **34**, 12077–12090 (2021)
32. Xie, S., Zhang, Z., Lin, Z., Hinz, T., Zhang, K.: Smartbrush: text and shape guided object inpainting with diffusion model. In: Proceedings of the IEEE/CVF Conference on Computer Vision and Pattern Recognition (CVPR), pp. 22428–22437 (2023)
33. Yang, C., Li, H., Lin, F., Jiang, B., Zhao, H.: Constrained R-CNN: a general image manipulation detection model. In: 2020 IEEE International Conference on Multimedia and Expo (ICME), pp. 1–6. IEEE (2020)
34. Zhang, J., Liu, H., Yang, K., Hu, X., Liu, R., Stiefelhagen, R.: CMX: cross-modal fusion for RGB-X semantic segmentation with transformers. IEEE Trans. Intell. Transp. Syst. 1–16 (2023). https://doi.org/10.1109/TITS.2023.3300537
35. Zhang, J., et al.: Delivering arbitrary-modal semantic segmentation. In: Proceedings of the IEEE/CVF Conference on Computer Vision and Pattern Recognition (CVPR), pp. 1136–1147 (2023)
36. Zhang, X., et al.: DE-GAN: domain embedded GAN for high quality face image inpainting. Pattern Recogn. **124**, 108415 (2022)
37. Zhou, P., Han, X., Morariu, V.I., Davis, L.S.: Learning rich features for image manipulation detection. In: Proceedings of the IEEE Conference on Computer Vision and Pattern Recognition (CVPR), pp. 1053–1061 (2018)

# Appearance-Motion Dual-Stream Heterogeneous Network for VideoQA

Feifei Xu[1], Zheng Zhong[1(✉)], Yitao Zhu[2], Yingchen Zhou[1], and Guangzhen Li[1]

[1] Shanghai University of Electric Power, Shanghai, China
`zhongzheng@mail.shiep.edu.cn`
[2] China Techenergy Co., Ltd., Shanghai, China

**Abstract.** Capturing spatio-temporal information in videos related to the question remains a key challenge in video question answering task (VideoQA). Though great success has been achieved in VideoQA, most of the existing methods do not sufficiently consider the correlation among appearance, motion, and object features, making it difficult to fully exploit the spatio-temporal relationships at different granularities. Besides, recent researches typically use the same interaction method when different features in the video interact with the question features separately, which ignores the spatio-temporal characteristics of the appearance and motion features in the video which leads to the problem of spatio-temporal mismatch. In this paper, we propose an Appearance-Motion Dual-stream Heterogeneous Network for VideoQA (AMHN), which pays attention to the synergy among three different features by heterogeneous interactions in terms of their spatio-temporal characteristics. AMHN unites object features with appearance features and motion features respectively to obtain two high-level visual representations containing object information. Then they are fed into the object-relational reasoning module to acquire relation-aware visual features. We use a bilinear attention network for appearance and put forward a Video-Text Symmetric Attention Network (VTSAN) for motion to achieve diverse features, which are fused under the guidance of the question to predict the final answer. We evaluate the performance of AMHN on two VideoQA benchmark datasets and perform an extensive ablation study. The experimental results demonstrate its state-of-the-art.

**Keywords:** Video Question Answering · Heterogeneous Network · Dual-stream

## 1 Introduction

Video question answering (VideoQA) is one of the most prominent Vision-Language tasks and has received increasing attention for its wide applications in video retrieval, intelligent question-and-answer systems, and autonomous driving. In comparison with image question answering [1], VideoQA is more challenging. In addition to modeling the semantic correlation between the question and

© The Author(s), under exclusive license to Springer Nature Switzerland AG 2024
S. Rudinac et al. (Eds.): MMM 2024, LNCS 14556, pp. 212–227, 2024.
https://doi.org/10.1007/978-3-031-53311-2_16

each image, it is further necessary to accurately extract the dynamic interactions between the question and the video.

Most of the existing methods [2] use recurrent neural networks (RNNs) and their variants to embed text and apply convolutional neural networks (CNNs) to extract video features. Video features mainly consider object features, single-frame static global appearance features [3], and multi-frame dynamic global motion features [4]. However, they usually build models based on only global appearance-motion features or only object features, which ignores the correlation between the object and appearance-motion features. Actually, the appearance and motion of objects also contain critical information in the video, from which we can exploit the synergy between the object and appearance-motion features.

On the other hand, most of the current research employs the same interaction method for both appearance-question and motion-question, which neglects the spatio-temporal differences in the appearance and motion features of videos. More attention should be paid to the spatial information during the appearance-question interaction, while more attention should be paid to the temporal information of the motion during the motion-question interaction. We need to use different interaction methods based on the spatio-temporal characteristics of different features to focus on the related information more comprehensively.

Given the above findings, we propose an Object-based Appearance-Motion Heterogeneous Network for video question answering which consists of four modules: the Video-Text Feature Extraction Module (VTFE), the Object Relation Inference Module (ORI), the V-Q Heterogeneous Interaction Module (VQHI), and the Answer Preference module. The VTFE module is designed to extract and encode the features of videos and questions, and we employ ResNet-101 [5] and 3D ResNeXt-101 [6] to extract the frame-level appearance and motion features of videos respectively. We acquire object features with RoIAlign on the convolutional layer of ResNet and on the feature map of ResNeXt. Then in the ORI module, we perform object-guided appearance and motion features as the input to the object-relation graph, which is utilized to obtain the relation-aware visual features. The VQHI module aims to embed rich cross-modal representations. In this module, we construct a heterogeneous network combining BAN [7] with our proposed VTSAN to compute the interaction between the two modalities by taking the output of the GCN in the ORI module and the question features as the inputs. Being guided by the semantics of the question, the answer preference module finally fuses the two features extracted by the VQHI module to perform answer prediction.

Our main contributions can be summarized as follows:

1. We propose an Appearance-Motion Dual-stream Heterogeneous Network for VideoQA (AMHN), which can better handle the relationship between appearance-motion features and object features of the video.
2. We design a heterogeneous interaction framework, which is more coherent with the spatio-temporal characteristics of video features and can capture more comprehensive semantic information.

3. We exploit an attention network for motion-question features interaction, Video-Text Symmetric Attention Network (VTSAN), which achieves more adequate interaction of video-text features by mutual guidance of the two modalities, and adds raw inputs from both features to improve the richness of features and the inference ability of the model.
4. We conduct a comprehensive evaluation of the MSVD-QA, and MSRVTT-QA datasets, achieving better performance than the state-of-the-art methods, and validate the effectiveness of each part of our approach.

## 2 Related Work

### 2.1 Video Question Answering

Video question answering aims to answer a given question related to the video content automatically. To better represent videos, earlier video representation methods [8,9] used VGG [10] to capture static information from frames as frame-level features and C3D [11] to acquire dynamic information as fragment-level features or video-level features. Most recent work [2,12] extracts global appearance and motion features of videos and designs different interactions for video and text features, such as question-guided attention [8,13] and Co-attention [14]. Although these approaches focus on the overall understanding of the video content, they ignore the importance of object features, which results in low accuracy rates. Combining global features with object features and exploiting the synergy between them may lead to a better visual representation. In this regard, our AMHN combines object features with appearance features and motion features of videos respectively to model spatio-temporal inference.

### 2.2 Visual-Linguistic Interaction

Visual-linguistic interaction is the projection of multimodal information into a common space for interactive retrieval [16]. Spatial attention models are designed to focus on significant spatial dimensions rather than the whole image to acquire more effective semantic cues [1]. In contrast to the spatial attention model, the temporal attention model focuses on temporal modeling of the context among video frames and aims to find the sequence of visual features relevant to the question [17]. To unify the visual semantic cues among appearance, motion, and questions, [12] first combines the spatial attention model with the temporal attention model and proposes a spatio-temporal attention model for the VideoQA model. Due to its superior performance, attention models such as multimodal fused memory [2], co-memory attention [14], multimodal attention [7], hierarchical attention [9] and multi-step progressive attention [18] are soon derived to capture a wider range of visual-linguistic interaction capabilities. In our visual-linguistic interaction framework, we use different interaction methods to interact with text features based on the different characteristics of video appearance and motion features. For motion features and text features of video, we design a new visual-linguistic interaction method, Video-Text Symmetrical Attention Network (VTSAN), which obtains the interaction between two modalities through question-guided visual features and visual-guided question features, which can help to deal with the long-term dependency problem in previous studies.

## 3   Method

In this section, we detail the architecture of AMHN, shown in Fig. 1. How to extract appearance and motion features is illustrated in Sect. 3.1. Section 3.2 describes the construction of object-relational graphs in the object-relational reasoning module which captures relation-aware visual features using GCN. The V-Q heterogeneous interaction module encoding the visual features combined with the questions, is given in Sect. 3.3. Finally, Sect. 3.4 explains how the answer preference module fuses the output of the VQHI module under the guidance of question semantics by determining whether the question semantics focus more on appearance or motion information, or a combination of both.

### 3.1   Video-Text Feature Extraction Module

**Question Representation.** VideoQA requires the model to understand the semantic information of the questions deeply, we apply all sentences to fine-tune BERT-base. For question, the 768-dimensional word embeddings of the words in

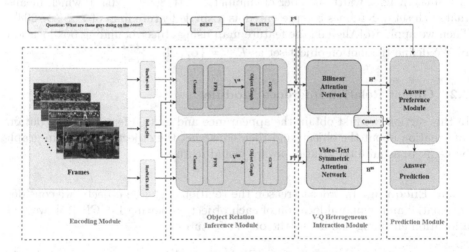

**Fig. 1.** The overview of AMHN. Appearance features with object information and motion features with object information are used as dual streams of the heterogeneous framework respectively. Take appearance features as an example. The appearance features with object information are input to the ORI module, which uses GCN to obtain the relation-aware visual representation. Then the visual representation and the question features are fed into the VQHI module to compute the interaction between the two modalities. The same is true for the stream of motion features with object information. Finally, the output of the VQHI module is fused to predict the final answer.

each question are extracted. The final question embedding representation and last hidden units from the bi-LSTM are denoted by $F^q \in \mathbb{R}^{M \times d}$ and $q \in \mathbb{R}^d$ respectively. Following the same question representation, the answer is represented as $\{a_c \mid a_c \in \mathbb{R}^d, c \leq 5\}$.

**Video Representation.** We extract the global appearance features $v_{global}^a \in \mathbb{R}^{T \times d}$ and motion features $v_{global}^m \in \mathbb{R}^{T \times d}$ from the video frames. Specifically, we receive the feature maps from the Conv5 layer of ResNet-101 and apply linear projection [14,16] to capture global appearance features $v_{global}^a \in \mathbb{R}^{T \times d}$, where d is the size of the hidden dimension. The global motion features $v_{global}^m \in \mathbb{R}^{T \times d}$ are obtained by the feature map in the last convolution layer of 3D ResNeXt-101 in which adaptive average pooling and linear projection are applied.

Object features are also extracted to form a high-level visual representation in conjunction with appearance and motion features respectively. Specifically, in terms of appearance, we use Faster R-CNN [19] to capture N objects from each frame, so that there are $K = N \times T$ objects in a video containing T frames. Then we apply RoIAlign [21] to obtain regions of interest from the convolutional layer of ResNet-101. We denote the appearance-object set as $\mathcal{R}^a = \{o_{t,n}^a, b_{t,n}\}_{t=1,n=1}^{t=T,n=N}$, where $o$ and $b$ mean object feature and bounding box location respectively. For motion features, we use the last convolutional layer of 3D ResNeXt-101 to acquire the feature maps with dimensions (number of frames, height, width, number of channels) $= (t/8, 7, 7, 2048)$, which means that each set of 8 frames is represented as a single feature map of $7 \times 7 \times 2048$. Then we apply RoIAlign on the feature map using object bounding box location $b$. We define the motion-object set as $\mathcal{R}^m = \{o_{t,n}^m, b_{t,n}\}_{t=1,n=1}^{t=T,n=N}$.

### 3.2    Object Relation Inference Module

In this section, we first obtain the appearance and motion features that contain information about the object, and then construct the object relationship graph by GCN to acquire the relation-aware visual features.

**Local Encoding.** In order to reason the relations between objects, we consider the spatial and temporal location of each object. Following L-GCN [15], we add a location encoding and define the object features as:

$$v_{object}^a = \text{FFN}\left([o^a; d^s; d^t]\right), \tag{1}$$

$$v_{object}^m = \text{FFN}\left([o^m; d^s; d^t]\right), \tag{2}$$

where $\mathbf{d}^s = \text{FFN}(\mathbf{b})$ and $\mathbf{d}^s$ is obtained by positional encoding based on the index of each frame. FFN is the feed-forward layer. Similar to object features, the location encoding information $\mathbf{d}^s$ is added to the global features. Then we concatenate object features with global features as a high-level visual representation to reflect the frame-level context of the object.

$$\mathbf{v}^a = \text{FFN}\left(\left[\mathbf{v}^a_{\text{object}}; \mathbf{v}^a_{\text{global}}\right]\right), \tag{3}$$

$$\mathbf{v}^m = \text{FFN}\left(\left[\mathbf{v}^m_{\text{object}}; \mathbf{v}^m_{\text{global}}\right]\right). \tag{4}$$

where $\mathbf{v}^a \in \mathbb{R}^{K \times d}$ and $\mathbf{v}^a \in \mathbb{R}^{K \times d}$ signify the object-guided appearance representation and object-guided motion representation respectively.

**Object Graph.** we construct the object graph $\mathcal{G} = (\mathcal{V}, \mathcal{E})$ to capture the spatio-temporal relationships between objects. $\mathcal{V}, \mathcal{E}$ denote the set of nodes and the set of edges of the graph respectively. The nodes of the graph $\mathcal{G}$ are given by $X^{a/m} \in \mathbb{R}^{K \times d}$. $X^{a/m}$ refers to the visual features $V^m$ and $V^a$ above, which are used as the input of the object-relationship graph, and the edges are given by $\left(v_i^{a/m}, v_j^{a/m}\right)$, which denote the relationship between two nodes. We perform a graph convolution operation on the graph $\mathcal{G}$ to obtain the relation-aware object features, and we get the similarity scores of the nodes by computing the dot product of the input features after projecting them into the interaction space. The adjacency matrix $A^{a/m} \in \mathbb{R}^{K \times K}$ is defined as follows:

$$A^{a/m} = \text{softmax}\left(\left(X^{a/m}W_1\right)\left(X^{a/m}W_2\right)^{\mathsf{T}}\right), \tag{5}$$

where $W_1$ and $W_2$ are trainable parameters. We denote the two-layer graph convolution with adjacency matrix $A$ on input $X^{a/m}$ as:

$$\text{GCN}\left(X^{a/m}; A^{a/m}\right) = \text{ReLU}\left(A^{a/m}\,\text{ReLU}\left(A^{a/m}X^{a/m}W_3\right)W_4\right), \tag{6}$$

$$F^{a/m} = \text{LayerNorm}\left(X^{a/m} + \text{GCN}\left(X^{a/m}; A^{a/m}\right)\right). \tag{7}$$

Here we use the ReLU activation function, where $W_3$ and $W_4$ denote trainable parameters, $F^a$ and $F^m$ mean appearance and motion visual features respectively. To alleviate the degradation problem of the model and to increase the generalization ability of the network, we add a residual connection to promote residual learning between object nodes.

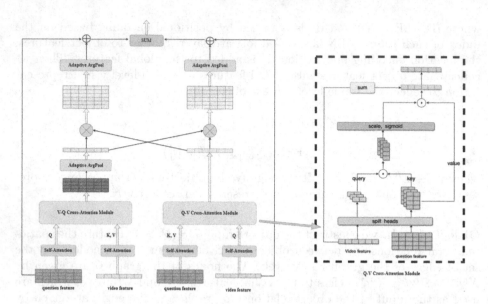

**Fig. 2.** Video-Text Symmetric Attention Network(VTSAN). Video features and question features are the inputs to VTSAN. Symmetric mutual guidance is performed by Q-V Cross-Attention Module and V-Q Cross-Attention Module through operations such as element-wise multiplication and global pooling.

### 3.3 V-Q Heterogeneous Interaction Module

This module is our main contribution. For the different spatio-temporal characteristics of video features, we propose heterogeneous networks. BAN is used as the interaction network when appearance features and question features interact. In the interaction between motion features and problem features we propose VTSAN, which uses symmetric interaction, where video and text features guide each other to make the interaction more adequate, and the original inputs of the features are added in the final feature fusion. Meanwhile, in VTSAN we propose a new cross-modal attention network: Q-V/V-Q cross-modal attention, which maximizes the original input information without changing the feature dimensions.

Specifically, we use the bilinear attention network (BAN) to compute the interaction between appearance features $F^a$ and question features $F^q$.

$$H_i = \text{BAN}_i \left( H_{i-1}, F^a; \mathcal{A}_i \right)^\top + H_{i-1}, \tag{8}$$

where $H_0 = F^q, 1 \leq i \leq g$ and $A$ denotes the attention map. In the above equation, the result $\text{BAN}(H, F^a; \mathcal{A}) \in \mathbb{R}^d$ is calculated and added to $H$ and repeated $g$ times. Eventually, $H^a \in \mathbb{R}^d$ represents the combined visual and linguistic features in the question space, incorporating different aspects of the two modalities.

**Video-Text Symmetric Attention Network.** To compute the interaction between motion features $F^m$ and question features $F^q$, we propose the Video-Text Symmetric Attention Network(VTSAN), as shown in Fig. 2. The motion feature $F^m$ interacts with the question feature $F^q$ in a symmetric structure. On account of the large difference between video and text modalities, for a more adequate video-text interaction, we use a symmetric approach for the two modalities to guide each other to acquire the video-guided question features and the question-guided video features.

In order to extract the features of the two modalities guided by each other, we propose **Q-V** and **V-Q Cross-Attention Network** with motion features $F^m$ and question features $F^q$ as inputs. Take the Q-V cross-attention network as an example, as shown in Fig. 2. Here our Q-V and V-Q Cross-Attention modules differ from self-attention in Transformer in two ways: 1) Our query, key, and value are derived from different modalities and they are not mapped to the same dimension. In this way, we can maximize the preservation of the original input information and the follow-up will not be affected since the final output is in the same dimension as the query. 2) We fuse the intra-frame information representation again through the element-wise addition operation to prevent it from being forgotten.

Both $F^{a/m}$ and $F^q$ are taken as inputs to the Q-V /V-Q Cross-Attention Network after Self-Attention, which we also denote by $F^{a/m}$ and $F^q$ here. In the Q-V cross-attention module, we take $F^m$ as the query and $F^q$ as the value and key, then add the original input of the motion features. The detailed equation is shown below:

$$P^{Q-V} = Attention\left(F^m, LN\left(F^q\right), LN\left(F^q\right)\right), \tag{9}$$

$$Attention(Q, K, V) = \text{softmax}\left(\frac{QK^\top}{\sqrt{d_k}}\right)V, \tag{10}$$

$$F^{Q-V} = LN\left(P^{Q-V} + F^m\right), \tag{11}$$

where $F^{Q-V} \in \mathbb{R}^d$ denotes the video-guided question features. Then in Eq. 9 and Eq. 11, we swap the left and right positions to obtain the question-guided video feature $F^{V-Q}$ with $F^q$ as the query and $F^m$ as the value and key. However, there is still a gap between the two modalities. In order to further interact with the information, we map $F^{Q-V}$ and $F^{V-Q}$ to the same low-dimensional space before multiplying each other's transpose matrix, which can be expressed as:

$$H^{Q-V} = \left(F^{Q-V}\right)^T \otimes \left(\text{ AdaptiveAvgPool }\left(F^{V-Q}\right)\right), \tag{12}$$

$$H^{V-Q} = \left(\text{ AdaptiveAvgPool }\left(F^{V-Q}\right)\right)^T \otimes F^{Q-V}, \tag{13}$$

where $\otimes$ denotes multiplication operation. Even though the dimensionality of the features is elevated in this process, it allows the elements of features to interact effectively. A high degree of similarity is generated between the two features. However, the high dimensionality is not conducive to the final prediction, so we use the AdaptiveAvgPool operation for both $H^{Q-V} \in \mathbb{R}^{d*d}$ and $H^{V-Q} \in \mathbb{R}^{d*d}$

features to reduce the computational overhead and retain the information they possess.

Finally, we add the respective original inputs to each of the two features to enrich the information representation of the model for better cross-modal inference. Then we combine the two features in an element-wise addition operation as the final video-text interaction features $H^m \in \mathbb{R}^d$. The detailed equation is shown below:

$$F^{Q-V} = LN \left( P^{Q-V} + F^m \right), \tag{14}$$

$$F^{Q-V} = LN \left( P^{Q-V} + F^m \right), \tag{15}$$

$$H^m = LN \left( \text{AdaptiveAvgPool} \left( H^{Q-V} \right) + F^m \right) \\ + LN \left( \text{AdaptiveAvgPool} \left( H^{V-Q} \right) + F^q \right). \tag{16}$$

## 3.4  Answer Preference Module

The fusion module finally fuses the two features, $H^a$ and $H^m$, based on the question context vector $q$. Depending on what the question ultimately asks, the model should determine whether the question context is more relevant to the appearance or motion information, or a combination of both.

We use the fusion method of MASN [22] to generate three features, appearance-centered, action-centered, and mixed appearance-motion. Then we aggregate them according to the question context. First, a regular scaled dot-product attention is defined to focus on different aspects of the features, which is used to obtain motion-centered, appearance-centered, and mixed attention. The attention scores of the three features are obtained under the guidance of the question context vector $q$. The scores can be interpreted as the importance of each feature in the question context. The fusion matrix of the question guidance is acquired based on the attention scores. The final output vector $f$ used to predict the answer is obtained by aggregating the information on the fusion matrix through the attention mechanism.

## 3.5  Answer Prediction and Loss Function

VideoQA tasks can generally be classified into three categories based on question types: multiple-choice, open-ended and counting tasks. Here we set up different answer predictors depending on the question types.

For the multiple-choice task, following previous work [12], we treat each candidate answer individually. Specifically, we attach an answer to the question and get $L$ candidates. For each candidate answer, we receive a score $S_l$ for each candidate answer by a linear transformation of the final output vector $f$.

$$S_l = W_5 f + b_1. \tag{17}$$

where $W_5$ denotes the trainable parameter, $b_1$ denotes the learnable bias. We minimize the hinge loss in each candidate pair, $\max(0, 1 + s_n - s_p)$, where $s_n$ and $s_p$ are the scores from incorrect and correct answers respectively.

The open-ended word task can be considered as a multi-classification problem that requires selecting the most appropriate option from the answer banks. In this paper, the classifier we designed generates scores for all the answers in the answer set by applying a linear classifier and a softmax function to the final output vector $f$, where the highest score is considered to be our predicted answer.

Finally, for the counting task, we obtain an integer result following Eq. 15 above, where linear regression is used instead of a classification function and non-integer results are rounded. The loss function uses Mean Square Error (MSE).

## 4  Experiments

In this section, we compare our proposed model with other state-of-the-art models on two benchmark datasets while conducting extensive ablation experiments to demonstrate the validity of our model.

### 4.1  Experimental Details

We apply Faster R-CNN [21] pre-trained on Visual Genome [23] to obtain local features. The number of extracted objects is N = 10. For global features, we use ResNet-101 pre-trained on ImageNet [24]. In the case of motion features, we apply 3D ResNeXt-101 pre-trained on the Kinetics action recognition dataset [25]. For the input of 3D ResNeXt-101, we concatenate a set of 8 frames around the sampled frame mentioned above. In terms of training details, we use Adam [26] as the optimizer with an initial learning rate of 1e-5, which decays by 0.1 after 25 epochs and stops training after 35 epochs. The number of BAN glimpse g is 4. We set the batch sizes for the multi-choice and open-ended tasks to 8 and 16 respectively. For multi-choice and open-ended tasks, we use accuracy to evaluate the performance of a VideoQA method.

### 4.2  Datasets

**MSVD-QA.** MSVD-QA [27] is derived from 1970 video clips in the MSVD video dataset [28], where the average video length is about 10 s. It contains 50,505 QA pairs with five types of questions, including what, how, when, where and who, with an average question length of 6 words.

**MSRVTT-QA.** Compared with MSVD-QA, MSRVTT-QA [27] has more videos and more complex scenarios. It consists of 10,000 untrimmed videos from the MSR-VTT dataset [29], including 243,000 QA pairs in the same five categories as the MSVD-QA. The average video length is 15 s, and the average question length is 7 words.

## 4.3   Comparison with State-of-the-Arts

**MSVD-QA**  To further explore the capability of AMHN in analyzing spatio-temporal information, we compare with the following models, including Co-Mem [14], STCA [30], CAN [31], HME [2], HGA [32], QueST [14], FAM [33], GMINI [34], VLCN [35], DSAVS [36]. As shown in Table 1 : our model AMHN achieves competitive results on both MSVD-QA and MSRVTT-QA datasets.

**MSRVTT-QA.**  To further explore the capability of AMHN in analyzing spatio-temporal information, we compare with the following models, including Co-Mem [14], STCA [30], CAN [31], HME [2], HGA [32], QueST [14], FAM [33], GMINI [34], VLCN [35], DSAVS [36]. As shown in Table 1 : our model AMHN achieves competitive results on both MSVD-QA and MSRVTT-QA datasets.

Among the five question words, "What" and "Who" need to identify and understand the objects, persons and scenes in the video, which focus more on the spatial information in the video. "Where" and "When" need to locate the position of objects or tasks in the video, so we need to consider the change of position on the timeline. As a result, "Where" and "When" should be more concern with the temporal information in the video. "How" needs to be inferred from both temporal and spatial information.

As shown in Table 1, AMHN achieves the best results in the"what", "how" and "when" tasks. These three types of questions account for the largest proportion in the dataset, and most of them need to deal with both temporal and spatial information rather than one or the other, benefiting from the adequate interaction between motion and question information, and the bilinear structure of the BAN network capturing spatial details through complex and huge computational effort, which results in a large improvement in the overall accuracy of all of them. As for the "Where" task, we could not achieve optimal results on the MSVD-QA Dataset. As in Table 1, there is a huge gap between the results obtained by the CoMem model on the two datasets. However, in the MSRVTT-QA dataset our model also achieves optimal results on the "where" task. On the "Who" task, the performance of AMHN can be improved compared to DSAVS, because DSAVS iterates several attention models on spatial information, while our model is much ahead of DSAVS in terms of time cost of training and speed of convergence. Overall, AMHN achieves competitive results on the MSVD-QA and MSRVTT-QA datasets.

**Table 1.** State-of-the-art comparison on dataset MSVD-QA and dataset MSVRTT-QA. The bolded font has the best result.

| Method | MSVD - QA | | | | | | MSRVTT - QA | | | | | |
|--------|------|------|------|------|-------|------|------|------|------|------|-------|------|
| | What | Who | How | When | Where | All | What | Who | How | When | Where | All |
| CoMem | 19.6 | 48.7 | 81.6 | 74.1 | 31.7 | 31.7 | 23.9 | 43.5 | 74.1 | 69.0 | 42.9 | 32.0 |
| HME | 22.4 | 50.1 | 73 | 70.7 | 42.9 | 33.7 | 26.5 | 43.6 | 74.1 | 69.0 | 28.6 | 33.0 |
| STCA | 24.3 | 49.6 | 83 | 74.1 | 53.6 | 35 | 27.4 | 45.4 | 83.7 | 74.0 | 33.2 | 34.2 |
| CAN | 21.1 | 47.9 | 84.1 | 74.1 | **57.1** | 32.4 | 26.7 | 43.4 | 83.7 | 75.3 | 35.2 | 33.2 |
| HGA | 23.5 | 50.4 | 83 | 72.4 | 46.4 | 34.7 | 29.2 | 45.7 | 83.5 | 75.2 | 34.0 | 35.5 |
| QuEST | 24.5 | 52.9 | 79.1 | 72.4 | 50 | 36.1 | 27.9 | 45.6 | 83.0 | 75.7 | 31.6 | 34.6 |
| FAM | 23.1 | 51.6 | 82.2 | 71.4 | 51.9 | 34.5 | 26.9 | 43.9 | 82.8 | 70.6 | 31.1 | 33.2 |
| GMINN | 24.8 | 49.9 | 84.1 | 75.9 | 53.6 | 35.4 | 30.2 | 45.4 | 84.1 | 74.9 | 34.2 | 36.1 |
| VLCN | 28.42 | 51.29 | 81.08 | 74.13 | 46.43 | 38.06 | 30.69 | 44.09 | 79.82 | 78.29 | 36.80 | 36.01 |
| DSAVS | 25.6 | **53.5** | 85.1 | 75.9 | 53.6 | 37.2 | 29.5 | **46.1** | 84.3 | 75.5 | 35.6 | 35.8 |
| AMHN | **28.6** | 51.5 | **85.4** | **76.3** | 50.9 | **40.3** | **31.2** | 44.9 | **85.7** | **79.6** | **43.1** | **39.3** |

## 4.4   Ablation Study

**Effect of the Main Components in AMHN.** To further verify each component in our AMHN, We select the MSRVTT-QA dataset with a large amount of data, and we analyze the role of the ORI module and the VQHI module through ablation experiments.And we also do an additional set of experiments to ablate the two interaction models within VQHI one by one while retaining the ORI module, we split VQHI into two components, BAN and VTSAN. the results are shown in Table 2.

**Table 2.** Ablation study on the MSRVTT-QA dataset.

| Methods | What | Who | How | When | Where |
|---------|------|------|------|------|-------|
| ORI | 28.5 | 37.7 | 81.5 | 75.6 | 39.9 |
| VQHI | 30.3 | 41.6 | 82.4 | 74.4 | 40.5 |
| BAN | 28.7 | 43.2 | 81.6 | 74.8 | 41.2 |
| VTSAN | 24.1 | 41.8 | 84.1 | 78.3 | 40.9 |
| AMHN | **31.2** | **44.9** | **85.7** | **79.6** | **43.1** |

When using only the ORL module to aggregate object node relationships, our approach resembles a traditional inference model that does not distinguish between temporal and spatial dual streams. The results obtained on all five question types are hardly satisfactory and have significant drawbacks compared to the refined interaction model. In contrast, when using only the VQHI module, the model performs approximately similarly to the previous dual-stream model due to the loss of the aggregation of object relations. It means that spatio-temporal fine-grained information plays an important role in video understanding, while the use of inter-object relations enables better inference tasks. The large effect on the "When", "How" and "Where" tasks when using only BAN demonstrates that focusing only on appearance information, which is temporal information, results in the loss of motion information, making it difficult to solve the challenges of VideoQA.The opposite problem occurs when using only VTSAN, which focuses only on temporal information, which is motion information, making the model perform poorly on the "What" and "Where" tasks. Our model AMHN achieves the most optimal results on all tasks, capable of exploiting the heterogeneous interaction structure to capture spatial details while focusing on timeline changes.

**Effect of the Video-Text Symmetric Attention Network.** The main purpose of our proposed VTSAN is to achieve a more adequate video-text interaction. For this purpose, we use symmetric structure where video and text are guided mutually. Compared to images, video has more complex spatio-temporal information,and the processing of spatio-temporal information should not be standardized. In order to demonstrate the effectiveness of VTSAN in processing spatio-temporal information, we replace the video-text interaction method in the appearance module or the motion module, and the replacement methods and experimental results are shown in Table 3.

**Table 3.** The results of VTSAN and BAN on the MSRVTT-QA dataset.

| Methods | What | Who | How | When | Where |
|---------|------|-----|-----|------|-------|
| BAN+BAN | **31.6** | 42.1 | 81.5 | 76.3 | 41.8 |
| BAN+CMA | 29.2 | 40.0 | 39.3 | 74.6 | 42.5 |
| VTSAN+BAN | 28.7 | 36.5 | 79.2 | 77.4 | 38.1 |
| VTSAN+CMA | 25.5 | 38.4 | 77.4 | 69.5 | 35.6 |
| VTSAN+VTSAN | 28.7 | 39.9 | 80.1 | 75.2 | 40.2 |
| BAN+VTSAN w/o (OI) | 28.7 | 39.9 | 80.1 | 75.2 | 40.2 |
| BAN+VTSAN | 31.2 | **44.9** | **85.7** | **79.6** | **43.1** |

We use different combinations of each to replace the two video-text interaction mechanisms in the VQHI module. As shown in Table 3. The results of the different model combinations can show the superiority of our model. In the "What" and "Who" tasks, BAN as an interaction model of appearance features and question features achieves a greater advantage than VTSAN, which mainly depends on the spatial information of the objects in the video frames is more obvious, and BAN interacts with bilinear interaction in the input two-channel clusters, and extracts the joint representations of the two channels by low-rank bilinear pooling, which is helpful for extracting the spatial information of the static ones. VTSAN performs better in locating temporal information, requiring a deeper understanding of the question and a better alignment of the question semantics with the inter-frame context of the video for better cross-modal inter-actions in the "When", "Who", and "How" tasks. This involves not only a deeper understanding of the question, but also a contextual alignment between the question semantics and the frame-to-frame context of the video to achieve a more effective cross-modal interaction. Our proposed VTSAN makes the interaction of the two modalities more adequate by making the features of the video and the question guide each other through a symmetric structure. Moreover, by adding the original inputs of the two features, the fifth row of Table 3 shows the accuracy achieved by VTSAN without adding the original inputs, while the added VTSAN improves on all the tasks, which indicates that the added original inputs improve the modeling ability and inference of the cross-modal relationship.

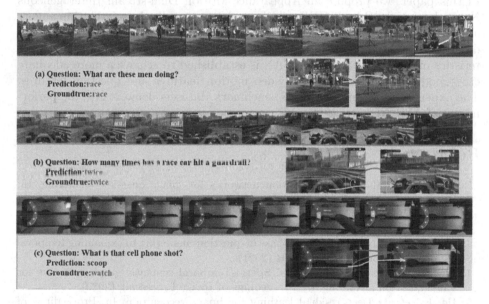

**Fig. 3.** Some qualitative results produced by our model. The ground-truth answers are in red and the incorrect answers are in blue. (Color figure online)

### 4.5   Qualitative Results

To qualitatively analyze the effectiveness of our proposed model AMHN, sev-eral typical examples selected from the test sets of the two datasets are visu-alized in Fig. 3. Examples (a) and (b) are examples from the MSVD-QA and MSRVTT-QA datasets respectively, which can be correctly predicted, and our model AMHN can accurately infer the answer. Specifically, the most frequently asked questions use "what" as the question word, they usually examine what the objects are doing in the video, and basically every frame of the video con-tains object information, and AMHN can achieve a high accuracy rate on these questions as it recognizes the object information accurately. the answers to some questions are reflected in a small number of video frames, so the model is required to capture the critical information accurately. As in example (b), Our model is competent for this kind of problem. Figure 3 also shows instances in which our model fails to infer the correct answer. The video shows the shooter holding a phone to shoot a watch. Since the environment is homogeneous and the object features remain unchanged throughout the video, the model is only capable of inferring from the appearance information of the object, thus generating similar answers.

## 5    Conclusion

In this paper, we propose an Appearance-Motion Dual-stream Heterogeneous Network for VideoQA (AMHN), which can better handle the relationship among appearance features, motion features, and object features of videos. Moreover, different interaction methods are designed based on the spatio-temporal characteristics of different features. VTSAN is established to achieve a more adequate cross-modal interaction between video motion features and question features. The experimental results on three benchmark datasets demonstrate the performance of AMHN.

## References

1. Yang, Z., et al.: Stacked attention networks for image question answering. In: Proceedings of CVPR (2016)
2. Fan, C., et al.: Heterogeneous memory enhanced multimodal attention model for video question answering. In: Proceedings of CVPR (2019)
3. Xiao, J., et al. :NExT-QA: next phase of question-answering to explaining temporal actions. In: Proceedings of CVPR (2021)
4. Liu, Y., et al.: Cross-Attentional Spatio-Temporal semantic graph networks for video question answering. In: Proceedings of Image Processing (2022)
5. He, K., et al.: Deep residual learning for image recognition. In: Proceedings of CVPR (2016)
6. Hara, K., et al.: Can spatiotemporal 3D CNNs retrace the history of 2D CNNs and imageNet? In: Proceedings of CVPR (2017)
7. Kim, J-H., et al.: Bilinear Attention Networks. In: Proceedings of NeurIPS (2018)
8. Xu, D., et al.: Video question answering via gradually refined attention over appearance and motion. In: Proceedings of ACM (2017)
9. Zhao, Z., et al.: Video question answering via hierarchical Dual-Level attention network learning. In: Proceedings of ACM (2017)
10. Simonyan, K., Zisserman, A.:Very deep convolutional networks for Large-Scale image recognition. CoRR abs/1409.1556 (2014)
11. Tran, D., et al.: Learning spatiotemporal features with 3D convolutional networks. In: Proceedings of ICCV (2015)
12. Jang, Y., et al.: TGIF-QA: Toward Spatio-Temporal reasoning in visual question answering. In: Proceedings of CVPR (2017)
13. Jiang, J., et al.: Divide and conquer: Question-Guided Spatio-Temporal contextual attention for video question answering. In: Proceedings of AAAI (2020)

14. Gao, J. et al.: Motion-Appearance Co-memory networks for video question answering. In: Proceedings of CVPR (2018)
15. Huang, D., et al.: Location-Aware graph convolutional networks for video question answering. In: Proceedings of AAAI (2020)
16. Zeng, K-H et al.: Leveraging video descriptions to learn video question answering. In: Proceedings of AAAI (2016)
17. Zhu, L., et al.: Uncovering the temporal context for video question answering. In: Proceedings of IJCV (2017)
18. Zhao, Z., et al.: Multi-Turn video question answering via Multi-Stream hierarchical attention context network. In: Proceedings of IJCAI (2018)
19. Ren, S., et al.: Faster R-CNN: towards Real-Time object detection with region proposal networks. In: Proceedings of TPAMI (2015)
20. Devlin, J., et al.: BERT: Pre-training of deep bidirectional transformers for language understanding. In: Proceedings of NAACL-HLT (2019)
21. He, K., et al.: Mask R-CNN. In: Proceedings of TPMAI (2017)
22. Seo, A., et al.: Attend what you need: Motion-Appearance synergistic networks for video question answering. In: Proceedings of ACL (2021)
23. krishna, R., et al.: Visual Genome: connecting language and vision using crowd-sourced dense image annotations. Int. J. Comput. Vis. **123**(1), 32–73 (2017). https://doi.org/10.1007/s11263-016-0981-7
24. Deng, J., et al. ImageNet: a large-scale hierarchical image database. In: Proceedings of CVPR (2009)
25. Kay, W., et al.: The kinetics human action video dataset. ArXiv abs/1705.06950 (2017)
26. Kingma, D., et al.: Adam: a method for stochastic optimization. CoRR abs/1412.6080 (2014)
27. Ye, Y., et al.: Video question answering via Attribute-Augmented attention network learning. In: Proceedings of SIGIR (2017)
28. Chen, D.L., William B.D.: Collecting Highly Parallel Data for Paraphrase Evaluation. In: Proceedings of ACL (2011)
29. Xu, J., et al.; MSR-VTT: a large video description dataset for bridging video and language. In: Proceedings of CVPR (2016)
30. Zha, Z., et al.: Spatiotemporal-Textual Co-Attention network for video question answering. In: Proceedings of TOMM (2019)
31. Yu, T., et al.: Compositional attention networks with Two-Stream fusion for video question answering. In: Proceedings of TIP (2020)
32. Jiang, P., Han, Y.: Reasoning with heterogeneous graph alignment for video question Answering. In: Proceedings of AAAI (2020)
33. Cai, J., et al.: Feature augmented memory with global attention network for VideoQA. In: Proceedings of IJCAI (2020)
34. Gu, M., et al.: Graph-Based Multi-Interaction network for video question answering. In: Proceedings of TIP(2021)
35. Abdessaied, A., et al.: Video language Co-Attention with multimodal Fast-Learning feature fusion for VideoQA. In: Proceedings of RepL4NLP (2022)
36. Liu, Y., et al.: Dynamic self-attention with vision synchronization networks for video question answering. In: Proceedings of ICPR (2022)
37. Li, X., et al.: Complementary spatiotemporal network for video question answering. In: Proceedings of Multimed Syst (2021)

# Adaptive Token Selection and Fusion Network for Multimodal Sentiment Analysis

Xiang Li[1], Ming Lu[1], Ziming Guo[1], and Xiaoming Zhang[2(✉)]

[1] School of Cyber Science and Technology, Beihang University, Beijing, China
{xlggg,luming,guoziming}@buaa.edu.cn
[2] State Key Laboratory of Software Development Environment, Beihang University, Beijing 100191, People's Republic of China
yolixs@buaa.edu.cn

**Abstract.** Multimodal sentiment analysis aims to predict human sentiment polarity with multiple modalities. Most existing methods focus on directly integrating original modal features into multimodal fusion, ignoring the redundancy and heterogeneity across modalities. In this paper, we propose a simple but efficient Adaptive Token Selection and Fusion Network (ATSFN) to mitigate the effect of redundancy and heterogeneity. ATSFN employs adaptive trainable tokens to extract unimodal informative tokens and perform dynamic multimodal token fusion. Specifically, we first integrate critical information from original features into adaptive selection tokens through token selection transformers. Sentiment features flow through these smaller sequences of tokens to capture important information while reducing redundancy. Next, we introduce a token fusion transformer to fuse multimodal features dynamically. It adaptively estimates the unique contribution of each modality to sentiment tendencies through learnable fusion tokens. Experiments on two benchmark datasets demonstrate that our proposed approach achieves competitive performance and significant improvements.

**Keywords:** Redundancy · Heterogeneity · Adaptive token · Multimodal sentiment analysis · Multimodal fusion

## 1 Introduction

With the rapid spread of online videos and movies on social media platforms, there is a growing interest in integrating and comprehending multimodal data. Multimodal Sentiment Analysis (MSA) aims to identify sentiment polarity using multimodal features segmented from video clips [31]. MSA involves time-series data from various modalities, e.g., natural language, facial expressions, and vocal behavior. Different modalities can provide complementary and heterogeneous semantic information to help humans make more accurate sentiment predictions. The crucial issue in multimodal sentiment analysis is effectively integrating heterogeneous data. A suitable fusion method should extract and integrate critical

© The Author(s), under exclusive license to Springer Nature Switzerland AG 2024
S. Rudinac et al. (Eds.): MMM 2024, LNCS 14556, pp. 228–241, 2024.
https://doi.org/10.1007/978-3-031-53311-2_17

information from multiple modalities while considering their independent contributions to sentiment prediction.

Various intermediate fusion techniques have been proposed to leverage multimodal information in MSA. The mainstream methods are roughly divided into representation-based methods and interaction-based methods. The former methods aim to acquire holistic representation by incorporating specific constraints like multi-task learning [8], self-supervised learning [27], and mutual information maximization [7] within the feature representation subspace. The latter methods employ complex fusion strategies to directly capture correlations and interactions between discrete elements across modalities. These fusion approaches encompass operation-based [16,18], tensor-based [11,28], attention-based [12,25], transformer-based [5,6,22], and MLP-based methods [20], among others.

Despite the achievements of mainstream MSA methods, they directly integrate original features into multimodal fusion. However, redundancy and heterogeneity across modalities still increase the difficulty of multimodal fusion. Firstly, the original features of each modality contain irrelevant and redundant information, such as superfluous frames in video, non-emotional features in text, and extraneous noise in speech. Therefore, directly integrating raw features without considering redundancy cannot capture meaningful information, negatively impacting multimodal fusion. Secondly, inherent heterogeneity makes various contributions of different modalities to sentiment tendencies. Different modalities express emotions from distinct viewpoints and contain independent and complementary information. Many emotions are more accessible to recognize via language, while some are easier by vision and audio. Hence, it is crucial to assess different modalities' unique contributions dynamically.

Based on the above analysis, we aim to extract critical features in the original modalities to reduce redundancy and dynamically integrate independent and complementary information from different modalities. To this end, we propose a simple and flexible model for multimodal sentiment analysis named Adaptive Token Selection and Fusion Network (ATSFN) in this paper. Motivated by attention bottlenecks [15] and trainable queries [2,10] in the transformer, we propose employing adaptive tokens to capture critical information from original modal features and achieve cross-modal dynamic fusion. Firstly, we introduce a token selection transformer that adaptively integrates meaningful information from the original token sequences through a small number of learnable selection tokens. In this way, the original features are compressed into fewer tokens to restrict redundant and noisy information. Secondly, to reduce the impact of heterogeneity, we advocate dynamically adjusting the contribution of each modality during the fusion process. Specifically, we employ a token fusion transformer to dynamically integrate independent and complementary information from different modalities. Trainable fusion tokens are used to estimate the combined weights of each modality and adaptively fuse them. Finally, we employ adaptive pooling to generate the robust and refined multimodal fusion representation and feed it into a classifier for sentiment prediction. We conduct a series of experiments to evaluate the effectiveness of our model. The experimental results demonstrate that ATSFN achieves competitive results and significant improvements on two MSA benchmark datasets.

To summarize, the main contributions of this paper can be summarized as follows:

- We propose a novel adaptive token selection and fusion network for multimodal sentiment analysis, effectively reducing the negative impact of redundancy and heterogeneity during multimodal fusion.
- We utilize fewer adaptive tokens to compress critical features and fuse independent information across modalities. It shows that multimodal features can be integrated using fewer tokens without performance loss.
- We conduct a series of experiments to validate the effectiveness of our model, and experimental results show that our model achieves significant improvements on two public datasets.

## 2 Related Work

Multimodal sentiment analysis aims to determine sentiment polarity from various modalities, such as text descriptions, visual expressions, and audio tones. Multimodal features can provide heterogeneous information from multiple perspectives to express human sentiment tendencies. How to effectively integrate multimodal features has attracted increasing attention in recent years. Fusion methods for MSA can be roughly divided into early, late, and hybrid fusion occurring in the intermediate stage.

Early fusion approaches [14,18] extract features of various modalities and combine all modalities directly at the feature level. In contrast, late fusion methods [17,26] first make decisions according to each modality and then integrate different modalities at the prediction level. Although these works outperform the single-modality methods, they do not explicitly consider the inherent cross-modal dependencies and can not effectively model cross-modal interactions.

Many hybrid fusion methods have recently attempted to exploit cross-modal interactions to explore more flexible intermediate fusion. Tensor Fusion Network [28] calculates the outer product of multimodal features to model intra- and inter-modal interactions. Liu et al. [16] introduce a Low-rank Multimodal Fusion method to perform multimodal fusion with modality-specific low-rank factors. Hazarika et al. [8] project each modality into modality-invariant and modality-specific subspaces with multi-task learning to explore consistency and complementarity between multiple modalities. With the significant progress of Transformer [24], Tsai et al. [22] propose the Multimodal Transformer to learn inter-modal correlations by directional pairwise cross-modal attention. Lin et al. [13] propose a novel hierarchical graph contrastive learning method to explore the complex relationships between multiple modalities. CubeMLP [20] introduces a fully MLP-based processing framework to fuse multimodal features with a much lower computing cost. However, most of the above methods do not consider the redundancy and heterogeneity across modalities, which negatively affects multimodal fusion. Unlike their methods, we focus on extracting critical information in the original model and achieving cross-modal adaptive interactions to mitigate the impact of redundancy and heterogeneity.

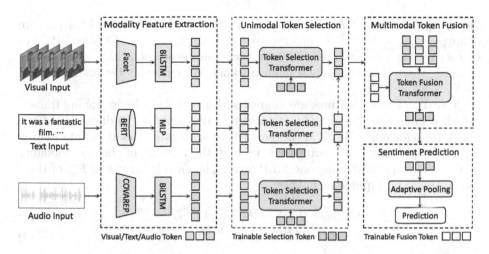

**Fig. 1.** The framework of Adaptive Token Selection and Fusion Network.

# 3   Methodology

The overview of our proposed model is shown in Fig. 1. It consists of the following primary parts: (1) modality feature extraction utilizes modality-specific encoders to convert raw multimodal inputs into sequential token features. (2) unimodal token selection takes the original modal features as input and compresses critical information through fewer trainable selection tokens. (3) multimodal token fusion employs adaptive fusion tokens to measure the contributions of different modalities and dynamically integrate multimodal features. (4) sentiment prediction uses fused multimodal features to predict sentiment scores. The following subsections discuss the details of our model.

## 3.1   Problem Definition

Multimodal sentiment analysis aims to predict the sentiment polarity using multimodal features extracted from video clips. The input comprises three sequences of encoded features from text ($t$), visual ($v$), and audio ($a$) modalities. These raw features are represented as $X_m \in \mathbb{R}^{l_m \times d_m}$, where $m \in \{t, v, a\}$, $l_m$ denotes the sequence length, and $d_m$ represents the embedding dimension of modality $m$. The corresponding sentiment label is denoted as $y$, and the goal of MSA is to utilize multimodal features to predict the sentiment strength $\hat{y}$.

## 3.2   Modality Feature Extraction

This paper adopts the pre-trained BERT-base-uncased model [4] as the text encoder. The BERT encoder takes the raw text transcribed from the video as input and outputs the last hidden layer representation $X_t \in \mathbb{R}^{l_t \times d_t}$. For the visual modality, we utilize Facet [1] to extract facial expression features from continuous video clips, including facial action units, head poses, etc. The visual features are denoted as $X_v \in \mathbb{R}^{l_v \times d_v}$. To process audio frames, we rely on COVAREP [3] to extract various low-level audio features, including speech polarity, glottal closure instants, spectral envelope, and pitch tracking. We represent the audio features as $X_a \in \mathbb{R}^{l_a \times d_a}$.

Visual and audio features are aligned with text tokens by averaging frames. We then input visual and audio features into two separate Bi-directional Long Short-Term Memory (BiLSTM) networks [9] to aggregate contextual information from multiple time steps. After that, we map the features of the text modality to the common dim $d$ using one MLP layer. The encoded features $E_m$ of three modalities are computed as follows:

$$E_{\{v,a\}} = BiLSTM(X_{\{v,a\}}) \in \mathbb{R}^{l_{\{v,a\}} \times d} \tag{1}$$

$$E_t = MLP(X_t) \in \mathbb{R}^{l_t \times d} \tag{2}$$

### 3.3  Unimodal Token Selection

Due to the abundance of redundant information in the original modal features, it is crucial to compress task-related information prior to multimodal fusion. To this end, we propose a transformer-based unimodal token selection module, which enhances the features of each modality by adaptively extracting critical information from the original modalities.

Inspired by the idea of trainable queries [2], we employ a small number of trainable selection tokens $S_m \in \mathbb{R}^{s \times d}$ to capture important features for sentiment prediction, where $s \ll l_m$ denotes the length of a short sequence. These selection tokens compress critical information and reduce redundancy by iterative querying with the original modal features.

We employ the token selection transformer to implement it, which is similar to the transformer decoder. As shown in Fig. 2, each token selection layer consists of token selection, layer normalization (LN), and feed-forward network (FFN) blocks, with residual connections applied. Token selection block adopts the selection attention to compress useful information in original modalities into the selection tokens. We define the formula of attention as $Attn(S_m, E_m) = softmax(\frac{QK^\top}{\sqrt{d}})V$, where queries $Q = W_Q S_m$, $K = W_K E_m$, and $V = W_V E_m$ with the learnable weights $W_{\{Q,K,V\}}$. For the sake of simplicity, we represent token selection as:

$$Selection(S_m, E_m) = softmax(\frac{S_m E_m^\top}{\sqrt{d}})E_m \tag{3}$$

Generally speaking, tokens with smaller attention values are task-irrelevant features. To filter out the minimum values and focus only on the more important tokens, we use a threshold $\tau$ during the selection attention calculation. Tokens with attention values lower than $\tau/n$ will be considered redundant features and ignored, where $n$ is the number of tokens in the original modality. The calculation process of selecting attention is shown on the right side of Fig. 2. In this paper,

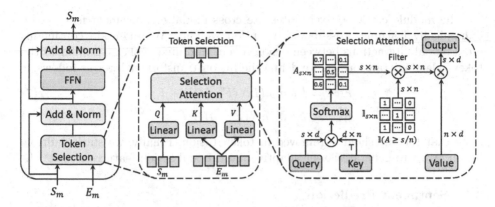

**Fig. 2.** The architecture of the Token Selection Transformer.

the threshold size is consistent with the selection token length $s$, and detailed discussions are introduced in the ablation experiment.

Specifically, each token selection layer used to extract the critical information from the original modality can be expressed as:

$$\hat{S}_m^{l+1} = Selection(LN(S_m^l, E_m^l)) + S_m^l \tag{4}$$

$$S_m^{l+1} = FFN(LN(\hat{S}_m^{l+1})) + \hat{S}_m^{l+1} \tag{5}$$

Here, $l$ denotes the layer number. Each time a transformer layer is passed, the selection token $S_m$ is updated and enhanced using original features $E_m$. Finally, we stack such layers and output the results $S_m \in \mathbb{R}^{s \times d}$ of the last layer as redundant-reduced features for each modality, $m \in \{t, v, a\}$.

## 3.4 Multimodal Token Fusion

The noise and redundancy in the original modal features are filtered out through the unimodal token selection module. However, information exchange among multiple modalities still needs to be explored effectively. Since different modalities often express distinct sentiment tendencies in diverse manners, it is crucial to consider the unique contributions of heterogeneous modalities dynamically.

To this end, we introduce fusion token $F \in \mathbb{R}^{s \times d}$ to adaptively aggregate heterogeneous information from different modalities. For simplicity, we set the number of fusion tokens to be the same as the selection tokens. Specifically, we concatenate all three modal tokens together and treat them as a mixer modality $M = \{S_t, S_v, S_a\} \in \mathbb{R}^{3s \times d}$. Fusion tokens $F$ are used to adaptively estimate the combined weight of each modal token and fuse multimodal features with cross-modal attention. The token fusion function can be defined as:

$$Fusion(F, M) = softmax(\frac{FM^\top}{\sqrt{d}}) \tag{6}$$

The module can adaptively integrate cross-modal enhancement features by dynamically learning cross-modal attention weights. We maintain continuous cross-modal interactions between fusion tokens and mixer tokens. Similar to the token selection transformer, we define the fusion transformer layer as follows:

$$\hat{F}^{l+1} = Fusion(LN(F^l, M^l)) + F^l \tag{7}$$

$$F^{l+1} = FFN(LN(\hat{F}^{l+1})) + \hat{F}^{l+1} \tag{8}$$

Note that the threshold is removed in token fusion. Finally, we stack multiple fusion layers to generate latent adaptive multimodal fusion features $F \in \mathbb{R}^{s \times d}$.

## 3.5 Sentiment Prediction

**Adaptive Pooling.** The multimodal fusion features $F \in \mathbb{R}^{s \times d}$ can be regarded as a semantic sequence $\{h_1, \ldots, h_j, \ldots, h_s\}$. However, not all time-step elements contribute equally to the final sentiment prediction. Therefore, we introduce adaptive pooling to dynamically assign weights to each time step based on its importance and generate an enhanced fused representation.

Specifically, we first perform a nonlinear transformation on $F$, and then use one learnable shared attention vector $q \in \mathbb{R}^{d \times 1}$ to calculate the attention value $\alpha_j$ for each multimodal vector $h_j \in \mathbb{R}^{1 \times d}$ as follows:

$$\alpha_j = q^T \cdot \tanh(W \cdot h_j^T + b) \tag{9}$$

where $W \in \mathbb{R}^{d \times d}$ is the weight matrix, and $b \in \mathbb{R}^{d \times 1}$ is the bias vector. We then normalize the attention values $\alpha_j$ with the softmax function to get the final weights:

$$\beta_j = \frac{\exp(\alpha_j)}{\sum_{j=1}^{s} \exp(\alpha_j)} \tag{10}$$

A large $\beta_j$ implies that the corresponding representation is important. The attended multimodal features are then calculated as the weighted average over the sequence to obtain the final representation:

$$Z = \sum_{j=1}^{s} \beta_j \odot h_j \tag{11}$$

Next, the refined multimodal representation $Z$ is supplied as the input to pass through the fully connected layers to predict the sentiment strength $\hat{y}$.

**Objective Optimization.** After obtaining the final prediction $\hat{y}$, along with the true value $y$, we use the mean absolute error (MAE) as the loss function during training:

$$\mathcal{L}_{MAE} = \frac{1}{N} \sum_{i=1}^{N} |y_i - \hat{y}_i| \tag{12}$$

where $N$ is the number of samples. The mean absolute errors between the prediction and the ground is a common practice in multimodal sentiment analysis.

# 4    Experiments

## 4.1    Datasets

CMU-MOSI [29] and CMU-MOSEI [30] are two challenging multimodal datasets widely utilized in multimodal sentiment analysis. Each dataset comprises videos segmented into several video clips. We evaluate our model on these datasets, and their detailed descriptions are presented below.

**CMU-MOSI.** The CMU-MOSI dataset is a widely used benchmark dataset in multimodal sentiment analysis research. It includes 2,199 subjective video segments collected from 93 YouTube movie reviews. 1,284, 229 and 686 samples are used as training, validation and testing sets.

**CMU-MOSEI.** The CMU-MOSEI dataset is an improved version of the MOSI dataset, with a higher number of video clips and greater variety in samples, speakers, and topics. It comprises 22,856 annotated review video segments sourced from YouTube, with a predefined data split of 16,326 training, 1,871 validation, and 4,659 testing samples.

## 4.2    Evaluation Metrics

The CMU-MOSI and CMU-MOSEI datasets are annotated with a continuous sentiment score from -3 (strongly negative) to 3 (strongly positive). Following recent studies [20,28], we use mean absolute error (MAE) and Pearson correlation (Corr) as evaluation metrics for the regression task. The continuous sentiment tendency can also be transformed into a classification task. For the classification task, we convert the regression outputs into categorical values and use binary classification accuracy (Acc-2) and F1-score (F1) as evaluation metrics.

## 4.3    Training Details

As the lengths of the samples vary, we pad smaller sequences with zeros and truncate longer sequences to match the max length of 100 during multimodal feature extraction. For the MOSI and MOSEI benchmark datasets, the common dim and batch size are {64, 128} and {16, 64}. The number of transformer layers is selected from 1 to 6, while the number of adaptive tokens is {4, 6} for each one, respectively. We use different learning rates to train BERT {2e-5, 1e-5} and other parts of the model {1e-4, 5e-4} and adopt learning rate decay strategies during training. Our model is implemented on one NVIDIA GeForce RTX 3090 GPU using the PyTorch framework and the Adam optimizer. All hyperparameters are determined via the validation set.

## 4.4    Baselines

We compare the performance of our model with baselines for sentiment polarity prediction. The previous models we compare are as follows.

**Table 1.** Comparison of MOSI and MOSEI benchmarks. Models with † indicate that results are retrieved from previous papers. Results with ‡ are reproduced under the same conditions. ↑ means higher is better and ↓ means lower is better. The best results are highlighted in bold.

| Datasets | CMU-MOSI | | | | CMU-MOSEI | | | |
|---|---|---|---|---|---|---|---|---|
| Metrics | MAE(↓) | Corr(↑) | ACC-2(↑) | F1(↑) | MAE(↓) | Corr(↑) | ACC-2(↑) | F1(↑) |
| TFN† | 0.901 | 0.698 | 80.8 | 80.7 | 0.593 | 0.700 | 82.5 | 82.1 |
| LMF† | 0.917 | 0.695 | 82.5 | 82.4 | 0.623 | 0.677 | 82.0 | 82.1 |
| MFM† | 0.877 | 0.706 | 81.7 | 81.6 | 0.568 | 0.717 | 84.4 | 84.3 |
| ICCN† | 0.862 | 0.714 | 83.0 | 83.0 | 0.565 | 0.713 | 84.2 | 84.2 |
| MulT† | 0.861 | 0.711 | 84.1 | 83.9 | 0.580 | 0.703 | 82.5 | 82.3 |
| MISA† | 0.804 | 0.764 | 82.1 | 82.0 | 0.568 | 0.724 | 84.2 | 83.9 |
| MAG-BERT † | 0.808 | 0.761 | 83.1 | 83.1 | 0.552 | 0.756 | 84.6 | 84.6 |
| CubeMLP† | 0.770 | 0.767 | **85.6** | **85.5** | **0.529** | 0.760 | 85.1 | 84.5 |
| CubeMLP‡ | 0.800 | 0.741 | 83.7 | 83.8 | 0.539 | 0.757 | 83.9 | 84.1 |
| ATSFN | **0.746** | **0.782** | 84.8 | 84.7 | 0.534 | **0.761** | **86.1** | **86.0** |

**TFN [28].** TFN (Tensor Fusion Network) models intra and inter-modal dynamics concurrently with local feature extraction network and outer product.

**LMF [11].** LMF (Low-rank Multimodal Fusion) performs multimodal fusion using low-rank tensors to improve efficiency and achieves competitive results

**MFM [23].** MFM (Multimodal Factorization Model) learns to decompose multimodal features into discriminative and modality-specific generative factors.

**ICCN [21].** ICCN (Interaction Canonical Correlation Network) accomplishes the fusion process by calculating the outer product of the text and the other two modalities, and the results are then fed into the CCA network.

**MulT [22].** MulT (Multimodal Transformer) extends standard transformer architecture with cross-modal attention to integrate multimodal sequences.

**MISA [8].** MISA (Modality-Invariant and -Specific Representations) projects each modality into modality-invariant and modality-specific subspaces to learn effective multimodal representations.

**MAG-BERT [19].** MAG-BERT (Multimodal Adaptation Gate for Bert) applies multimodal adaptive gate alignment to the BERT backbone so that they can receive audio and visual information during fine-tuning.

**CubeMLP [20].** CubeMLP introduces three independent MLP units to integrate all modality features and performs mixing operations along three axes.

## 5    Results and Analysis

### 5.1    Summary of Results

The comparison results of our ATSFN model on CMU-MOSI and CMU-MOSEI benchmarks are shown in Table 1. We observe that the proposed model

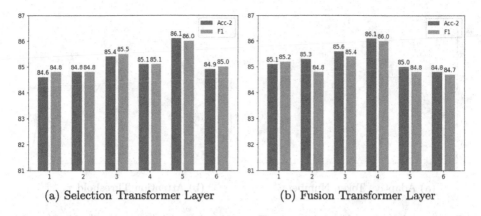

(a) Selection Transformer Layer          (b) Fusion Transformer Layer

**Fig. 3.** Ablation results of Transformer Layer on CMU-MOSEI dataset.

**Table 2.** Ablation analysis of Adaptive Pooling on CMU-MOSEI dataset.

| Pooling Method | MAE($\downarrow$) | Corr($\uparrow$) | ACC-2($\uparrow$) | F1($\uparrow$) |
|---|---|---|---|---|
| Head Token | 0.541 | 0.756 | 85.1 | 84.8 |
| Max Pooling | 0.566 | 0.761 | 85.6 | 85.5 |
| Average Pooling | 0.556 | 0.757 | 85.6 | 85.6 |
| Adaptive Pooling | **0.534** | **0.761** | **86.1** | **86.0** |

performs well compared to the adopted baselines. Compared with tensor-based methods TFN and LMF, the ATSFN adaptively extracts critical information to learn effective multimodal representations while reducing redundancy. Our model performs better than the representation-based method MISA, illustrating the effectiveness of dynamic interactions across modalities. ATSFN consistently improves over transformer-based MulT and MAG-BERT models, demonstrating the feasibility of adaptive token selection and fusion for MSA. Furthermore, our model performs slightly worse on the MOSI dataset than CubeMLP. This may be attributed to the small size of the MOSI, making it difficult for the transformer to capture important multimodal features adaptively.

## 5.2   Ablation Analysis

**Effect of Transformer Layer.** This paper adopts the transformer as the main architecture to achieve multimodal adaptive token selection and fusion. Consequently, we investigate the impact of the number of transformer layers on model performance. Figure 3a and 3b present the performance of our method with varying numbers of transformer layers. An increase in layers results in improved performance due to the increased information capture. However, excessively large numbers of layers lead to performance degradation, indicating potential overfitting. Hence, employing an appropriate number of layers is necessary.

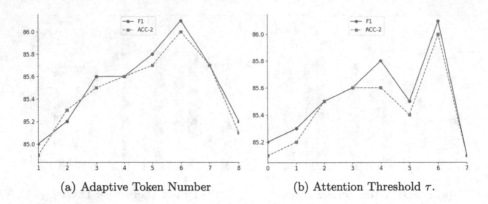

(a) Adaptive Token Number          (b) Attention Threshold $\tau$.

**Fig. 4.** Ablation results of Adaptive Token Number and Attention Threshold $\tau$.

**Effect of Adaptive Token Number.** This paper uses adaptive tokens to extract critical task features and perform cross-modal dynamic fusion. We evaluate our model with different adaptive token numbers from 1 to 8 on the MOSEI dataset to quantify the impact. As shown in Fig. 4a, the model with smaller tokens can lead to poor performance due to the loss of useful information. In contrast, too large one makes the model focus on too many features, resulting in redundancy and performance degradation. Therefore, selecting a middle number of adaptive tokens is more advantageous.

**Effect of Attention Threshold $\tau$.** We use a threshold $\tau$ to filter out redundant tokens and focus on selecting the most important parts of the attention map. We study the impact of attention threshold $\tau$. As shown in Fig. 4b, the model's performance is relatively poor when no threshold is used. As the threshold increases, the model's performance steadily improves, suggesting that filtering out irrelevant information in the original modality is beneficial. Interestingly, when $\tau$ equals the number of selection tokens $s$, the model performs best on the MOSEI dataset. Continuing to increase the threshold will cause model performance to degrade. The possible reason is that an excessive threshold will cause critical information to be discarded.

**Effect of Adaptive Pooling.** We explore the impact of adaptive pooling on sentiment prediction. See the experimental results of the head token, max pooling, and average pooling strategies on the MOSEI dataset in Table 2. Head token refers to using the head token in the multimodal feature sequence as the global representation, but it does not provide sufficient sentiment information, leading to poor performance. Average pooling and max pooling perform equally well, and both can provide more sentiment features. Adaptive pooling performs best compared to several pooling methods mentioned above, which shows that it helps capture the features most relevant for sentiment prediction.

**Table 3.** Case study with some sample inputs and predictions.

| Case | Text | Visual | Audio | Pred | Truth |
|------|------|--------|-------|------|-------|
| A | *I thought the other two were pretty exciting.* | Smiling face | Rising tone | 1.1316 | 1.0000 |
| B | *It's a film currently in theaters.* | No expression | Peaceful tone | 0.0012 | 0.0000 |
| C | *(umm) I really really wish he could have done a better job.* | Frown & Stress | Pitch variation | -1.3300 | -1.3333 |
| D | *So, I would highly recommend that you save your money (uhh).* | No expression | Peaceful tone | 0.9938 | -1.6666 |

## 5.3 Case Study

Some predicted and true values are shown in Table 3, along with the corresponding raw data inputs (for visual and audio, we only illustrate literally). In case A, the model can capture sentiment words such as "exciting" and combine them with the visual modality "Smiling face" to make a positive prediction. Case B is a neutral case; the emotional tendencies of the three modalities are consistent, and our model predicts it well. For case C, no clear prediction can be made by considering text modality alone. What is exciting is that our model is able to combine audio and visual cues to dynamically capture cross-modal associations to correctly infer negative sentiment, which is consistent with our motivation.

In case D, the text description "highly recommend" confuses the model, while neither visual nor audio clues are provided. The model focuses on the positive emotional tendencies of the text but ultimately misleads the incorrect predictions due to insufficient contextual and auxiliary cues. It can be seen that text modality plays an indispensable role in MSA, and it is crucial to integrate independent and complementary information from multiple modalities.

## 6    Conclusion

This paper proposes a novel model named ATSFN for multimodal sentiment analysis. Compared with existing methods, we adopt fewer trainable adaptive tokens to filter the redundant information of the original modal features and achieve adaptive multimodal fusion to mitigate cross-modal heterogeneity. Experimental results on two benchmark datasets show the superiority of the ATSFN model. Furthermore, ablation experiments further validate the effectiveness of our proposed framework and modules.

In future work, we will expand ATSFN to more multimodal fusion tasks and explore how to use text modality to guide multimodal feature fusion effectively.

**Acknowledgments.** This work was supported by the National Natural Science Foundation of China (No.62272025 and No.U22B2021) and the Fund of the State Key Laboratory of Software Development Environment.

# References

1. Baltrušaitis, T., Robinson, P., Morency, L.P.: Openface: an open source facial behavior analysis toolkit. In: 2016 IEEE Winter Conference on Applications of Computer Vision (WACV). IEEE (2016)
2. Carion, Nicolas, Massa, Francisco, Synnaeve, Gabriel, Usunier, Nicolas, Kirillov, Alexander, Zagoruyko, Sergey: End-to-End Object Detection with Transformers. In: Vedaldi, Andrea, Bischof, Horst, Brox, Thomas, Frahm, Jan-Michael. (eds.) ECCV 2020. LNCS, vol. 12346, pp. 213–229. Springer, Cham (2020). https://doi.org/10.1007/978-3-030-58452-8_13
3. Degottex, G., Kane, J., Drugman, T., Raitio, T., Scherer, S.: Covarep-a collaborative voice analysis repository for speech technologies. In: 2014 IEEE International Conference on Acoustics, Speech and Signal Processing (ICASSP). IEEE (2014)
4. Devlin, J., Chang, M.W., Lee, K., Toutanova, K.: Bert: Pre-training of deep bidirectional transformers for language understanding. ArXiv preprint. arXiv:1810.04805 (2018)
5. Du, Pengfei, Gao, Yali, Li, Xiaoyong: Bi-attention Modal Separation Network for Multimodal Video Fusion. In: Þór Jónsson, Björn., Gurrin, Cathal, Tran, Minh-Triet., Dang-Nguyen, Duc-Tien., Hu, Anita Min-Chun., Huynh Thi Thanh, Binh, Huet, Benoit (eds.) MMM 2022. LNCS, vol. 13141, pp. 585–598. Springer, Cham (2022). https://doi.org/10.1007/978-3-030-98358-1_46
6. Han, W., Chen, H., Gelbukh, A., Zadeh, A., Morency, L.P., Poria, S.: Bi-bimodal modality fusion for correlation-controlled multimodal sentiment analysis. In: Proceedings of the 2021 International Conference on Multimodal Interaction, pp. 6–15 (2021)
7. Han, W., Chen, H., Poria, S.: Improving multimodal fusion with hierarchical mutual information maximization for multimodal sentiment analysis. In: Proceedings of the 2021 Conference on Empirical Methods in Natural Language Processing (Nov 2021)
8. Hazarika, D., Zimmermann, R., Poria, S.: Misa: Modality-invariant and-specific representations for multimodal sentiment analysis. In: Proceedings of the 28th ACM International Conference on Multimedia (2020)
9. Hochreiter, S., Schmidhuber, J.: Long short-term memory. Neural Comput. **9**(8), 1735–1780 (1997)
10. Iashin, V., Xie, W., Rahtu, E., Zisserman, A.: Sparse in space and time: Audio-visual synchronisation with trainable selectors. In: 33rd British Machine Vision Conference 2022, BMVC 2022, London, UK, November, pp. 21–24, 2022. BMVA Press (2022)
11. Jin, T., Huang, S., Li, Y., Zhang, Z.: Dual low-rank multimodal fusion. In: Findings of the Association for Computational Linguistics: EMNLP 2020, pp. 377–387 (2020)
12. Kumar, A., Vepa, J.: Gated mechanism for attention based multi modal sentiment analysis. In: ICASSP IEEE International Conference on Acoustics, Speech and Signal Processing (ICASSP). IEEE (2020)
13. Lin, Z., et al.: Modeling intra-and inter-modal relations: Hierarchical graph contrastive learning for multimodal sentiment analysis. In: Proceedings of the 29th International Conference on Computational Linguistics (2022)
14. Morency, L.P., Mihalcea, R., Doshi, P.: Towards multimodal sentiment analysis: Harvesting opinions from the web. In: Proceedings of the 13th International Conference on Multimodal Interfaces, pp. 169–176 (2011)

15. Nagrani, A., Yang, S., Arnab, A., Jansen, A., Schmid, C., Sun, C.: Attention bottle-necks for multimodal fusion. Advances in Neural Information Processing Systems 34 (2021)
16. Nguyen, D., Nguyen, K., Sridharan, S., Dean, D., Fookes, C.: Deep spatio-temporal feature fusion with compact bilinear pooling for multimodal emotion recognition. Comput. Vis. Image Underst. **174**, 33–42 (2018)
17. Nojavanasghari, B., Gopinath, D., Koushik, J., Baltrušaitis, T., Morency, L.P.: Deep multimodal fusion for persuasiveness prediction. In: Proceedings of the 18th ACM International Conference on Multimodal Interaction (2016)
18. Poria, S., Chaturvedi, I., Cambria, E., Hussain, A.: Convolutional MKL based multimodal emotion recognition and sentiment analysis. In: 2016 IEEE 16th International Conference on Data Mining (ICDM). IEEE (2016)
19. Rahman, W., et al.: Integrating multimodal information in large pretrained transformers. In: Proceedings of the Conference. Association for Computational Linguistics. Meeting. vol. 2020. NIH Public Access (2020)
20. Sun, H., Wang, H., Liu, J., Chen, Y.W., Lin, L.: CubeMLP: An MLP-based model for multimodal sentiment analysis and depression estimation. In: Proceedings of the 30th ACM International Conference on Multimedia (2022)
21. Sun, Z., Sarma, P., Sethares, W., Liang, Y.: Learning relationships between text, audio, and video via deep canonical correlation for multimodal language analysis. In: Proceedings of the AAAI Conference on Artificial Intelligence. vol. 34 (2020)
22. Tsai, Y.H.H., Bai, S., Liang, P.P., Kolter, J.Z., Morency, L.P., Salakhutdinov, R.: Multimodal transformer for unaligned multimodal language sequences. In: Proceedings of the Conference. Association for Computational Linguistics. Meeting vol. 2019. NIH Public Access (2019)
23. Tsai, Y.H.H., Liang, P.P., Zadeh, A., Morency, L.P., Salakhutdinov, R.: Learning factorized multimodal representations. In: International Conference on Learning Representations (2019)
24. Vaswani, A., Shazeer, et al.: Attention is all you need. In: Advances in Neural Information Processing Systems. vol. 30. Curran Associates, Inc. (2017)
25. Wang, Y., Shen, Y., Liu, Z., Liang, P.P., Zadeh, A., Morency, L.P.: Words can shift: Dynamically adjusting word representations using nonverbal behaviors. In: Proceedings of the AAAI Conference on Artificial Intelligence. vol. 33 (2019)
26. Wu, C.H., Liang, W.B.: Emotion recognition of affective speech based on multiple classifiers using acoustic-prosodic information and semantic labels. IEEE Trans. Affect. Comput. **2**(1), 10–21 (2010)
27. Yu, W., Xu, H., Yuan, Z., Wu, J.: Learning modality-specific representations with self-supervised multi-task learning for multimodal sentiment analysis. In: Proceedings of the AAAI Conference on Artificial Intelligence. vol. 35 (2021)
28. Zadeh, A., Chen, M., Poria, S., Cambria, E., Morency, L.P.: Tensor fusion network for multimodal sentiment analysis. In: Proceedings of the 2017 Conference on Empirical Methods in Natural Language Processing (2017)
29. Zadeh, A., Zellers, R., Pincus, E., Morency, L.P.: Multimodal sentiment intensity analysis in videos: facial gestures and verbal messages. IEEE Intell. Syst. **31**(6), 82–88 (2016)
30. Zadeh, A.B., Liang, P.P., Poria, S., Cambria, E., Morency, L.P.: Multimodal language analysis in the wild: CMU-MOSEI dataset and interpretable dynamic fusion graph. In: Proceedings of the 56th Annual Meeting of the Association for Computational Linguistics (Volume 1: Long Papers) (2018)
31. Zhu, L., Zhu, Z., Zhang, C., Xu, Y., Kong, X.: Multimodal sentiment analysis based on fusion methods: a survey. Inf. Fusion **95**, 306–325 (2023)

# Exploring Imperceptible Adversarial Examples in $YC_bC_r$ Color Space

Pei Chen, Zhiyong Feng, Meng Xing[✉], Yiming Zhang, and Jinqing Zheng

College of Intelligence and Computing, Tianjin University, Tianjin, China
{chenpei,zyfeng,xingmeng,kevin_zhangym,cszjq}@tju.edu.cn

**Abstract.** Numerous studies have shown that well-designed perturbations can easily fool deep neural networks. Existing attacks are mainly conducted on the low-level pixels of $RGB$ images, resulting in noise-like perturbations distributed over the entire image, highly vulnerable and low attack transferability. Furthermore, they delve into the data space with point-wise perturbation, which may neglect the geometric characteristics and fail to study the role and impact of various image components. Compared with $RGB$ images, $YC_bC_r$ images can express various image components more intuitively. In this paper, we propose generating semantically preserved adversarial examples by perturbing the frequency band energy corresponding to inconspicuous colors and textures in the $YC_bC_r$ color space. Specifically, we first transform clean images from spatial to frequency domain, followed by applying a fusion module to indirectly inject perturbations. Moreover, the low-frequency constraint and luma-chroma optimization strategy are further introduced to ensure visual imperceptibility. Extensive experiments on multiple datasets indicate that our attack retains a high attack success rate while significantly improving visual quality.

**Keywords:** Adversarial examples · $YC_bC_r$ · Imperceptibility

## 1 Introduction

Recent studies [2,9,20,21] have shown that Deep Neural Networks (DNNs) can be susceptible to adversarial examples created by adding subtle perturbations to normal images, thereby deceiving the DNNs. In practice, a successful adversarial example needs to satisfy a high attack success rate while possessing low perceptibility to the human visual system (HVS), i.e., the adversarial example retains its semantic characteristics.

Many attacks search for pix-level invasions [2,3,18,21] within a bounded $L_p$ norm ($p = 0,\ 2,\ \infty$) in the $RGB$ color space. However, the distance between points in the $RGB$ color space has a weak correlation with the perceptual difference between the colors they represent [31]. Furthermore, the perception of human eyes to the error of different image regions is different and image processing methods have different effects on the different areas of the image, the

© The Author(s), under exclusive license to Springer Nature Switzerland AG 2024
S. Rudinac et al. (Eds.): MMM 2024, LNCS 14556, pp. 242–256, 2024.
https://doi.org/10.1007/978-3-031-53311-2_18

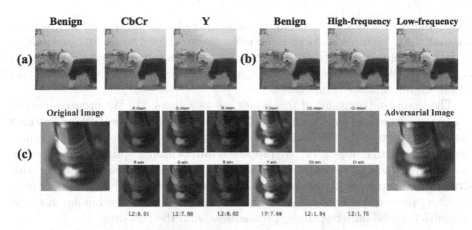

**Fig. 1.** (a): Visual comparison of chrominance and luminance channels distortions. (b): Visual comparison of low-frequency and high-frequency components distortions. Both using the same settings and exaggerated perturbations to better illustrate the effects. (c): Qualitative comparison of $L2$ distances each channel between the original and adversarial images in both $RGB$ and $YC_bC_r$ spaces. We use FGSM attack($\epsilon = 0.04$) to create adversarial images.

$RGB$ space is not perceptually uniform and lacks semantic clues [9] to achieve a good trade-off between attacking strength and stealthiness, resulting in low transferability as well as vulnerability to existing defense methods. [28] directly manipulated spatial pixel flows without $L_p$ constraints, but it lacks explicit visual representation. [31] leverage unrestricted color attacks, which may be easily distinguishable by humans. Recently, it has been demonstrated that removing existing information from the original image is an effective method for generating adversaries [6]. However, AdvDrop [6] may randomly discard certain visually perceptible features and blur the smooth region of images. These challenges raise an important question: *Is it feasible to achieve imperceptible perturbations while maintaining a high attack success rate and robustness?*

In this paper, we utilize a perturbation module to modify the energy of inconspicuous color and texture corresponding frequency bands in the $YC_bC_r$ color space via the adversarial loss. The indirect injection of adversary into frequency domain avoids the redundant noise of attacks in the pixel level, leading to a more invisible attack. Compared with the $RGB$ space, we discover that the human perceptual mechanism has significant sensitivity differences for different channels in the $YC_bC_r$ space, as shown in Fig. 1(a). HVS is more sensitive to information loss in the $Y$ channel (i.e., luminance component) than the $C_b$ and $C_r$ channels (i.e., chrominance components), which means that significant modifications of the $C_b$ and $C_r$ channels can effectively ensure attack strength with less semantic information loss. Therefore, we design a luma-chroma optimization strategy to achieve a better trade-off between attacking strength and stealthiness. Furthermore, signal-processing techniques [19] have demonstrated that low-frequency components maintain the basic structures, while high-frequency components correspond to the texture and important details of objects. In Fig. 1(c), we can see that the $Y$ channel provides more information related to texture and shape. As HVS is more sensitive to distortions in low-frequency components, as depicted in Fig. 1(b). We introduce a low-frequency constraint to restrict perturbations within high-frequency components.

The main contributions of this paper are as follows: (1) We propose a novel frequency adversarial attack using the $YC_bC_r$ color space by perturbing features that are important for DNNs but insensitive to HVS while semantically representing the original object. (2) We propose a luma-chroma optimization strategy to balance attacking strength and stealthiness. (3) A low-frequency constraint strategy is introduced to further boost the imperceptible adversarial samples. (4) Extensive experiments demonstrate the efficiency of our technique in terms of transferability, imperceptibility, robustness, and the run time.

## 2    Related Work

**Adversarial Attacks.** A host of adversarial attack techniques have been developed to attack DNNs. [9] propose FGSM, which generates adversarial examples instantly. After that, BIM [18] and PGD [21] were proposed. They are multi-step attacks. Applying the uniform intensity of perturbations to the entire image will destroy the smooth region and is easily perceptible. In addition, the C&W [2] method achieves a powerful attack effect by constructing upper bounds on the robustness of neural networks, but it is computationally exhausted. AutoAttack [3] is proposed by integrating multiple attack methods. Zhao et al. [33] introduce PerC-AL to improve inconspicuousness by constraining perturbations through minimizing the perceptual color distance. [7,31] generate adversarial examples by modifying lighting, color, and other factors. [12] exploit motion blurring to lose details in a pixel-wise way. [28,31] propose unrestricted attacks by leveraging certain attributes of the images. [14] manually distinguishes semantic classes that are sensitive to the human eye, but it does not scale to complex datasets.

**Frequency-Based Analysis and Attacks.** Prior to our work, several papers have explored DNNs in the frequency domain. Wang et al. [26] observe that DNNs could capture high-frequency components of images that are imperceptible. Dong et al [30] find that naturally trained models are highly sensitive to additive perturbations in high frequencies. Guo et al. [10] propose a low-frequency random search method to achieve black-box attacks. Although low-frequency perturbations are effective, the visual quality of adversarial images is noticeably diminished. Duan et al [6] propose the AdvDrop attack which generates adversarial examples by dropping existing features of clean images. However, due to the limited and reduced information in the generated adversarial examples, it is challenging to improve the confidence of predictions and there may be quantization artifacts present. [20] propose a semantic similarity attack, but it requires more time to generate adversarial examples. [15] perturb transform coefficients in the graph spectral domain to 3D Point Cloud. [1] create adversarial images via swapping image components.

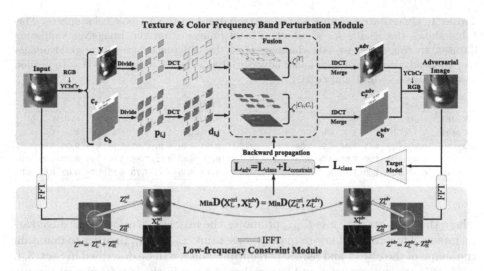

**Fig. 2.** The pipeline of the proposed method. Top: Texture & Color Frequency Band Perturbation Module; Bottom: Low-frequency Constraint Module.

## 3 Methodology

### 3.1 Overview

Consider an image $x$ denotes the original image and an image classifier $C$. An adversary aims to compute a small distortion $\delta$ into a clean image for generating $x_{adv}$ so that the classifier predicts $C(x_{adv}) \neq C(x)$ (untargeted attack) or $C(x_{adv}) = t$ (targeted attack), where $t$ is the target class. Typically, an adversarial example $x_{adv}$ is bounded by the $L_p$ norm. Our goal is to quickly generate adversarial examples that are both insensitive to the human visual system (HVS) and robust. Specifically, the first step is to convert the $RGB$ image to $YC_bC_r$ space and then transform channels into the frequency domain through discrete cosine transform(DCT). The second step involves perturbation fusion for each channel to modify the energy of the relative high-frequency components of the transformed image. Finally, we transform the image back to the $RGB$ domain. Now our constrained optimization problem becomes:

$$\min_{\zeta^{[c]}} \mathcal{L}_{adv}\left(x_{adv}, x, y\right), \text{where } x_{adv} = \mathcal{T}^{-1}\left(\mathcal{F}\left(\mathcal{T}(x), \zeta^{[c]}\right)\right)$$
$$\text{s.t. } \left\|\zeta^{[c]}\right\|_{\infty} < \epsilon^{[c]}, c \in \{Y, C_b, C_r\} \tag{1}$$

where $\mathcal{T}(\cdot)$ and $\mathcal{T}^{-1}(\cdot)$ denote DCT and IDCT, $\mathcal{F}$ represents the fusion module to modify the energy in the spectral domain, and $\mathcal{L}_{adv}(\cdot)$ is a loss function connecting the model. Meanwhile, $\zeta^{[c]}$ and $\epsilon^{[c]}$ denote the perturbation that is directly optimized and its perturbation budget. We further define our adversarial loss $\mathcal{L}_{adv}\left(x_{adv}, x, y\right)$ as follows:

$$\mathcal{L}_{adv}\left(x_{adv}, x, y\right) = \mathcal{L}_{class}\left(x_{adv}, y\right) + \beta \cdot \mathcal{L}_{constrain}\left(x_L^{ori}, x_L^{adv}\right) \tag{2}$$

**Table 1.** Quantitative comparison of perturbations in the $YCbCr$ color spaces. This table shows the results for a subset of 5000 images from the ImageNet validation dataset. In this table, we take the average of the difference between original images and their adversarial versions under adversarial attacks FGSM and PGD with epsilon values between $\epsilon=0.01$ and $\epsilon=0.04$.

| MODELS | FGSM | | | | | | | | | PGD | | | | | | | | |
|---|---|---|---|---|---|---|---|---|---|---|---|---|---|---|---|---|---|---|
| | e=0.01 | | | e=0.02 | | | e=0.04 | | | e=0.01 | | | e=0.02 | | | e=0.04 | | |
| YCbCr | Y | Cb | Cr | Y | Cb | Cr | Y | Cb | Cr | Y | Cb | Cr | Y | Cb | Cr | Y | Cb | Cr |
| Vgg | **6.30** | 4.19 | 4.15 | **7.48** | 4.76 | 4.68 | **8.51** | 5.22 | 5.11 | **6.60** | 4.40 | 4.46 | **7.95** | 5.13 | 5.19 | **9.26** | 5.80 | 5.94 |
| Resnet | **6.40** | 4.24 | 4.20 | **7.58** | 4.78 | 4.70 | **8.60** | 5.20 | 5.10 | **6.50** | 4.40 | 4.52 | **7.86** | 5.16 | 5.21 | **9.09** | 5.98 | 6.10 |
| Densenet | **6.40** | 4.27 | 4.30 | **7.13** | 4.85 | 4.88 | **8.10** | 5.33 | 5.35 | **6.30** | 4.40 | 4.52 | **7.73** | 5.19 | 5.28 | **9.02** | 6.01 | 6.16 |

where the cross-entropy loss $\mathcal{L}_{\text{class}}$ promotes the misclassification of the adversarial image. The proposed low-frequency constraint, $\mathcal{L}_{\text{constrain}}$, in order to maintain the shape of the object and reduce perceptible noise, will be described in Sect. 3.4. $x_L^{adv}$, $x_L^{ori}$ are the reconstructed images based on only the low-frequency components of $x_{adv}$, $x$, respectively. The overview of the proposed method is illustrated in Fig. 2.

### 3.2 Luma-Chroma Optimization

Typically, adversarial perturbations are computed using the $RGB$ color space. However, using different color spaces can be beneficial in terms of some fields [31]. For example, the JPEG compression algorithm [25] leverages the $YC_bC_r$ color space, downsampling the chroma channels to reduce the storage size of images. In the $YC_bC_r$ and $RGB$ color spaces, we calculate the per-channel $L2$ difference. Our results, shown in Fig. 1(c), qualitatively demonstrate that each channel in the $RGB$ space presents similar amounts of perturbations. However, in the $YC_bC_r$ color space, the $Y$ channel exhibits more substantial disturbance than the rest of channels, indicating that the energy of invasions is more concentrated on the $Y$ than the $C_b$ and $C_r$ channels. We further quantitatively illustrate this conclusion in Table 1. Based on the analysis of perturbations discussed above, as an alternative to existing attacks in the $RGB$ domain, we are motivated to perform adversarial attacks in the $YC_bC_r$ space. To mitigate the visual distortion caused by information loss by leveraging perceptual features, we propose assigning larger values for $\epsilon^{[C_b]}$, $\epsilon^{[C_r]}$ than $\epsilon^{[Y]}$.

### 3.3 Spectrum Transformation

We transform the already obtained $YC_bC_r$ channels to the corresponding frequency domain for facilitating subsequent processing. In particular, we use the DCT, which expresses the image as a set of cosine functions oscillating at different frequencies. DCT has better energy concentration than the discrete Fourier transform (DFT). In practice, we divide the images into a group of disjoint $K \times K$ blocks with $K = 8$. Larger block sizes lead to the computation time-consuming, while smaller block sizes result in severe image distortion. The definition of two-dimensional DCT is:

$$F(u,v) = C_u C_v \sum_{i=0}^{N-1} \sum_{j=0}^{N-1} f(i,j)$$

$$\cos\left[\frac{\pi u(2i+1)}{2N}\right] \cos\left[\frac{\pi v(2j+1)}{2N}\right]$$

$$(3)$$

where $f(i,j)$ is the value on the coordinate $(i,j)$ of image, $C_u$, $C_v$ can be considered as the compensation coefficient for making the DCT matrix become an orthogonal matrix, and $N$ is the block size.

We deploy a trainable perturbation $\zeta^{[c]}$ to perturb the spectral representation of the image. To adaptively adjust the perturbation $\zeta^{[c]}$ and improve the success rate of the proposed spectral attack, we solve the optimization problem in Eq. 1 by leveraging the gradients of the target model $f(\cdot)$ through backward propagation. The complete fusion module is defined as:

$$\mathcal{F}\left(x, \zeta^{[c]}\right) = \mathcal{T}(x) + \mathcal{T}\left(\zeta^{[c]}\right)$$

$$= \mathcal{T}\left(x + \zeta^{[c]}\right)$$

$$(4)$$

During the adaptive learning process, we iteratively adjust the perturbation $\zeta^{[c]}$ to increase the success rate of the attack. Specifically, we update the perturbation $\zeta^{[c]}$ with the gradients and learn the perturbation $\zeta^{[c]}$ as:

$$\zeta_{(k+1)}^{[c]} = \zeta_{(k)}^{[c]} + lr \cdot \text{sign}\left(\nabla_{\zeta^{[c]}} \mathcal{L}_{adv}\left(x_{adv}, x, y\right)\right), c \in \{Y, C_b, C_r\} \quad (5)$$

where $lr$ represents the learning rate, and $k$ denotes the number of iterations.

In conclusion, we apply the above methods to $x^{C_b,C_r}$ and $x^Y$ to obtain $x_{abv}^{C_b,C_r}$ and $x_{abv}^Y$, followed by concatenating the input image adversarial chrominance $x_{abv}^{C_b,C_r}$ and luminance $x_{abv}^Y$ channels to obtain the adversarial example $x_{abv}^{Y,C_b,C_r}$.

### 3.4 Low-Frequency Constraint

Although the perturbations are added and optimized in the frequency, these perturbations still have the potential to distort certain frequency bands, resulting in visually perceptible distortions in local regions. On the other hand, traditional constraints may lead to disturbances randomly distributed over the image. Therefore, we seek a new constraint to limit them in regions less sensitive to HVS.

The low-frequency components correspond to the main part of the image, characterized by large and flat regions which are more noticeable to HVS. In contrast, the high-frequency components typically consist of the edges, noise, and detailed parts of the image, which are not easily perceived. First, we use Fourier transformer to convert the image from spatial domain to frequency domain to get the spectrum $Z$. Then we decompose the spectrum to obtain the high frequency component $Z_H$ and low frequency component $Z_L$. The discrete Fourier transform (DFT) $\mathscr{F} = F(u, v)$ can be formulated as follows:

$$F(u, v) = \sum_{x=0}^{M-1} \sum_{y=0}^{N-1} f(x, y) \left( \cos 2\pi \left( \frac{ux}{M} + \frac{vy}{N} \right) - i \sin 2\pi \left( \frac{ux}{M} + \frac{vy}{N} \right) \right) \quad (6)$$

where $M$ and $N$ denote the height and width of the image respectively; $f(x, y)$, the value of the image at position $(x, y)$ in the spatial domain; $i$ denotes the imaginary unit; $u = 0, ..., M - 1, v = 0, ..., N - 1$. $F(u, v)$ is the complex value of the spectrum.

After obtaining the spectrum, we shift frequency components with low values to the center, and then extract the high frequency component $Z_H$ and low frequency component $Z_L$ from the shifted spectrum. To do this, we calculate the Euclidean distance between each point and the center point of the spectrum. We then compare this distance to a predetermined threshold value that has been set. We can define the decomposition function as follows:

$$\begin{aligned} z_L(i, j) &= \begin{cases} z(i, j), & \text{if } d\left((i, j), (c_i, c_j)\right) \leq r \\ 0, & \text{otherwise} \end{cases} \\ z_H(i, j) &= \begin{cases} 0, & \text{if } d\left((i, j), (c_i, c_j)\right) \leq r \\ z(i, j), & \text{otherwise} \end{cases} \end{aligned} \quad (7)$$

where $d(\cdot, \cdot)$ is the function to calculate the Euclidean distance. If the input is a color image, the above processing is performed independently for each color channel of the image. Using inverse Fourier transform $\mathscr{F}^{-1}$, we can obtain $X_L$ from $Z_L$, the low-frequency component of the image.

Inspired by Focal Frequency Loss (FFL) [16] used in image reconstruction and synthesis tasks, we utilize it to constrain low-frequency component and calculate the distance between adversarial and original examples. Since the result of the Fourier transform is a complex number, $Z$ can be denoted as $Z = a + bi$. Let the $Z_L^{ori} = a_{ori} + b_{ori}i$ be the low frequency component of the original image spectrum, and the corresponding $Z_L^{adv} = a_{adv} + b_{adv}i$ with the same meaning w.r.t. the adversarial image spectrum. The distance between the original image $Z_L^{ori}$ and the adversarial attack $Z_L^{adv}$ can be calculated as: $D\left(z_L^{ori}, z_L^{adv}\right) = d\left((a_{ori}, b_{ori}), (a_{adv}, b_{adv})\right)$.

## 4     Experiments

### 4.1     Experimental Setup

We evaluate the performance of the comparison methods on CIFAR-10 [17], CIFAR-100 [17], and ImageNet-1K [24]. The CIFAR-10 dataset consists of 60,000 images in 10 classes of size 32×32; CIFAR-100 contains 100 classes with the same number of samples as CIFAR-10. ImageNet dataset comprises 1.28 million images from 1000 classes.

**Table 2.** Attack success rate (ASR) and five evaluation metrics by different attacks on the CIFAR-10 and CIFAR-100 datasets in the untargeted scenario. ↓ means the value is lower the better, and vice versa. The best results are marked in bold.

| Dataset | Attack | $l_2$ ↓ | $l_\infty$ ↓ | FID↓ | SSIM↑ | LF↓ | ASR(%)↑ |
|---------|--------|---------|--------------|------|-------|-----|---------|
| CIFAR-10 | StepLL [18] | 0.85 | **0.03** | 14.12 | 0.982 | 0.23 | 98.10 |
| | MIM [5] | 1.90 | **0.03** | 25.57 | 0.956 | 0.48 | 98.64 |
| | PGD [21] | 1.28 | **0.03** | 26.88 | 0.962 | 0.34 | 99.94 |
| | C&W [2] | 0.78 | 0.08 | 10.61 | 0.970 | 0.18 | **100.00** |
| | AdvDrop [6] | 1.13 | 0.08 | 15.19 | 0.973 | 0.42 | 99.36 |
| | AutoAttack [3] | 1.49 | **0.03** | 28.28 | 0.965 | 0.40 | **100.00** |
| | PerC-AL [33] | 0.86 | 0.18 | 9.58 | 0.978 | 0.15 | 98.29 |
| | Ours | **0.55** | 0.05 | **6.74** | **0.988** | **0.11** | 99.56 |
| CIFAR-100 | StepLL [18] | 0.83 | **0.03** | 12.67 | 0.976 | 0.32 | 97.73 |
| | MIM [5] | 1.87 | **0.03** | 22.63 | 0.942 | 0.65 | 98.26 |
| | PGD [21] | 1.29 | **0.03** | 24.77 | 0.958 | 0.42 | 99.62 |
| | C&W [2] | 0.92 | 0.08 | 12.41 | 0.969 | 0.25 | **100.00** |
| | AdvDrop [6] | 1.09 | 0.08 | 15.56 | 0.970 | 0.43 | 99.40 |
| | AutoAttack [3] | 1.52 | **0.03** | 26.01 | 0.952 | 0.56 | **100.00** |
| | PerC-AL [33] | 1.47 | 0.20 | 12.91 | 0.954 | 0.39 | 99.66 |
| | Ours | **0.72** | 0.05 | **8.13** | **0.983** | **0.19** | 99.89 |

**Table 3.** Attack success rate (ASR) and evaluation metrics by different attacks on the ImageNet-1K datasets in the untargeted scenario.

| Dataset | Attack | $l_2$ ↓ | $l_\infty$ ↓ | FID↓ | SSIM↑ | LPIPS↓ | RunTime(s)↓ | ASR(%)↑ |
|---------|--------|---------|--------------|------|-------|--------|-------------|---------|
| ImageNet-1K | StepLL [18] | 26.85 | **0.03** | 39.61 | 0.948 | 0.1443 | 3998 | 98.76 |
| | PGD [21] | 54.97 | **0.03** | 37.21 | 0.891 | 0.2155 | **3451** | 99.56 |
| | C&W [2] | 6.81 | 0.05 | 12.14 | 0.977 | 0.0617 | > 100000 | 99.27 |
| | AdvDrop [6] | 14.95 | 0.07 | 9.68 | 0.977 | 0.0639 | 48355 | 99.76 |
| | AutoAttack [3] | 66.49 | **0.03** | 53.55 | 0.885 | 0.2549 | 27312 | **99.99** |
| | SSAH [20] | 4.97 | **0.03** | **5.22** | 0.991 | 0.0352 | 38018 | 98.01 |
| | PerC-AL [33] | **4.65** | 0.12 | 11.56 | 0.992 | 0.0439 | > 100000 | 98.78 |
| | Ours | 4.77 | **0.03** | 5.31 | **0.995** | **0.0328** | 7847 | 99.24 |

We use Attack Success Rate (ASR) to evaluate the performance of the attacks and employ different metrics to evaluate the quality of the generated adversarial images, including conventional $L_2$ norm, $L_\infty$ norm and Fréchet Inception Distance (FID) [13], which measures the perceptual similarity, average distortion of low-frequency components $\left(LF = \frac{1}{N}\sum_{i=1}^{N}\left\|\mathscr{F}^{-1}\left(z_{i_L}^{ori}\right) - \mathscr{F}^{-1}\left(z_{i_L}^{adv}\right)\right\|_2\right)$, learned perceptual image patch similarity (LPIPS) [32] and SSIM [27] that assesses the structural similarity between the two images.

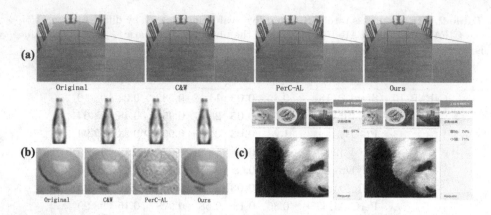

**Fig. 3.** Qualitative analysis of our approach. (a) Adversarial images generated by different methods on ImageNet-1K. (b) Adversarial images generated by different methods on CIFAR-100. (c) Classification results of Tencent cloud classifier. Left: origin image. Right: adversarial image generated by ours. Please zoom in for a better view.

We set the same perturbation budget $\epsilon = 8/255$ with respect to $L_\infty$ norm for StepLL [18], MIM [5], PGD [21] and AutoAttack [3]. This budget is specified with the iterative step size $\alpha = 1/255$. Adam is chosen as the optimizer for both PerC-AL [33] and C&W [2] with the learning rate of 0.01. The learning rate for SSAH [20] is set to 0.001. For ImageNet-1K, pre-trained ResNet-50 with the top-1 error of 23.85% is employed. We have also evaluated the performance of all comparison methods on the testing set of CIFAR-10 and CIFAR-100. For the classifier, we use pre-trained ResNet-20 with 7.4% and 30.4% top-1 error on CIFAR-10 and CIFAR-100, respectively. For CIFAR-100 and CIFAR-10, the hyperparameter $\beta$ is set to 0.01, while the default value on ImageNet-1K is 0.1. The learning rate $lr$ in Eq. 5 is set as $\epsilon^{[c]}/k$ for all the three datasets. The radius $r$ for dividing high and low frequency was set to $(M + N)//8$. we set larger perturbation budget for $\epsilon^{[C_b]} = \epsilon^{[C_r]} = 6$ than $\epsilon^{[Y]} = 2.5$ in order to craft more imperceptible adversarial examples.

## 4.2   White-Box Attacks

Table 2 and Table 3 shows the attack performances of various methods on CIFAR-10, CIFAR-100 and ImageNet-1K, as well as the quality of the adversarial images evaluated with different metrics. It demonstrates that our attack achieves competitive results not only in terms of traditional $L_p$ norm metrics compared to baseline methods. More importantly, our approach achieves the remarkable performance regarding SSIM, FID, and LF, accompanied by a significant success rate, specifically in terms of FID, which is considered to be in agreement with human judgment and can effectively reflect the degree of disturbance. On ImageNet-1K, compared to C&W, PerC-AL, and AdvDrop, our method reduces the FID scores by 6.83, 6.25, and 4.37, respectively, indicating that adversarial

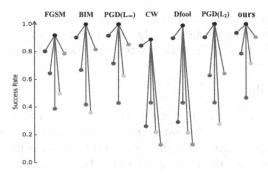

**Fig. 4.** Robustness of FGSM($L_\infty$), BIM($L_\infty$), PGD($L_\infty$), C&W($L_2$), Dfool($L_2$), PGD($L_2$) and Ours against Feature Squeeze(MF-3) (•), Feature Squeeze(Bit-6) (•), JPEG-30 (•), AT (•) and PD (•). (•) is no defend. The vertical axis represents the success rate of attacks under the corresponding defense methods.

images crafted by our framework have higher structural and perceptual similarity than those generated by the comparative method. Although the SSAH method is slightly better than ours in terms of FID, it requires more time and is not as good as ours in terms of structural similarity and LPIPS. Our method is better able to achieve a balance between the various metrics. Moreover, our perceptual metrics are also superior on CIFAR-10 and CIFAR-100.

Figure 3(a) and (b) illustrate the adversarial examples crafted using different methods on ImageNet-1K and CIFAR-100 datasets, respectively. We can observe that the proposed method is capable of generating visually imperceptible adversarial examples compared to other methods. This may be attributed to our strategy of modifying the energy of channels corresponding to color and texture that are insensitive to HVS in order to create realistic adversarial examples.

### 4.3 Robustness

We evaluate our attack and other methods by attacking a test set of 2000 randomly selected correctly classified images from ImageNet [24] across four defenses - JPEG defense [4], feature squeezing [29], adversarial training [21] and pixel deflection [23]. As adversarial training requires too much computation resource, we adopt a black-box setting and use pretrained adversarial model ResNet50 [8] to evaluate various attacks on adversarial training. We set attacks on the $L_\infty$ norm with $\epsilon = 4$ and on the $L_2$ norm with $\epsilon = 0.06$, which are commonly used attack settings in previous defense methods [11].

Specifically, among all these defenses, adversarial training remains the most effective defense because it allows the model to learn more robust feature representations. On the other hand, our method achieves a success rate of 71.30% under the JPEG defense mechanism and an 78.50% success rate under feature squeezing with reduced bit depth, largely outperforming other approaches (see Fig. 4). The possible explanation is that our method focuses on disturbing specific areas in the image. The modifications in these areas are difficult to be erased by image processing techniques, which further improve the robustness. For example, JPEG compression greatly influence the smooth area of the image but have little influence on the complex texture area and edge.

### 4.4   Transferability

To investigate the transferability of our framework in a black-box setting, we assess the transferability of adversarial images across architectures and datasets. We use the ImageNet-compatible dataset containing 1000 images with a resolution of $299\times299\times3$ to conduct experiments. As baselines, we select SSA [20], Gaussian noise, and GD-UAP [22] perturbations, with the $L_\infty$ norm constraint set as 10/255. The experimental results in Table 4 indicate that our attack is more easily transferable to other models.

**Table 4.** The attack success rates (%) of transferring adversarial examples across different architectures and datasets.

| Surrogate | Training set | Attack | ResNet-18 | VGG-16 |
|---|---|---|---|---|
| - | - | Gaussian Noise | 9.18 | 10.20 |
| ResNet-20 | CIFAR-10 | GD-UAP [22] | 14.09 | 11.44 |
| | | SSA [20] | 17.66 | 16.87 |
| | | Ours | **20.84** | **18.01** |
| VGG-11 | CIFAR-10 | GD-UAP [22] | 14.63 | 12.38 |
| | | SSA [20] | 16.92 | 18.52 |
| | | Ours | **21.04** | **24.71** |

**Table 5.** Ablation study of attack success rates (%) on different channels shows that the indispensability of each channel in our attack mechanism.

| Dataset | Only Y | Only CbCr | Ours |
|---|---|---|---|
| Cifar-10 | 94.19 | 73.94 | **99.56** |
| Cifar-100 | 95.41 | 85.94 | **99.89** |
| ImageNet-1K | 95.71 | 83.71 | **99.24** |

**Table 6.** Ablation study on the effectiveness of LFC on three dataset. CIEDE2000 [33] is the metric to measure perceptual color difference, other metrics have been stated in the setup.

| Dataset | Attack | $l_2\downarrow$ | $l_\infty\downarrow$ | FID↓ | SSIM↑ | LF↓ | CIEDE2000↓ |
|---|---|---|---|---|---|---|---|
| Cifar-10 | Ours w/o LFC | 0.75 | 0.06 | 9.50 | 0.981 | 0.32 | 93.30 |
| | Ours | **0.55** | **0.05** | **6.74** | **0.988** | **0.11** | **70.77** |
| Cifar-100 | Ours w/o LFC | 0.78 | 0.06 | 9.55 | 0.980 | 0.37 | 99.01 |
| | Ours | **0.72** | **0.05** | **8.13** | **0.983** | **0.19** | **82.44** |
| ImageNet-1K | Ours w/o LFC | 5.42 | 0.05 | 7.57 | 0.991 | 2.18 | 172.72 |
| | Ours | **4.77** | **0.03** | **5.31** | **0.995** | **1.15** | **137.50** |

To explore the practical potential of our method, we test the adversarial example generated by our method on a real-world image recognition system, i.e., Tencent Cloud API. Since we have no knowledge about the training data and model architecture of Tencent Cloud Classifier, the attack conducted in the real world becomes a black-box attack. As shown in Fig. 3(c), when the test bear image without adding perturbation is fed into the classifier, the classifier identifies it as a bear with a confidence score of 97%. However, when the adversarial example is input into the classifier, the classifier recognizes it as grass with 74% confidence, demonstrating the convincingness of our attack method.

## 4.5 Ablation Study

To study the impact of different constraints $\epsilon^{[C_b]}$, $\epsilon^{[C_r]}$, and $\epsilon^{[Y]}$, we adopt *lpips* [32], an effective measure of image similarity, to evaluate the perceptual loss of our method under different constraints and compare it with PGD, one of the strongest attacks. As shown in Fig. 5, our method perceptual loss is below 0.1 for different feature extraction networks, and the adversarial images generated by our method are more consistent with clean images compared to PGD.

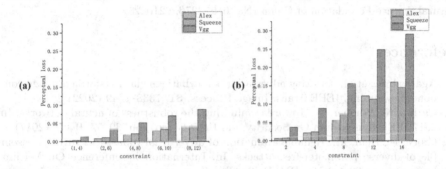

**Fig. 5.** (a) The perceptual loss of our method under different constraint conditions using the AlexNet, SqueezeNet, and VGG feature extraction networks. e.g., $(2, 6)$ denote perturbation budget of $\epsilon^{[Y]}$ as 2 and the values of $\epsilon^{[C_b]}$ and $\epsilon^{[C_r]}$ as 6. (b) The perceptual loss of PGD on $L_\infty$ settings using the same networks.

Table 6 shows the performance of our method with the low-frequency constraint(LFC) and without LFC. This indicates that LFC is an indispensable component and can effectively preserve the object structure and improve the image quality, which achieves less perceptible adversarial examples.

To quantify the impact of each channel, we conduct ablation experiments on both chrominance and luminance channels under the same working mechanism. The results presented in Table 5 show evidence that color and texture both play important roles in classification. The results also indicate that if we only use certain channels in our mechanism, the success rate of attack will be affected to varying degrees, and may yield visually perceptible adversarial images.

# 5    Conclusion

In this paper, we propose a frequency domain adversarial attack using $YC_bC_r$ color space. It aims to perturb images by disturbing the energy of frequency bands corresponding to relative high-frequency texture and color features. Considering that the $YC_bC_r$ color channels have different sensitivities to human visual perception, larger perturbations are allocated to the less perceptually sensitive $C_b$ and $C_r$ channels. Furthermore, introducing the low-frequency constraint to preserve the semantics of the adversarial image and reduce distortions in sensitive regions. We also conduct comparative experiments to clearly illustrate the superiority of our approach in terms of transferability, imperceptibility, and robustness against various defense techniques.

**Acknowledgements.** This work was supported in part by the National Natural Science Key Foundation of China (No. 61832014, 62032016), and in part by the National Natural Science Foundation of China (No. 61972276, 62102281).

# References

1. Agarwal, A., et al.: Crafting adversarial perturbations via transformed image component swapping. IEEE Trans. Image Process. **31**, 7338–7349 (2022)
2. Carlini, N., Wagner, D.: Towards evaluating the robustness of neural networks. In: 2017 IEEE Symposium On Security and Privacy (sp), pp. 39–57. IEEE (2017)
3. Croce, F., Hein, M.: Reliable evaluation of adversarial robustness with an ensemble of diverse parameter-free attacks. In: International Conference On Machine Learning, pp. 2206–2216. PMLR (2020)
4. Das, N., et al.: Shield: Fast, practical defense and vaccination for deep learning using jpeg compression. In: Proceedings of the 24th ACM SIGKDD International Conference on Knowledge Discovery & Data Mining, pp. 196–204 (2018)
5. Dong, Y., et al.: Boosting adversarial attacks with momentum. In: Proceedings of the IEEE Conference on Computer Vision and Pattern Recognition, pp. 9185–9193 (2018)
6. Duan, R., Chen, Y., Niu, D., Yang, Y., Qin, A.K., He, Y.: Advdrop: adversarial attack to dnns by dropping information. In: Proceedings of the IEEE/CVF International Conference on Computer Vision, pp. 7506–7515 (2021)
7. Duan, Ret al.: Adversarial laser beam: Effective physical-world attack to dnns in a blink. In: Proceedings of the IEEE/CVF Conference on Computer Vision and Pattern Recognition, pp. 16062–16071 (2021)
8. Engstrom, L., Ilyas, A., Salman, H., Santurkar, S., Tsipras, D.: Robustness (python library). https://github.com/MadryLab/robustness **4**(4), 4-3 (2019)
9. Goodfellow, I.J., Shlens, J., Szegedy, C.: Explaining and harnessing adversarial examples. arXiv preprint arXiv:1412.6572 (2014)
10. Guo, C., Frank, J.S., Weinberger, K.Q.: Low frequency adversarial perturbation. arXiv preprint arXiv:1809.08758 (2018)
11. Guo, C., Rana, M., Cisse, M., Van Der Maaten, L.: Countering adversarial images using input transformations. arXiv preprint arXiv:1711.00117 (2017)
12. Guo, Q., et al.: Watch out! motion is blurring the vision of your deep neural networks. Adv. Neural. Inf. Process. Syst. **33**, 975–985 (2020)

13. Heusel, M., et al.: Gans trained by a two time-scale update rule converge to a local nash equilibrium. In: Advances in Neural Information Processing Systems, vol. 30 (2017)
14. Hosseini, H., Poovendran, R.: Semantic adversarial examples. In: Proceedings of the IEEE Conference on Computer Vision and Pattern Recognition Workshops, pp. 1614–1619 (2018)
15. Hu, Q., Liu, D., Hu, W.: Exploring the devil in graph spectral domain for 3D point cloud attacks. In: European Conference on Computer Vision, pp. 229–248. Springer (2022). https://doi.org/10.1007/978-3-031-20062-5_14
16. Jiang, L., Dai, B., Wu, W., Loy, C.C.: Focal frequency loss for image reconstruction and synthesis. In: Proceedings of the IEEE/CVF International Conference on Computer Vision, pp. 13919–13929 (2021)
17. Krizhevsky, A., Hinton, G., et al.: Learning multiple layers of features from tiny images (2009)
18. Kurakin, A., Goodfellow, I., Bengio, S.: Adversarial machine learning at scale. arXiv preprint arXiv:1611.01236 (2016)
19. Li, Y., Liu, B.: Improved edge detection algorithm for canny operator. In: 2022 IEEE 10th Joint International Information Technology and Artificial Intelligence Conference (ITAIC), vol. 10, pp. 1–5. IEEE (2022)
20. Luo, C., Lin, Q., Xie, W., Wu, B., Xie, J., Shen, L.: Frequency-driven imperceptible adversarial attack on semantic similarity. In: Proceedings of the IEEE/CVF Conference on Computer Vision and Pattern Recognition, pp. 15315–15324 (2022)
21. Madry, A., Makelov, A., Schmidt, L., Tsipras, D., Vladu, A.: Towards deep learning models resistant to adversarial attacks. arXiv preprint arXiv:1706.06083 (2017)
22. Mopuri, K.R., Ganeshan, A., Babu, R.V.: Generalizable data-free objective for crafting universal adversarial perturbations. IEEE Trans. Pattern Anal. Mach. Intell. 41(10), 2452–2465 (2018)
23. Prakash, A., Moran, N., Garber, S., DiLillo, A., Storer, J.: Deflecting adversarial attacks with pixel deflection. In: Proceedings of the IEEE Conference on Computer Vision and Pattern Recognition, pp. 8571–8580 (2018)
24. Russakovsky, O., et al.: Imagenet large scale visual recognition challenge. Int. J. Comput. Vision 115, 211–252 (2015)
25. Wallace, G.K.: The jpeg still picture compression standard. Commun. ACM 34(4), 30–44 (1991)
26. Wang, H., et al.: High-frequency component helps explain the generalization of convolutional neural networks. In: Proceedings of the IEEE/CVF Conference on Computer Vision and Pattern Recognition, pp. 8684–8694 (2020)
27. Wang, Z., Bovik, A.C., Sheikh, H.R., Simoncelli, E.P.: Image quality assessment: from error visibility to structural similarity. IEEE Trans. Image Process. 13(4), 600–612 (2004)
28. Xiao, C., Zhu, J.Y., Li, B., He, W., Liu, M., Song, D.: Spatially transformed adversarial examples. arXiv preprint arXiv:1801.02612 (2018)
29. Xu, W., Evans, D., Qi, Y.: Feature squeezing: Detecting adversarial examples in deep neural networks. arXiv preprint arXiv:1704.01155 (2017)
30. Yin, D., Gontijo Lopes, R., Shlens, J., Cubuk, E.D., Gilmer, J.: A fourier perspective on model robustness in computer vision. In: Advances in Neural Information Processing Systems, vol. 32 (2019)
31. Yuan, S., Zhang, Q., Gao, L., Cheng, Y., Song, J.: Natural color fool: towards boosting black-box unrestricted attacks. arXiv preprint arXiv:2210.02041 (2022)

32. Zhang, R., Isola, P., Efros, A.A., Shechtman, E., Wang, O.: The unreasonable effectiveness of deep features as a perceptual metric. In: Proceedings of the IEEE Conference on Computer Vision and Pattern Recognition, pp. 586–595 (2018)
33. Zhao, Z., Liu, Z., Larson, M.: Towards large yet imperceptible adversarial image perturbations with perceptual color distance. In: Proceedings of the IEEE/CVF Conference on Computer Vision and Pattern Recognition, pp. 1039–1048 (2020)

# Fractional-Order Image Moments and Applications

Liyun Xu(✉) ⒾⒹ and Min Zhang

Institute of Big Data Science and Industry, School of Computer and Information
Technology, Shanxi University, Taiyuan 030006, China
xuliyun@sxu.edu.cn

**Abstract.** Image moments, as a global feature descriptor for images, have become a powerful tool for pattern recognition and image analysis. Most of the currently existing fractional-order image moments are polynomial-based. Three novel moments, namely Zernike fractional Fourier moment, Merlin fractional Fourier moment, and exponential fractional Fourier moment, combined with classical moments and fractional Fourier transform, are introduced based on angular functions. Additionally, we propose a zero watermarking algorithm based on these moments. We present robustness and comparative analysis experiments, including the effects of noise, filtering, rotation, and scaling. The experimental results demonstrate that the zero watermarking algorithm utilizing fractional-order moments can effectively withstand image processing attacks and geometric attacks. In both cases, their performance surpasses that of their corresponding integer-order image moments.

**Keywords:** Orthogonal moment · Fractional Fourier transform · Zero watermarking · Information security

## 1 Introductory

As a global feature descriptor for images, image moments have become powerful tools for pattern recognition and image analysis [1]. Image moments can be classified into orthogonal and non-orthogonal moments based on the orthogonality of the transform kernel function. Orthogonal moments are more suitable for image reconstruction than non-orthogonal moments because non-orthogonal moments inevitably result in information redundancy [2]. In recent years, various new orthogonal moment concepts have been proposed and applied in various fields of image processing [3]. Ren Heping et al. [4] proposed Jacobi Fourier moments, which exhibited good image descriptive capabilities and noise immunity when reasonable parameters were chosen. They extended the construction of a particular function alone to the construction of a family of functions, thus expanding the

Supported by the National Natural Science Foundation of China (Project No. 61901248); and the Shanxi Province Basic Research Programme (Project No. 202303021211023).

ⓒ The Author(s), under exclusive license to Springer Nature Switzerland AG 2024
S. Rudinac et al. (Eds.): MMM 2024, LNCS 14556, pp. 257–269, 2024.
https://doi.org/10.1007/978-3-031-53311-2_19

horizons of moment descriptors. Zernike moments (ZM), which are orthogonal and rotationally invariant moments, effectively address spatial pattern matching problems [5]. Yap et al. [6] were the first to propose the polar harmonic transform based on trigonometric functions. Unlike classical Zernike moments and pseudo-Zernike moments, the kernel calculation of the polar harmonic transform is straightforward and does not involve numerical stability issues [7]. The polar harmonic transform combines the advantages of orthogonality and invariance of Zernike and pseudo-Zernike moments while avoiding their inherent limitations. Hosny et al. [8] introduced orthogonal fractional-order Legendre Fourier moments. Omar et al. [9] proposed fractional-order Jacobi moments and verified that fractional-order polynomials outperformed their corresponding integer-order counterparts in terms of representing detailed functions. However, they were only robust against noise and were not effective in resisting other geometric attacks. Therefore, fractional-order image moments may outperform their corresponding integer-order image moments in certain aspects.

In recent years, orthogonal moments have also found wide applications in the field of image encryption [3,10,11], especially in the domain of image watermarking excellence. To preserve the integrity of images, Wen et al. [12] introduced the concept of a zero watermarking algorithm. The zero watermarking algorithm utilizes image features for extraction and watermark embedding without altering any image information. This algorithm has undergone extensive research in various fields [13]. Gao et al. [14] conducted a robustness analysis of image zero watermarking algorithms using Bessel-Fourier moments, which could effectively withstand various signal processing and geometric transformations, albeit with relatively complex computations. Wang et al. [15] proposed a ternary number of radial harmonic Fourier moments and applied them to the comprehensive processing of three-dimensional images. They also introduced a robust zero watermarking algorithm for stereo images that could resist various symmetric and asymmetric attacks. In a different approach, Wang et al. [16] developed a zero watermarking algorithm for color images using quaternionic exponential moments, which demonstrated robustness against geometric attacks but exhibited weakness in resisting large-scale cropping attacks. Meanwhile, Xia et al. [17] employed quaternionic polar harmonic Fourier moments for simultaneous copyright protection, enhancing the watermarking system's efficiency and conserving storage space. However, this method showed limitations in its robustness against stronger geometric attacks.

While researchers have proposed many orthogonal moments, most of the current fractional-order image moments rely on polynomial functions for fractional orders. In this paper, we introduce fractional-order image moments based on angular functions, designed to withstand image processing and geometric attacks. Compared to the Fourier transform, the fractional Fourier transform offers an additional free parameter, making it a valuable tool in digital watermarking and image encryption. Hence, we propose Zernike fractional Fourier moments (ZFFM), Merlin fractional Fourier moments (MFFM), and exponential fractional Fourier moments (EFFM), in combination with classical moments.

Subsequently, we apply these moments to a zero watermarking algorithm and conduct a robustness analysis through experiments involving noise, filtering, rotation, and scaling. The experimental results demonstrate that the proposed fractional-order image moments outperform their corresponding integer-order counterparts in terms of robustness against attacks.

## 2   Related Work

In this section, we will begin by introducing the fractional Fourier transform, a generalized form of the classical Fourier transform. We will then list three commonly used moments. These preliminary concepts will serve as motivation for the moments proposed in the next section.

### 2.1   Fractional Fourier Transform

The fractional Fourier transform is a representation of a signal in the time-frequency plane achieved by counterclockwise rotation of the coordinate axes around the origin at an arbitrary angle. It is a generalized form of the Fourier transform, essentially representing the signal as a superposition of chirp signals. The $p$ order fractional Fourier transform of a one-dimensional signal $x(t)$ is defined as follows

$$X_p(u) = F^p[x](u) = \int_{-\infty}^{+\infty} x(t)K_p(t, u)dt, \tag{1}$$

where $F^p$ is the fractional Fourier transform operator and the kernel function $K_p(t, u)$ has the following expression

$$K_p(t, u) = A_\alpha \exp\left\{i\left(\frac{\cot\alpha}{2}t^2 - tu\csc\alpha + \frac{\cot\alpha}{2}u^2\right)\right\}, \tag{2}$$

where $A_\alpha = \sqrt{(1 - i\cot\alpha)/2\pi}$ , $\alpha = p\pi/2$ represents the rotation angle of the time-frequency plane. The fractional Fourier transform reduces to the classical Fourier transform when $p = 1$, i.e. $\alpha = \pi/2$.

### 2.2   Zernike Moment

The order $n$ with repetition $m$ Zernike moment of a two-dimensional function $f(x, y)$ is defined as follows:

$$Z_{nm} = \frac{n+1}{\pi} \iint\limits_{x^2+y^2 \leq 1} f(x, y)V_{nm}^*(x, y)dxdy = \frac{n+1}{\pi} \int_0^{2\pi}\int_0^1 f(r, \theta)V_{nm}^*(r, \theta)rdrd\theta, \tag{3}$$

where $*$ denotes the conjugate and $V_{nm}$ is the Zernike polynomial defined as

$$V_{nm}(x, y) = V_{nm}(r, \theta) = R_{nm}(r)\exp(jm\theta), x^2 + y^2 \leq 1, \tag{4}$$

where $R_{nm}(r)$ is defined as

$$R_{nm}(r) = \sum_{s=0}^{(n-|m|)/2} \frac{(-1)^s (n-s)^2 r^{n-2s}}{s!(\frac{n+|m|}{2} - s)!(\frac{n-|m|}{2} - s)!}. \tag{5}$$

## 2.3    Fourier-Merlin Moment

The Fourier- Merlin moment (OFFM) of an image $f(r,\theta)$ in polar coordinates is defined as

$$\phi_{nm} = \frac{1}{2\pi a_n} \int_0^{2\pi} \int_0^1 f(r,\theta) Q_n(r) \exp(-jm\theta) r dr d\theta, \tag{6}$$

where $a_n = \frac{1}{2(n+1)}$ is the normalisation factor and the radial basis function polynomial $Q_n(r)$ is defined as

$$Q_n(r) = \sum_{s=0}^n a_{ns} r^s, \tag{7}$$

where $a_{ns} = (-1)^{n+s} \frac{(n+s+1)!}{(n-s)!s!(s+1)!}$ is the coefficient of the polynomial.

## 2.4    Exponential Fourier Moment

The exponential moment(EM) of an image $f(r,\theta)$ in polar coordinates is defined as

$$E_{nm} = \frac{1}{4\pi} \int_0^{2\pi} \int_0^1 f(r,\theta) A_n^*(r) \exp(-jm\theta) r dr d\theta, \tag{8}$$

where $A_n^*(r)$ is the conjugate of $A_n(r) = \sqrt{2/r} \exp(j2n\pi r)$.

# 3    Fractional-Order Moments

## 3.1    Zernike Fractional Fourier Moment

The family of functions $B_{nm}(r,\theta) = R_{nm}(r) K_p(\theta, m)$ is obtained by taking the Zernike polynomials as radial functions and the kernel functions of the fractional Fourier transform as angular functions. Thus the definition of the Zernike fractional $p$ order Fourier moments of the $n(n > 0)$ order repetition $m(|m| >= 0)$ of the image $f(r,\theta)$ can be expressed as:

$$Z_{nmp} = \frac{n+1}{\pi} \int_0^{2\pi} \int_0^1 f(r,\theta) B_{nm}^*(r,\theta) r dr d\theta, \tag{9}$$

where the kernel function $K_p(\theta, m)$ is

$$K_p(\theta, m) = \exp i \left\{ \frac{\cot \alpha}{2} \theta^2 - m\theta \csc \alpha + \frac{\cot \alpha}{2} m^2 \right\} \qquad (10)$$

To avoid ambiguity due to order, the Zernike fractional Fourier moments are denoted as $(n, m, p)$-order Zernike fractional Fourier moments. Since the kernel function of the fractional Fourier transform has orthogonality, i.e.

$$\int_{-\infty}^{+\infty} K_p(\theta, m) K_p^*(\theta, m') dt = \delta(u - u'). \qquad (11)$$

Thus the family of functions is known to be orthogonal on the unit circle,

$$\int_0^{2\pi} \int_0^1 B_{nm}(r, \theta) B_{kl}^*(r, \theta) r dr d\theta = \chi \delta_{nmkl}, \qquad (12)$$

where $\chi$ is the normalisation factor and $B_{kl}^*(r, \theta)$ is the conjugate of $B_{nm}(r, \theta)$.

## 3.2 Merlin Fractional Fourier Moment

The Merlin polynomials are taken as radial functions and the kernel functions of the fractional Fourier transform as angular functions. Thus the definition of the Merlin fractional $p$ order Fourier moments of the $n(n > 0)$ order repetition of $m(|m| >= 0)$ for image $f(r, \theta)$ can be expressed as

$$\phi_{nmp} = \frac{1}{2\pi a_n} \int_0^{2\pi} \int_0^1 f(r, \theta) Q_n(r) K_p(\theta, m) r dr d\theta. \qquad (13)$$

To avoid ambiguity due to order, denoted as $(n, m, p)$-order Merrill fractional Fourier moments.

## 3.3 Exponential Fractional Fourier Moment

The exponential polynomial is taken as a radial function and the kernel function of the fractional Fourier transform as an angular function. The definition of the exponential fractional $p$ order Fourier moments of the $n(+\infty > n > -\infty)$ order repetition $m(+\infty > m > -\infty)$ of the image $f(r, \theta)$ can be expressed as:

$$E_{nmp} = \frac{1}{4\pi} \int_0^{2\pi} \int_0^1 f(r, \theta) A_n^*(r) K_p(\theta, m) r dr d\theta. \qquad (14)$$

To avoid ambiguity due to order, it is denoted as $(n, m, p)$-order exponential fractional Fourier moments.

# 4  Zero Watermarking Algorithm

In a zero watermarking algorithm, copyright protection is achieved by extracting essential image features and storing them in an intellectual property database. Zero watermarking algorithms offer two primary advantages over traditional watermarking methods. Firstly, they do not degrade the quality of the original image, making them highly imperceptible. Secondly, these algorithms strike a balance between the resistance to digital watermarking attacks, the amount of embedded information in the watermark, and imperceptibility, showcasing strong resistance. A zero watermarking algorithm typically consists of two processes: generating a zero watermarked image and extracting watermark verification. In this paper, various image moments will be employed for image feature extraction.

## 4.1  Watermark Embedding

The watermark image generation process consists of four steps: computing orthogonal moments, randomly selecting suitable moments, generating a binary feature map, and embedding the watermark. The flowchart is depicted in Fig. 1

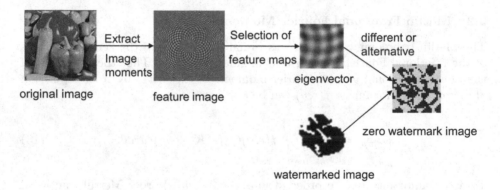

**Fig. 1.** Watermark embedding process.

The algorithm is described as follows:

Step 1: Calculate the orthogonal moments

Select the parameters, $(k_1, k_2, k_3, k_4)$ are generated as an encrypted sequence by using logical mapping, and the order image moments of the image $f(r, \theta)$ are computed using the corresponding formula.

Step 2: Random selection of moments

Use $k_3$ , $k_4$ to select the feature map $M$(size $P \times Q$) from the above $(k_1, k_1, k_2)$ order image moments (where $(k_3, k_4)$ are the coordinates of the left vertex of the feature map).

Step 3: Binary feature map generation

The pixels $L_A$ of the feature map $M$ are globally thresholded to obtain a binarised feature map

$$L_f(x,y) = \begin{cases} 1, & L_A(x,y) \geq Threshold \\ 0, & L_A(x,y) \leq Threshold \end{cases} \qquad (15)$$

where $Threshold$ is the average value of the image pixels.

Step 4: Watermark Embedding

The binarised feature map $L_f$ and the watermarked image $L_w$ are subjected to the dissimilarity operation to obtain the zero watermarked image $w$, i.e.

$$w = XOR(L_f, L_w). \qquad (16)$$

## 4.2  Watermark Extract

The primary objective of the watermark extraction process is to establish the copyright ownership of the image. This extraction process comprises four steps: calculating orthogonal moments, selecting specific moments, generating a binary feature image, and extracting the watermark. Watermark extraction is the inverse process of watermark embedding. The critical task is to obtain $(k1, k2, k3, k4)$ from the encrypted sequence. Steps one to three are identical to the watermark embedding process, while step four is described as follows:

Step 4: Watermark Extraction

The binarised feature map $L_f$ and the zero-watermarked image $w$ are dissimilar to each other to obtain the watermarked image $L_w$,

$$L_w = XOR(L_f, w). \qquad (17)$$

The pixels $L_w$ will be globally thresholded to obtain the final watermarked image.

# 5  Experiments and Analysis

In order to evaluate the performance of the zero watermarking algorithm, this paper verifies it through experiments. The grey scale image used for the experiment is $256 * 256$.

## 5.1  Watermark Capacity

The watermarking capacity is an important metric for evaluating the performance of a zero watermarking algorithm. This capacity depends on the number of image moments, and the number of orthogonal moments is determined by the maximum order of image moments. The relationship between the watermark capacity $N$ and the maximum order of orthogonal moments $n_{max}$ is given by $N = (n_{max}+1)(2n_{max}+1)$. As shown in Fig. 2, for the watermark image, an order greater than 22 should be selected. Therefore, in the subsequent experiments, orthogonal moments of order $(25, 25, p)$ will be chosen for feature extraction.

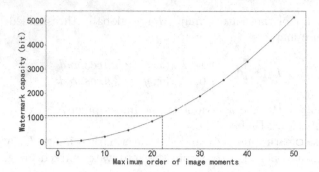

**Fig. 2.** Relationship between watermark capacity and maximum order of image moments.

## 5.2 Robustness

In this paper, different attacks are performed on the image to verify the robustness of the fractional order moments. The robustness of the algorithm is evaluated using the accuracy ratio $(AR)$, which is expressed as

$$AR = \frac{bits_c}{bits_t} \times 100\%, \tag{18}$$

where $bits_c$ denotes the number of bits retrieved correctly and $bits_t$ denotes the total number of bits in the watermarked image. The larger the $AR$, the closer the extracted watermarked image to the embedded watermarked image, and the better the robustness of the algorithm.

## 5.3 Experiments

**Noise Attacks.** To assess the algorithm's noise robustness, this section subjects the image to Gaussian and pretzel noise with various standard deviations. The experiments involve embedding the original grayscale image with different fractional orders and then extracting the corresponding orders for the image after the noise attack. From Fig. 3, it is evident that Zernike fractional Fourier moments perform best against Gaussian noise when $p = 0.5$ and against pretzel noise when $p = 0.3$. When $p = 0.8$, Merlin fractional Fourier moments exhibit superior resistance to Gaussian noise, while at $p = 0.3$, they are optimal against pretzel noise. As for exponential fractional Fourier moments, they excel against Gaussian noise at $p = 0.7$ and against pretzel noise at $p = 0.6$.

**Filtering Attacks.** To assess the algorithm's robustness against filtering, this section subjects the image to various filters, including average filtering, Gaussian filtering, bilateral filtering, and 2D convolutional filtering. The experiments involve embedding the original grayscale image with different fractional orders and then extracting the corresponding orders for the image after the filter attack.

From Fig. 4, it is evident that Zernike fractional Fourier moments perform best against filtering attacks when $p = 0.5$. When $p = 0.5$, Merlin fractional Fourier moments exhibit superior resistance to filtering attacks, while at $p = 0.2$, exponential fractional Fourier moments excel in countering filtering attacks.

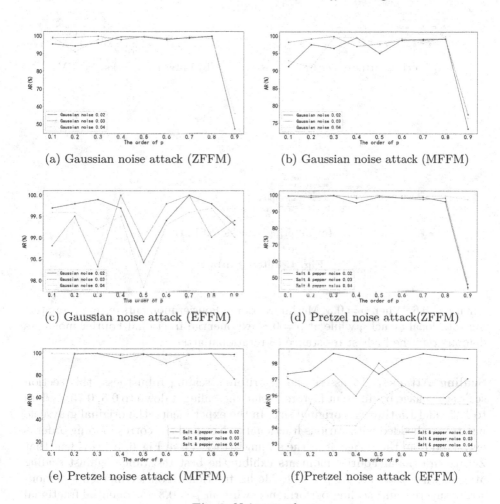

(a) Gaussian noise attack (ZFFM)          (b) Gaussian noise attack (MFFM)

(c) Gaussian noise attack (EFFM)          (d) Pretzel noise attack(ZFFM)

(e) Pretzel noise attack (MFFM)          (f)Pretzel noise attack (EFFM)

**Fig. 3.** Noise attacks

**Rotation Attacks.** To assess the algorithm's robustness against rotation, the images are rotated by $5°$, $15°$, $30°$, and $45°$. This paper embeds the original grayscale image with different fractional orders and then extracts the corresponding order of the image after the rotation attack. From Fig. 5, it is evident that Zernike fractional Fourier moments perform best against rotational attacks

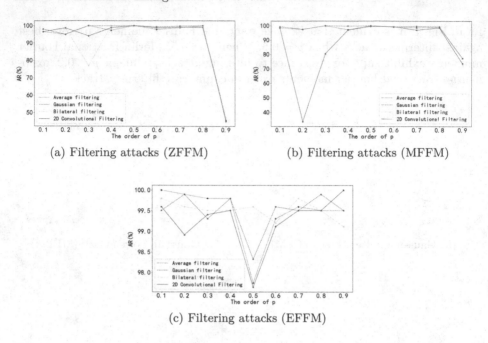

(a) Filtering attacks (ZFFM)          (b) Filtering attacks (MFFM)

(c) Filtering attacks (EFFM)

**Fig. 4.** Filtering attacks

when $p = 0.6$. When $p = 0.3$, Merlin fractional Fourier moments excel in countering rotational attacks, while at $p = 0.8$, exponential fractional Fourier moments demonstrate the highest resistance to rotational attacks.

**Scaling Attacks.** To assess the algorithm's scaling robustness, this section scales the image by different factors, including scaling it down to 0.5, 0.75, and up to 1.25, and 1.5 times its original size. In the experiments, the original grayscale image is embedded with various fractional orders, and the corresponding order is extracted from the image after the scaling attack. From Fig. 6, it is evident that Zernike fractional Fourier moments exhibit the best resistance against scaling attacks when $p = 0.8$. At $p = 0.1$, Merlin fractional Fourier moments demonstrate superior anti-scaling performance, while at $p = 0.3$, exponential fractional Fourier moments excel in resisting scaling attacks.

**Comparisons.** To further demonstrate the algorithm's superiority, we compare and analyze Zernike fractional Fourier moments with Zernike moments (ZM), Merlin fractional Fourier moments with Merlin Fourier moments (OFMM), and exponential fractional Fourier moments with exponential Fourier moments (EM). From the results in Table 1, it is clear that the Zernike fractional Fourier moments and the Merlin fractional Fourier moments generally outperform their counterparts, except that the performance is not optimal in resisting very few attacks. The exponential fractional Fourier moments outperform the exponential Fourier

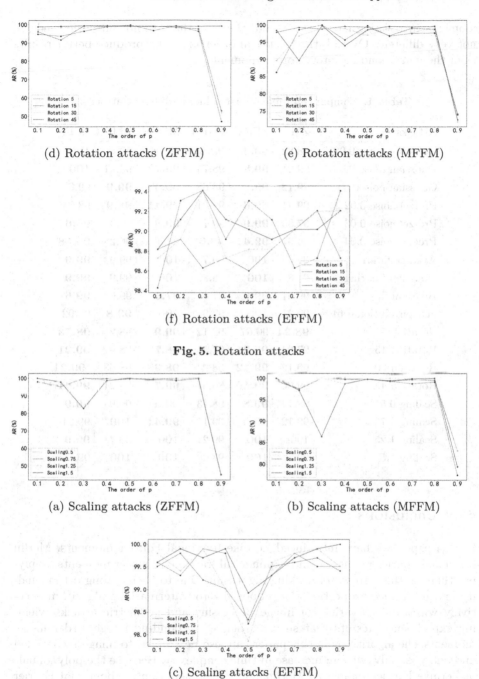

(d) Rotation attacks (ZFFM)              (e) Rotation attacks (MFFM)

(f) Rotation attacks (EFFM)

**Fig. 5.** Rotation attacks

(a) Scaling attacks (ZFFM)              (b) Scaling attacks (MFFM)

(c) Scaling attacks (EFFM)

**Fig. 6.** Scaling attacks

moments in resisting most attacks and the ARs that are not outperformed are not very different. Overall, the fractional order moments produce better results than the corresponding integer order moments.

Table 1. Comparison table of $AR$ values of different attacks.

| Attacks | ZM | ZFFM | OMFF | MFFM | EM | EFFM |
|---|---|---|---|---|---|---|
| Gaussian noise 0.02 | 98.82 | **99.6** | **99.6** | 99.21 | 99.8 | **100** |
| Gaussian noise 0.03 | 99.21 | **99.8** | 98.73 | **99.12** | 99.31 | **100** |
| Gaussian noise0.04 | 99.12 | **99.8** | 98.63 | **99.9** | **99.9** | 99.6 |
| Pretzel noise 0.02 | 99.41 | **99.6** | 99.21 | **99.41** | 98.92 | 98.24 |
| Pretzel noise 0.03 | 97.85 | **99.9** | 99.7 | **99.8** | 95.41 | **96.67** |
| Pretzel noise 0.04 | 96.67 | **99.41** | **99.6** | 99.51 | 96.28 | **96.58** |
| Average filtering | 99.6 | **100** | 99.7 | **100** | 99.9 | **99.9** |
| Gaussian filtering | 98.82 | **100** | 98.82 | **100** | 99.9 | **99.9** |
| Bilateral filter | 99.31 | **99.6** | 97.6 | **97.6** | 99.6 | **99.6** |
| 2D convolutional filter | 99.21 | **100** | 97.9 | **100** | **99.8** | 98.92 |
| Rotation 5 | **98.24** | 96.67 | 99.12 | **99.9** | 98.2 | **98.73** |
| Rotation 15 | 97.94 | **99.21** | 97.55 | **98.73** | 98.92 | **99.21** |
| Rotation 30 | 99.12 | **99.12** | 98.24 | **98.98** | 98.63 | **99.21** |
| Rotation 45 | 99.60 | **99.60** | 86.91 | **99.7** | 99.21 | **99.41** |
| Scaling 0.5 | 98.73 | **99.8** | 98.73 | **99.41** | 99.9 | **99.9** |
| Scaling 0.75 | 99.12 | **99.9** | 99.12 | **99.41** | 100 | 99.21 |
| Scaling 1.25 | 100 | **100** | 99.21 | **100** | 99.41 | **99.9** |
| Scaling 1.5 | 99.7 | **100** | 97.36 | **100** | **100** | 99.7 |

# 6    Conclusions

In this paper, we have introduced Zernike fractional Fourier moments, Merlin fractional Fourier moments, and exponential fractional Fourier moments, applying them to the zero watermarking algorithm. Due to their strong orthogonality and invariance properties, the proposed zero watermarking algorithm effectively withstands a variety of image processing and geometric attacks, yielding experimental results that surpass their corresponding integer-order image moments.The algorithms in this paper can also be applied to image recognition and analysis, only with an increase in time complexity. Because the polynomials of Zernike Fourier moments, Merlin Fourier moments and exponential Fourier moments involve numerical stability, their corresponding fractional-order image moments will be less efficient at certain fractional orders, and it is hoped that such problems will be solved through more in-depth analytical studies in future work.

# References

1. Yang, L.L., Ye, D.Y.: Moment and texture based algorithm for text detection in natural scene images. J. Chinese Comput. Syst. **37**(6), 1313–1317 (2016)
2. Jian, L.Q.: Research on image local feature extraction based on orthogonal moments. Electron. Technol. Softw. Eng. **21**, 184–188 (2022)
3. Khalid M.Hosny., Mohamed M., Darwish.: New set of quaternion moments for color images representation and recognition. J. Math. Imag. Vision, **60**, 717–736 (2018)
4. Ren, H.P., Ping, Z.L., Fu, W.R.G.: Jacobi-Fourier moment is used to describe the image. J. Opt. **01**, 5–10 (2004)
5. Chandan Singh., Anu Bala.: A local Zernike moment-based unbiased nonlocal means fuzzy C-Means algorithm for segmentation of brain magnetic resonance images. Expert Systems With Applications, 118: 625–639 (2019)
6. Pew-Thian Yap., Jiang, X.D.: Alex Chichung Kot.: Two-dimensional polar harmonic transforms for invariant image representation. IEEE Trans. Pattern Anal. Mach. Intell. **32**(7), 1259–1270(2010)
7. Hosny, K.M., Darwish, M.M. and Aboelenen, T.: New Fractional-order Legendre-Fourier moments for pattern recognition applications. Pattern Recogn. **103**, 107324 (2020)
8. El Ogri, O., et al.: Novel fractional-order Jacobi moments and invariant moments for pattern recognition applications. Neural Comput. Appl. **33**(20), 13539–13565 (2021). https://doi.org/10.1007/s00521-021-05977-w
9. Wang, Y.Z., Sun, H.B., Ma, Y.K.: Image robust hashing algorithm based on quaternion harmonic transformation moment and salient features. Comput. Appl. Softw. **38**(3), 210–217 (2021)
10. Karthick, S., Sankar, S.P., Prathab, T.R.: An approach for image encryption / decryption based on quat-ernion fourier transform. In: Proceedings of 2018 International Conference on Emerging Trends and Innovations in Engineering and Technological Research (ICETIETR) (2018)
11. Liu, X.L., Han, G.N., Wu, J.S.: Fractional Krawtchouk transform with an application to image watermarking. IEEE Trans. Signal Process. **65**(7), 1894–1908 (2017)
12. Wen, Q., Sun, T. F., Wang, S. X.: Based zero-watermark digital watermarking technology, in: the Third China Information Hiding and Multimedia Security Workshop (CIHW). Xidian University Press. Xian, China, pp. 102–109 (2001)
13. Long, M., Peng, F., Du, Q.Z.: Zero-watermarking for authenticating 2D engineering graphics based on optimal binary searching-tree. J. Chin. Comput. Syst. **33**(6), 1296–1299 (2012)
14. Gao, G.Y., Jiang, G.P.: Bessel-Fourier moment-based robust image zero-watermarking. Multimedia Tools Appl. **74**, 841–858 (2015)
15. Wang, C.P., Wang, X.Y., Xia, Z.Q.: Ternary radial harmonic Fourier moments based robust stereoimage zero-watermarking algorithm. Inf. Sci. **470**, 109–120 (2019)
16. Wang, C.P., Wang, X.Y., Xia, Z.Q.: Geometrically resilient color image zero-watermarking algorithm based on quaternion Exponent moments. J. Vis. Commun. Image Represent. **41**, 247–259 (2016)
17. Xia, Z.Q., Wang, X.Y., Li, X.X.: Efficient copyright protection for three CT images based on quaternion polar harmonic Fourier moments. Signal Process. **164**, 368–379 (2019)

# Time-Quality Tradeoff of MuseHash Query Processing Performance

Maria Pegia[1,3]([✉]) [iD], Ferran Agullo Lopez[2] [iD], Anastasia Moumtzidou[1] [iD],
Alberto Gutierrez-Torre[2] [iD], Björn Þór Jónsson[3] [iD],
Josep Lluís Berral García[2,4] [iD], Ilias Gialampoukidis[1] [iD], Stefanos Vrochidis[1] [iD],
and Ioannis Kompatsiaris[1] [iD]

[1] Information Technologies Institute, Centre for Research and Technology Hellas,
Thessaloniki, Greece
{mpegia,moumtzid,heliasgj,stefanos,ikom}@iti.gr
[2] Barcelona Supercomputing Center, Barcelona, Spain
{ferran.agullo,alberto.gutierrez}@bsc.es
[3] Reykjavik University, Reykjavík, Iceland
{ mpegia22,bjorn}@ru.is
[4] Universitat Politècnica de Catalunya, Barcelona, Spain
josep.ll.berral@upc.edu

**Abstract.** Nowadays, large quantities of multimedia data are generated by various applications on smartphones, drones and other devices. Facilitating retrieval from these multimedia collections requires (a) effective media representation and (b) efficient indexing and query processing approaches. Recently, the MuseHash approach was proposed, which can effectively represent a variety of modalities, improving on previous hashing-based approaches. However, the interaction of the MuseHash approach with existing indexing and query processing approaches has not been considered. This paper provides the first systematic evaluation of the time-quality tradeoff that arises when MuseHash media representation is combined with state-of-the-art approximate nearest-neighbour indexes along multi-core and GPU processing.

**Keywords:** MuseHash · Query processing performance · Approximate nearest neighbour indexes · High-Performance Computing

## 1 Introduction

Joint representations of media modalities have recently come into focus within the multimedia community. The most common and successful approaches are based on supervised learning of hash functions, where a model is trained to encode the different modalities into a joint representation [18,27], typically a hash code based in Hamming space. At query time, the given query modalities are then hashed into the same representation, and the media collection is ranked

---

M. Pegia and F. A. Lopez—Contributed equally to the research reported in this paper

© The Author(s), under exclusive license to Springer Nature Switzerland AG 2024
S. Rudinac et al. (Eds.): MMM 2024, LNCS 14556, pp. 270–283, 2024.
https://doi.org/10.1007/978-3-031-53311-2_20

based on similarity to the hash code of the query. So far, in this work, the focus of the evaluation has been on the *effectiveness* of the hash codes, as represented by retrieval accuracy over relatively small benchmark collections. As media collections become increasingly large and complex, however, we must consider the *efficiency* of query processing with these representations. This paper represents the first step in this direction.

Query processing using such hash codes is one instance of the nearest neighbour problem, which is a well-studied problem in the literature. Challenges for nearest-neighbour queries [29] arise when dealing with large-scale and high-dimensional collections. This has been termed "curse of dimensionality" since, as dimensionality or dataset size increases, using indexes to return exact nearest neighbour results becomes intractable and a brute-force scan of the collection is necessary [30]. To make retrieval feasible at large scale, a multitude of approximate indexing methods have been proposed, based on a variety of approaches, including tree-based structures, graph-based structures, and hashing-based structures. Such approximate indexes yield a *time-quality tradeoff* between efficiency and effectiveness, more precisely between query processing performance and result quality, that must be evaluated. For some tasks, approximate indexes can even improve quality compared to the brute-force scan. Furthermore, with the increase in available processing power, multi-core and GPU processing can be applied to facilitate hash-code creation and also, depending on the index structure, similarity computations.

In this paper, we present an analysis of query processing performance with the MuseHash approach [27], a recent state-of-the-art multimodal hashing approach for media representation. The main contributions of this paper are:

- We have combined MuseHash media representations with a large set of approximate indexes implemented in the ANN Benchmarks system.
- We have explored the impact of multi-core and GPU processing, using High-Performance Computing (HPC) infrastructures.
- We present experiments for two benchmark collections, a small aerial dataset and a much larger lifelog collection, as well as synthetic collections.

The results indicate that MuseHash can be combined with approximate indexes for efficient query processing, in particular the graph-based HNSW structure, which is considered state-of-the-art in the indexing community, outperforming a full sequential scan in terms of result quality. We also show that while multi-core and GPU processing can improve the performance of the evaluated approach, this impact alone is smaller than the impact of the indexing structure.

The remainder of this paper is organised as follows. Section 2 presents the state-of-the-art in hash-based media representation, including the MuseHash approach. Section 3 then presents the relevant state-of-the-art approximate indexes, while Sect. 4 presents multi-core processing and HPC. Section 5 then offers our analysis of the experimental results. Finally, Sect. 6 concludes.

## 2    Hash-Based Representation and MuseHash

Hash-based representations play a crucial role in efficiently indexing and retrieving multimedia data. They condense complex feature vectors into compact hash codes, streamlining storage and retrieval processes. Various hashing methods utilize adversarial learning [4], some leverage deep learning networks [7], and others employ an approach based on a similarity matrix [9]. Some are designed specifically for utilizing a single modality [4,9], while others are tailored for combining multiple modalities [7].

MuseHash [27] is a recent multimodal hashing algorithm which excels in handling multimedia data with diverse modalities, leveraging a combination of modalities in its queries, such as visual and temporal aspects, to deliver highly relevant results. In brief, MuseHash extracts features independently for each modality, utilizing Bayesian ridge regression to learn hash functions that map features to the Hamming space. This separate computation for each modality enables support for both unimodal and multimodal queries.

In more detail, the method involves three main phases: training, offline, and querying. In the training phase, hash functions are generated from the training collection using Bayesian ridge regression. These functions map feature vectors from each modality to the Hamming space. Affinity matrices are created based on ground truth labels, and semantic probabilities are derived from these matrices. During the offline phase, features are extracted from the retrieval set for each modality. Using the learned hash functions, hash codes are computed and stored in a database, ensuring efficient storage and retrieval of multimedia data. Finally, in the querying phase, hash functions learned in the previous steps are applied to a given query. Unified hash codes are generated from query-specific hash codes using the XOR operation. The database is queried using Hamming distances, leading to the retrieval of top-k relevant results. Overall, MuseHash combines supervised learning, Bayesian regression, and Hamming distance-based retrieval to significantly enhance the accuracy and efficiency of multimedia data retrieval.

Overall, MuseHash as a hash-based approach gives compact representations of data and uses less memory for data storage. However, the fast similarity search for the ranking procedure is too slow, when you use brute-force and you have to query large collections. Thus the motivation to use optimization techniques in the ranking procedure drive us to decide approximate nearest neighbors approaches.

## 3    Approximate Indexes and ANN Benchmarks

The nearest neighbors problem is crucial in computer science, involving finding the closest data points to a given query within a dataset [1]. It is applied in pattern recognition, data mining, image retrieval, and recommendation systems. Exact solutions to this problem are computationally challenging, particularly with large datasets. Approximate nearest neighbors algorithms [3] offer a practical compromise, providing reasonably accurate matches while significantly improving computational efficiency, making them ideal for large-scale applications. This involves balancing retrieval accuracy and faster query processing, ideal for large-scale datasets. Evaluating approximate nearest neighbor algorithms includes assessing accuracy and efficiency using metrics such as precision, recall, query time, and index construction time.

Previous works, such as [2,19], use pretrained models to extract features for each modality on diverse datasets. They apply methods from [3], evaluating performance based on exact kNN points rather than provided dataset labels. However, our emphasis is on utilizing dataset labels for measuring method performance. We selected the following approximate nearest neighbors similar to the cited works on the current research [3]: tree-based structures, graph-based structures, pruning techniques, brute-force approaches and baseline methods.

**Tree-based methods:** BallTree [5] uses hyper-spheres to create a tree hierarchy, while CKDTree [24] extends the KD-trees for multiple dimensions with hyper-rectangles. Random Projection Tree (RPT) methods, like Annoy (Approximate Nearest Neighbors Oh Yeah) [21], utilize random projections to split data points and build index structures for fast approximate nearest neighbor search. On the contrary, PyNNDescent [12] employs randomized k-d trees, combining randomized partitioning and nearest neighbor search to efficiently navigate the tree structure and find approximate nearest neighbors.

**Graph-based methods:** The HNSW (Hierarchical Navigable Small World) [22] arranges the dataset into small-world graphs for efficient approximate nearest neighbor search with minimal memory usage. SW-graph (Small World Graph) [23] combines small-world graph and locality-sensitive hashing. It balances retrieval accuracy and efficiency by leveraging the data's local and global structures.

**Pruning methods:** SCANN (Scalable Nearest Neighbors) [14] uses locality-sensitive hashing for fast approximate nearest neighbor search with single-threaded and multi-threaded implementations for large-scale datasets.

**Brute-force methods:** BruteForce and BruteForce-BLAS [10] are brute-force methods for solving the nearest neighbor search. They calculate the distances between the query point and all data points to find the closest neighbors. Although ensuring accuracy, these methods can be computationally expensive, especially for large datasets. BruteForce-BLAS improves efficiency using the BLAS library for faster distance calculations, serving as a baseline for calculating advanced approximate indexes.

**Baseline methods:** Dummy-Algo-MT [13] and Dummy-Algo-ST offer simpler and generic implementations. Dummy-Algo-MT is a multi-threaded brute-force search implementation with parallel processing for improved performance. Dummy-Algo-ST is a single-threaded version, provides a basic implementation for comparison. These baseline methods serve as reference points for evaluating advanced approximate indexes.

## 4   Multi-Core and GPU Processing

Current hardware capabilities can be exploited to improve the speed of a variety of tasks, including feature retrieval. For example, when dealing with data that cannot be accommodated in memory, methods such as DiskANN [17] or SPANN [8] utilise data locality and fast SSD storage. Also, multi-core processing has been exploited in retrieval, with examples such as SCANN [14], or through parallelism at the level of query and data processing. This allows an algorithm to make use of all the available hardware resources, drastically improving performance, even though a linear speed-up may not be achievable. On the other hand, making use of Graphic Processing Units (GPUs) has been popularized, specially in the image processing field and in intensively parallel problems. In the case of the retrieval task, the SONG [31] algorithm, for example, has been co-designed to make use of GPUs, beating similar algorithms that are CPU-based. This highlights that when developing new indexing structures, it is advantageous to keep in mind the capabilities of the underlying hardware.

In this section we define how the MuseHash algorithm is optimized to use the available hardware as efficiently as possible and how it is transformed to increase its time performance when more resources are available (scalability). Concretely, we are focusing on the offline and querying phases of the algorithm. Our approach is driven by the usage of parallel computation using multi-core and GPU processing.

In the offline phase, the focus is on feature extraction. As this implies extracting features from the latent space of a Neural Network, the most viable optimization is to use GPUs alongside specialized accelerated software when performing the required forward pass (CUDA [25] and cuDNN [11]). Moreover, the task can be done simultaneously by multiple GPUs as there are no dependencies between samples. Notice that the proposed optimization can also be applied in the querying phase if the incoming sample has no precomputed features.

In the querying phase, however, we propose to improve query processing performance through two different strategies: query parallelism and data parallelism. Both strategies are designed for high-capability environments, enabling the scalability of algorithms with increased resources. Nevertheless, they are also applicable to smaller devices with multi-core possibilities.

On one hand, the **query parallelism** strategy divides the incoming queries into a set of processes, henceforth query processors, where each of them has a different copy of the overall data. The query processors can be located in different machines and can work independently due to the absence of dependencies between queries. On the other hand, the **data parallelism** strategy splits the workload of a single query into multiple processes that divide the full set of data in equal parts to perform the subsequent similarity comparison. These processes need to synchronize to produce a common result.

The proposed strategies are not contradictory and can be integrated together for an increased performance, making use of multiple machines where different query processors are located (query parallelism) and utilizing all the available resources in each of them (data parallelism). Similarly to query parallelism, the data parallelism strategy can be applied to processes located in different machines, but its implementation is much harder and the subsequent communication latency could be detrimental. In addition, we propose using GPU processing to speed up the similarity comparison, as this accelerated hardware can increase the performance of the internal computations of the algorithms.

## 5   Experiments

In this section, we report results from a set of experiments that explore query processing performance of state-of-the-art methods with the MuseHash media representations. In the offline phase, the focus is on feature extraction and the implementation options are relatively straight-forward: using 4 GPUs simultaneously resulted in 20-fold speed-up compared to the CPU-only version. In the querying phase, however, we have multiple choices in approximate indexing and in hardware-based implementations, so due to space limitations the focus here is on a detailed analysis of the querying phase. We start by investigating the impact of applying approximate high-dimensional indexes from the ANN Benchmarks collection. In the remaining four experiments, we study the impact of hardware on the brute-force scan and on one particular approximate indexing strategy.

### 5.1   Datasets

We utilize two benchmark collections and three synthetically generated collections. The benchmark collections provide us with diverse modalities and rich content for realistic analysis of results quality. While the relevant properties and modalities are summarized in Table 1, they are:

**AU-AIR** The AU-AIR dataset [6] consists of eight aerial traffic surveillance video clips at an intersection in Aarhus, Denmark. Captured on windless days, the videos depict varied lighting conditions due to the time of day and weather. With a resolution of $1920 \times 1080$ pixels, it comprises 32,823 frames extracted at five frames per second to avoid redundancies.

**LSC'23** This dataset was generated by an active lifelogger over the course of 18 months [15]. The primary resource of the collection is an image dataset featuring fully redacted and anonymized wearable camera images. Captured using a Narrative Clip device, these images have a resolution of $1024 \times 768$ pixels. Due to huge imbalance on the dataset, we filtered the original dataset with using only the data that include label information with label frequency greater than 257. This preprocessed dataset used in our experiments comprises 40926 images.

To evaluate the scaling of the various approaches, three other synthetic datasets are randomly generated using the uniform distribution. The number of samples in the training sets are 28000 (**small synthetic**), 112000 (**medium synthetic**) and 448000 (**large synthetic**), whereas the testing size is constant at 450 samples.

**Table 1.** Summary of two benchmark datasets used in experiments.

| Dataset | Ground Truth Labels | Modalities | | | | Collection Sizes | | | |
|---|---|---|---|---|---|---|---|---|---|
| | | Image | Text | Time | Location | Whole | Retrieval | Training | Test |
| AU-AIR | 8 | ✓ | ✗ | ✓ | ✓ | 32283 | 32183 | 2000 | 100 |
| LSC'23 | 135 | ✓ | ✓ | ✓ | ✓ | 40926 | 40676 | 4000 | 250 |

## 5.2  Experimental Settings

We have run a large-set of experiments, but due to space limitations we report the results of a representative sample in this section. For example, since the ANN Benchmark collection contains a large set of algorithm implementations, we have chosen to omit algorithms that (a) are alternative implementations of the same approach or (b) perform poorly, leaving 5 indexing strategies and one brute-force approach.

In our benchmark experiments, we have assessed the retrieval performance of MuseHash using several evaluation metrics, including mean Average Precision (mAP), precision, recall, and F-score, but due to space limitations we focus on F-score in our presentation. We have assessed query performance using latency (milliseconds per query) and throughput (reported as thousand queries per second). In the following we report on throughput, sometimes represented as speed-up over a baseline implementation.

For the benchmark collections, the following feature vectors from each modality are used as input representations for our evaluation:

**Image** 4096-D vector from the fc-7 layer of pre-trained VGG16 network.[1]
**Textual** 768-D vector from pre-trained BERT model.[2]
**Temporal** 191-D vector representation for LSC'23 and 203-D vector for AU-AIR [27]. The first four coordinates represent the year, the next 12 are for month (one-hot-encoded), the next 31 for day (one-hot-encoded), the next 24 for hours (one-hot-encoded), the next 60 for minutes (one-hot-encoded), and the next 60 for seconds (one-hot-encoded). AU-AIR has an additional 12 digits for microseconds (binary encoded).
**Spatial** 3-D vector with values (altitude, longitude, altitude).

We compute hash codes using MuseHash for different bit lengths $d_c = 16, 32, 64,$ 128, 256, 512, 1024, 2048. Moreover, we conduct experiments using both single modalities and a combination of all modalities.

The first experiment with ANN Benchmarks was performed using a server with Intel(r) Core(TM) i9-10920X CPU @ 3.50 GHz (12 cores, 2 threads/core) with 134.25 GB of RAM. The remaining experiments were performed on a MareNostrum IV accelerated cluster from the Barcelona Supercomputing Center, where each machine has 2 IBM Power9 8335-GTH @ 2.4 GHz (20 cores,

---

[1] https://github.com/Leo-xxx/pytorch-notebooks/blob/master/Torn-shirt-classifier/VGG16-transfer-learning.ipy.
[2] https://github.com/maknotavailable/pytorch-pretrained-BERT.

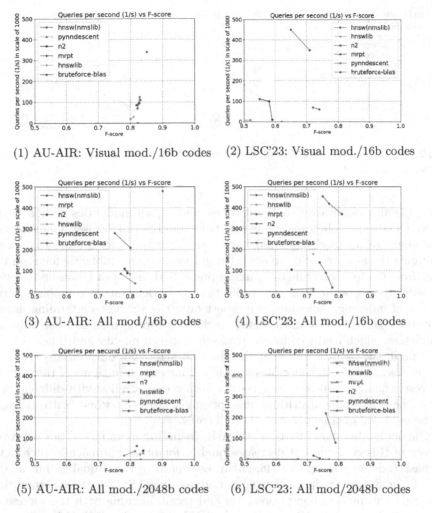

(1) AU-AIR: Visual mod./16b codes     (2) LSC'23: Visual mod./16b codes

(3) AU-AIR: All mod/16b codes     (4) LSC'23: All mod./16b codes

(5) AU-AIR: All mod./2048b codes     (6) LSC'23: All mod/2048b codes

**Fig. 1.** CPU experiment results: AU-AIR (1st column), LSC'23 (2nd column). Rows show visual modality with 16-bit codes, all modalities with 16-bit codes, and all modalities with 2048-bit codes.

4 threads/core) with 512 GB of RAM and 4 NVIDIA V100 GPU with 16 GB RAM. In each case, 5 executions are averaged, with error margins calculated with the standard deviation.

## 5.3    Experiment 1: Impact of Approximate Indexes

In this first experiment, we evaluate retrieval results on the AU-AIR and LSC'23 datasets. Figure 1 shows a representative sample of the results; similar observations hold with other settings. The first column of Fig. 1 displays results for the AU-AIR collection, covering from top to bottom (1) visual modality with 16-bit

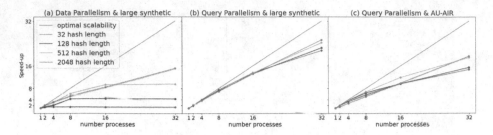

**Fig. 2.** Speed-up of data parallelism vs query parallelism for different hash lengths and datasets.

codes, (3) all modalities with 16-bit codes, and (5) all modalities with 2048-bit codes. The second column presents results for the LSC'23 dataset, following the same structure of modalities and hash code lengths. In each case, the $x$-axis represents the F-score value (to make the graphs more readable, we focus on the range from 0.5 to 1.0), while the $y$-axis represents throughput (thousand queries per second). As outlined above, while the ANN Benchmark contains a large number of indexing algorithms, we report only the six best-performing implementations here. As seen in the figure, each such approach offers some different parameters, which lead to different tradeoffs between quality and time.

Overall, the figures show that the Hnswlib algorithm outperforms all other approaches across both collections and all settings, both in terms of throughput and result quality. This confirms results from recent studies with different settings. The brute-force algorithm, unsurprisingly, performs worst, with throughput below 1,000 queries per second in all cases.

The figure also shows that for AU-AIR, including more modalities improves F-score of Hnswlib, while it decreases quality for other approaches. Increasing the hash code length, up to 2048, enhances overall retrieval quality, but at the cost of significantly reduced throughput. Regarding LSC'23, the incorporation of textual information boosts precision and recall, aligning with the success of textual queries using CLIP models in the VBS competition [20]. Using all available modalities further enhances retrieval results, again at the cost of reduced throughput.

### 5.4 Experiment 2: Query Parallelism Vs. Data Parallelism

In this experiment, we focus on comparing data parallelism (different processes contribute to the computation of each query) and query parallelism (different processes answer different queries) strategies for scaling the brute-force algorithm (using an implementation taken from the scikit-learn package [26]) which traverses the full length of the dataset to find the exact match ($O(n)$ complexity). Figure 2 shows a representative sample of results, showing (a) data parallelism and (b) query parallelism for the large synthetic dataset, and (c) query parallelism for the AU-AIR dataset with all modalities, for a variety of hash lengths

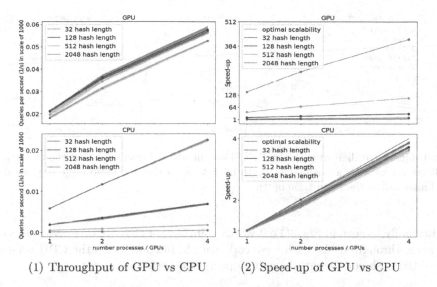

(1) Throughput of GPU vs CPU    (2) Speed-up of GPU vs CPU

**Fig. 3.** Comparison of query parallelism using GPU and CPU for the large synthetic dataset. Speed-up is computed relative to a single CPU-based process.

In all graphs, the speed-up represents throughput with varying number of processors compared to the throughput of a single process; the optimal performance would be that with $X$ processes, the speed-up should be a factor of $X$.

As Fig. 2(a) shows, data parallelism can yield considerable increase in performance, but only when considering longer hash-codes; with smaller hash-codes the speed-up reaches a plateau and the same presumably holds even for hash-code lengths of 2048. Figure 2(b) indicates that query parallelism scales better than data parallelism, obtaining a speed-up factor of 25x, which is very close to the optimal of 32x, for the largest synthetic collection. For both data and query parallelism, however, throughput remains in the order of a few thousands of queries per second, or much lower than seen with approximate indexes. Finally, Fig. 2(c) shows that query parallelism results are somewhat worse for the AU-AIR dataset, presumably because of the Amdahl's law [16].

### 5.5    Experiment 3: Comparison of GPU and CPU Processing

In this experiment, we explore the advantages of specialized hardware, concretely GPUs, with a brute-force scan implementation taken from the cuML package [28]. Figure 3 presents (1) throughput for GPU and CPU respectively, on the left side, and (2) the resulting speed-up on the right side, compared to a single CPU-based process.

As the figure shows, GPU processing provides much better performance than the standard CPU multi-core processing. With the largest synthetic dataset and the longest hash length, a speed-up of 144x is obtained with only one GPU, and with four GPU processors an outstanding speed-up of 419x is achieved.

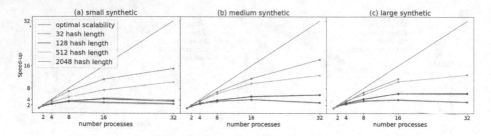

**Fig. 4.** Query parallelism with the PyNNDescent algorithm. (Note that the execution of 32 processes with a hash-code length of 2048 with the synthetic large dataset did not finish within the fixed time of 2 h.)

Additionally, as seen in Fig. 3(1), it worth pointing out that GPUs provide almost the same throughput for each hash-code length, in contrast to the CPU version where the hash-code length heavily impacts the performance.

### 5.6 Experiment 4: Query Parallelism with PyNNDescent Indexing

Having established that query parallelism outperforms data parallelism, we consider whether we can effectively combine it with a state-of-the-art approximate indexing algorithm. For this purpose, we chose the PyNNDescent method [12], with default parameters. Besides, note that query parallelism can be applied straight-forwardly to any indexing algorithm, in contrast to data-parallelism, which involves modifying the internals of the algorithm.

Figure 4 shows the speed-up results of scaling PyNNDescent using multi-core processing in a query parallelism scenario with the synthetic datasets. The algorithm scales similarly with all datasets, with speed-up between 3x and 8x when using 8 simultaneous processes. From that point on, employing more parallel processes does not substantially improve the speed-up, especially in the case of small hash-code lengths. In addition, these results show that the PyNNDescent algorithm is quite constant with regard to the number of dataset samples, obtaining similar results with different dataset sizes.

### 5.7 Experiment 5: Curse of Dimensionality

As discussed in the introduction, high-dimensional retrieval suffers from the curse of dimensionality, but the impact on different indexing strategies varies [30]. In these experiments we compare the performance of a classical tree method (BallTree), which we expect to perform poorly, and a state-of-the-art indexing method (PyNNDescent), which we expect to perform better than a brute-force approach, when increasing the dimensionality through the length of the hash-codes. Figure 5 shows the results for the largest synthetic dataset, with the hash-code length on the $x$-axis and the throughput on the $y$-axis. We can see that the BallTree algorithm already degrades to brute-force performance when increasing

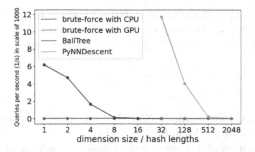

**Fig. 5.** Curse of dimensionality of brute-force, BallTree (with $leaf\_size = 100$) and PyNNDescent (with default parameters) algorithms in the large synthetic dataset with different hash lengths. Note that the $x$-axis is not proportional.

the hash-code length to 8. In contrast, the PyNNDescent algorithm covers a much larger range of hash-code lengths, only degrading to brute-force performance after increasing the length beyond 512.

### 5.8   Discussion

In both benchmark datasets, we observed from Experiment 1 that the Hnswlib, a fast approximate nearest neighbor search method, outperformed all other methods, both in terms of throughput and result quality. Overall, many indexing approaches performed well, yielding a general enhancement of the MuseHash algorithm. As explored in Experiment 2 and 4, these indexing algorithms can be combined with multi-core processing to speed-up the computations, making an efficient use of all the available hardware. Experiment 3 showed the potential of GPUs to heavily enhance the throughput, but the tested method was still heavily outperformed by a CPU-based indexing method in medium-sized datasets.

## 6   Conclusion

In this paper, we have considered query performance for the multi-modal hashing codes produced by MuseHash, using both state-of-the-art approximate indexing approaches and hardware-based approaches. Our results indicate that these techniques improve performance to a varying extent, but that they can also often be combined for further performance improvements. Future work includes applying a combination of the top-performing Hnswlib indexing algorithm, along with parallel hardware-based strategies.

**Acknowledgment.** The work reported here was supported by the EU's Horizon 2020 research and innovation programme under grant agreements H2020-101070250 CALLISTO, H2020-101070262 WATERVERSE, and H2020-101080090 ALLIES, and by the Spanish Ministry of Science (MICINN), the Research State Agency (AEI) and European Regional Development Funds (ERDF/FEDER) under grant agreements PID2021-

126248OB-I00, MCIN/AEI/10.13039/ 501100011033/FEDER, UE, Severo Ochoa Center of Excellence CEX2021-001148-S-20-3 and the Generalitat de Catalunya (AGAUR) 2021-SGR-00478.

# References

1. Abbasifard, M.R., Ghahremani, B., Naderi, H.: A survey on nearest neighbor search methods. Int. J. Comput. Appl. **95**(25) (2014)
2. Arulmozhi, P., Abirami, S.: A comparative study of hash based approximate nearest neighbor learning and its application in image retrieval. Artif. Intell. Rev. **52**(1), 323–355 (2017). https://doi.org/10.1007/s10462-017-9591-1
3. Aumüller, M., Bernhardsson, E., Faithfull, A.: ANN-benchmarks: a benchmarking tool for approximate nearest neighbor algorithms. In: International Conference on Similarity Search and Applications, pp. 34–49. Springer (2017). https://doi.org/ 10.1007/978-3-319-68474-1_3
4. Bai, C., Zeng, C., Ma, Q., Zhang, J., Chen, S.: Deep adversarial discrete hashing for cross-modal retrieval. In: Proceedings of the 2020 International Conference on Multimedia Retrieval (2020)
5. Boytsov, L., Naidan, B.: Engineering efficient and effective non-metric space library. In: Brisaboa, N., Pedreira, O., Zezula, P. (eds.) SISAP 2013. LNCS, vol. 8199, pp. 280–293. Springer, Heidelberg (2013). https://doi.org/10.1007/978-3-642-41062-8_28
6. Bozcan, I., Kayacan, E.: AU-AIR: A multi-modal unmanned aerial vehicle dataset for low altitude traffic surveillance. In: 2020 IEEE International Conference on Robotics and Automation (ICRA), pp. 8504–8510. IEEE (2020)
7. Cao, Y., Long, M., Liu, B., Wang, J.: Deep cauchy hashing for hamming space retrieval. In: Proceedings of the IEEE Conference on Computer Vision and Pattern Recognition (2018)
8. Chen, Q., et al.: Spann: highly-efficient billion-scale approximate nearest neighborhood search. Adv. Neural. Inf. Process. Syst. **34**, 5199–5212 (2021)
9. Chen, S., et al.: Hadamard Codebook based Deep Hashing. arXiv Preprint arXiv:1910.09182 (2019)
10. Chen, X., Güttel, S.: Fast Exact Fixed-Radius Nearest Neighbor Search based on Sorting. Preprint at arXiv. 1048550/ARXIV2212 (2022)
11. Chetlur, S., et al.: cuDNN: Efficient Primitives for Deep Learning (2014)
12. Dong, W., Moses, C., Li, K.: Efficient K-nearest neighbor graph construction for generic similarity measures. In: Proceedings of the 20th International Conference on World Wide Web, pp. 577–586 (2011)
13. Geiger, M.J.: A Multi-threaded local search algorithm and computer implementation for the multi-mode, resource-constrained multi-project scheduling problem. Eur. J. Oper. Res. **256**(3), 729–741 (2017)
14. Guo, R., et al.: Accelerating large-scale inference with anisotropic vector quantization. In: International Conference on Machine Learning, pp. 3887–3896. PMLR (2020)
15. Gurrin, C., et al.: Introduction to the sixth annual lifelog sarch challenge, LSC'23. In: Proceedings of the 2023 ACM International Conference on Multimedia Retrieval, pp. 678–679 (2023)
16. Hill, M.D., Marty, M.R.: Amdahl's law in the multicore era. Computer **41**(7), 33–38 (2008). https://doi.org/10.1109/MC.2008.209

17. Jayaram Subramanya, S., Devvrit, F., Simhadri, H.V., Krishnawamy, R., Kadekodi, R.: Diskann: fast accurate billion-point nearest neighbor search on a single node. In: Advances in Neural Information Processing Systems, vol. 32 (2019)
18. Jing, L., Vahdani, E., Tan, J., Tian, Y.: Cross-Modal Center Loss for 3D Cross-Modal Retrieval. In: Proceedings of the IEEE/CVF Conference on Computer Vision and Pattern Recognition (2021)
19. Li, W., et al.: Approximate nearest neighbor search on high dimensional data-experiments, analyses, and improvement. IEEE Transactions on Knowledge and Data Engineering (2019)
20. Lokoč, J., et al.: Interactive video retrieval in the age of effective joint embedding deep models: lessons from the 11th VBS. Multimedia Syst. **29**(6), 3481–3504 (2023)
21. Luna, A.: Using Annoy in Package C++ Code
22. Malkov, Y.A., Yashunin, D.A.: Efficient and robust approximate nearest neighbor search using hierarchical navigable small world graphs. IEEE Trans. Pattern Anal. Mach. Intell. **42**(4), 824–836 (2018)
23. Malkov, Y., Ponomarenko, A., Logvinov, A., Krylov, V.: Approximate nearest neighbor algorithm based on navigable small world graphs. Inf. Syst. **45**, 61–68 (2014)
24. Narasimhulu, Y., Suthar, A., Pasunuri, R., Venkaiah, V.C.: CKD-Tree: an improved kd-tree construction algorithm. In: ISIC, pp. 211–218 (2021)
25. NVIDIA, Vingelmann, P., Fitzek, F.H.: CUDA, Release: 10.2.89 (2020). https://developer.nvidia.com/cuda-toolkit
26. Pedregosa, F., et al.: Scikit-learn: machine learning in Python. J. Mach. Learn. Res. **12**, 2825–2830 (2011)
27. Pegia, M., et al.: MuseHash: supervised bayesian hashing for multimodal image representation. In: ICMR, pp. 434–442 (2023)
28. Raschka, S., Patterson, J., Nolet, C.: Machine Learning in Python: Main Developments and Technology Trends in Data Science, Machine Learning, and Artificial Intelligence. arXiv Preprint arXiv:2002.04803 (2020)
29. Ukey, N., Yang, Z., Li, B., Zhang, G., Hu, Y., Zhang, W.: Survey on Exact KNN Qeries over High-Dimensional Data Space. Sensors (2023)
30. Weber, R., Schek, H.J., Blott, S.: A quantitative analysis and performance study for similarity-search methods in high-dimensional spaces. In: VLDB. vol. 98, pp. 194–205 (1998)
31. Zhao, W., Tan, S., Li, P.: Song: approximate nearest neighbor search on GPU. In: 2020 IEEE 36th International Conference on Data Engineering (ICDE), pp. 1033–1044. IEEE (2020)

# Dual-Fisheye Image Stitching
# via Unsupervised Deep Learning

Zhanjie Jin[1,2], Anming Dong[1,2]($\boxtimes$) $\textcircled{\tiny ID}$, Jiguo Yu[3] $\textcircled{\tiny ID}$, Shuxiang Dong[4],
and You Zhou[5]

[1] Key Laboratory of Computing Power Network and Information Security,
Ministry of Education, Shandong Computer Science Center (National Supercomputer
Center in Jinan), Qilu University of Technology (Shandong Academy of Sciences),
Jinan 250353, China
anmingdong@qlu.edu.cn
[2] School of Information Science and Technology, Qilu University of Technology
(Shandong Academy of Sciences), Jinan 25353, China
[3] Big Data Institute, Qilu University of Technology (Shandong Academy
of Sciences), Jinan 25353, China
[4] Advanced Technology Research Institute, Beijing Institute of Technology,
Jinan, China
[5] echnology Department, Shandong HiCon New Media Institute Co., Ltd,
Jinan, China

**Abstract.** Constructing panoramic images from a dual-fisheye lens has
been increasingly used along with the recent booming of new computer
vision applications, such as virtual reality (VR) and augmented real-
ity(AR). The recent development of deep learning (DL) techniques has
shed new light on the field of image stitching, but little research has
been conducted on DL-based dual-fisheye image stitching. In this work,
we propose an unsupervised deep learning method for dual-fisheye image
stitching. Specifically, we construct a stitching system consisting of fisheye distortion correction, unsupervised image reconstruction, and image
edge rectangularization blocks. Experiment results show that the proposed scheme can perform accurate and natural stitching of two images,
and exceed the traditional method in PSNR, SSIM, RMSE, MSE, and
other performance indicators.

This work is supported in part by the National Natural Science Foundation of China
(NSFC) under Grant 62272256, the Shandong Provincial Natural Science Foundation
under Grants ZR2021MF026 and ZR2023MF040, the Innovation Capability Enhance-
ment Program for Small and Medium-sized Technological Enterprises of Shandong
Province under Grants 2022TSGC2180 and 2022TSGC2123, the Innovation Team Cul-
tivating Program of Jinan under Grant 202228093, the Piloting Fundamental Research
Program for the Integration of Scientific Research, Education and Industry of Qilu
University of Technology (Shandong Academy of Sciences) under Grants 2021JC02014
and 2022XD001, the Talent Cultivation Promotion Program of Computer Science and
Technology in Qilu University of Technology (Shandong Academy of Sciences) under
Grants 2021PY05001 and 2023PY059.

© The Author(s), under exclusive license to Springer Nature Switzerland AG 2024
S. Rudinac et al. (Eds.): MMM 2024, LNCS 14556, pp. 284–298, 2024.
https://doi.org/10.1007/978-3-031-53311-2_21

**Keywords:** Dual fisheye · Panoramic image · Image stitching · Deep learning

# 1  Introduction

Through panoramic vision technology, people can understand the overall situation of a scene more intuitively, enhance their spatial perception and immersion experience, and provide a more convenient and efficient way of visual interaction. Achieving panoramic vision requires the acquisition of panoramic images containing all surrounding scene information, but this requires expensive professional equipment and complex post-processing [1]. Leveraging the wide-angle capabilities of fisheye lenses facilitates the camera in capturing a broader field of view, addressing the current demand for more accessible panoramic imagery [2]. It has important application prospects in the fields of national defense, unmanned vehicles [3,4], security monitoring, medical care, and so on.

The dual fisheye lens combines two fisheye lenses to provide a wider field of view and a more realistic viewing experience for building panoramas. However, the images obtained with dual fisheye lenses must be precisely stitched together to form a panoramic image with a wider field of view [5]. However, due to the fisheye image edge details after correction may be lacking or lost, reducing the resolution of the edges [6]. Therefore, the traditional image stitching method may be effective for conventional images, but it is not satisfactory for fisheye images that lack image features. Therefore, fisheye image stitching needs to consider not only translation and rotation transformation in distortion correction but also the effect of the quality of corrected image features on the stitching [7].

There are many methods and techniques in the field of dual fisheye image stitching, which are mainly divided into feature-based methods, grid-based methods, and deep learning methods.

The feature-based methods are mainstream and most widely used, which mainly extract feature key points through feature detection algorithms such as SIFT, SURF, and ORB, match them through the correspondence between key points, and use the RANSAC algorithm to estimate homography. Finally realize the alignment of the two images [8–13]. The grid-based methods are to insert mesh into the image and realize the image stitching through mesh warp. The realization of mesh warp is to find the corresponding feature points of the mesh by the feature extraction method and apply a constraint to it to make the mesh warp [5,14,15]. This method also needs to extract the feature points of the grid, which has relatively high-quality requirements for the feature points. However, the features of the corrected fisheye image may be missing or lost, resulting in insufficient extraction of feature points and affecting the stitching effect, which is undoubtedly disastrous for the above method. There are few types of research in this field on methods based on deep learning, which mainly use the powerful representation ability of CNN to quickly learn and extract image features, and the features learned are more flexible than those extracted by traditional feature detection methods [16–18]; It is also achieved by weakly supervised learning, which takes multiple fisheye images as input and creates 360-degree output images in isometric column projection format [19]. The former also needs to extract key feature points, but the way of neural network is used to extract more quickly. The latter proposes weak supervision to train the network without the need for real ground truth images, but the generation of pictures also needs to provide additional images of the same as the stitching scene to act as weak supervision.

To solve these problems, Using deep learning-based unsupervised image stitching, the unsupervised image reconstruction network from feature to pixel is adopted to overcome the difficulty of traditional methods of stitching low-feature images with low resolution. We also adopt an unsupervised training scheme to constrain the unsupervised homography network with an ablation-based loss so that the input images are roughly aligned. Then, in the image reconstruction network, the low-resolution deformation branch and high-resolution thinning branch are used to learn the deformation rule and improve the resolution respectively. Compared with weakly supervised learning methods, our method is more efficient and convenient and does not require additional image input to act as weakly supervised, and since unsupervised learning is not limited by label quality or quantity, it is generally more adaptive and can adapt to different quality and diversity of data. In addition, an image rectangle operation is proposed to eliminate the irregular edges generated by image stitching, which can provide a more compact and efficient picture.

The experimental results show that the image is natural and realistic without irregular edges, which improves the user's visibility and the adaptability of the machine vision system to the image and is superior to the traditional methods in PSNR, SSIM, MSE, and other performance indexes. This paper has the following contributions:

- To the best of our knowledge, it is the first time that we have designed a dual-fisheye image stitching method based on unsupervised learning, which is composed of a combination of distortion correction, an unsupervised learning image stitching network, and image edge rectangularization.
- We verify the effectiveness of unsupervised learning on image stitching through experiments, and the generated images are of better quality and the pages are more efficient and compact.

## 2    Related Work

In this section, we summarize three types of approaches: feature-based method, grid-based method, and deep learning-based method.

### 2.1    Feature-Based Method

Harris corner detection is a pixel-based feature information detection method proposed by Harris in 1998. On this basis, a large number of corner point-related alignment methods have been proposed [20]. David Lowe published the SIFT algorithm in 1999 and then perfected it in 2004. It finds the extreme point in the spatial scale and extracts its position, scale, and rotation invariants [21]. In 2006, Bay et al. improved the SIFT algorithm to improve its efficiency and proposed the SURF algorithm [22]. In 2006, Rosten et al. proposed a FAST feature that is fast but does not have scale invariance and rotation invariance [23]. In In 2018, Lo et al. used an isometric column projection transformation to correct the overlapping area of two fisheye cameras. Local distortion is also used for image alignment and feature matching to improve stitching quality [11]. In 2018, Souza proposed an adaptive stitching technology that can extract feature cluster templates from the stitching region and align them using template matching, thereby reducing discontinuity in the panorama [9]. In 2020 Roberto et al., method used a feature detection algorithm to extract features for global alignment and then applied local optimization based on rigid least squares transformation and slit-based transformation line estimation [24]. In 2021, Xue et al. extracted SURF feature points from the corrected left and right images, and then proposed a weighted fusion algorithm AW based on DW in the image fusion part, which improved the linear change of weights to nonlinear change, and solved the ghosts in the overlapping region of the stitching images. In 2022, Dong et al. proposed a video stitching method based on multiple homography, in which a key point detector and descriptor (SIFT) obtained feature points from two video images and directly aligned wide-angle images based on multiple homography alignment [12].

## 2.2 Grid-Based Method

In 2019, Lo et al. proposed a video stitching method that includes mesh deformation operations to minimize geometric distortion, and adaptive near-line cutting operations [14]. In 2021, Lo et al. adopted a mesh deformation model and proposed an adaptive slicing method for image stitching to reduce geometric distortion and ensure optimal spatiotemporal alignment [5]. In 2022, Cheng et al. used image alignment and image blending of deformable meshes to optimize interlens parallax and ensure high alignment accuracy [15].

## 2.3 Deep Learning-Based Method

In 2017, Ufer et al. proposed a new approach for semantic matching based on pretrained CNN features using a heuristic feature pyramid and activation-guided feature selection [16]. The method proposed by Rocco in 2017 addresses the problem of estimating geometric transformations between images by using a CNN-based model. [17]. Lai et al. proposed a video stitching method based on a push-scan stitching network, which converted all undistorted fisheye images into a common cylinder through column projection, and then completed the stitching using CNN-based push-scan stitching network. The video stitching problem is transformed into a spatial view interpolation problem. Song et al., in 2022, developed a weakly supervised learning mechanism to train a stitching model without real ground-truth images [19].

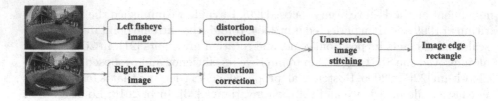

**Fig. 1.** The overall framework diagram of the system.

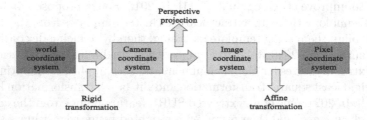

**Fig. 2.** Coordinate system transformation

# 3  Unsupervised Deep Learning Dual-Fisheye Image Stitching System

Figure 1 shows the overall framework diagram of the system in this paper. It mainly contains three parts: fisheye distortion correction, image stitching, and image rectangularization.

## 3.1  Fisheye Image Distortion Correction

By applying a polynomial function to map pixel coordinates in a fisheye image to the corrected coordinate system, the distortion correction is realized by solving the polynomial function [25].

**Principle of Coordinate Transformation.** First, a physical point is selected as the origin of the world coordinate system to determine the 3D coordinates of each pixel; then rigid body transformation and projection are performed to convert the 3D coordinates to pixel coordinates; finally, the image position is determined, and the image center is transformed to the final pixel coordinate system. Figure 2 shows the transformation of the coordinate system.

$$Z \begin{pmatrix} u \\ v \\ 1 \end{pmatrix} = \underbrace{\underbrace{\begin{pmatrix} \frac{1}{dX} & -\frac{\cot\theta}{dX} & u_0 \\ 0 & \frac{1}{dY\sin\theta} & v_0 \\ 0 & 0 & 1 \end{pmatrix}}_{\text{Affine transformation}} \underbrace{\begin{pmatrix} f & 0 & 0 & 0 \\ 0 & f & 0 & 0 \\ 0 & 0 & 1 & 0 \end{pmatrix}}_{\text{Perspective projection}}}_{\text{Intrinsic matrix}} \underbrace{\begin{pmatrix} R & T \\ 0 & 1 \end{pmatrix}}_{\substack{\text{Rigid transformation} \\ \text{Extrinsic matrix}}} \begin{pmatrix} U \\ V \\ W \\ 1 \end{pmatrix} \quad (1)$$

**Fig. 3.** Checkerboard pictures, take a picture to obtain the coordinate information of corner calculation.

Equation 1 illustrates the transformation relationship between the coordinate systems. Where $(U, V, W)$ are the physical coordinates of a point, $(u, v)$ are the corresponding pixel coordinates of the point in the pixel coordinate system, and $Z$ is the scale factor. $f$ is the image distance and $dX$, $dY$ denotes the physical length of a pixel in the $X$ and $Y$ direction on the camera's light-sensitive plate, respectively. $u_0$, $v_0$ denote the coordinates of the center of the camera's light-sensitive plate in the pixel coordinate system, respectively. $\theta$ Indicates the angle between the horizontal and vertical edges of the photographic plate. $R$ denotes the rotation matrix, and $T$ denotes the translation vector.

**Polynomial Distortion Correction.** Polynomial distortion correction is used to correct the distortion effect in the image [26]. The method is based on the mapping principle of polynomial functions, which maps the pixel coordinates in an image to a corrected coordinate system, usually described approximately using the first few terms of a Taylor series polynomial expanded at $r = 0$, and then calculates the corrected pixel positions. The coordinate relationship between radial distortion and tangential distortion before and after correction is shown in Eq. 2 and Eq. 3 below:

$$\hat{x} = x \left(1 + k_1 r^2 + k_2 r^4 + k_3 r^6\right),$$
$$\hat{y} = y \left(1 + k_1 r^2 + k_2 r^4 + k_3 r^6\right), \tag{2}$$

$$\hat{x} = x + \left(2 p_1 y + p_2 (r^2 + 2x^2)\right),$$
$$\hat{y} = y + \left(2 p_2 x + p_1 (r^2 + 2y^2)\right), \tag{3}$$

where $(x, y)$, $(\hat{x}, \hat{y})$ are the ideal normalized image coordinates without distortion, and the normalized image coordinates after distortion, respectively. $k_1$, $k_2$, $k_3$, $p_1$, $p_2$ are the parameters to be obtained.

The checkerboard calibration board as shown in Fig. 3 was adopted for the calibration method, and the pixel coordinates $(u, v)$ of each corner were obtained by the image detection algorithm. Then calculate its physical coordinates $(U, V, W = 0)$. Using this information, the camera can be calibrated (Fig. 4).

**Fig. 4.** The two images on the left are the fisheye images to be corrected, and the two images on the right are the images after polynomial model correction.

## 3.2   Image Stitching

Traditional image stitching technology needs to extract feature points from the image, but because the edge resolution of the image will be reduced after fisheye correction, this leads to the difficulty of feature point extraction, which is not acceptable to the traditional method. They mainly use mathematical methods such as gradient features and local maximum or minimum grayscale to describe the feature information of an image [11]. However, some features that determine image properties are abstract and cannot be described by simple mathematical formulas [27]. However, deep learning technology has advanced performance in image processing, which can learn the feature properties of images without representation.

At present, there are few feature extraction methods based on deep learning, and they need to extract the feature information of the image, which performs poorly in the case of low image feature quality. The unsupervised deep learning method does not need to display and extract image feature information, which means that it can overcome the impact of low quality of image features on stitching technology. Therefore, the unsupervised depth image stitching framework proposed by Nie [28] is adopted to solve our problem. The specific image stitching framework is shown in Fig. 5, which includes two stages: unsupervised coarse image stitching and unsupervised image reconstruction. In the coarse alignment stage, an ablation-based strategy is designed to constrain the large baseline unsupervised single-strain estimation and the stitching domain transform layer. In the reconstruction stage, the warped image is used to reconstruct the stitched image from features to pixels.

**Unsupervised Coarse Image Alignment.** In the first stage, an ablation-based strategy was designed to constrain the unsupervised single-strain estimation of large baselines, and the input images can be warped to roughly align with each other in the proposed suture domain transformation layer. Specifically, the first stage takes the complete image as input to ensure that all overlapping regions are included. When the distorted target image is brought close to the reference image, the reference image clears the content of the distorted target image where invalid pixels are located. The objective function for unsupervised homography response by Eq. 4.

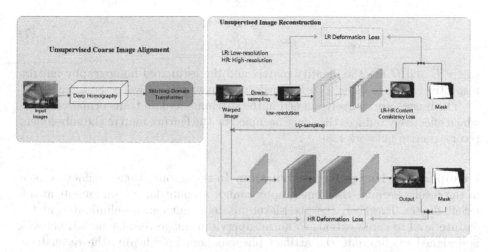

**Fig. 5.** Image stitching structure diagram, the left half is the coarse alignment stage, while the right half is the image reconstruction to realize the image reconstruction from features to pixels.

$$L_{PW} = \left\| \mathcal{H}(N) \otimes I^a - \mathcal{H}\left(I^b\right) \right\| \tag{4}$$

where $I^a$ and $I^b$ denote the reference image and the target image, respectively, $\mathcal{H}(.)$ uses the estimated homography to warp an image to align with another image. $\otimes$ is the pixel-wise multiplication and $N$ is an all-one matrix with identical size with $I^a$ .

In addition, the transform layer is introduced to distort the input image in the stitching-domain space [29], stitching-domain transformer layer is implemented in the following way. First, we compute the four vertices in the target image by Eq. 5.

$$(x_i^w, y_i^w) = (x_i^b, y_i^b) + (\Delta x_i, \Delta y_i), i \in \{1,2,3,4\}, \tag{5}$$

where $(x_i^w, y_i^w)$, $(x_i^b, y_i^b)$ denote the warped target image and the ith vertex coordinates of the target image, respectively. $(\Delta x_i, \Delta y_i)$ denotes the offset of the ith vertex. The size $(W^* \times H^*)$ of the warped image can be obtained by Eq. 6.

$$W^* = \max_{i \in \{1,2,3,4\}} \{x_i^w, x_i^a\} - \min_{i \in \{1,2,3,4\}} \{x_i^w, x_i^a\},$$
$$H^* = \max_{i \in \{1,2,3,4\}} \{y_i^w, y_i^a\} - \min_{i \in \{1,2,3,4\}} \{y_i^w, y_i^a\}, \tag{6}$$

where $(x_i^a, y_i^a)$, $(x_i^b, y_i^b)$ is the same, $(x_i^a, y_i^a)$ said the vertex coordinates of the reference image. Assign a specific value to the pixels of the distorted image B, which can be denoted by Eq. 7.

$$I^{aw} = \sqsupseteq (I^a, E),$$
$$I^{bw} = \sqsupseteq (I^b, H), \tag{7}$$

where $E$ and $H$ are the identity matrix and the estimated homography matrix, respectively. $\sqsupseteq()$ is using a $3 \times 3$ transformation matrix to warp the image.

The images are coarsely aligned in the stitching domain transformation layer, which effectively reduces the space occupied by the feature map in the subsequent reconstruction network [28].

**Unsupervised Image Reconstruction.** In the second stage, artifacts appear in the results because the homography cannot account for all alignments at different depths. However, the pixel-level misalignment can be eliminated at the feature level to some extent, so an unsupervised image reconstruction network is designed to eliminate the artifact phenomenon. Specifically, the reconstruction network can be implemented by two branches: a low-resolution deformation branch and a high-resolution refinement branch to learn the deformation rules of image stitching and to improve the resolution, respectively.

Since the perceptual field shrinks with increasing resolution and cannot adequately perceive the unaligned regions, a low-resolution branch is designed to learn the deformation rules of image stitching. Another high-resolution fine branch is designed to increase the resolution and refine the stitched images. An encoder-decoder network is used to reconstruct the stitched images. At the beginning of the encoder stage, the network focuses only on the overlapping regions. As the resolution decreases, deeper semantic features are extracted and reconstructed. At the decoder stage, it starts to focus on non-overlapping regions. As the resolution is recovered, a more accurate feature map is reconstructed. Finally, pixel-level reconstruction is performed on the stitched image. The image stitching results are shown in Fig. 6.

### 3.3 Image Rectangularization

Image stitching gives the desired large field of view but introduces irregular borders that affect the visual effect. The traditional approach is to obtain regular rectangular borders by cropping and stitching. However, the former may reduce the important content of the image, while the latter may fill in the wrong information and cause the original image information to change. These methods can appear unreliable in practical applications. The first solution–rectangling, proposed by He [30], forms a two-stage mesh deformation by searching the initial mesh and optimizing the target mesh without changing the original image content. Then, the irregular image is mapped into rectangles by mesh deformation. However, this is less robust to nonlinear images with complex scenes and may lead to severe distortion [31].

In this article, the first deep learning image rectangle solution proposed by Nie [32] is used to solve these problems. The concrete image rectangle framework is shown in Fig. 7. Specifically, the stitched image and mask are used as input.

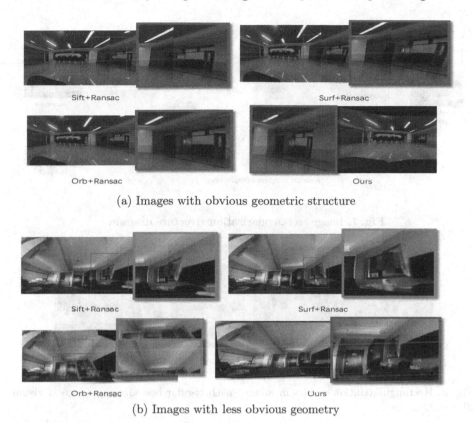

(a) Images with obvious geometric structure

(b) Images with less obvious geometry

**Fig. 6.** We enlarge the overlapping area of the stitching image, (a) and (b) are scene images with relatively obvious geometric features and not obvious geometric features respectively. The results of sift, surf, orb, and ransac are shown respectively, and the last one is our method. Intuitively, it can be seen that our method is superior to others

Simple convolutional stitching blocks are stacked to form a predefined feature extractor to extract high-level semantic features from the input. After feature extraction, an adaptive pooling layer is used to fix the resolution of the feature image.

The horizontal and vertical motion of each vertex is predicted by using the fully convolutional architecture as a regression variable. Then, a residual asymptotic regression strategy is used to gradually estimate the exact grid motion with distortion of the intermediate feature maps. The performance of the model is improved with a slight increase in computational effort, and the warped results are again used as network inputs. Then, through two regression machines with the same structure, the initial reticular motion and the residual reticular motion are predicted, respectively. Although they have the same structure, they are designated for different tasks because of their different input characteristics. Finally,

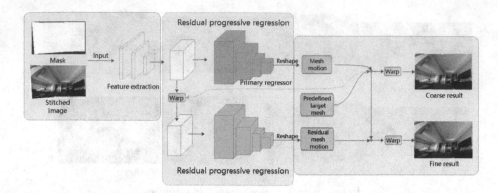

**Fig. 7.** Image rectangularization structure diagram

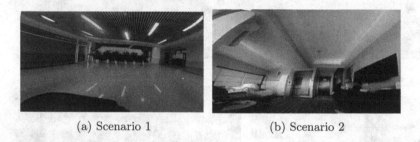

(a) Scenario 1                    (b) Scenario 2

**Fig. 8.** Rectangularization results in images with regular boundaries and good visual effects.

**Table 1.** PSNR, SSIM, RMSE, and MSE of the overlapping regions on images with indistinct geometric structure.

| Area | Index / Method | PSNR ↑ | SSIM ↑ | RMSE ↓ | MSE ↓ |
|---|---|---|---|---|---|
| Top | SIFT+RANSAC | 15.36 | 0.34 | 43.45 | 1888.49 |
| | SURF+RANSAC | 14.55 | 0.32 | 47.70 | 2276.08 |
| | ORB+RANSAC | 15.51 | 0.35 | 42.75 | 1828.35 |
| | Ours | **22.16** | **0.58** | **19.86** | **394.60** |
| Middle | SIFT+RANSAC | 10.39 | 0.16 | 77.05 | 5937.70 |
| | SURF+RANSAC | 10.52 | 0.09 | 75.87 | 5756.47 |
| | ORB+RANSAC | 10.67 | 0.12 | 74.60 | 5566.04 |
| | Ours | **19.33** | **0.58** | **27.51** | **757.17** |
| Bottom | SIFT+RANSAC | 10.34 | 0.12 | 77.53 | 6011.59 |
| | SURF+RANSAC | 10.13 | 0.14 | 79.41 | 6307.37 |
| | ORB+RANSAC | 11.70 | 0.17 | 66.27 | 4391.72 |
| | Ours | **19.51** | **0.48** | **26.95** | **726.42** |
| Average | SIFT+RANSAC | 12.03 | 0.20 | 66.01 | 4,612.59 |
| | SURF+RANSAC | 11.74 | 0.18 | 67.66 | 4,779.97 |
| | ORB+RANSAC | 12.62 | 0.21 | 61.21 | 3,928.70 |
| | Ours | **20.34** | **0.55** | **24.77** | **626.06** |

there is a predefined rigid target mesh, which estimates only one initial mesh to form mesh deformation. The initial grid is predicted by a complete convolution network and a residual asymptotic regression strategy. Figure 8 shows the result of a rectangle of our image.

**Table 2.** PSNR, SSIM, RMSE, and MSE analysis were performed on the overlapping regions of images with more obvious geometric structures.

| Area | Method | PSNR↑ | SSIM↑ | RMSE↓ | MSE↓ |
|---|---|---|---|---|---|
| Top | SIFT+RANSAC | 14.12 | 0.34 | 50.15 | 2515.80 |
| | SURF+RANSAC | 14.24 | 0.27 | 49.46 | 2446.77 |
| | ORB+RANSAC | 13.29 | 0.27 | 55.16 | 3042.95 |
| | Ours | **16.04** | **0.39** | **40.19** | **1615.52** |
| Middle | SIFT+RANSAC | 19.50 | 0.56 | 26.99 | 728.80 |
| | SURF+RANSAC | 17.38 | 0.44 | 34.47 | 1188.51 |
| | ORB+RANSAC | 16.73 | 0.44 | 37.14 | 1379.64 |
| | Ours | **22.15** | **0.61** | **19.88** | **395.60** |
| Bottom | SIFT+RANSAC | 24.18 | 0.75 | 15.75 | 248.35 |
| | SURF+RANSAC | 21.13 | 0.68 | 22.36 | 500.15 |
| | ORB+RANSAC | **24.00** | **0.76** | **15.39** | **236.97** |
| | Ours | 23.27 | 0.74 | 17.49 | 306.10 |
| Average | SIFT+RANSAC | 19.26 | 0.55 | 108.50 | 1086.78 |
| | SURF+RANSAC | 17.58 | 0.46 | 35.43 | 1378.48 |
| | ORB+RANSAC | 18.13 | 0.49 | 35.90 | 1553.19 |
| | Ours | **20.49** | **0.58** | **25.85** | **772.41** |

# 4 Experiment and Analysis

In order to verify the validity and reliability of the method, two sets of images were taken with the same fisheye lens, which is a non-zoom lens, and the two sets of images were taken panning. Then we compare our method with the traditional method.

## 4.1 Dataset

In terms of datasets, we use the UDIS-D unsupervised depth image stitching dataset provided in [28], which contains a variety of real images with different disparity degrees and different environments to enhance the robustness of the model.

## 4.2  Result Analysis

Relevant experiments were carried out according to the above methods. Using the same non-autozoom fisheye lens, the effect of a dual fisheye lens with constant relative position is simulated by panning instead of rotating the lens. The experimental data can be seen in Table 1 and Table 2. In order to make the results more accurate, we enlarged the overlapping area and divided it into three parts to calculate separately. It can be seen that our method outperforms the others in PSNR, SSIM, RMSE and other performance indicators.

## 5  Conclusion

The use of a dual fisheye lens can better identify object information, obtain object depth-of-field information, and realize 3D stereo images based on image differences, which has good practical value. Therefore, an unsupervised deep learning-based dual fisheye image stitching method is proposed in this paper. The experimental results show that the method solves the limitation of traditional methods limited by the quality of image features. It realizes the accurate stitching of two fisheye images, improves the visual coverage, and removes irregular boundaries without cutting the content. However, there are still some problems, such as how to solve the real-time problem, which will greatly promote future applications.

## References

1. Chalfoun, J., et al.: Mist: accurate and scalable microscopy image stitching tool with stage modeling and error minimization. Sci. Rep. **7**(1), 1–10 (2017)
2. Duan, H., Min, X., Sun, W., Zhu, Y., Zhang, X-P., Zhai, G.: Attentive deep image quality assessment for omnidirectional stitching, IEEE J. Se. Top. Sign. Process. 2023
3. Jia, Q., Feng, X., Liu, Y., Fan, X., Latecki, L.J.: Learning pixel-wise alignment for unsupervised image stitching. Network **1**(1), 1 (2023)
4. Lo, I-C., Chen,H.H. : Acquiring 360∘ light field by a moving dual-fisheye camera, IEEE Transactions on Image Processing, 2023
5. Lo, I.-C., Shih, K.-T., Chen, H.H.: Efficient and accurate stitching for 360∘ dual-fisheye images and videos. IEEE Trans. Image Process. **31**, 251–262 (2021)
6. Kweon, H., Kim, H., Kang, Y., Yoon, Y., Jeong, W., Yoon, K.-J.: Pixel-wise warping for deep image stitching. Proc. AAAI Conf. Artif. Intell. **37**(1), 1196–1204 (2023)
7. Lo, Y.C. Huang, C.C., Tsai, Y.F., Lo, I.C., Wu, A.Y.A., Chen, H.H.: Face recognition for fisheye images, in 2022 IEEE International Conference on Image Processing (ICIP). IEEE, 2022, pp. 146–150
8. Li, Y.-H., Lo, I.-C., Chen, H.H.: Deep face rectification for 360 dual-fisheye cameras. IEEE Trans. Image Process. **30**, 264–276 (2020)
9. Souza, T., et al.: 360 stitching from dual-fisheye cameras based on feature cluster matching, In: 31st SIBGRAPI Conference on Graphics, Patterns and Images (SIBGRAPI). IEEE **2018**, pp.313–320 (2018)

10. Ni, G., Chen, X., Zhu, Y., He, L., "Dual-fisheye lens stitching and error correction, In : 10th International Congress on Image and Signal Processing, BioMedical Engineering and Informatics (CISP-BMEI). IEEE **2017**, pp .1–6 (2017)

11. Lo, I.-C., Shih, K.-T., Chen, H.H.: Image stitching for dual fisheye cameras, In: 2018 25th IEEE International Conference on Image Processing (ICIP). IEEE, 2018, pp. 3164–3168

12. Dong, Y., Pei, M., Zhang, L., Xu, B., Wu, Y., Jia, Y.: Stitching videos from a fisheye lens camera and a wide-angle lens camera for telepresence robots. Int. J. Soc. Robot. **14**(3), 733–745 (2022)

13. Brown, M., Lowe, D.G.: Automatic panoramic image stitching using invariant features. Int. J. Comput. Vision **74**, 59–73 (2007)

14. Lo, I.-C., Shih, K.-T., Chen, H.H.: 360o video stitching for dual fisheye cameras, In 2019 IEEE International Conference on Image Processing (ICIP). IEEE, 2019, pp. 3522–3526

15. Cheng, H., Xu, C. Wang, J. Zhao,L. : Quad-fisheye image stitching for monoscopic panorama reconstruction, In: Computer Graphics Forum, vol. 41, no. 6. Wiley Online Library, 2022, pp. 94–109

16. Ufer, N., Ommer, B.: Deep semantic feature matching, In: Proceedings of the IEEE Conference on Computer Vision and Pattern Recognition, 2017, pp. 6914–6923

17. Rocco,I. Arandjelovic, R. Sivic,J.: Convolutional neural network architecture for geometric matching, In: Proceedings of the IEEE Conference on Computer Vision and Pattern Recognition, 2017, pp. 6148–6157

18. Jeon, Sangryul, Kim, Seungryong, Min, Dongbo, Sohn, Kwanghoon: PARN: Pyramidal Affine Regression Networks for Dense Semantic Correspondence. In: Ferrari, Vittorio, Hebert, Martial, Sminchisescu, Cristian, Weiss, Yair (eds.) ECCV 2018. LNCS, vol. 11210, pp. 355–371. Springer, Cham (2018). https://doi.org/10.1007/978-3-030-01231-1_22

19. Song, D.-Y., Lee, G., Lee, H., Um, G.-M., Cho, D.: Weakly-supervised stitching network for real-world panoramic image generation, In: European Conference on Computer Vision. Springer, 2022, pp. 54–71

20. Harris, C., Stephens, M., et al.,: A combined corner and edge detector, In; Alvey Vision Conference, vol. 15, no. 50. Citeseer, 1988, pp. 10–5244

21. Lowe, D.G.: Distinctive image features from scale-invariant keypoints. Int. J. Comput. Vision **60**(2), 91–110 (2004)

22. Bay, H., Ess, A., Tuytelaars, T., Van Gool, L.: Speeded-up robust features (surf). Comput. Vis. Image Underst. **110**(3), 346–359 (2008)

23. Rosten, Edward, Drummond, Tom: Machine Learning for High-Speed Corner Detection. In: Leonardis, Aleš, Bischof, Horst, Pinz, Axel (eds.) ECCV 2006. LNCS, vol. 3951, pp. 430–443. Springer, Heidelberg (2006). https://doi.org/10.1007/11744023_34

24. Roberto, R., et al.: Using local refinements on 360 stitching from dual-fisheye cameras. In: VISIGRAPP (5: VISAPP), 2020, pp. 17–26

25. Zhang, Z.: Flexible camera calibration by viewing a plane from unknown orientations, In: Proceedings of the seventh IEEE International Conference on Computer Vision, vol. 1. IEEE, 1999, pp. 666–673

26. —, A flexible new technique for camera calibration, IEEE Trans. Pattern Anal. Mach. Intell. vol. 22, no. 11, pp. 1330–1334, 2000

27. Tareen, S.A.K., Saleem, Z.: A comparative analysis of sift, surf, kaze, akaze, orb, and brisk, in,: International Conference on Computing, Mathematics and Engineering Technologies (iCoMET). IEEE **2018**, pp. 1–10 (2018)

28. Nie, L., Lin, C., Liao, K., Liu, S., Zhao, Y.: Unsupervised deep image stitching: reconstructing stitched features to images. IEEE Trans. Image Process. **30**, 6184–6197 (2021)
29. Jaderberg, M., Simonyan, K., Zisserman, A., et al.: Spatial transformer networks, Adv. Neural Inf. process. syst. vol. 28, 2015
30. He, K., Chang, H., Sun, J.: Rectangling panoramic images via warping. ACM Trans. Graph. (TOG) **32**(4), 1–10 (2013)
31. Li, D., He, K., Sun, J., Zhou, K., : A geodesic-preserving method for image warping, In: Proceedings of the IEEE Conference on Computer Vision and Pattern Recognition, 2015, pp. 213–221
32. Nie, L., Lin, C., Liao, K., Liu, S., Zhao, Y.: Deep rectangling for image stitching: a learning baseline, In: Proceedings of the IEEE/CVF Conference on Computer Vision and Pattern Recognition, 2022, pp. 5740–5748

# CA-GAN: Conditional Adaptive Generative Adversarial Network for Text-to-Image Synthesis

Junpeng Liu[1(✉)] and Hengkang Bao[2]

[1] School of Film, Xiamen University, Xiamen 361005, China
liujunpeng@stu.xmu.edu.cn
[2] School of Economics, Peking University, Beijing 100871, China

**Abstract.** Text-to-image synthesis has been a popular multimodal task in recent years, which faces two major challenges: the semantic consistency and the fine-grained information loss. Existing methods mostly adopt either a multi-stage stacked architecture or a single-stream model with several affine transformations as the fusion block. The former requires additional networks to ensure the semantic consistency between text and image, which is complex and results in poor generation quality. The latter simply extracts affine transformation from Conditional Batch Normalization (CBN), which can not match text features well. To address these issues, we propose an effective Conditional Adaptive Generative Adversarial Network. Our proposed method (i.e., CA-GAN) adopts a single-stream network architecture, consisting of a single generator/discriminator pair. To be specific, we propose: (1) a conditional adaptive instance normalization residual block which promotes the generator to synthesize high quality images containing semantic information; (2) an attention block that focuses on image-related channels and pixels. We conduct extensive experiments on CUB and COCO datasets, and the results show the superiority of the proposed CA-GAN in text-to-image synthesis tasks compared with previous methods.

**Keywords:** Multi-modal · Text-to-Image · GAN

## 1 Introduction

The emergence of Generative Adversarial Networks (GANs) [1] in recent years has greatly advanced the field of text-to-image synthesis [7,14,16,20,23], making it possible to generate realistic images with semantic consistency. Text-to-image synthesis is a research direction that attracts much attention in the fields of computer vision (CV) and natural language processing (NLP). Its core goal is to transform natural language descriptions into realistic image content. The development of this field enables computer systems to understand and generate images, providing potential opportunities for applications such as computer-aided design, photo editing and virtual scene construction.

© The Author(s), under exclusive license to Springer Nature Switzerland AG 2024
S. Rudinac et al. (Eds.): MMM 2024, LNCS 14556, pp. 299–312, 2024.
https://doi.org/10.1007/978-3-031-53311-2_22

Since Reed et al. [11] first introduced Generative Adversarial Networks (GANs) to text-to-image synthesis, much of the ensuing work has focused on GAN as the primary method. There are two main types of methods for text-to-image synthesis nowadays. The first type is represented by AttnGAN [16], which adopts a multi-stage stacked network architecture to enhance the image resolution (see, e.g., [7,22,23]). The second type is represented by DF-GAN [14], which only uses a single-stream network architecture to directly generate high-resolution images (see, e.g., [17,21]). Although these methods are helpful for text-to-image synthesis, they still have problems. First, the entanglement between generators in the stacked network architecture affects the quality of the final generated images, and the introduction of additional networks to ensure semantic consistency of text images increases the training complexity. Second, the single-stream model with multiple affine transformations as fusion blocks only extracts affine transformations from Conditional Batch Normalization (CBN) [18] and neglects to normalise individual images in the batch, which leads to a mismatch in the feature space of the generated images and text vectors, and does not allow a good establishment of the semantic relationship between the two modalities. Third, as the feature map resolution increases, the over-deep residual network structure may cause some fine-grained word-level information to be blurred or lost in the subsequent layers of the network.

To address these issues, we propose an effective Conditional Adaptive Generative Adversarial Network (CA-GAN). For the first issue, we adopt a single-stream network architecture consisting of a single generator/discriminator, which avoids the entanglement between generators in a stacked network and can directly generate high-quality images. For the second issue, we propose a conditional adaptive instance normalization residual block (CARBlock) to help generate images fusing textual semantic information. In CARBlock, inspired by adaptive instance normalization (AdaIN) [5], we propose Conditional instance normalization (CAdaIN), which applies adaptive instance normalization of text vector conditions in image generation to ensure that the generated image is visually consistent with the conditional description. Moreover, some of the original features can be retained through residual connection, which helps to preserve details and structural information. For the third issue, we propose an attention block (ABlock), which can improve the network's attention to important features and reduce the response to irrelevant information through channel attention and spatial attention mechanisms. This helps to retain more fine-grained information in subsequent layers of the network, thus mitigating the problem of losing fine-grained information due to an overly deep network structure.

The contributions of this paper are the following:

- We propose an effective Conditional Adaptive Generative Adversarial Network for Text-to-Image Synthesis, which can guide the generation of high-quality images.
- Conditional AdaIN (CAdaIN) is introduced to assist the generator in flexibly controlling the texture of the generated images.

- We propose a new Conditional AdaIN Residual Block, which provides better control over the semantic consistency between the generated images and the given text conditions.
- We propose an attention block that helps to generate images focused on textual content from the channel level and pixel level to retain more fine-grained information.
- Extensive experiments on the CUB and COCO datasets show that the proposed CA-GAN outperforms state-of-the-art text-to-image models.

## 2  Related Work

Text-to-image Synthesis is one of the multimodal tasks, aiming to generate images that are consistent with text descriptions and highly realistic, so that computers can understand and generate visual content related to natural language descriptions. In recent years, with the introduction of Generative Adversarial Networks (GANs) [1], neural networks generate images close to real images and find a way to solve the Text-to-image problem. The advantage of GAN lies in its ability to learn and simulate high-dimensional, complex distributions of real data [6,19]. The model consists of two core components, the generator G and the discriminator D. The main goal of the generator is to generate realistic images that are indistinguishable from real images, while the task of the discriminator is to accurately distinguish real images from those generated by the generator. These two models compete with each other and are balanced by alternate training. Through this adversarial learning process, the generator gradually improves its ability to generate realistic images, while the discriminator becomes more adept at distinguishing between real and generated images. However, the method suffers from the problem of pattern collapse. Based on this, conditional GAN was first proposed by Reed et al. [11] to generate an accurate image from a textual representation. StackGAN [20] introduces a multi-stage stack network architecture to improve image resolution. AttnGAN [16] proposes a novel attention-generating network for producing fine-grained text-to-image results. MirrorGAN [10] first generates an image from the text and then re-generates the text from the generated image to achieve consistency between the two text descriptions. DM-GAN [23] introduces memory network to solve the problem of image quality during initialization. DF-GAN [14] proposes a new single-stage text-to-image backbone, which can synthesize high-resolution images with higher image quality. DTGAN [21] is the first to propose the use of conditional normalization functions and dual attention modules to fine-tune each feature map scale. MF-GAN [17] introduces triplet loss for the first time in text-to-image synthesis. ASIN-GAN [4] proposes an Adaptive Semantic Instance Normalisation (ASIN) structure for text-to-image synthesis tasks that takes into account individual differences in image features and better incorporates the textual information given during the generation process.

Different from the above methods, CA-GAN repeatedly and deeply fuses text information into the process of image generation through Conditional AdaIN Residual Block to obtain more fine-grained image features. In addition, the attention module helps the generator focus only on the text generated content. In the

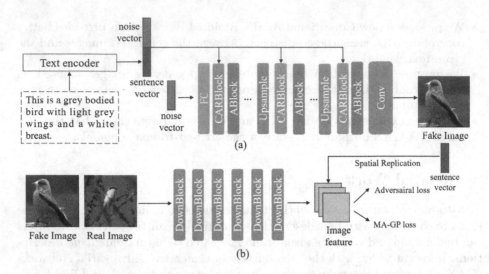

**Fig. 1.** The architecture of the proposed CA-GAN. In (a), Generator consists of a fully connected layer, multiple CARBlocks, ABlocks and upsampling layers to help generate images that fuse the semantic information of the text. The discriminator contains multiple DownBlocks. In (b), By applying Matching-aware Gradient Penalty (MA-GP), our model can synthesize text-matched images.

exploration of related work, we can see that remarkable progress has been made in the field of text-to-image synthesis, but there are still some challenges and limitations. Through our research, we expect to provide a new and effective solution to the semantic consistency and fine-grained information loss problems in text-to-image synthesis.

## 3    Method

In this section, we will introduce the proposed CA-GAN, including the architecture, proposed components (CARBlock and ABlock), and network objectives, as shown in Fig. 1.

### 3.1    Model Overview

CA-GAN is mainly composed of three main components: a text encoder [16], a generator, and a discriminator.

The text encoder uses a pre-trained bi-directional Long Short-Term Memory (LSTM) network [13], which is responsible for extracting sentence-level semantic features from the input text information, where each word corresponds to two hidden states, one in each direction. These hidden states capture the information of the word in its preceding and following contexts. To represent the semantics of a word, two hidden states are concatenated together. The feature matrix, denoted as $e \in \mathbb{R}^{D \times T}$, represents the features of all words in the text description where $D$ is the dimension of the word vector and $T$ is the number of words in the description. Each column of the feature matrix $e_i$ represents the feature vector of the $i^{th}$ word. In addition, the final hidden state of the bidirectional LSTM captures the summary information of the entire text description and connects them to form a global sentence vector. The vector, denoted as $e \in \mathbb{R}^D$, is used to represent the semantic information of the whole sentence.

The generator has two inputs: a sentence vector $s$ generated from the text encoder and a noise vector $z$ sampled from a Gaussian distribution. We pass $z$ through the Fully Connected Layer layer reshape and then through a series of upsampling layers, conditional adaIN residual blocks (CARBlock) and attention blocks (ABlock) for deep fusion of text semantic information and image features. Similar to DF-GAN [14], we splice sentence vector $s$ and noise vector $z$ as control vector $c$, acting on CARBlock to help generate high-quality images that are not only rich in textual information but also more diverse.Mathematically,

$$\hat{z} = F(z) \tag{1}$$

$$h_0 = C_0(\hat{z}, c) \tag{2}$$

$$\widehat{h_0} = A_0(h_0) \tag{3}$$

$$h_i = C_i(u_i(\widehat{h_{i-1}}), c) \quad i = 1, 2, \ldots, n \tag{4}$$

$$\widehat{h_i} = A_i(h_i) \tag{5}$$

$$o - G(\widehat{h_n}) \tag{6}$$

where $z$ is the random noise from a Gaussian distribution, $F$ is a fully connected layer, $c$ is the control vector after stitching the sentence vector $s$ and the noise vector $z$, $C_i$ is the CARBlock, $A_i$ is the ABlock, $u_i$ is the upsampling layer, $G$ is the final convolution layer, and $o$ is the final generated image output.

We use the one-way discriminator proposed by DF-GAN [14], which has proven to be very effective. The discriminator consists of multiple DownBlocks, each of which contains a downsampling layer and a residual network. We convert the generated images into image features by a series of DownBlock modules, which are then connected with text vectors to compute the adversarial loss. By combining Matching-aware zero-centered Gradient Penalty (MA-GP) loss and one-way output, the target-aware discriminator can guide the generator to synthesize more real and text-matched images.

## 3.2   CARBlock

The process of the CARBlock is shown in Fig. 2, which takes two inputs: the feature map $x$ and the global sentence vector $c$. The aim of CARBlock is to facilitate the deep fusion of generated images with semantic information, by summing the input feature maps and the features fused with semantic conditions through residual concatenation, enabling the network to train and learn the details and textures of the images more efficiently. Most existing text-to-image generation

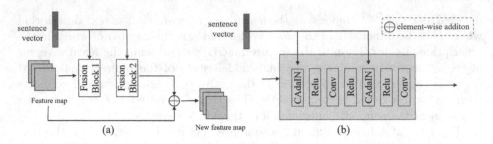

**Fig. 2.** Overview of the proposed Conditional AdaIN Residual Block. CARBlock fuses text and image features through two fusion blocks.

models [14,17] employ multiple individual affine transformations to fuse text information and image features, but neglect to normalize individual image features in batch processing to match the semantic relationships of text information. ASIN-GAN [4] proposes ASIN to be one of the components of its stacked network. Unlike previous work, we adopt CAdaIN as one of the main components of the Fusion Block, which uses semantic information as an additional conditional input for learning affine parameters during style migration, and normalises the feature map to match the semantic information. In the forward propagation, the conditional vector $c$ is passed to two fully connected layers, which are processed by linear transformations and nonlinear activation functions to compute the weight and bias. These learned parameters are used to scale and offset the normalized feature maps to obtain the final image features. Mathematically,

$$\gamma(c) = W_1(c), \beta(c) = W_2(c) \tag{7}$$

$$\text{CAdaIn}(x, c) = \gamma(c)\left(\frac{x - \mu(x)}{\sigma(x)}\right) + \beta(c) \tag{8}$$

where $x$ denotes the feature map, $\mu(x)$ and $\sigma(x)$ denote the mean and standard deviation of the channels on the feature map, respectively, $\gamma$ and $\beta$ are determined by the sentence vector $c$, and $W_1$ and $W_2$ are two fully connected layers that convert the sentence vector $c$ into linguistic cues.

In addition, as shown in Fig. 2, after each CAdaIN layer, we adopt the ReLU to introduce nonlinear properties, and insert convolutional modules between the two CAdaIN layers, aiming to extract richer and useful feature representations. The experimental results show that our CARBlock can promote the fusion of semantic information and image.

### 3.3   Attention Block

While CARBlock can help generate deeply fused images rich in textual semantic information, the stacking of excessively deep residuals may cause unnecessary features to be retained or amplified, introducing redundant information. To solve this problem, we design a plug-and-play Attention Block, which aims to further

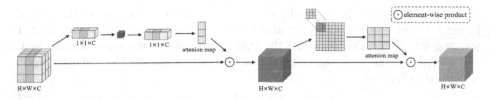

**Fig. 3.** Overview of the proposed attention block. ABlock includes a channel-aware attention module and a pixel-aware attention module.

adjust the feature representation in the generator network to pay more attention to the feature channels and spatial locations that fuse the text information. The Attention Block is inspired by Squeeze-and-Excitation Networks [3], but different from previous work, our proposed Attention Block not only focuses on the feature channel, but also expects to extract the feature from the spatial region of the generated image to capture the image features rich in text semantic information (Fig. 3).

First, we downscale the feature map $h$ by an adaptive global averaging pooling operation to obtain global feature information, generating a channel attention map $X_h \in \mathbb{R}^{C \times 1 \times 1}$. This step reduces the spatial dimension of the feature map to $1 \times 1$ and averages the values of each channel for pooling. Mathematically,

$$X_h = GAP(h) \tag{9}$$

where $X_h$ denotes the feature map and $GAP$ denotes the global average pooling.

Next, we use a $1 \times 1$ convolutional layer to process the reduced-dimensional feature map to reduce the number of channels and control the complexity of the model. This convolutional layer reduces the number of channels of the input feature map from a larger value to a smaller value by a linear transformation, resulting in a feature map $X_c \in \mathbb{R}^{\frac{C}{16} \times 1 \times 1}$. To introduce nonlinear properties and enhance feature representation, we apply a ReLU activation function on the reduced feature map. This activation function sets the negative values to zero and keeps the positive values constant to introduce the nonlinear transformation. Mathematically,

$$X_c = \sigma(W_a(X_h)) \tag{10}$$

where $W_a$ is the convolution of $1 \times 1$ and $\sigma$ is a nonlinear function, such as ReLU.

Subsequently, we again process the feature map $X_c$ after the activation operation using a $1 \times 1$ convolutional layer to recover its channel number and make it consistent with the original feature map. It is defined as:

$$\hat{X}_h = W_b(X_c) \tag{11}$$

where $W_b$ is the convolution of $1 \times 1$ and $\hat{X}_h \in \mathbb{R}^{C \times 1 \times 1}$.

Finally, we apply the sigmoid function to compress the output of the convolution layer to a range between 0 and 1. This can be considered as the channel weights, which are used to adjust the channel weights of the input feature map and multiply them with the original feature map to update the feature map.It is defined as follows:

$$\alpha = \sigma(\hat{X}_h) \tag{12}$$

$$o = \alpha \odot h \tag{13}$$

where $\sigma$ denotes the sigmoid function and $\odot$ is the element-wise multiplication.

After fusing the feature channels of text information, we capture the overall spatial region features of the generated image $h$ by a $1 \times 1$ convolutional layer with a sigmoid function to obtain a spatial weight attention map, and we multiply them with the feature map $o$ to update the feature map. Mathematically,

$$\beta = \sigma(W_c(o)) \tag{14}$$

$$x = \beta \odot o \tag{15}$$

where $\beta$ is the spatial weight attention map, $W_c$ is the $1 \times 1$ convolutional layer, $x$ is the final output image, and $\odot$ is the element-wise multiplication.

It is demonstrated that the generator can better combine the semantic information of text and image features through Attention Block to extract more semantically relevant feature representations, thus generating a better image after fusing text information.

### 3.4 Objective Function

**Adversarial Loss.** We use hinge loss [8,14,19] to match the generated sample with the input text semantics to stabilize the training process. The adversarial loss of the discriminator is denoted as:

$$
\begin{aligned}
L^D_{\text{adv}} = & \mathbb{E}_{x \sim p_{\text{data}}} [\max(0, 1 - D(x, s))] \\
& + \frac{1}{2} \mathbb{E}_{x \sim p_G} [\max(0, 1 + D(\hat{x}, s))] \\
& + \frac{1}{2} \mathbb{E}_{x \sim p_{\text{data}}} [\max(0, 1 + D(x, \hat{s}))]
\end{aligned}
\tag{16}
$$

where $x$ is the real image, $\hat{x}$ is the fake image, $s$ is the matching text description, $\hat{s}$ is the mismatching text description.

The corresponding generator loss is:

$$L^G_{\text{adv}} = -\mathbb{E}_{x \sim p_G}[D(\hat{x}, s)] \tag{17}$$

**MA-GP Loss.** We use the match-aware Gradient Penalty (MA-GP) [14] to ensure that the discriminator smooths real data and matching data points. It is as follows:

$$L_M = \mathbb{E}_{x \sim p_{\text{data}}} [(\|\nabla_x D(x, s)\|_2 + \|\nabla_s D(x, s)\|_2)^p] \tag{18}$$

**Generator Loss.** The generator loss consist of adversarial loss:

$$L_G = L_{\text{adv}}^G = -\mathbb{E}_{x \sim p_G}[D(\hat{x}, s)] \tag{19}$$

**Discriminator Loss.** The discriminator loss contains adversarial loss and MA-GP loss:

$$L_D = L_{\text{adv}}^D + \lambda_M L_M \tag{20}$$

## 4   Experiment

**Datasets.** We conduct our experiments on two popular datasets, CUB [15] and MS COCO [9].The CUB dataset contains 200 bird species with 10 language descriptions per image, with a total of 11,788 images, including 8,855 training images and 2,933 test images.The MS COCO dataset contains 123,287 images, with 10 language descriptions per image, with 8,2783 training images and 40,504 validation images. 82783 and validation images 40504 images.

**Training Details.** We use Adam to optimise the network, where $\beta_1 = 0.0$, $\beta_2 = 0.9$. We follow the two time scale updating rule (TTUR), with the learning rate of the generator set to 0.0001 and the learning rate of the discriminator set to 0.0004. We use Pytorch to train the CA-GAN on CUB for 1000 epochs, and on the COCO dataset set for Evaluation Details.

**Evaluation Details.** As in our previous work [14], we used two metrics, Inception Score (IS) [12] and Frechet Inception Distance (FID) [2], to evaluate the performance of text-generated image models.

*IS.* IS is obtained by calculating the Kullback-Leibler (KL) scatter between the conditional and marginal distributions.The higher the IS, the higher the quality of the generated image.It's defined as:

$$I = \exp\left(\mathbb{E}_x\left[D_{KL}(p(y \mid x)\|p(y))\right]\right) \tag{21}$$

where $x$ denotes the data sampled from the generated images and $y$ denotes the image labels predicted by the pre-trained Inception v3 network. $p(y|x)$ denotes the conditional distribution and $p(y)$ denotes the marginal distribution.

*FID.* The FID is calculated by calculating the distance between the generated image distribution and the real image distribution at the feature level. the lower the FID, the closer the generated image is to the real image.It's defined as:

$$\text{FID} = \|\mu_r - \mu_g\|^2 + \text{Tr}\left(\Sigma_r + \Sigma_g - 2\left(\Sigma_r \Sigma_g\right)^{1/2}\right) \tag{22}$$

where $\mu_r$ is the mean of the true image feature distribution, $\mu_g$ is the mean of the generated image feature distribution, $\Sigma_r$ is the covariance of the true image feature distribution, and $\Sigma_g$ is the covariance of the generated image feature

**Table 1.** The outcomes of evaluating our CA-GAN on the CUB and COCO datasets, in comparison to state-of-the-art techniques, based on the IS and FID metrics, with bold typeface indicating the optimal results.

| Methods | CUB | | COCO |
|---|---|---|---|
| | IS↑ | FID↓ | FID↓ |
| StackGAN [20] | 3.70 ± 0.04 | 51.89 | 74.05 |
| AttnGAN [16] | 4.36 ± 0.03 | 23.98 | 35.49 |
| ControlGAN [7] | 4.58 ± 0.09 | – | – |
| DM-GAN [23] | 4.75 ± 0.07 | 16.09 | 32.64 |
| DF-GAN(nf=32) [14] | 4.86 | 14.81 | 21.42 |
| DTGAN [21] | 4.88 ± 0.03 | 16.35 | 23.61 |
| **Ours(nf=32)** | **4.88 ± 0.06** | **12.21** | **17.64** |

distribution. The FID extracts the feature vectors by a pre-trained Inception v3 network.

We randomly generated 30k images from the test dataset to evaluate the IS and FID of the models. In particular, it was observed by previous work [14] that the IS on the COCO dataset was not able to evaluate the quality of the synthesized images for the text-to-image model. Therefore, we did not compare IS on the COCO dataset.

### 4.1 Quantitative Evaluation

We compared our CA-GAN with current state-of-the-art models StackGAN [20], AttnGAN [16], ControlGAN [7], DM-GAN [23], DF-GAN [14], and DTGAN [21] on CUB and COCO datasets.

As shown in Table 1, IS and FID of our proposed CA-GAN and other comparative methods are reported on the CUB dataset, and FID is reported on the COCO dataset. On the CUB dataset, we observe that our model achieves the highest IS of 4.88 and the lowest FID of 12.21. On the COCO dataset, our model achieves the lowest FID of 17.64. Compared to AttnGAN, which fuses textual and image features and has cross-modal focus, our CA-GAN improves the IS from 4.36 to 4.88 on the CUB dataset, reduces the FID from 23.98 to 12.21, and reduces the FID from 35.49 to 17.64 on the COCO dataset. In comparison to the current state-of-the-art single-stream model, DF-GAN, it improves the IS from 4.86 to 4.88 on the CUB dataset, reduces the FID from 14.81 to 12.21, and reduces the FID from 21.42 to 17.64 on the COCO dataset. Quantitative experiments demonstrate that CA-GAN performs well on the COCO dataset with more complex images.

### 4.2 Qualitative Evaluation

As shown in Fig. 4, we present a qualitative comparison of the generated images from AttnGAN, DF-GAN, and our proposed CA-GAN. On both the CUB

**Fig. 4.** Examples of images synthesized by AttnGAN, DF-GAN, and our proposed CA-GAN conditioned on text descriptions from the test set of CUB and COCO datasets.

dataset and the more challenging COCO dataset, we evaluate the quality and semantic consistency of text-generated images. On the CUB dataset, we observe that the bird images synthesized by the AttnGAN model lack a significant amount of detail, including feathers, shape, and color. This limitation can be attributed to the complex stacked network architecture of AttnGAN. The single-stage network architecture proposed by DF-GAN is simpler than that of AttnGAN and results in more realistic bird images. However, it still falls short in capturing fine details in the generated images. In contrast, our CA-GAN exhibits a richer level of bird details and a more consistent background. Moving on to the COCO dataset, we notice that images generated by AttnGAN are already significantly distorted, making it challenging to generate images from textual descriptions. DF-GAN, while generating two skiers in the 5th column, deviates from the textual input, which specifies a single skier. Additionally, the keyword 'tomato' is missing from the generated pizza image in the 7th column, failing to fulfill the textual requirements. In contrast, our CA-GAN not only produces high-quality images with semantic consistency with the text but also incorporates a more extensive range of image details.

### 4.3   Ablation Study

In this section, we perform an ablation study on the CUB dataset to verify the effectiveness of each component of CA-GAN, which includes the Conditional AdaIN Residual Block (CARBlock), the Attention Block (ABlock), and the Conditional AdaIN (CAdaIN). We first replace CAdaIN with Affine to test the validity of CAdaIN, then we remove the CARBlock module to test the validity of CARBlock, then we retain the CARBlock module remove the ABlock module to test the validity of the ABlock module. Finally we remove CARBlock

**Table 2.** The performance of different components of our model on the CUB dataset.

| Methods | IS↑ | FID↓ |
|---|---|---|
| CA-GAN without (ABlock + CARBlock) | $3.56 \pm 0.02$ | 50.35 |
| CA-GAN without CARBlock | $3.95 \pm 0.04$ | 43.27 |
| CA-GAN without ABlock | $4.67 \pm 0.03$ | 14.35 |
| CA-GAN relpace CAdaIN with Affine | $4.72 \pm 0.04$ | 15.82 |
| **CA-GAN** | $\mathbf{4.88 \pm 0.06}$ | **12.21** |

and ABlock. we tested their performance on the CUB dataset and the results are shown in Table 2.

After replacing CAdaIN with Affine, IS decreases from 4.88 to 4.72, and FID increases from 12.21 to 15.82, which shows that CAdaIN is able to better ensure that the generated images are visually consistent with the conditional descriptions. After we remove ABlock, IS decreases from 4.88 to 4.67 and FID increases from 12.21 to 14.35, which suggests that ABlock can help to generate deep fusion images rich in textual semantic information, and finally we proceed to remove CARBlock, and IS decreases from 4.67 to 3.56, which suggests that CARBlock can help to improve the generated image quality.

## 5    Conclusions

In this paper, we introduce a Conditional Adaptive Generation Adversarial Network (CA-GAN) for text-to-image synthesis, providing a simple and effective approach for multi-modal text-to-image generation. Conditional AdaIN (CAdaIN) allows flexible control over the semantic consistency of the generated image with respect to the provided text condition. Additionally, we propose a novel CAdaIN Residual Block that more deeply integrates textual and image features. Furthermore, an attention module is employed to guide the generator's focus towards the relevant channels and pixels associated with the input text. Extensive experiments demonstrate that our CA-GAN surpasses other state-of-the-art methods on the CUB and COCO datasets.

## References

1. Goodfellow, I., et al.: Generative adversarial nets. Adv. Neural Inf. Process. Syst. 27 (2014)
2. Heusel, M., Ramsauer, H., Unterthiner, T., Nessler, B., Hochreiter, S.: GANs trained by a two time-scale update rule converge to a local Nash equilibrium. Advances in Neural Information Processing Systems 30 (2017)
3. Hu, J., Shen, L., Sun, G.: Squeeze-and-excitation networks. In: Proceedings of the IEEE Conference on Computer Vision and Pattern Recognition, pp. 7132–7141 (2018)

4.  Huang, S., Chen, Y.: Generative adversarial networks with adaptive semantic nor-
    malization for text-to-image synthesis. Digital. Signal Proc. **120**, 103267 (2022)
5.  Huang, X., Belongie, S.: Arbitrary style transfer in real-time with adaptive instance
    normalization. In: Proceedings of the IEEE International Conference on Computer
    Vision, pp. 1501–1510 (2017)
6.  Karras, T., Laine, S., Aila, T.: A style-based generator architecture for generative
    adversarial networks. In: Proceedings of the IEEE/CVF Conference on Computer
    Vision and Pattern Recognition, pp. 4401–4410 (2019)
7.  Li, B., Qi, X., Lukasiewicz, T., Torr, P.: Controllable text-to-image generation.
    Adv. Neural Inf. Process. Syst. 32 (2019)
8.  Lim, J.H., Ye, J.C.: Geometric GAN. arXiv preprint. arXiv:1705.02894 (2017)
9.  Lin, Tsung-Yi., Maire, Michael, Belongie, Serge, Hays, James, Perona, Pietro,
    Ramanan, Deva, Dollár, Piotr, Zitnick, C. Lawrence.: Microsoft COCO: Common
    Objects in Context. In: Fleet, David, Pajdla, Tomas, Schiele, Bernt, Tuytelaars,
    Tinne (eds.) ECCV 2014. LNCS, vol. 8693, pp. 740–755. Springer, Cham (2014).
    https://doi.org/10.1007/978-3-319-10602-1_48
10. Qiao, T., Zhang, J., Xu, D., Tao, D.: MirrorGAN: Learning text-to-image genera-
    tion by redescription. In: Proceedings of the IEEE/CVF Conference on Computer
    Vision and Pattern Recognition, pp. 1505–1514 (2019)
11. Reed, S., Akata, Z., Yan, X., Logeswaran, L., Schiele, B., Lee, H.: Generative adver-
    sarial text to image synthesis. In: International Conference on Machine Learning,
    pp. 1060–1069. PMLR (2016)
12. Salimans, T., Goodfellow, I., Zaremba, W., Cheung, V., Radford, A., Chen, X.:
    Improved techniques for training GANS. Adv. Neural Inf. Process. Syst. 29 (2016)
13. Schuster, M., Paliwal, K.K.: Bidirectional recurrent neural networks. IEEE Trans.
    Signal Process. **45**(11), 2673–2681 (1997)
14. Tao, M., Tang, H., Wu, F., Jing, X.Y., Bao, B.K., Xu, C.: DF-GAN: A simple and
    effective baseline for text-to-image synthesis. In: Proceedings of the IEEE/CVF
    Conference on Computer Vision and Pattern Recognition, pp. 16515–16525 (2022)
15. Wah, C., Branson, S., Welinder, P., Perona, P., Belongie, S.: The Caltech-UCSD
    birds-200-2011 dataset (2011)
16. Xu, T., et al.: Attngan: Fine-grained text to image generation with attentional gen-
    erative adversarial networks. In: Proceedings of the IEEE Conference on Computer
    Vision and Pattern Recognition, pp. 1316–1324 (2018)
17. yang, Y., et al.: MF-GAN: Multi-conditional Fusion Generative Adversarial Net-
    work for Text-to-Image Synthesis. In: Þór Jónsson, Björn., Gurrin, Cathal, Tran,
    Minh-Triet., Dang-Nguyen, Duc-Tien., Hu, Anita Min-Chun., Huynh Thi Thanh,
    Binh, Huet, Benoit (eds.) MMM 2022. LNCS, vol. 13141, pp. 41–53. Springer,
    Cham (2022). https://doi.org/10.1007/978-3-030-98358-1_4
18. Yin, G., Liu, B., Sheng, L., Yu, N., Wang, X., Shao, J.: Semantics disentangling
    for text-to-image generation. In: Proceedings of the IEEE/CVF Conference on
    Computer Vision and Pattern Recognition, pp. 2327–2336 (2019)
19. Zhang, H., Goodfellow, I., Metaxas, D., Odena, A.: Self-attention generative adver-
    sarial networks. In: International Conference on Machine Learning, pp. 7354–7363.
    PMLR (2019)
20. Zhang, H., et al.: Stackgan: Text to photo-realistic image synthesis with stacked
    generative adversarial networks. In: Proceedings of the IEEE International Con-
    ference on Computer Vision, pp. 5907–5915 (2017)
21. Zhang, Z., Schomaker, L.: DTGAN: Dual attention generative adversarial networks
    for text-to-image generation. In: 2021 International Joint Conference on Neural
    Networks (IJCNN), pp. 1–8. IEEE (2021)

22. Zhu, J., Li, Z., Ma, H.: TT2INet: Text to photo-realistic image synthesis with transformer as text encoder. In: 2021 International Joint Conference on Neural Networks (IJCNN). pp. 1–8. IEEE (2021)
23. Zhu, M., Pan, P., Chen, W., Yang, Y.: Dm-gan: Dynamic memory generative adversarial networks for text-to-image synthesis. In: Proceedings of the IEEE/CVF Conference on Computer Vision and Pattern Recognition, pp. 5802–5810 (2019)

# RDC-YOLOv5: Improved Safety Helmet Detection in Adverse Weather

Dexu Yao[1], Aimin Li[1(✉)], Deqi Liu[2], and Mengfan Cheng[2]

[1] Key Laboratory of Computing Power Network and Information Security,
Ministry of Education, Shandong Computer Science Center (National Supercomputer
Center in Jinan), Qilu University of Technology (Shandong Academy of Sciences),
Jinan, China
10431210601@stu.qlu.edu.cn, lam@qlu.edu.cn
[2] Shandong Provincial Key Laboratory of Computer Networks,
Shandong Fundamental Research Center for Computer Science, Jinan, China
{10431210382,10431210366}@stu.qlu.edu.cn

**Abstract.** Outdoor construction sites are frequently affected by fog and
various adverse weather conditions, resulting in a decline in the quality of
the captured images. This deterioration ultimately leads to a significant
drop in the performance of helmet-wear detection systems at construction sites. To address this challenge, we proposed an improved YOLOv5
model. Firstly, we set up a restoration network to effectively restore
hazy image quality and enhance intricate details. Secondly, we introduced a micro-scale detection layer, enabling more efficient capture of
smaller objects and effectively mitigating the adverse weather-induced
variations in object scale. Finally, we added a cross-layer connection to
meticulously amplify the fine-grained features of objects within the network's shallower layers. Our comprehensive evaluation of the enhanced
YOLOv5 model, conducted on two distinct datasets, yielded a final mean
average precision (mAP) value of 95.1%. This substantial improvement
effectively reduces both false detections and missed detections, enhances
overall robustness, and significantly improves detection performance in
challenging adverse weather conditions.

**Keywords:** Safety helmet detection · Adverse weather · YOLOv5

## 1 Introduction

Object detection plays a vital role in computer vision, and recent years have
witnessed significant advancements driven by the rapid progress in deep learning
techniques [1]. This technology has applications in many domains, including
autonomous driving, where it helps vehicles perceive and intelligently navigate
through traffic signs, pedestrians, and other vehicles [2]. Additionally, object
detection is vital in the medical field, enabling automated analysis and diagnosis
of lesions in medical images, thereby improving disease early detection rates [3].
Furthermore, its utility extends to image segmentation, drones [4], agriculture [5],
and many other domains.

© The Author(s), under exclusive license to Springer Nature Switzerland AG 2024
S. Rudinac et al. (Eds.): MMM 2024, LNCS 14556, pp. 313–326, 2024.
https://doi.org/10.1007/978-3-031-53311-2_23

In recent years, Convolutional Neural Networks (CNN) have revolutionized object detection [6]. CNN-based detectors have become dominant and have single-stage and two-stage detection models. Single-stage detection, exemplified by the R-CNN (Region-based Convolutional Neural Networks) family, which includes R-CNN, Fast R-CNN [7], and Faster R-CNN [8], generates candidate object regions through selective search or the Region Proposal Network (RPN) [9], then employs CNNs for region classification and localization.

In addition to the traditional two-stage methods, single-stage detection methods have gained prominence for their superior ability to perform object detection. YOLO series is a widely recognized single-stage object detection model. YOLO transforms object detection into a regression problem by predicting an object's class, location, and confidence within a grid. Several iterations of the YOLO model have been introduced, including YOLOv1 [10], YOLOv2 [11], YOLOv3 [12], YOLOv4 [13], and YOLOv5 [14], among others. These iterations have realized substantial improvements through enhancements in network architectures, utilization of more powerful feature extractors, and implementation of more efficient training strategies. The YOLOv5 has garnered attention for its high accuracy and faster processing speed, achieving exceptional results across various object detection datasets and real-world applications.

Detecting wearing safety helmets at construction sites is essential for mitigating accident risks and fostering secure working environments. Waranusast et al. [15] employed object detection algorithms to accurately identify helmet usage among construction workers. Li et al. [16] used deep learning techniques to achieve real-time helmet detection at construction sites, providing immediate safety alerts. However, helmet detection in real-world outdoor construction scenarios differentiates it from other detection tasks. Outdoor construction sites are often affected by weather-related interference, such as foggy conditions or dusty and sandy weather, which can decrease image quality and detection performance.

This paper aims to improve safety helmet detection performance at construction sites for adverse weather. Our contributions can be summarized as follows:

- Set up A Restoration Network: This network enhances original image quality, improves visibility, and reduces the pressure for the detection network.
- Add A Micro-Scale Detection Layer(152 × 152): The micro-scale detection layer improves the ability of models to detect smaller objects.
- Add the Cross-Layer Connection: The Cross-Layer Connection to capture intricate object features, especially within the shallower network layers, facilitates finer and more context-rich information extraction.

These contributions address the challenges of adverse weather in helmet detection at construction sites. Our experimental results also validate the effectiveness of our improvements.

## 2    Related Work

### 2.1    Object Detection in Adverse Weather

Detecting helmets in adverse weather presents substantial challenges, primarily due to low image quality, lighting changes, and background interference. Adverse weather conditions, such as haze, reduce the contrast between objects and the background, leading to blurred edges and details of small objects. Unstable lighting conditions introduce further complexity, with factors like shadows and glare altering the appearance of small objects. Additionally, background interference, often characterized by blurriness and interfering objects, can confound the features of small objects and result in a higher likelihood of false and miss detection.

Researchers have proposed many methods, including multimodal sensor fusion [17], where multiple sensors like visible cameras, infrared thermal cameras [18], and LIDAR [19] work together to improve helmet detection accuracy in adverse weather. However, these methods may have detection accuracy and hardware requirements limitations.

Deep learning algorithms [20] and image enhancement techniques have gained prominence in recent research. These techniques involve preprocessing images to enhance quality by removing elements like raindrops, snowflakes, and fog. Training models on these improved images significantly enhance detection accuracy.

### 2.2    YOLOv5 Model

YOLOv5 stands out as a prominent model in object detection, offering four versions: YOLOv5s, YOLOv5m, YOLOv5l, and YOLOv5x.

In this paper, we use the YOLOv5s version mainly because of its lightweight network architecture, which better meets the real-time detection requirements of the device. YOLOv5 employs CSPNet (Cross-Stage Partial Network) as its infrastructure, building a streamlined design that makes it ideal for real-time object detection. CSPNet offers excellent feature extraction efficiency and few parameters, improving inference speed while ensuring accuracy. It is worth emphasizing that YOLOv5 excels in multi-scale object detection, effectively detecting objects with different feature scales. Furthermore, we implement an adaptive training strategy that can be dynamically adjusted based on object confidence levels. This method effectively tackles the class imbalance challenge in object detection, thereby improving the model's ability to detect smaller, more challenging samples.

## 3    Our Proposed Method

### 3.1    Set up a Restoration Network

In adverse weather conditions, the quality of captured images reduced pixel quality, diminished sharpness, blurred edges, loss of detail, and a significant presence of noise and interference, all of which adversely affect detection performance.

To address these challenges, we set up a restoration network. The primary objective of this network is to mitigate the adverse effects of noise and blurring present in these foggy images, revealing the underlying features and structural information akin to clear images. This transformation process effectively converts a blurred or distorted image into a more visually coherent and interpretable form. Post-restoration, the image is forwarded to the detection network for further processing. This method restores detailed feature information, providing valuable cues for subsequent detection tasks.

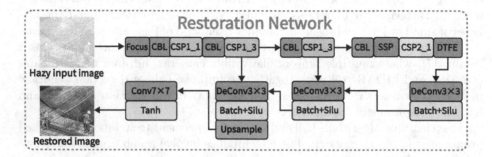

**Fig. 1.** The architecture of restoration network.

The architecture of the restoration network is shown in Fig. 1. The restoration network commences with feature extraction, and complex latent features inherent to the input image are captured. After extracting features, we introduce the Dynamic Transformer Feature Enhancement (DTFE) module to enhance feature extraction and representation capabilities. The DTFE utilizes three DeConv operations, one Upsample operation, and one Tanh activation function to restore clean image features. These operations generate a final clean image, subsequently input into the detection network for object detection.

**Dynamic Transformer Feature Enhancement Module.** The DTFE module comprises two deformable conv3 × 3 layers and a vision transformer component, as shown in Fig. 2. These elements play critical roles in extending the adaptive shape-aware domain of the model.

The two deformable conv3 × 3 layers perform dynamic feature transformations, facilitating adaptive shape-aware modelling. Deformable convolution introduces filters capable of dynamic shape and position adjustments, aligning with the feature distribution of the input data. This operation significantly enhances the network's feature transformation capabilities. By introducing this deformable conv3 × 3, we effectively extend the receptive domain, accommodating adaptive shapes and reinforcing the model's feature extraction capabilities. This adaptability improves the precision in capturing object shape, size, and location information in adverse weather.

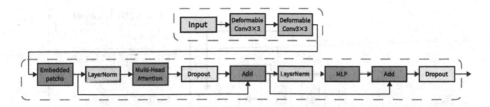

**Fig. 2.** The architecture of DTFE.

The vision transformer achieves a broader contextual understanding through global awareness and self-attention mechanisms. These mechanisms enhance feature representation capabilities, enabling the comprehensive capture of global features and relationships within the input image. This capability is particularly critical in adverse weather where atmospheric factors may lead to object blurriness or occlusion, necessitating a greater reliance on contextual information for accurate detection. Furthermore, the vision transformer improves the model's resilience to interference, making it better equipped to handle image disruptions caused by adverse weather conditions such as fog.

In summary, the restoration network enhances key and detailed features in the grayscale image, highlighting features that facilitate object detection. The clean image generated by the restoration network is then fed into the detection network to improve the detection accuracy further. This cooperation between the restoration and detection networks is essential for object detection in adverse weather.

### 3.2   Adding a 152 × 152 Detection Layer for Micro-Scale

In object detection networks, comprehending the capabilities of shallow and deep networks holds great significance. Shallow networks specialize in low-level feature extraction, characterized by compact receptive fields and high spatial resolutions. These attributes enable them to effectively capture fine details, including object contours and subtle colour variations, making them proficient at detecting smaller objects. However, as network depth increases, the capacity to capture detailed features related to smaller objects diminishes. Conversely, deep networks in higher layers possess broader receptive fields, rendering them more adept at detecting medium and large objects.

YOLOv5 can be divided into three detection layers: 19 × 19, 38 × 38, and 76 × 76. The smallest object scales these three layers can detect are large-scale with 32 × 32, medium-scale with 16 × 16, and small-scale with 8 × 8. Consequently, micro-scale objects with less than 8 × 8 pixels cannot be detected. However, safety helmets commonly found on construction sites are generally small objects, which can pose a significant detection challenge.

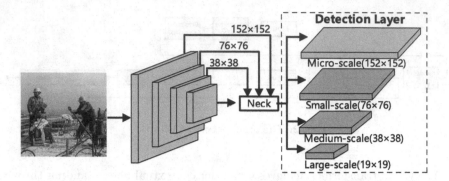

**Fig. 3.** Add a $152 \times 152$ detection layer for micro-scale.

To address this issue, we add a $152 \times 152$ detection layer for micro-scale objects, 608 divided by 152 is 4, as shown in Fig. 3. This adjustment effectively reduces the minimum detectable object size from $8 \times 8$ pixels to $4 \times 4$ pixels, significantly improving the performance of detecting smaller objects.

### 3.3    Adding a Cross-Layer Connection

In object detection within neural networks, as information travels from shallower to deeper layers, there is a gradual reduction in the scale of the feature map, leading to the blurring of feature and gradient information. This phenomenon gives rise to a critical issue known as "feature loss". The problem can result in misclassification of objects and the loss of crucial object information during detection.

To tackle this challenge, we add the cross-layer connection. This connectivity mechanism enables the fusion and sharing of information across distinct network layers, fostering a seamless integration of lower-level and upper-level features. This integration provides a more comprehensive understanding of an object's structural and contextual characteristics. Furthermore, incorporating cross-layer connections enhances model responsiveness by expediting model convergence and decision-making processes, which is particularly beneficial for real-time object detection in adverse weather.

The cross-layer connection structure is realized through the density block module, as illustrated in Fig. 4. The density block module plays a crucial role in facilitating the transfer of feature information. Its image features are directly transmitted to the feature fusion network, where they are spliced and fused. The splicing layer then conveys the fused feature information to the subsequent layer, augmenting the feature representation capability. Within this structure, the splicing layer is responsible for merging two feature maps from different layers, each with dimensions $K_1 \times H_1 \times W_1$ and $K_2 \times H_1 \times W_1$, respectively. The output feature information of the splicing layer can be expressed as:

$$Z_{concat} = (K_1 + K_2) \times H_1 \times W_2 \tag{1}$$

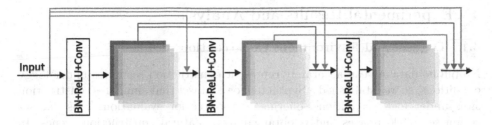

**Fig. 4.** The architecture of the density block module.

$K_1$ and $K_2$ represent the number of channels in the input feature maps from different layers, while $H_1$ and $W_1$ denote the height and width of the input feature maps required for the splicing layer. Equation 1 demonstrates that combining features from different layers in the splicing layer enhances subsequent network layers' feature representation. The cross-layer connection then transmits the new feature information $K_3 \times H_1 \times W_1$ to the corresponding splicing layer, resulting in the following feature information:

$$Z_{concat} = (K_1 + K_2 + K_3) \times H_1 \times W_2 \tag{2}$$

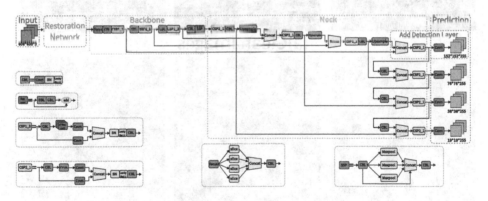

**Fig. 5.** The structure of the improved YOLOv5 model.

The height and width of the output feature maps of the fifth and seventh layers in YOLOv5's feature extraction network match the input scales of the two splicing layers. The addition of cross-layer connections in these two layers combines feature information extracted from shallower layers with that from deeper layers, enriching feature information for small and medium-sized objects. This improvement allows the network to better capture features of objects at different scales in subsequent stages, thereby improving the accuracy and robustness of the model. The structure of the final improved YOLOv5 model is shown in Fig. 5.

## 4    Experimental Results and Analysis

### 4.1    Dataset and Environment Construction

No public datasets can accurately represent construction scenarios under foggy conditions, so we established a Synthetic Fog Dataset that simulates construction sites under foggy conditions, offering a resource for evaluation. This dataset comprises 6,000 images and is obtained from natural construction scenes. In addition, to test our model more realistically, we also used the real-world Foggy Driving Dataset [21] for comparison tests with other models. The synthetic fog dataset applied an atmospheric scattering model to replicate foggy conditions to generate synthetic foggy images, as shown in Fig. 6.

(a)Clean images

(b)Synthetic foggy images

**Fig. 6.** Example images in the proposed fog dataset.

Specifically, atmospheric scattering models are employed to simulate the scattering and absorption of light in fog. The following equation quantifies the scattering effect:

$$I(x) = J(x) \cdot t(x) + A \cdot (1 - t(x)) \tag{3}$$

$I(x)$ represents the image under foggy conditions, $J(x)$ is the original image, $A$ stands for the estimated atmospheric light, and $t(x)$ denotes the transmittance, which dictates the extent of light attenuation within the fog. $t(x)$ is calculated using the following equation:

$$t(x) = \exp(-\beta \cdot d(x)) \tag{4}$$

$\beta$ represents the fog concentration coefficient, indicating the fog's concentration level. $d(x)$ signifies the distance between the camera and the object, defined as:

$$d(x) = \sqrt{(x_2 - x_1,)^2 + (y_2 - y_1)^2} \tag{5}$$

$(x_1, y_1)$and $(x_2, y_2)$ denote the coordinates of two pixel points. Our experiments standardized the global atmospheric light parameter $A$ to 0.6. We introduced variability in the atmospheric scattering parameter $\beta$ to manipulate the fog level, selecting random values from 0.08 to 0.12.

Our synthetic fog dataset was divided into two subsets: 5,000 images for training and 1,000 images for thorough testing. Additionally, we employed a systematic approach by categorizing the dataset into four scales based on object pixel size: micro-objects, small-objects, medium-objects, and large-objects. This segmentation enabled a comprehensive assessment of the model's detection performance across a spectrum of object sizes, ensuring a thorough evaluation of its capabilities.

Our improved YOLOv5 model was built upon the PyTorch 1.8.1 framework, and both training and testing phases were executed on NVIDIA A100 GPUs. We adhered to the hyperparameter settings defined in the YOLOv5 model architec ture throughout the training process. To combat overfitting, we applied a weight decay coefficient of 0.0005, and to prevent the model from converging to local optima, we set the momentum value at 0.937. The initial learning rate was 0.01, with termination occurring at 0.2. After 200 training iterations, the model attained optimal performance with the specified weight configuration.

## 4.2  Evaluation Criteria

This paper employs widely accepted evaluation criteria in object detection, mean Average Precision (mAP). The mathematical expressions for mAP are detailed as follows:

$$P = Precision = \frac{TP}{TP + FP} \tag{6}$$

$$R = Recall = \frac{TP}{TP + FN} \tag{7}$$

$$AP = \int_0^1 P(R)dR \tag{8}$$

$$mAP = \frac{\sum_{i=0}^n AP(i)}{n} \tag{9}$$

$P$ is precision, $TP$ is true positives, $FP$ is false positives, and $FN$ is false negatives. $AP$ represents the average accuracy for a specific category, where $n$ is the total number of categories, and $mAP$ is the overall average value of $AP$ across all categories.

## 4.3   Ablation Studies

To test the impact of each component in the improved model, we conducted ablation experiments to analyze different components, including the restoration network, the detection layer(152 × 152), and the cross-layer connections. We used YOLOv5 as the baseline model and added the different components to the baseline model step by step as follows:

1. **F1: Restoration Network**
2. **F2: Detection Layer(152 × 152)**
3. **F3: Cross-Layer Connection**

**Table 1.** The Results of Ablation Experiments

| Variants | Micro | Small | Medium | Large | mAP(%) |
|---|---|---|---|---|---|
| YOLOv5 | 92.6 | 92.9 | 93.2 | 91.7 | 92.6 |
| YOLOv5+F1 | 93.1 | 93.2 | 93.5 | 91.8 | 92.9 |
| YOLOv5+F2 | 92.8 | 92.9 | 93.5 | 92.2 | 92.9 |
| YOLOv5+F3 | 93.0 | 93.1 | 93.7 | 92.6 | 93.1 |
| YOLOv5+F1+F2 | 93.5 | 93.6 | 93.6 | 93.1 | 93.5 |
| YOLOv5+F1+F3 | 92.8 | 93.6 | 94.0 | 93.1 | 93.4 |
| YOLOv5+F2+F3 | 93.2 | 93.6 | 94.5 | 93.2 | 93.6 |
| **YOLOv5+F1+F2+F3** | **93.8** | **94.6** | **96.0** | **95.2** | **95.1(↑2.5)** |

The experimental results are shown in Table 1, that each component of the improved YOLOv5 model helps to improve the object detection performance. The data shows that the restoration network and 152 × 152 Detection Layer are more helpful in detecting small objects, and the cross-layer connection is more inclined to detect large objects' performance improvement. Through this experiment, we verified the effectiveness of these improved methods. Our improved YOLOv5 model improved the performance of helmet detection in adverse weather by 2.5% compared to the YOLOv5 model.

### 4.4   Comparison Experiments

**Comparison Experiments of Synthetic Fog Dataset.** In order to more fully test the performance of the improved YOLOv5 model, we compared it with the state-of-the-art model. The results of the comparison experiments are shown in Table 2.

The table results show that the improved YOLOv5 model is much better than the other models in terms of mAP and achieves the best detection results. What is more noteworthy is that the results of our improved YOLOv5 model in the scale of "Micro", compared with the detection results of other scales,

**Table 2.** Comparative Experimental Results Of Synthetic Fog Dataset

| Models | Micro | Small | Medium | Large | mAP(%) |
|---|---|---|---|---|---|
| YOLOv5 | 92.6 | 92.9 | 93.2 | 91.7 | 92.6 |
| DENet [22] | 92.8 | 93.3 | 93.5 | 93.6 | 93.3 |
| DS-Net [23] | 93.1 | 94.0 | 93.9 | 94.2 | 93.8 |
| CF-YOLO [24] | 93.4 | 93.9 | 94.1 | 94.2 | 93.9 |
| R-YOLO [25] | 93.2 | 93.9 | 93.5 | 93.8 | 93.7 |
| FFA-YOLOv5 [26] | 93.2 | 94.4 | 94.6 | 93.8 | 94.0 |
| **RDC-YOLOv5(Ours)** | **93.8** | **94.6** | **96.0** | **95.2** | **95.1** |

achieved a more obvious advantage, further indicating that our improvement in the detection of small objects is a significant effect, more conducive to the detection of small helmet objects in adverse weather. In addition, our improved model frames per second (FPS) reach 32, which can meet the real-time inspection requirements at construction sites.

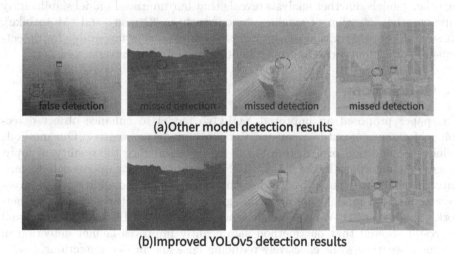

(a)Other model detection results

(b)Improved YOLOv5 detection results

**Fig. 7.** Comparison of the Detection Results of the Improved YOLOv5 Model with Other Models.

We show the detection results of the improved YOLOv5 model with other models, as shown in Fig. 7. Our improved model effectively mitigates the leakage and false detection problems of other models compared to other models in the case of dense foggy, and the improved model has better robustness.

**Comparison Experiments of Foggy Driving Dataset.** We conducted comparative experiments with other models on a real-world foggy driving dataset.

Our goal was to demonstrate the effectiveness of our improved YOLOv5 model. The results are shown in Table 3.

**Table 3.** Comparative Experimental Results Of Foggy Driving Dataset

| Models | Person | Motorbike | Bicycle | Car | Bus | mAP(%) |
|---|---|---|---|---|---|---|
| YOLOv5 | 89.8 | 89.9 | 90.5 | 91.4 | 91.9 | 90.7 |
| DENet [22] | 90.5 | 90.6 | 91.3 | 91.1 | 92.6 | 91.2 |
| DS-Net [23] | 90.6 | 91.0 | 91.4 | 92.4 | 92.9 | 91.7 |
| CF-YOLO [24] | 90.4 | 90.9 | 91.1 | 91.2 | 90.9 | 90.3 |
| R-YOLO [25] | 91.0 | 91.2 | 91.7 | 92.6 | 92.9 | 91.9 |
| FFA-YOLOv5 [26] | 91.2 | 91.4 | 91.8 | 92.5 | 93.2 | 92.0 |
| **RDC-YOLOv5(Ours)** | **92.1** | **92.0** | **92.3** | **93.2** | **93.6** | **92.6** |

According to the experimental data, our improved YOLOv5 model performs best on the foggy driving dataset, and its mAP value is significantly higher than the other models. Further analysis reveals that our improved model significantly improves the detection of smaller objects such as "Person" and "Motorbike". These findings confirm that our improved method is more favourable for detecting small objects and can improve the detection effect in adverse weather.

## 5    Conclusions

This paper proposed an improved YOLOv5 model to enhance object detection performance at outdoor construction sites in adverse weather. Our methods include the addition of a restoration network to enhance the quality of input images, the incorporation of a $152 \times 152$ detection layer for micro-scale to mitigate the challenges posed by varying object sizes, and the addition of a cross-layer connection module to enrich fine-grained object features within the network's shallow layers. We conducted experiments on two different datasets, and the result showed that our method significantly improves helmet detection in inclement weather while effectively reducing false and missed detections.

**Acknowledgement.** This work was supported by the Key R&D Plan of Shandong Province, China (No.2021CXGC010102).

## References

1. Xu, J.: Deep learning for object detection: a comprehensive review. Towards Data Science (2017)
2. Niranjan, D., VinayKarthik, B.: Deep learning based object detection model for autonomous driving research using Carla simulator. In: 2nd International Conference on Smart Electronics and Communication (ICOSEC), pp. 1251–1258. IEEE (2021)

3. Wang, J., Zhu, H., Wang, S.-H., Zhang, Y.-D.: A review of deep learning on medical image analysis. Mobile Netw. Appl. **26**, 351–380 (2021)
4. Xiao, Y., et al.: A review of object detection based on deep learning. Multimedia Tools Appl. **79**, 23-729–23-791 (2020)
5. Hasan, A.M., Sohel, F., Diepeveen, D., Laga, H., Jones, M.G.: A survey of deep learning techniques for weed detection from images. Comput. Electron. Agric. **184**, 106067 (2021)
6. LeCun, Y., Bottou, L., Bengio, Y., Haffner, P.: Gradient-based learning applied to document recognition. Proc. IEEE **86**(11), 2278–2324 (1998)
7. Girshick, R., Donahue, J., Darrell, T., Malik, J.: Rich feature hierarchies for accurate object detection and semantic segmentation. In: Proceedings of the IEEE Conference on Computer Vision and Pattern Recognition, pp. 580–587 (2014)
8. Girshick, R.: Fast R-CNN. In: Proceedings of the IEEE International Conference on Computer Vision, pp. 1440–1448 (2015)
9. Ren, S., He, K., Girshick, R., Sun, J.: Faster R-CNN: towards real-time object detection with region proposal networks. In: Advances in Neural Information Processing Systems, vol. 28 (2015)
10. Redmon, J., Divvala, S., Girshick, R., Farhadi, A.: You only look once: unified, real-time object detection. In: Proceedings of the IEEE Conference on Computer Vision and Pattern Recognition, pp. 779–788 (2016)
11. Redmon, J., Farhadi, A.: YOLO9000: better, faster, stronger. In: Proceedings of the IEEE Conference on Computer Vision and Pattern Recognition, pp. 7263–7271 (2017)
12. Redmon, J., Farhadi, A.: YOLOv3: an incremental improvement. arXiv preprint arXiv:1804.02767 (2018)
13. Bochkovskiy, A., Wang, C.-Y., Liao, H.-Y.M.: YOLOv4: optimal speed and accuracy of object detection. arXiv preprint arXiv:2004.10934 (2020)
14. Huang, J., et al.: Speed/accuracy trade-offs for modern convolutional object detectors. In: Proceedings of the IEEE Conference on Computer Vision and Pattern Recognition, pp. 7310–7311 (2017)
15. Waranusast, R., Bundon, N., Timtong, V., Tangnoi, C., Pattanathaburt, P.: Machine vision techniques for motorcycle safety helmet detection. In: 28th International Conference on Image and Vision Computing New Zealand (IVCNZ 2013), vol. 2013, pp. 35–40. IEEE (2013)
16. Li, Y., Wei, H., Han, Z., Huang, J., Wang, W.: Deep learning-based safety helmet detection in engineering management based on convolutional neural networks. Adv. Civil Eng. **2020**, 1–10 (2020)
17. Zhang, C., et al.: Robust-FusionNet: deep multimodal sensor fusion for 3-D object detection under severe weather conditions. IEEE Trans. Instrum. Meas. **71**, 1–13 (2022)
18. Bhadoriya, A.S., Vegamoor, V., Rathinam, S.: Vehicle detection and tracking using thermal cameras in adverse visibility conditions. Sensors **22**(12), 4567 (2022)
19. Kim, T.-L., Arshad, S., Park, T.-H.: Adaptive feature attention module for robust visual-lidar fusion-based object detection in adverse weather conditions. Remote Sens. **15**(16), 3992 (2023)
20. Sharma, T., Debaque, B., Duclos, N., Chehri, A., Kinder, B., Fortier, P.: Deep learning-based object detection and scene perception under bad weather conditions. Electronics **11**(4), 563 (2022)
21. Sakaridis, C., Dai, D., Van Gool, L.: Semantic foggy scene understanding with synthetic data. Int. J. Comput. Vis. **126**, 973–992 (2018)

22. Qin, Q., Chang, K., Huang, M., Li, G.: DENet: detection-driven enhancement network for object detection under adverse weather conditions. In: Proceedings of the Asian Conference on Computer Vision, pp. 2813–2829 (2022)
23. Huang, S.-C., Le, T.-H., Jaw, D.-W.: DSNet: joint semantic learning for object detection in inclement weather conditions. IEEE Trans. Pattern Anal. Mach. Intell. **43**(8), 2623–2633 (2020)
24. Ding, Q., et al.: CF-YOLO: cross fusion yolo for object detection in adverse weather with a high-quality real snow dataset. IEEE Trans. Intell. Transp. Syst. (2023)
25. Wang, L., Qin, H., Zhou, X., Lu, X., Zhang, F.: R-YOLO: a robust object detector in adverse weather. IEEE Trans. Instrum. Meas. **72**, 1–11 (2022)
26. Qin, X., Wang, Z., Bai, Y., Xie, X., Jia, H.: FFA-Net: feature fusion attention network for single image dehazing. In: Proceedings of the AAAI Conference on Artificial Intelligence, vol. 34, no. 07, pp. 11-908–11-915 (2020)

# Sustainable Commercial Fishery Control Using Multimedia Forensics Data from Non-trusted, Mobile Edge Nodes

Aril Bernhard Ovesen[1]([✉]), Tor-Arne Schmidt Nordmo[1],
Michael Alexander Riegler[1,2,3], Pål Halvorsen[2,3], and Dag Johansen[1]

[1] UiT The Arctic University of Norway, Tromsø, Norway
aril.b.ovesen@uit.no
[2] Simula Metropolitan Center for Digital Engineering, Oslo, Norway
[3] Oslo Metropolitan University, Oslo, Norway

**Abstract.** Uncontrolled over-fishing has been exemplified by the UN as a serious ecological challenge and a major threat to sustainable food supplies. Emerging trends within governing bodies point towards digital solutions by deploying CCTV-based video monitoring systems on a large scale. We conjecture that such systems are not feasible when reliant on satellite broadband in remote areas, and expose workers aboard fishing vessels to unneeded manual surveillance. To facilitate this, we propose Dorvu, a AI-based multimedia distributed storage system designed for edge environments, with a specific focus on commercial fishery monitoring. Dorvu addresses the challenges of secure data storage, fault tolerance, availability, and remote access in hostile edge environments. The system employs a novel data distribution scheme involving sensor readings and AI video content extraction to ensure the preservation of forensic evidence even in unstable conditions. Experimental evaluations demonstrate the feasibility of real-time multimedia data collection, analysis, and distribution in networks of edge devices on-board active fishing vessels. Dorvu offers a practical alternative to current governmental surveillance trends that compromise data security and privacy, and we propose it as a solution for edge-based forensic data management in commercial fisheries and similar applications.

**Keywords:** multimedia storage · edge computing · privacy · digital forensics · file systems

## 1 Introduction

Sustainability issues resulting from criminal over-fishing has resulted in several governments around the world proposing and implementing remote video surveillance as a form of continuous monitoring of fishing trawlers. Fishery video monitoring is for example mandated by law in Denmark, and it is expected that the majority of Danish trawlers will be equipped with CCTV monitoring systems during 2023 [28]. Similar policies are expected to be implemented in New

© The Author(s), under exclusive license to Springer Nature Switzerland AG 2024
S. Rudinac et al. (Eds.): MMM 2024, LNCS 14556, pp. 327–340, 2024.
https://doi.org/10.1007/978-3-031-53311-2_24

Zealand [22] and has been proposed in Norway [15]. In the European Union, compulsive video surveillance has been mandated in cases where trawlers are known to catch beyond their assigned quota [17]. These policies have been characterized as antagonizing by some of those working aboard fishing vessels [14,25]. At the same time, the dependence on food from the ocean is growing worldwide [5], while over-fishing is increasingly becoming a threat to global resources [29]. The combination of criminal over-fishing, increased global reliance on fish, and effects of climate change, can have significant consequences both for life in the ocean and the global population [26].

Our work focuses on providing an alternative to proposed solutions for commercial fishery surveillance programs in Norway, but this foundation does not limit our work from being deployed in other areas or scenarios. Ongoing going governmental proposals for deployments in fishing vessels include video streaming, manual surveillance, and automatic logging of catch [15]. We conjecture that edge systems for surveillance in the commercial fishery domain should attempt to (i) minimize the negative impact from introducing workplace surveillance [25], (ii) provide the ability to automatically process video information without human input [1] (iii) provide mechanisms to limit the consumption of video data, even for privileged users (i.e. federal fishery authorities) [21], and (iv) be specifically designed for the difficult edge environments found onboard active fishing vessels [18] with regards to limited connectivity, risk of system faults, and tampering from malicious actors.

To that end, we have developed the Dorvu[1] multimedia distributed storage system, designed for robust and secure data collection, analysis and storage, in hostile edge environments. Dorvu's primary objective is 24/7 reliable data collection, real-time analysis, and protection in non-trusted off-shore edge environments where adversaries might be local to the data monitored, analyzed, and stored. Due to very limited connectivity from the isolated and remote Arctic ocean, a secondary objective is to support reliable satellite transmission of locally analyzed event data of potential relevance to cloud-connected mainland governmental surveillance nodes. The threat model includes physical attacks and environmental instability, emphasizing system robustness. We propose an architecture containing a distributed storage layer that is configured in real-time by automatic multimedia processing and data extraction, in addition to input from privileged remote clients. The Dorvu system is designed for deployment and data collection on-board active fishing vessels, while clients can access the system remotely from mainland nodes through satellite connections. Clients are able to issue system configuration patches in the form of data policies, denoting filters for data that is important for the remote operation, for example as forensic evidence of fishery activities.

---

[1] *Dorvu* is a Northern Sámi term, meaning *security* or *trust*.

## 1.1   Related Work

The storage and analysis of visual data on specialized systems at edge as been implemented in previous works. Winkler et al. [32] define a class of visual sensor networks that enable reactive monitoring for enforcement, for example traffic monitoring. While reactive networks trigger data recording after certain events, Bischof et al. [3] note that an additional solution is to analyze a continuous stream of data to detect these events as part of the same on-board data pipeline. Privacy-preservation is a topic in certain visual sensor networks; Fitwi et al. [6] define that privacy in video surveillance is achieved when people cannot be identified by human observers without their knowledge and consent. Collection and storage of sensitive data at the edge have been explored in [2,32]. While traditional secure storage systems rely on keys for static confidentiality and authorization, when producing potentially privacy-sensitive information, the actual contents of data or meta-data can be important to enforce a principle of least privilege while maintaining the ability to extract information [9,33].

Traditional distributed storage systems typically do not optimize for low bandwidth and fluctuating computation capabilities, while dedicated low bandwidth systems are aimed at compression or minimizing the transfer of redundant data [16], i.e., data that already exists at the recipient, likely already delivered in a different context. The contents of certain data objects at clients operating at the recipient is of no concern to the underlying file system, because generalized distributed storage systems are designed to provide a storage layer for numerous applications whose critical operations may rely on insertion and retrieval of arbitrary data objects [8]. Distributed fault tolerance schemes aim to allow systems to continue all of these operations despite failures of certain components [13]. A domain-specific system may utilize domain knowledge to make intelligent decisions in the face of different faults. They can evaluate failures and allow them to persist in order to prioritize other components, or dynamically schedule operations based on their usefulness to specialized clients, sacrificing the general usefulness of the system to retain critical operations.

## 2   System Overview

Dorvu is a distributed multimedia storage system for the edge, that is designed as a supplement to ongoing efforts of digitization and remote surveillance of commercial fishery operations in the European Economic Area (EEA). It is developed as a middleware system that can be deployed on numerous compute nodes, and provides robust, privacy-preserving multimedia storage and access for data collection in low-bandwidth and remote areas.

The system is deployed with a pre-configured number of compute nodes, each of which with its own storage device attached. Nodes can be connected to their own video input devices as their main source of data input. These nodes are deployed in a network onboard a fishing vessel, connected to each other through either Ethernet of Wi-Fi, with power being supplied from the vessel. We assume that each node is reasonably physically sealed and tamper-proofed, but also that adversaries with physical possession of a node may be able to compromise it given enough time. Clients that want to access the system during the vessel's voyage may do so through satellite broadband communication, which is present on the vessel as part of the system deployment. Client access involves querying data and computation results, and communicating new policies to configure system behaviour during run time. Additional sensor input can be used to annotate video data; for the remote fishery monitoring scenario described in this paper, we utilize a data stream from an Automatic Identification System (AIS). This is an automated tracking system present on virtually all commercial vessels in the EEA [7], and can be used to provide geo-location annotations for video data.

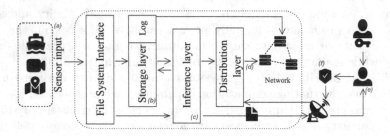

**Fig. 1.** An overview of the system architecture of Dorvu. The colored area indicate components that are internal to the system.

Figure 1 shows an overview of the data flow across several layers in Dorvu, which will be further elaborated in the following sections. The figure shows that (a) video data is generated at active fishing vessels, where it is (b) sent to a node running Dorvu and stored (Sect. 2.1), logged (Sect. 2.3), and (c) processed (Sect. 2.4). This processing involves AI inference to extract information about ongoing events on the fishing vessel. The output of this layer represents the semantic transformation of the input video file, which can be (e) requested and accessed by remote clients. The purpose of this transformation is to reduce the bandwidth cost of transferring information, and eliminate information that is irrelevant for the fishery operation, with respect to the privacy of fishery workers [21]. After data is processed, it is (d) distributed among nodes in the edge system (Sect. 2.2). Finally, clients can update the configuration of the system in real-time by (f) patching in new policies that determine how newly generated data is handled (Sect. 2.5).

Dorvu is implemented in 3648 lines of Rust code, covering the File System in Userspace (FUSE) overlay file system [27] and encryption scheme, and 2105 lines of C# code, covering most of the remaining components, while AI inference modules are implemented in Python. A FUSE file system API is used to access data in Dorvu. This is used to implement a POSIX-like interface, in order to provide interoperability with most tools a client would use to consume data from the storage system.

## 2.1 Secure Storage Platform

A key component of Dorvu is the ability to secure data stored locally at the edge. The off-shore edge environment introduces particular needs for strict security properties, as adversaries may gain physical possession of parts of the system. The purpose of providing local storage is to generate a plausible view of events occurring in proximity to the system, i.e. observed by any of its multiple sensors.

Usually, distributed storage systems can rely on several techniques to ensure that all data is captured, to give a view of events that is as close to the reality as the system can possibly provide. To prevent data loss, cloud-connected systems can elastically scale out their capacity to increase storage space and redundancy, while edge systems may deploy data sinks to secure data and free up local storage devices [23]. If real-time observation is required, data can be processed on-the-fly, and be streamed to remote users [3,32]. These traditional techniques are not available in the deployment environment of Dorvu; bandwidth limitations and unstable networks prevent access to cloud resources and most data streaming, while the remoteness of the system deployment prevents any manual intervention and reliance on data sinks. This means that the system must continuously adapt to observed faults, and allow degradation, in terms of storage space, compute capability and availability, to occur gracefully. In addition to this, a question of access control and privacy rights arises in the context of remote surveillance. Data must be handled with fine granularity, and access to the system does not necessarily imply access to all its stored and continuously generated data, even in cases of fishing vessel inspections conducted by federal authorities. Thus the system must provide real-time unsupervised management of access rights and encryption keys, to prevent unauthorized access and privacy infringement.

Because of the requirements outlined above, data should be written to Dorvu in a way where the following guarantees are upheld: First, newly created data must be annotated with access policies based on their contents before files are persisted to disk, where these policies prevent unauthorized access, even for privileged users and adversaries obtaining physical control of the system. This is elaborated in Sect. 2.6. Second, once written, any modifications to the data will compromise its integrity and must be transparent to data consumers. As physical control of the system will imply the ability to destroy existing data and prevent the creation of new data, the goal of the storage platform is to guarantee the integrity of existing data.

## 2.2 Data Distribution Layer

Data is stored locally at the receiving node, and distributed across the edge network to achieve fault tolerance through replication. The aim is to ensure that the failure of a single component does not compromise the integrity of the remaining components. Figure 2 illustrates the data flow between edge nodes in Dorvu, with the leftmost part of the illustration showing replication between nodes through the data distribution layer. Distribution of data in Dorvu is deterministic based on each individual node's view on the network. Nodes transmit

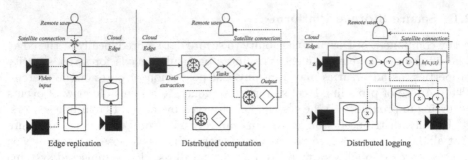

**Fig. 2.** An overview of three layers of data movement internal to the Dorvu system.

challenges to others to prove that data replicated from themselves exists and is unmodified on other nodes in the network. Cryptographic and algebraic challenges has been explored as a means to prove the existence of files in previous works, most notably related to provable data possession [30] and provable data retrievability [11]. In Dorvu, due to fear of tampering, nodes have limited *trust* in other nodes, and data movement is independent of decisions made by other nodes. Node $A$ will reduce its *trust* in Node $B$ if it observes Node $B$ failing a challenge on data that originated in $A$. A challenge for file $F$ at $B$ is initiated an arbitrary amount of time after $F$ was transmitted from $A$, where $A$ requests a cryptographic hash of an arbitrary range of the encrypted $F$. The use of *trust* to manage the network is further detailed in Sect. 2.5.

### 2.3 Distributed Log Layer

Figure 2 (right) illustrates logging mechanisms in Dorvu. Names of newly created files along with hash digests of file contents are appended to a sequential log stored at each node. Storing digests or signatures alongside file meta-data to detect changes is a common approach for file system integrity checking [12]. Nodes in Dorvu distribute this local log to the rest of the network on fixed intervals. The number of files included in each interval depends on the frequency of file generation, and in the case of remote surveillance, the length of individual video chunks. Files within the same interval can be appended to calculate a common hash digest, as an optimization, because integrity checking can only be performed at the granularity of which the data is transmitted from their origin nodes. The purpose of this logging mechanism is to store and provide forensic evidence of data tampering in cases where some (but not all) nodes are byzantine or compromised by an adversary. Other approaches to similar problems of nodes with limited trust in a remote distributed storage system include implementing a blockchain within the network [4]. Our approach is simpler, as nodes does not represent individual stakeholders; data distribution is used as a fault tolerance mechanism to ensure data survival.

## 2.4   Inference Layer

Video data that is input to Dorvu is automatically processed through inference in order to extract information about its contents. Details regarding AI inference and training are out of scope of this paper, but is further detailed in [19]. In short, we have previously developed a dataset from fishery surveillance videos from an active voyage of a collaborating fishing trawler [20] that provides us with annotations to detect people, fishing equipment, and various events of interest. This procedure is conducted on each node for its individual data input, independent of other nodes. A checkpointing scheme is utilized in order to resume processing from replicas, in case of node failure [1]. This is illustrated in Fig. 2 (middle), in which the topmost node has failed at the third and final step of computation, and the right node resumes the work. For a file $f$ originating at node $A$, replicated at node $B$, $B$ will poll $A$ for checkpoints of the computation on $f$, and resume data processing in cases where $A$ stops responding.

## 2.5   Policies and Trust

Data distribution, data processing, and remote client access, are tied together with data policies, which are used by privileged clients to modify the behaviour of the system during run-time in remote environments. Data policies are assigned to features that can be found in data, in addition to sensor readings. For our use-case, we focus on people, fishery equipment, fish, and special events [20]. For sensor input, we use AIS data points annotated in their corresponding video files. Remote clients can configure network behaviour by providing nodes with policies related to which files are of highest importance to retain. Challenges, as described in Sect. 2.2, are used to confirm that files are adequately replicated in the network. Classification of file importance through challenges is performed by providing relevant video features (output from the data processing layer) and sensor readings. This is useful data input for fishery surveillance, because AIS data points can be used to limit data to a certain geographical area. An additional benefit of this approach is that a sequence of AIS readings can be used to calculate vessel movement. Data policies are thus provided by remote clients during the Dorvu run-time to determine to what degree of robustness video data should be secured, based on their calculated content, and their context provided from sensor readings.

Robustness within the network is determined by *trust*, as described in Sect. 2.2. *trust* in the network is implemented in relation to data policies and video features, so that each node contains a $nxm$ matrix of 32 bit floating point values between 0 and 1, with $n$ being the number of nodes in the network and $m$ being the features extracted from video data. Additionally, each node maintains a list of active policies $p$ that inherits trust from one or more of the $m$ features. This means that nodes maintain a degree of *trust* in that other nodes are able to satisfy their active policies. Each node maintains their list of policies independently of other nodes, and must therefore calculate *trust* to identify nodes most likely to retain their files. This approach allows users to customize the inherent data distribution mechanisms in Dorvu to maximize the availability and survivability of the video data that is most relevant as forensic evidence.

## 2.6  Encryption

Data stored locally on disk in edge environments are subject to strict requirements of confidentiality, due to the risk of data leakage. Bhatnagar et al. [2] define an encryption scheme for storage in edge nodes in wireless sensor networks. This scheme has the aim of minimizing data leakage in case an adversary gains physical possession of a storage node. This is achieved by generating new keys for every data object, with a one-way function, such as a cryptographic hash function, applied on the old key. A trusted party has control of the initial key, and every subsequent key can as such be re-calculated when the node is in a trusted environment. Dorvu utilizes a modified version of this. For our scenario, the goal of minimizing data leakage is the same, but using the same one-way key for every file is infeasible if remote data access should be possible for certain, but not all, files. The scheme is modified to better fit this specific scenario, by adding a second initial seed used to generate not only keys, but also subsequent seeds. An initial seed $(S_1)$, as well as two cryptographic one-way functions ($F_1$ and $F_2$), are initialized on both an edge node, and a trusted main-land node. A key is generated by applying the function $F_1$ on the seed $S_1$, generating $K_1$. A second seed is generated by applying function $F_2$ on $S_1$, generating $S_2$. $S_1$ is subsequently deleted from the disk on the edge node. $K_1$ can now be used to encrypt data with a synchronous encryption scheme, without the possibility of this key being re-calculated by information present in the edge environment. After a file has been encrypted with $K_1$, a new key $K_2$ is generated by applying $F_1$ on $K_1$, and $K_1$ is deleted from the disk. This continues by generating $K_n$ for the $n$th file written to disk. After a pre-configured time interval, the key on the edge node is discarded, and a new key is generated by applying $F_1$ on $S_2$, and deleting $S_2$ after generating $S_3$.

As a result of this scheme, the trusted party that initialized the seed can grant access to an authorized third-party for a given time-interval, by calculating the key for that time period by applying $F_2$ on the initial seed $S_1$. At the same time, any adversaries that gain possession of the device, and any keys stored on it, will be unable to read any data written prior to the possession. This achieves the functionality of file retrieval for authorized users under certain conditions, while retaining the guarantees from physical attacks that were described in the original scheme [2]. The time interval defined in our encryption scheme determines how much data a user can access *at minimum* when granted an access key. This can be reduced to a single file, by never applying $F_1$ on a key $K$. As this scheme will be used to encrypt video data, it is assumed that files need a minimum length to be useful. There is a trade-off between encrypting the smallest units possible, each individual frame, with unique keys, and granting a potentially too large window of access because the time-interval is configured to be too long, opening up for data leaks.

# 3   Experiments and Results

In evaluating this system, we investigate individual components of the proposed data pipeline setup. The purpose of these experiments is to evaluate the feasibility of the system in an edge environment. As such, we investigate the capabilities of a satellite link and the embedded computing boards that we have used to deploy our prototype. For these experiments, we have utilized a Iridium Certus 200 broadband satellite service, an L-band non-geostationary satellite network, connected to a Thales VesselLink 200 maritime antenna. Service providers claim that the Certus network provides global coverage through a network containing 66 satellites, with transmission speeds up to 176 kilobits per second (kBps) [10]. The VesselLink router is connected over Ethernet to a Nvidia Jetson Xavier NX embedded computing board (NVIDIA Carmel ARMv8.2 CPU @ 1400MHz, 384-core Volta GPU @ 1100MHz), which was used as a test system for the following experiments. For many of these experiments, test data is collected from the Njord dataset [20], which contains 29 h of video footage recorded from cameras installed in an operational fishing trawler.

## 3.1   File Storage

The time to encrypt or digest a video file with additional metadata and write it to persistent storage was measured. This experiment encompasses our encryption scheme (Sect. 2.6), and hash functions. The video file is a 10 s snippet from the Njord dataset. A sample AIS signal indicating the location where the video originated from is included as metadata. Various encryption and hashing functions were utilized, and most of these OpenSSL implementations were selected because the hardware configuration on the Nvidia Jetson Xavier NX board denoted support for them through a specialized instruction set. This experiment included reading the video file from persistent storage, producing an encrypted file or digest, and writing the result to persistent storage again. The results from this experiment are shown in Fig. 3.

## 3.2   Policy-Based Data Replication

Data replication based on data policies in Dorvu is evaluated in unstable environments with comparison to hash-based replication schemes [31] and round robin schemes. An unstable environment is emulated by injecting probabilistic crashes in the network. The system is deployed with 8 nodes, with 100 video files streamed to each node. Each node is given a random failure rate $\lambda \in (0.0, 1.0)$ which is used to generate a timeline of failures under an exponential distribution. The system runs for 30 s, each data access increments the timeline, and each failure prevents the next data access. A separate monitor program continuously queries all files and tracks their availability across all nodes. If all replicas that contain a given file is unresponsive, the file is considered unavailable. For this experiment, Dorvu runs with an active policy that detects persons in video files. Of the 100 files in the test set, 33 contains detectable persons. The number

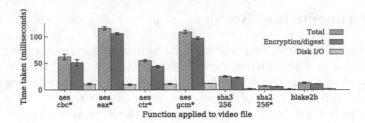

**Fig. 3.** Encryption and digest time of a tuple containing a video file and location metadata. Error bars represent standard deviation after 10 repetitions. (*) denotes hardware support.

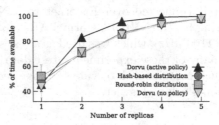

**Fig. 4.** Availability of files in an unstable network. Results show average availability after 10 runs, where error bars represent standard deviation.

of nodes that each file is replicated to is varied from 1 to 5 for this experiment. Results are shown in Fig. 4, where the availability of files that match the active policy, and the files that do not, are shown as separate data points.

### 3.3  Satellite Network Connection

The potential for video streaming or transfer over satellite broadband, as an alternative to local storage, is evaluated. Live video transfer from active vessels to mainland nodes could reduce risk of data loss and increase response time for system observers. For this experiment, a test video file is hosted on the test system, connected to the internet through satellite broadband. The video is served through the Python *ffmpeg* library implementation of Dynamic Adaptive Streaming over HTTP (DASH) [24]. The test system serves the test client a manifest (.mpd) file containing an index of all available video chunks (.m4s). The video is separated into both 2 and 10 s snippets from the original video file in various resolutions and bitrates. This allows a viewer of the DASH stream to select segments in different sizes depending on current network quality. For this experiment, the client used VLC media player 3.0.11 to receive the stream. This software was responsible for adapting its file requests according to the observed bandwidth. File system access was observed to indicate which files were read at a given time, and the total length of each segment transferred over the network was summed to get an indication of expected quality for DASH streaming under the network conditions in the testing environment. The results from this experiment are shown in Fig. 5 (left). Additionally, the files were downloaded by the client using the Linux *wget* utility. This utility allows for the documentation of both the total transfer time of a file, in addition to a running estimated transfer time, given the most recently observed network conditions extrapolated to the entire file size. The results from this experiment are shown in Fig. 5 (right).

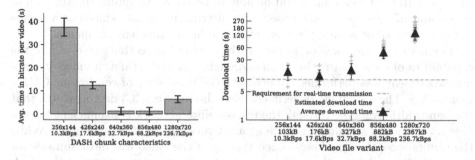

**Fig. 5. Left:** Time spent sending videos of different bitrates during DASH streaming over satellite broadband. **Right:** Time taken to download a 10 s video file of different qualities over satellite broadband. Grey markers indicate the estimated time to completion after every 50 kilobits of download. **Both:** Results show average results after ten repetitions, with error bars representing the standard deviation of observations.

## 4 Discussion

The solution presented in this article is intended for a specific and rapidly emerging domain of trusted edge computing where the edge nodes are mobile, potentially malicious, and error-prone. Because of this, direct points of comparisons to existing systems were hard to find. Therefore, through our experiments we aim to evaluate two main aspects. First, evaluate if the proposed solution is feasible given the constraints in a hostile edge environment. Second, compare the proposed solution to the alternative of full off-shore surveillance through CCTV video streaming described in some governmental propositions [15].

The first component we evaluate is the storage for locally collected video and sensor data. The experiment described in Sect. 3.1 investigates the time to encrypt and store data. The results show that, while speed-up can be achieved by using ciphers or functions with explicit hardware instruction support, the test system is capable of encrypting or digesting a file and writing it to disk in shorter than 10 s time using our scheme, with every function in the experiment. This experiment demonstrates that encryption and storage at edge devices in real time are possible to achieve, and that the choice of encryption scheme could affect performance if multiple video inputs were to be handled in parallel. The experiment described in Sect. 3.2 shows that the policy-based distribution in

Dorvu achieves favorable performance compared to traditional replication techniques with regards to file availability. An important detail is that the availability of non-policied files is not significantly worse than the alternative methods. Systems deployed in environments where elastic storage scaling or manual intervention and repairs are possible, would likely replace or ignore nodes that are demonstratively faulty or malicious, but in off-shore edge deployments, systems may have to rely on all available storage space, despite risks of data losses. In those scenarios, the system should be able to adapt to maximize the survival of important information. Policy-based data distribution is introduced as a means towards the goal of securing forensic evidence in unstable environments.

Further evaluation was performed with satellite connectivity to evaluate the capabilities of data transfer in an off-shore edge scenario. First, a video streaming experiment was performed, to evaluate the possibility of continuous CCTV monitoring. The streaming experiment described in Sect. 3.3 demonstrates that the bandwidth available through the test satellite network is insufficient to provide a continuous streaming service at a bit-rate of 17.6 kBps or higher. While the results show that the lowest sized variant of the video was most suitable for the network conditions, the experiment does not indicate anything of the quality of the service provided when streaming that particular video. The following bandwidth experiment indicate that the bandwidth supplied by the satellite connection is in most cases not sufficient to transfer entire files of any of these bit-rates in real-time, although it was occasionally observed for the smallest file sizes.

## 5   Conclusion

We have designed and implemented a multimedia distributed storage system for video and sensor data captured in edge environments, with specific focus on the emerging field of commercial fishery monitoring. Conflicting properties like security, fault tolerance, availability and remote access has been taken into account so that the system can be put to good practical use for securing forensic evidence in a compliant manner. This solution is opposed to the current governmental trend that mandates deployment of surveillance systems that violate some of these properties. Our experimental evaluation shows that traditional video streaming solutions are not always viable in off-shore edge environments, and that our proposed system makes desirable trade-offs for the targeted domains. This is accomplished by introducing AI inference into the storage system, combined with a novel data distribution scheme where individual node trust in the network, combined with user input over satellite links, determines data distribution behaviour.

# References

1. Alsile, J.A., et al.: Áika: a distributed edge system for AI inference. Big Data Cogn. Comput. **6**(2), 68 (2022)
2. Bhatnagar, N., Miller, E.L.: Designing a secure reliable file system for sensor networks. In: Proceedings of the 2007 ACM workshop on Storage security and survivability, pp. 19–24 (2007)
3. Bischof, H., Godec, M., Leistner, C., Rinner, B., Starzacher, A.: Autonomous audio-supported learning of visual classifiers for traffic monitoring. IEEE Intell. Syst. **25**(3), 15–23 (2010)
4. Conoscenti, M., Vetro, A., De Martin, J.C.: Blockchain for the internet of things: A systematic literature review. In: 2016 IEEE/ACS 13th International Conference of Computer Systems and Applications (AICCSA), pp. 1–6. IEEE (2016)
5. Costello, C.e.a.: The future of food from the sea. Nature (London) **588**(7836), 95 (2020)
6. Fitwi, A., Chen, Y., Zhu, S.: Lightweight frame scrambling mechanisms for end-to-end privacy in edge smart surveillance. IET Smart Cities **4**(1), 17–35 (2022)
7. Fournier, M., Casey Hilliard, R., Rezaee, S., Pelot, R.: Past, present, and future of the satellite-based automatic identification system: Areas of applications (2004–2016). WMU J. Marit. Aff. **17**, 311–345 (2018)
8. Ghemawat, S., Gobioff, H., Leung, S.T.: The google file system. In: Proceedings of the 19th ACM symposium on Operating systems principles, pp. 29–43 (2003)
9. Gu, B., Wang, X., Qu, Y., Jin, J., Xiang, Y., Gao, L.: Context-aware privacy preservation in a hierarchical fog computing system. In: ICC 2019–2019 IEEE International Conference on Communications (ICC), pp. 1–6, IEEE (2019)
10. Inc, I.C.: Iridium certus 200 datasheet. https://www.iridium.com/services/iridium-certus-200/ (2023). Accessed 8 Sept 2021
11. Juels, A., Kaliski Jr, B.S.: Pors: Proofs of retrievability for large files. In: Proceedings of the 14th ACM Conference on Computer and Communications Security, pp. 584–597 (2007)
12. Kim, G.H., Spafford, E.H.: The design and implementation of tripwire: a file system integrity checker. In: Proceedings of the 2nd ACM Conference on Computer and Communications Security, pp. 18–29 (1994)
13. Kuhl, J.G., Reddy, S.M.: Distributed fault-tolerance for large multiprocessor systems. In: Proceedings of the 7th Annual Symposium on Computer Architecture, pp. 23–30 (1980)
14. Martinussen, T.M.: Danske fiskere samler seg mot kameraovervåkning. Fiskeribladet (2020), https://www.fiskeribladet.no/nyheter/danske-fiskere-samler-seg-mot-kamera-overvakning-i-fiskeriene/2-1-839478. Accessed 5 Dec 2022
15. Ministry of Trade: Industry and Fisheries: Framtidens fiskerikontroll. NOU **19**, 21 (2019)
16. Muthitacharoen, A.: A low-bandwidth network file system. In: Proceedings of the 18th ACM Symposium on Operating Systems Principles, pp. 174–187 (2001)
17. Márcia Bizzotto: Fishing rules: Compulsory cctv for certain vessels to counter infractions. European Parliament Press Release (2021). https://www.europarl.europa.eu/news/en/press-room/20210304IPR99227/fishing-rules-compulsory-cctv-for-certain-vessels-to-counter-infractions
18. Nordmo, T.A.S., Ovesen, A.B., Johansen, H.D., Riegler, M.A., Halvorsen, P., Johansen, D.: Dutkat: A multimedia system for catching illegal catchers in a privacy-preserving manner. In: Proceedings of the 2021 Workshop on Intelligent Cross-Data Analysis and Retrieval, pp. 57–61 (2021)

19. Nordmo, T.A.S.: Dutkat: a privacy-preserving system for automatic catch documentation and illegal activity detection in the fishing industry (2023)
20. Nordmo, T.A.S., et al.: Njord: a fishing trawler dataset. In: Proceedings of the 13th ACM Multimedia Systems Conference, pp. 197–202 (2022)
21. Ovesen, A.B., Nordmo, T.A.S., Johansen, H.D., Riegler, M.A., Halvorsen, P., Johansen, D.: File system support for privacy-preserving analysis and forensics in low-bandwidth edge environments. Information 12(10), 430 (2021)
22. for Primary Industries, M.: On-board cameras for commercial fishing vessels (2023), https://www.mpi.govt.nz/fishing-aquaculture/commercial-fishing/fisheries-change-programme/on-board-cameras-for-commercial-fishing-vessels/ Accessed 20 Nov 2022
23. Rajagopalan, R., Varshney, P.K.: Data aggregation techniques in sensor networks: a survey (2006)
24. Sodagar, I.: The mpeg-dash standard for multimedia streaming over the internet. IEEE Multimedia 18(4), 62–67 (2011)
25. Solberg, R.R.: Bærekraftig fiskeri, governance og tid. Master's thesis, The University of Bergen (2022)
26. Sumaila, U.R., Tai, T.C.: End overfishing and increase the resilience of the ocean to climate change. Front. Mar. Sci. 7, 523 (2020)
27. Tarasov, V., Gupta, A., Sourav, K., Trehan, S., Zadok, E.: Terra incognita: on the practicality of user-space file systems. In: 7th {USENIX} Workshop on Hot Topics in Storage and File Systems (HotStorage 15) (2015)
28. Tornsberg, L.: Danske pelagiske fiskere indfører 100% dokumenteret fiskeri. Fiskerforum (2022), https://fiskerforum.dk/danske-pelagiske-fiskere-indfoerer-100-dokumenteret-fiskeri-%E2%80%A8/ Accessed 15 Jan 2023
29. UNODC: Fisheries crime: transnational organized criminal activities in the context of the fisheries sector (2016)
30. Wang, Q., Ren, K., Yu, S., Lou, W.: Dependable and secure sensor data storage with dynamic integrity assurance. ACM Trans. Sensor Netw. (TOSN) 8(1), 1–24 (2011)
31. Weil, S.A., Brandt, S.A., Miller, E.L., Maltzahn, C.: Crush: Controlled, scalable, decentralized placement of replicated data. In: Proceedings of the 2006 ACM/IEEE conference on Supercomputing, pp. 122-es (2006)
32. Winkler, T., Rinner, B.: Security and privacy protection in visual sensor networks: a survey. ACM Comput. Surv. (CSUR) 47(1), 1–42 (2014)
33. Xu, M., Xu, W., O'Kane, J.: Content-aware data dissemination for enhancing privacy and availability in wireless sensor networks. In: 2011 IEEE Eighth International Conference on Mobile Ad-Hoc and Sensor Systems, pp. 361–370. IEEE (2011)

# MC-TCMNER: A Multi-modal Fusion Model Combining Contrast Learning Method for Traditional Chinese Medicine NER

Shan Cao[1] and Qingfeng Wu[2(✉)]

[1] School of Informatics, Xiamen University, Xiamen, China
30920201153954@stu.xmu.edu.cn
[2] Xiamen University, Xiamen, China
qfwu@xmu.edu.cn

**Abstract.** Traditional Chinese Medicine (TCM) texts contain a wealth of knowledge accumulated over thousands of years, making the extraction of knowledge from these texts a pivotal concern. Named Entity Recognition (NER) can serve as an effective tool for extracting knowledge information from TCM texts. However, TCM texts contain a large number of rare characters and homophones, and the attributions of entities are also more complex, making TCMNER more challenging. In order to address this issue, this paper introduces MC-TCMNER, a novel method that leverages the multi-modal features of Chinese characters and incorporates a training strategy based on contrastive learning. Experiments have shown that our proposed method achieves an F1 score of 94.05% on the TCMNER dataset and 52.84% on the C-CLUE benchmark, demonstrating the effectiveness of MC-TCMNER. Furthermore, owing to the limited availability of a comprehensive dataset for TCMNER, we have taken the initiative to publicly release a TCMNER dataset that we meticulously collected and annotated.

**Keywords:** Named Entity Recognition · Traditional Chinese Medicine · Multi-Modal · Contrast Learning

## 1 Introduction

In recent years, Traditional Chinese Medicine (TCM) has garnered increasing attention due to its unique theoretical basis and clinical treatment methods [1]. Experienced through extended periods of refinement, TCM encompasses an expansive and intricate system of knowledge[1]. Applying Named Entity Recognition(NER) technology to the knowledge mining of TCM text can enable the prompt and accurate extraction of TCM entity information, which is of significant importance for the organization and mining of TCM knowledge [2,3].

---

[1] https://github.com/cshan-github/TCM_NER_datasets.

© The Author(s), under exclusive license to Springer Nature Switzerland AG 2024
S. Rudinac et al. (Eds.): MMM 2024, LNCS 14556, pp. 341–354, 2024.
https://doi.org/10.1007/978-3-031-53311-2_25

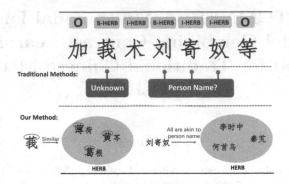

**Fig. 1.** The comparison of our method with others.

As a branch of Chinese NER, TCMNER not only faces the challenges of word segmentation [3–6,8] but also includes numerous obscure characters and terminologies. As depicted in Fig. 1, when confronted with the sentence "加莪术刘寄奴等"(Add curcuma zedoary, Artemisia anomala, and so on), traditional NER models struggle to understand the meaning of the character "莪"(curcuma) since it is not commonly used in modern Chinese. Additionally, the issue of semantic conflicts between Modern Chinese and TCM texts exists. As exemplified by "刘寄奴" (Artemisia anomala) in Fig. 1, which appears as a personal name in modern Chinese, making it challenging to establish a correlation with NER labels specific to TCM, such as diseases or symptoms. This underscores the limitations of conventional methods in the context of TCMNER.

In Chinese, there is a common saying: "A scholar recognizes only half of the characters." This saying implies that when encountering unfamiliar characters, we can infer their meaning by considering their glyph features. For example, in Fig. 1, even if we do not fully understand the exact meaning of the character "莪" (curcuma), we can roughly identify it as a type of herbal plant based on its structure, which includes the "艹" component. This is because Chinese characters with the "艹" radical are often associated with plants. Furthermore, there is also a strong semantic similarity between the labels in TCM text and modern Chinese. For instance, entities labeled as herb encompass numerous names similar to personal names, while entities labeled as symptom include entities related to temperature or weather, such as "湿"(damp) and "热"(hot).

Building upon this foundation, this study introduces MC-TCMNER, a multimodal fusion NER model that incorporates semantic, glyph, and phonetic features of Chinese characters. This design empowers the model to infer entity attributes even when faced with rare characters or homophones by leveraging the external features of the character. Furthermore, this paper adopts a training strategy that incorporates contrastive learning to enhance the fusion of external and semantic features and enable the model to learn commonalities among entities of the same attribute. Moreover, this study employed web scraping techniques to assemble a TCM text corpus from online sources and performed manual annotation in NER format to create an openly accessible dataset.

## 2   Related Work

As a subset of Chinese NER, TCMNER has garnered attention from various researchers: TCMKG-LSTM-CRF model proposed in [9], which incorporates a knowledge attention vector mechanism to establish attention between the neural network's hidden vector and the knowledge graph's candidate vector while taking into account the influence of previous words. Similarly, [8] proposed a back-tagging method for NER in TCM, which significantly improves the effect of remote supervision. They also discussed on this basis how to use distant supervision methods to achieve better performance of the NER model. Furthermore, [3] proposed a NER method for TCM books that integrates dictionary information into the presentation layer to enhance its semantic information. The authors also built a large-scale corpus to obtain pretraining vectors.

As a kind of hieroglyphics, the Chinese characters' glyph and pronunciation have been utilized by numerous researchers in various NLP tasks. [12] employed historical Chinese scripts to enhance the pictographic evidence within characters. Their method achieved state-of-the-art performance in the majority of NLP tasks. [13] proposed a Fusion Glyph Network for Chinese NER named FGN, except for encoding glyph information with a novel CNN, this method may extract interactive information between character distributed representation and glyph representation by a fusion mechanism. Experiments showing that FGN with LSTM-CRF as tagger achieves leading performance for Chinese NER. [14] integrated phonetic features of Chinese characters with the lexicon information, and proposed a novel multi-tagging-scheme learning method to alleviate the data sparsity and error propagation problems. Experiments demonstrated that the introduction of the phonetic feature and the multi-tagging-scheme has a significant positive effect. [15] proposed a Chinese text classification model derived from the construction methods and evolution of Chinese characters, and conducted on three datasets to demonstrate validity and plausibility.

Contrastive learning is a method aimed at developing powerful representations by bringing similar features closer together in a representation space. Some research teams have applied this approach to NLP tasks: [7] proposed a novel Multi-Granularity Contrastive Learning framework (MCL), that aims to optimize the intergranularity distribution distance and emphasize the critical matched words in the lexicon. This network can explicitly leverage lexicon information on the initial lattice structure, and further provide more dense interactions of across-granularity. [16] proposed a multi-level supervised contrastive learning framework for low-resource natural language inference, outperforming other methods on NLI datasets and cross-domain text classification tasks. [17] proposed Mutual Learning and Large-Margin Contrastive Learning (ML-LMCL), aiming to iteratively share knowledge between these two models. Experiments on three datasets show that ML-LMCL outperforms existing models and achieves new state-of-the-art performance. [18] presented a method to train multilingual Information Retrieval (IR) systems and designed a semantic contrastive loss to align representations of parallel sentences. This method can work well even with a small number of parallel sentences, and be used as an add-on module to any backbones and other tasks.

## 3    Method

### 3.1    Multi-modal Fusion NER Model

This paper introduces a multi-modal TCMNER model, the overall structure as illustrated in Fig. 2. For each Chinese character in the input text, we analyze it by distinguishing between semantic and external features. Semantic features refer to encoding characters based on their usage experience and semantic information. External features involve encoding Chinese characters based on their glyph and phonetics features. During the fusion stage of them, this paper combines the two by adding them together and performs feature fusion calculations using the stack of three layers of Transformer Blocks [23]. In the final Head part, dimension reduction is achieved using a linear layer, and a conditional random field (CRF) layer is combined to serve as the NER output. The remaining sections will provide detailed explanations of the methods for extracting features.

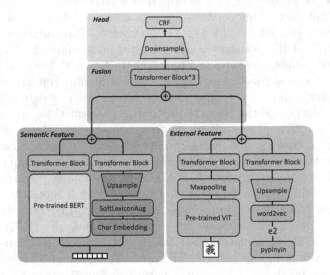

**Fig. 2.** The overall architecture of the multi-modal TCMNER method.

**Sematic Features**

This paper employs two methods to represent the semantic features of characters: pre-trained BERT and SoftLexicon augmentation. Pre-trained BERT, which has undergone extensive Chinese text training, has become one of the commonly used encoding methods in NLP [25]. Given that the task of this paper is TCMNER, which include a substantial amount of classical Chinese sentences, we have chosen GuwenBERT[2] as the pre-trained BERT model for our application. GuwenBERT was pre-trained on a corpus of 15,694 classical Chinese texts and has demonstrated excellent performance across various classical Chinese language tasks, aligning well with our application situation.

Additionally, numerous studies have shown that the lexicon augmentation can be beneficial for word segmentation [3,5]. The SoftLexicon Augmentation strategy proposed by [5] is widely used due to its adaptability, and this paper adopts the same method to enhance character features. The input sentence is treated as a character sequence, denoted as $s = \{c_1, c_2, ..., c_n\} \in V_c$, where $V_c$ is the character vocabulary. Each character $c_i$ is represented using monogram embeddings and bigram embeddings:

$$x_i^c = [e^c(c_i); e^b(c_i, c_{i+1})] \tag{1}$$

Here, $e_c$ represents the character embedding lookup table, and $e_b$ represents the bigram embedding lookup table. Next, all matched words of each character $c_i$ is categorized into four word sets "BMES" (Begin, Middle, End, Single). If a word set is empty, a special word "NONE" is added to the empty word set. After obtaining the "BMES" word sets for each character, each word set is then condensed into a fixed-dimensional vector. Subsequently, in order to facilitate the fusion of word sets, we utilize the occurrence frequency of each term as an indicator of its weight. The weighted representation of the term set S is defined as follows:

$$v_s(S) = \frac{4}{Z} \sum_{w \in S} z(w)e^w(w), \tag{2}$$

where,

$$Z = \sum_{w \in B \cup M \cup E \cup S} z(w). \tag{3}$$

$z(w)$ represents the frequency of the vocabulary word $w$ occurring in the statistical data, and $e^w$ denotes the word embedding lookup table. Here, we apply weight normalization to all words within the four word sets to enable a comprehensive comparison. Finally, the representations of the four sets of words are combined into a fixed-dimensional feature, which is subsequently added to the representation of each character:

$$e^s(B, M, E, S) = [v^s(B); v^s(M); v^s(E); v^s(s)], \tag{4}$$

$$x^c \leftarrow [x^c; e^s(B, M, E, S)]. \tag{5}$$

After extracting both of these features, this paper utilizes a Transformer Block to align and fuse the features. The advantage of this design is the ability to perform fine-grained fusion of the features, reducing the dimensionality

---

[2] https://github.com/ethan-yt/guwenbert.

of the fused features compared to concatenation. It is important to note that since the dimensions of the two features are different, the features generated by SoftLexicon need to be linearly upsample to achieve feature length matching.

**External Features**

Semantic features are commonly employed in Chinese NER due to the rich prior knowledge provided by our usage habits in Chinese. However, in TCMNER, rare characters and homophones are frequently encountered, this affects the performance of NER models that rely solely on semantic information. The introduction of external features can assist in cases of missing semantic information, enabling the model to infer entity attributes even when encountering unfamiliar characters. This paper uses the character's glyph and pronunciation as external features and encodes them separately using different methods.

In this paper, a pre-trained ViT [21] with MAE [20] was used as the image encoder for Chinese character structure. Compared to retraining an image encoder, pre-trained models possess enhanced feature extraction capabilities, making them more effective in extracting structural features of Chinese characters. For each Chinese character, the paper obtained the character's image using the PIL library's ImageDraw and resized it to [1024, 1024] in order to align with the input size of ViT. The output dimensions of the pretrained ViT's final layer were [64, 64, 768], to match the dimensions of semantic features, the paper reduced the features to [768] using Maxpooling.

For the extraction of phonetic features, this paper employed the same method as described in [14]. We utilized the Pypinyin library[3] to transliterate the characters into the Latinized pinyin system, representing the tone of each character with numerical notation. For instance, the character "莪" (curcuma) in the word of "莪术" (curcuma zedoary) would be represented as "e2" in pinyin. Once we obtained the pinyin representation for each character, we trained phonetic embeddings by providing the transliterated text, as follows:

$$v_{i,p,q}^{phn} = e^{phn}(Py(c_{p,q}^i)) \tag{6}$$

In the equation, $v_{i,p,q}^{phn}$ represents the dense vector embedding of the phonetics of character $c_{p,q}^i$, superscript $i$ denotes the $ith$ character in the character sequence $c_{p,q}$, and $Py(.)$ represents the operation of converting the character into the form of Pinyin, and $e^{phn}(.)$ is the phonetic embedding lookup table trained by the word2vec tool [24].

After extracting both of these features as mentioned above, similar to the semantic features, we perform feature alignment using Transformer Encoder Block and then sum them together. Just like the SoftLexicon features, phonetic features require upsample to achieve dimension matching.

### 3.2 Training Strategies Combining Contrastive Learning

The model proposed in this paper can comprehensively consider the features of Chinese characters. However, training this model using only a standard NER loss

---

[3] https://pypi.python.org/pypi/pypinyin.

may lead to the following issues: (1) alignment between the external features and semantic features is not established, impacting the final fusion results; (2) The model lacks the ability to cluster features within entities of the same attribute. In [22], the concept of semantic class was introduced, which represents the semantic category of entities without considering the positional roles within the entity. To further improve the performance of the multi-modal TCMNER method, two sets of additional contrastive losses are incorporated in addition to the traditional NER loss. These contrastive losses are the External Semantic Loss (ESL), which contrasts external features and semantic features, and the Semantic Class Loss (SCL), which contrasts entities belonging to the same semantic class. The calculation methods for these two types of losses are illustrated in Fig. 3. We will introduce these two types of losses in the following pages.

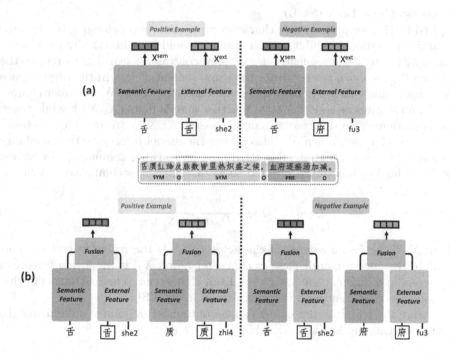

**Fig. 3.** Categorization diagram of contrast learning. (a) represents the External Semantic Loss (ESL), and (b) represents the Semantic Class Loss (SCL).

## External Semantic Loss (ESL)

Inspired by ALBEF [19], this paper aims to enhance the model's ability to create a smoother fusion between the external features and semantic features. To achieve this, the paper introduces the ESL, which aims to align the semantic features and external features of the same character while separating those of characters from different semantic classes. For character $i$, we consider its external features, $X_i^{ext}$, and semantic features, $X_i^{sem}$. We treat the external features and semantic features of the same character as positive sample instances. To construct negative samples, we select all $X_j^{ext}$, where character $j$'s semantic class is different from character $i$, along with character $i$'s $X_i^{sem}$ from the sentence. Next, the objective of the contrastive learning is to minimize the following loss, which can be expressed as:

$$L_{ESL} = -log\frac{e^{sim(X_i^{ext},X_i^{sem})/\tau_1}}{\sum_j^T e^{sim(X_j^{ext},X_i^{sem})/\tau_1}} \tag{7}$$

where, $sim()$ calculates the similarity of different characters. $T$ is the number of remaining characters after removing duplicate characters in a sentence. $\tau_1$ is the temperature.

### Semantic Class Loss (SCL)

Inspired by [22], we propose that character representations belonging to the same semantic class should exhibit greater similarity and be semantically relevant to their entity types. Consequently, we employ contrastive learning to reduce the distance between characters within the same semantic class in the target space while separating characters from different semantic classes. We consider characters from the same semantic class as positive sample pairs $(X_i, X_i^+)$, while token representations from different semantic classes $(X_i, X_i^-)$ are regarded as negative samples for each token $X_i$. This allows the model to acquire the capability to cluster characters with similar attributes by learning commonalities among entities. This loss function is referred to as SCL, and it is computed as follows:

$$L_{SCL} = -log\frac{e^{sim(X_i,X_i^+)/\tau_2}}{\sum_j^T e^{sim(X_i,X_i^-)/\tau_2}} \tag{8}$$

where $X_i$ is the fusion feature of character $i$, $X_i^+$ is the fusion feature of characters in the same semantic class as character $i$, and $X_i^-$ is the fusion feature of characters in a different semantic class from character $i$. The meanings of other variables are consistent with those explained in the ESL.

During the training, the model can be optimized by jointly minimizing the contrastive training loss and NER loss:

$$L = L_{NER} + L_{ESL} + L_{SCL} \tag{9}$$

## 4   Experiments

All experiments in this paper were conducted in a consistent hardware environment and framework. The experiments utilized an Intel(R) Core(TM) i12-700K CPU with 32 GB of memory, along with a GeForce RTX 3090 GPU, and were implemented exclusively within the PyTorch framework. The evaluation metrics for the model in the experiments are the commonly used metrics in NER tasks, including F1_score, Precision, and Recall. The hyperparameter settings and datasets for different experiments will be individually introduced in each section.

## 4.1 Compared with Commonly Used Chinese NER Methods in TCMNER

To validate the effectiveness of the method proposed in this paper, we compared it with several commonly used Chinese NER methods. Since there is currently a lack of publicly available large-scale TCMNER datasets, we collected and annotated a TCMNER dataset independently and used it to evaluate the model's performance. We first used web scraping techniques to extract disease-related information from the website[4], which includes brief descriptions of diseases, symptoms, causes, treatment methods, and others, as shown as Fig. 4. For the extracted text content, we initially cleaned it by removing duplicate sentences and characters that are not Chinese. Four clinical experts from China (chief physicians) used the YEDDA[5] annotation tool to individually annotate text for four types of information using BIO tagging. This dataset contains a total of 387,465 Chinese characters, with five types of entities: Symptoms, Causes, Herbs, Preparations (already prepared medicine), and Effects. The ratio of the training set, testing set, and validation set is 8:1:1. The entity counts for each attribute are shown in the Table 1.

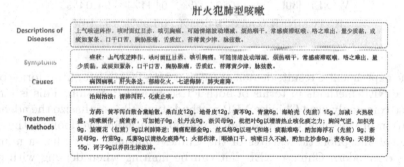

**Fig. 4.** TCM website text example.

The hyperparameters in experiments were set as Table 2. The methods we compared include LatticeLSTM [27], TENER [29], SoftLexicon [5], and $W^2$NER [30], with their hyperparameter settings identical to those in their original papers on the OntoNotes [28]. This is because the quantity of characters in the TCMNER dataset is similar to the OntoNotes dataset. All models utilized character embeddings identical to those in [27], and vocabulary embeddings identical to those in [3], which were derived from a pre-trained word embedding lexicon specifically tailored to the TCM domain[6]. The experimental results are shown in the Table 3.

---

[4] https://www.zhzyw.com/jbdq.html.
[5] https://github.com/jiesutd/YEDDA.
[6] https://github.com/Sporot/TCM_word2vec.

**Table 1.** Entity attributes on the TCM-NER dataset.

| Attribute | Quantity |
|---|---|
| SYM | 5221 |
| CAU | 1435 |
| HER | 675 |
| PRE | 1049 |
| EFF | 656 |

**Table 2.** Training Hyperparameters.

| Param. | Value |
|---|---|
| Batch_size | 32 |
| Epoch | 30 |
| Learning rate | $2*10^{-4}$ |
| Dropout | 0.5 |
| Optimizer | Adam |
| $\tau 1$ | 0.5 |
| $\tau 2$ | 0.03 |

**Table 3.** Comparison with commonly used Chinese NER methods on TCMNER dataset.

| | F1_score | Precision | Recall |
|---|---|---|---|
| LatticeLSTM [27] | 88.81% | 89.43% | 88.29% |
| TENER [29] | 91.20% | 91.18% | 91.42% |
| SoftLexicon [5] | 90.58% | 90.66% | 90.42% |
| W²NER [30] | 93.37% | 94.11% | **94.04%** |
| MC-TCMNER (Ours) | **94.05%** | **95.99%** | 93.10% |

The experimental results demonstrate that MC-TCMNER outperforms other Chinese NER methods in terms of both F1_score and Precision. This is largely because commonly used Chinese NER methods typically emphasize the adequate utilization of vocabulary and semantic information while overlooking the morphological and phonetic features of the Chinese language. This has a minimal impact on modern Chinese NER but can be limiting when dealing with complex TCM texts. Additionally, the inclusion of contrastive learning allows the model to better capture commonalities between entities with similar attributes, resulting in improved experimental results.

## 4.2   Compared the Performance on C-CLUE Benchmark

To assess the scalability of the proposed method in this paper, we also evaluated the model's performance on Classical Chinese NER (CCNER). This was done because Classical Chinese text contains a significant number of rare characters and homophones, which are similar to the core challenges we aim to address. The dataset used in the experiments is C-CLUE [26], which has been annotated using a crowdsourcing system by publishers for the compilation of "The Twenty-Four Histories" of ancient Chinese dynasties. The statistical information of the dataset is provided in Table 4 [26] conducted a benchmark test, evaluating the performance of different pre-trained models. The hyperparameters in the experiment are the same as those in [26], the character lookup table and vocabulary are identical to

**Table 4.** Entity attributes on the C-CLUE.

| Attribute | Quantity |
|-----------|----------|
| PER | 11435 |
| LOC | 3520 |
| ORG | 2064 |
| POS | 2131 |

**Table 5.** MC-TCMNER's performance on C-CLUE benchmark.

| | F1_score | Precision | Recall |
|---|----------|-----------|--------|
| BERT-base [26] | 29.82% | 35.59% | 32.12% |
| BERT-wwm [26] | 32.98% | 43.82% | 35.40% |
| Roberta-zh [26] | 28.28% | 34.93% | 31.09% |
| ZKY-BERT [26] | 33.32% | 42.71% | 36.16% |
| BiLSTM-CRF [10] | 49.76% | 54.77% | 52.15% |
| Lattice-LSTM [10] | 51.96% | 56.92% | 54.33% |
| AT-CCNER [10] | **53.37%** | **58.22%** | 55.69% |
| MC-TCMNER (Ours) | 52.84% | 54.37% | **55.91%** |

those in [27], and the remaining hyperparameter settings are consistent with the previous section. The experimental results are shown as Table 5.

The experimental results indicate that MC-TCMNER outperforms [26] on the C-CLUE dataset and approaches the performance level of [10], suggesting its potential for CCNER. Compared to [10], MC-TCMNER demonstrates better adaptability as it does not require additional Classical Chinese to Modern Chinese translation datasets. This indicates that the incorporation of external features can assist the model in better understanding the rare characters and homophones commonly found in Classical Chinese texts.

### 4.3    Ablation Experiment

To validate the effectiveness of the proposed multi-modal structure and contrastive learning, we conducted ablation experiments on the model. The hyperparameter for the experiments remained consistent with that in the first section, and the F1_score of the models on the TCMNER dataset was chosen as the evaluation criterion. The SoftLexicon without BERT features and external features was used as the baseline model, we compare the differences between various model architectures and training methods. The experimental results are shown in the Fig. 5.

The experimental results demonstrate that the inclusion of external features and the use of appropriate training strategies can lead to better performance for TCMNER. It's important to note that when fusing semantic features and external features through a simple sum, the resulting may not be as good as considering only semantic features. This indicates that choosing an appropriate fusion method to align the features is crucial.

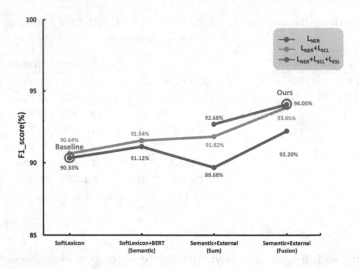

**Fig. 5.** Ablation Experiment. The X-axis represents different model structures, the curves of colors represent different training methods, and the Y-axis represents the F1_score of the models.

## 5    Conclusions

This paper introduces MC-TCMNER, a TCMNER method that combines the multi-modal features of Chinese characters with contrastive learning. Experiments demonstrate that our method can effectively utilize the external features and semantic characteristics of Chinese characters and cluster characters with similar attributes, resulting in superior performance on the TCMNER task and CCNER task. Furthermore, in this study, we have released a TCMNER dataset, contributing to the informatization of TCM.

**Acknowledgment.** This work was supported by Industry-University-Research Cooperation Project of Fujian Science and Technology Planning (No: 2022H6012), Science and Technology Research Project of Jiangxi Provincial Department of Education (No. GJJ 2206003), Natural Science Foundation of Fujian Province of China (No. 2021J011 169, No .2022J011224).

## References

1. Zhang, Q., Zhou, J., Zhang, B.: Computational traditional chinese medicine diagnosis: a literature survey. Comput. Biolo. Med. **133**, 104358 (2021)
2. Su, M.H., Lee, C.W., Hsu, C.L., Su, R.C.: RoBERTa-based traditional chinese medicine named entity recognition model. In: ROCLING, pp. 61–66 (2022)
3. Song, B., Bao, Z., Wang, Y.Z., Zhang, W., Sun, C.: Incorporating Lexicon for named entity recognition of traditional Chinese medicine books. In: Zhu, X., Zhang, M., Hong, Yu., He, R. (eds.) NLPCC 2020. LNCS (LNAI), vol. 12431, pp. 481–489. Springer, Cham (2020). https://doi.org/10.1007/978-3-030-60457-8_39

4. Xiaonan Li, Hang Yan, Xipeng Qiu, Xuanjing Huang: FLAT: Chinese NER Using Flat-Lattice Transformer. ACL 2020: 6836–6842

5. Ma, R., Peng, M., Zhang, Q., Huang, X.: Simplify the Usage of Lexicon in Chinese NER. ACL, pp. 5951–5960 (2020)

6. Xue, M., Bowen, Yu., Liu, T., Zhang, Y.: Erli Meng, pp. 3831–3841. Porous Lattice Transformer Encoder for Chinese NER. COLING, Bin Wang (2020)

7. Zhao, S., Wang, C., Hu, M., Yan, T., Wang, M.: MCL: Multi-granularity contrastive learning framework for Chinese NER. In: AAAI, pp. 14011–14019 (2023)

8. Zhang, D., et al.: Improving distantly-supervised named entity recognition for traditional chinese medicine text via a novel back-labeling approach. IEEE Access **8**, 145413–145421 (2020)

9. Jin, Z., Zhang, Y., Kuang, H., Yao, L., Zhang, W., Pan, Y.: Named entity recognition in traditional Chinese medicine clinical cases combining BiLSTM-CRF with knowledge graph. In: Douligeris, C., Karagiannis, D., Apostolou, D. (eds.) KSEM 2019. LNCS (LNAI), vol. 11775, pp. 537–548. Springer, Cham (2019). https://doi.org/10.1007/978-3-030-29551-6_48

10. Qi, Y., Ma, H., Shi, L., Zan, H., Zhou, Q.: Adversarial transfer for classical Chinese NER with translation word segmentation. In: NLPCC, vol. 1 pp. 298–310 (2022)

11. Liu, C.L., Chu, C.T., Chang, W.T., Zheng, T.Y.: When classical chinese meets machine learning: explaining the relative performances of word and sentence segmentation tasks. DH (2020)

12. Meng, Y., et al.: Glyce: glyph-vectors for Chinese character representations. In: NeurIPS, pp. 2742–2753 (2019)

13. Xuan, Z., Bao, R., Jiang, S.: FGN: fusion glyph network for Chinese named entity recognition. In: Chen, H., Liu, K., Sun, Y., Wang, S., Hou, L. (eds.) CCKS 2020. CCIS, vol. 1356, pp. 28–40. Springer, Singapore (2021). https://doi.org/10.1007/978-981-16-1964-9_3

14. Mai, C., et al.: Pronounce differently, mean differently: a multi-tagging-scheme learning method for Chinese NER integrated with lexicon and phonetic features. Inf. Process. Manag. **59**(5), 103041 (2022)

15. Yan-Xin, H., Bo, L.: A Chinese text classification model based on radicals and character distinctions. IEEE Access **11**, 45520–45526 (2023)

16. Li, S., Hu, X., Lin, L., Liu, A., Wen, L., Philip, S.Y.: A multi-level supervised contrastive learning framework for low-resource natural language inference. IEEE ACM Trans. Audio Speech Lang. Process. **31**, 1771–1783 (2023)

17. Cheng, X., Cao, B., Ye, Q., Zhu, Z., Li, H., Zou, Y.: ML-LMCL: Mutual learning and large-margin contrastive learning for improving ASR robustness in spoken language understanding. In: ACL (Findings), pp. 6492–6505 (2023)

18. Hu, X., et al.: Language agnostic multilingual information retrieval with contrastive learning. In: ACL (Findings), pp. 9133–9146 (2023)

19. Li, J., Selvaraju, R., Gotmare, A., Joty, S., Xiong, C., Hoi, S.C.H.: Align before fuse: vision and language representation learning with momentum distillation. In: NeurIPS, pp. 9694–9705 (2021)

20. He, K., Chen, X., Xie, S., Li, Y., Dollár, P., Girshick, R.: Masked autoencoders are scalable vision learn- ers. In: CVPR (2022)

21. Dosovitskiy, A., et al.: An image is worth 16x16 words: transformers for image recognition at scale. In: ICLR (2021)

22. Zhang, Z., et al.: NerCo: a contrastive learning based two-stage Chinese NER method. In: IJCAI, pp. 5287–5295 (2023)

23. Vaswani, A., Shazeer, N., Parmar, N., Uszkoreit, J., Jones, L., Gomez, A.N.: Lukasz Kaiser, pp. 5998–6008. Attention is All you Need. NIPS, Illia Polosukhin (2017)

24. Mikolov, T., Sutskever, I., Chen, K., Corrado, G. S., Dean, J.: Distributed representations of words and phrases and their compositionality. In: Proceedings of the Conference On Neural Information Processing Systems, pp. 3111–3119 (2013)
25. Devlin, J., Chang, M.W., Lee, K., Toutanova, K.: BERT: pre-training of deep bidirectional transformers for language understanding. In: NAACL-HLT, pp. 4171–4186 (2019)
26. Ji, Z., Shen, Y., Sun, Y., Yu, T., Wang, X.: C-CLUE: a benchmark of classical Chinese based on a crowdsourcing system for knowledge graph construction. In: Qin, B., Jin, Z., Wang, H., Pan, J., Liu, Y., An, B. (eds.) CCKS 2021. CCIS, vol. 1466, pp. 295–301. Springer, Singapore (2021). https://doi.org/10.1007/978-981-16-6471-7_24
27. Zhang, Y., Yang, J.: Chinese NER using lattice LSTM. In: Proceedings of the 56th Annual Meeting of the Association for Computational Linguistics (ACL), pp. 1554–1564 (2018)
28. Weischedel, R., et al.: Ontonotes release 4.0. LDC2011t03. Philadelphia: Linguistic Data Consortium (2011)
29. Yan, H., Deng, B., Li, X., Qiu, X: TENER: adapting transformer encoder for named entity recognition. CoRR abs/1911.04474 (2019)
30. Li, J., et al.: Unified named entity recognition as word-word relation classification. In: Proceedings of the AAAI Conference on Artificial Intelligence 36, 10965–10973 (2022)

# C3-PO: A Convolutional Neural Network for COVID Onset Prediction from Cough Sounds

Xiangyu Chen[1], Md Ayshik Rahman Khan[2] (ID), Md Rakibul Hasan[3] (ID), Tom Gedeon[1,3], and Md Zakir Hossain[1,3](✉) (ID)

[1] Australian National University, Canberra, ACT 2601, Australia
[2] La Trobe University, Bundoora, VIC 3083, Australia
[3] Curtin University, Perth, WA 6102, Australia
{Rakibul.Hasan,Tom.Gedeon,Zakir.Hossain1}@curtin.edu.au

**Abstract.** This study presents a novel approach to diagnosing the highly contagious COVID-19 respiratory disease. Traditional diagnosis methods, such as polymerase chain reaction (PCR) and rapid antigen test (RAT), have been found to be resource-intensive and expensive, prompting the need for alternative diagnostic methods. Existing machine learning-based diagnosis approaches, such as X-rays and CT scans, suffer from suboptimal performance, primarily due to data imbalance and data paucity. To this end, this study proposes **C3-PO**, **C**ough **s**ounds **o**n **C**onvolutional neural network (CNN) for COVID-19 Prediction. The framework utilises data augmentation and segmentation techniques to increase the volume of data to more than three times the original size. It includes an ensemble method to further mitigate the impacts of data paucity and data imbalances. Our CNN model was tested on the crowdsourced Coswara dataset and validated by the Russian dataset. It achieved an accuracy rate of 92.7% and an area under the receiver operating characteristics curve (AUC-ROC) of 98.1% on the Russian dataset, exceeding the existing works by 22% in terms of accuracy. On the Coswara dataset, the method achieved an accuracy rate of 72.3% and an AUC-ROC of 80.0%. Codes and evaluations are publicly available at https://github.com/ZakirANU/C3-PO-CovidCough-CNN.

**Keywords:** COVID-19 · Diagnosis · Cough sound · Convolutional neural network

## 1 Introduction

COVID-19 is a novel acute respiratory disease that emerged at the end of 2019 and spread worldwide. As of August 2023, more than 769 million people have been infected [25], with a death toll reaching 6.9 million. However, the statistics may fail to illustrate the severity of COVID-19 as there were many notable cases of under-reporting the infection and mortality data [19].

© The Author(s), under exclusive license to Springer Nature Switzerland AG 2024
S. Rudinac et al. (Eds.): MMM 2024, LNCS 14556, pp. 355–368, 2024.
https://doi.org/10.1007/978-3-031-53311-2_26

In addition to its impact on human health, the quarantine policies and travel bans implemented to address COVID-19 have dealt a massive blow to global society and economy. Although the impact of the pandemic on the world is now fading, COVID-19 has sounded the alarm for the world.

COVID-19 is typically diagnosed using polymerase chain reaction (PCR) and rapid antigen test (RAT). PCR tests require specialised equipment and medical personnel, while RAT tests require mass production and are less accurate than PCR. Traditional detection methods, when dealing with such highly contagious diseases, not only consume huge medical resources and delay the timely treatment of other patients but also are inefficient and cannot produce quick results. With the rapid development of machine learning (ML), several studies [5,10,13] have proposed using ML algorithms to classify biological features and achieved some success. Furkan et al. [5] utilised CT images combined with machine vision algorithms to identify COVID-19, reaching an area under the receiver operating characteristics curve (AUC-ROC) of 95%. Kassania et al. [10] used X-ray and CT scan images for diagnosis and achieved up to 99% accuracy. Other studies [1,15] used sound signals for diagnosis. To this end, we aim to diagnose COVID-19 through cough sounds since sounds are easier to collect and far more cost-effective than other alternative signals, such as chest X-rays [10] and CT scans [5].

This paper has three major contributions. (**1**) To address the challenges posed by limited data and imbalanced datasets, we employ data augmentation and segmentation techniques to augment the volume of data. (**2**) We employ an ensemble of models to effectively leverage the samples from the majority class. (**3**) We propose a novel convolutional neural network (CNN)-based framework, **C3-PO**, to diagnose COVID-19 using cough sounds. Our method yields promising results on the Coswara [21] and Russian [6] datasets.

## 2   Related Work

### 2.1   Feasibility

Authors in [24] revealed the difference between medical symptoms of COVID-19 and other respiratory diseases. They observed that even in patients exhibiting fever, the commonly employed pharyngeal swab PCR test may yield negative results. This is attributed to the potential absence of viruses in the upper respiratory tract despite the presence of pneumonia. This observation suggests that COVID-19 may predominantly impact the lower respiratory tract, thereby distinguishing it from other respiratory diseases that primarily affect the upper respiratory tract. Using cough sounds, our study leverages theoretical medical knowledge to diagnose COVID-19.

AI4COVID pipeline [9] uses a nonlinear dimension reduction technique, t-SNE (t-Distributed Stochastic Neighbor Embedding), to perform two-dimensional visualisation of the extracted MFCC (Mel-Frequency Cepstral Coefficients) features from cough sounds. This verifies the feasibility of using cough sounds for COVID-19 diagnosis.

## 2.2   Cough Classification

In recent years, diagnostic methods have been proposed using cough sounds or respiratory sounds, such as the algorithm proposed in [18]. To detect pertussis, they preprocessed audio using a 3-phase cough model. They leveraged linear regression models to classify audio features such as MFCC [7], Zero-Crossing Rate (ZCR), and crest factor and achieved 92% accuracy. Authors in [11] used a lightweight CNN model on three different modalities (voice, breath and cough) and achieved approximately 92% accuracy in predicting COVID-19. Furthermore, authors in [20] used an ensemble of CNN models to classify COVID-19 and reported an AUC of 80.7%.

Lack of data has always been an issue for diagnosing COVID-19 from cough sounds. To address this, the algorithm proposed in [14] uses data augmentation methods such as pitch shift and time stretch to increase the quantity of data and improve the model's robustness. The algorithm improved the AUC-ROC from 72.23 to 87.07 on the first DiCOVA competition [16] dataset, winning first place in the competition. [2] discovered that each audio clip in the dataset might contain multiple cough sounds. As a result, they significantly increased the number of samples by segmenting the cough audios using root mean square energy (RMSE). The AI4COVID-19 [9] algorithm first trained CNN on cough detection datasets and then fine-tuned them on COVID-19 datasets.

To enhance the model's reliability, several authors have used various techniques. For instance, the AI4COVID-19 [9] algorithm uses three models that require a unanimous agreement for it to provide a definitive result, reducing the misdiagnosis probability. Authors in [12] simulated the effects of muscular degradation on audio samples. They performed the prediction on three different pre-trained models. The models extract audio features and concatenate them for classification, which helped them achieve an accuracy of 97.1% on a verified dataset.

## 3   Method

### 3.1   Cough Segmentation

To address issues such as poor generalisation and overfitting caused by the lack of data, audio samples are segmented, extracting single coughs as samples instead of whole audio files. This not only increases the number of samples but also removes useless sound segments, like noise and silence parts. Following the COUGHVID project [17], we also used RMSE to determine valid audio regions:

$$\text{RMSE} = \sqrt{\frac{1}{N}\sum_{i=1}^{N}(x_i - \bar{x})^2} \tag{1}$$

where $x_i$ denotes each sample value in audio signal, $\bar{x}$ represents the average value of all sample values, and $N$ represents the number of all sample values.

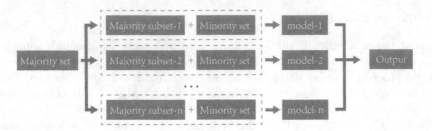

**Fig. 1.** The process of resampling majority set and ensemble models. The majority subset is divided into multiple subsets to match the minority subset to create multiple balanced datasets.

### 3.2 Data Augmentation

Data augmentation techniques are used to further enhance the model's generalisation capabilities. The algorithm generates a transformed sample for each original sample, with a variable rate between 0.7–1.40 and 0.9 probability of being stretched in time, 1 probability of pitch shifting (semitones ranging from −2 to 4), horizontal shift between −0.5 and 0.5 with 80% probability, trim and gain probability of 1, and 0.8 probability of polarity inversion. Note that data augmentation is performed on the training set only.

### 3.3 Split Majority Set and Ensemble Models

Common losses such as cross-entropy loss function do not consider the issue of dataset imbalance, which might result in a lack of generalisability and the model prioritising the majority class samples. To address these issues, a special method is proposed, an approach similar to [4]. The algorithm samples $n$ subsets from the majority data set; each subset has the same number of samples as the minority data set, and these subsets are combined with the minority set to form $n$ balanced data sets. Figure 1 illustrates the process.

The $n$ datasets are used to train $n$ independent models, and the results $y_i$ of the $n$ models are combined to generate the final output $y$:

$$y = \frac{1}{n} \sum_{i=0}^{n} y_i \tag{2}$$

It is worth noting that the segments in the test set may not come from the same audio source as those in the training set when splitting the training and test sets.

### 3.4 Model Architecture

The model architecture (Fig. 2) includes two convolutional layers and three fully connected layers, with cross-entropy as the loss function. In the network, dropout layers and batch normalisation layers are added to reduce overfitting. We implement the model using the Pytorch framework.

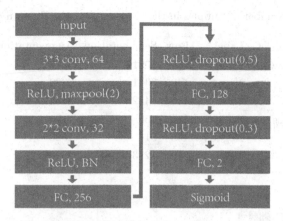

**Fig. 2.** An illustration of the network architecture. ReLU – Rectified Linear Unit, BN – Batch Normalisation, FC – Fully Connected.

### 3.5 Feature Selection

To ensure the effectiveness of the features, a forward selection method was used to screen the features. The specific steps are listed below:

1. Combine the k features to be adopted with the already adopted features.
2. Perform cross-validation to obtain the accuracy of the k combined features.
3. Compare the accuracy of the k combined features; if all are lower than the accuracy of the already adopted feature combination, stop selecting new features. Otherwise, select the to-be-adopted feature corresponding to the combination feature with the highest accuracy.
4. Return to the first step and iterate.

## 4   Experiments and Results

### 4.1   Dataset

This study primarily employs the Coswara dataset [21] and Russian dataset [6] to train and test the models. The Coswara dataset represents crowdsourced datasets and has been used in the second DiCOVA Challenge [22]. Each participant provided nine different kinds of sound samples in this dataset [3]. The Russian dataset, on the other hand, is a clinical dataset, implying that a portion of its data has been medically verified. However, the sample size is much smaller due to the higher cost of sample collection. The Russian dataset is a partially validated dataset, so we only take the validated data from the dataset, ensuring the labels are highly credible. The details of the datasets are provided in Table 1.

**Table 1.** Number of samples in the Coswara and the Russian datasets.

| Dataset | Label | Raw | Segmentation | Augmentation |
|---|---|---|---|---|
| Coswara | Positive | 589 | 1091 | 2182 |
| | Negative | 1405 | 3341 | 6682 |
| | Total | 1994 | 4432 | 8864 |
| Russian | Positive | 381 | 359 | 718 |
| | Negative | 438 | 914 | 1828 |
| | Total | 819 | 1273 | 2546 |

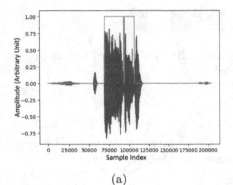
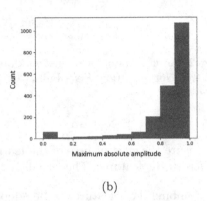

(a)　　　　　　　　　　　　　　　　　　　(b)

**Fig. 3.** (a) An example of segmenting an audio signal, where the segments marked by orange line are cropped as useful segments. (b) Histogram of the maximum absolute value of sounds in the Coswara dataset. (Color figure online)

## 4.2 Data Preprocessing

In the Russian dataset, some audio files are recorded in stereo channels. We retain the data from one of the channels based on the size of the RMSE of each channel's signal.

In the Coswara dataset, heavy cough data is chosen for diagnosing COVID-19 since coughing is a significant characteristic of such respiratory diseases, and heavier cough sounds can provide more detailed information. The samples labelled as 'healthy' are selected as COVID-negative samples, and samples labelled as 'positive mild' and 'positive moderate' are selected as COVID-positive samples. After data cleaning, the dataset consists of 589 positive samples and 1,405 negative samples, totalling 1,994 samples. The number of positive samples is very small, so resampling the dataset would reduce the total number of samples to only 1,198. Such a small number of data could easily result in inadequate generalisation capabilities of the model.

Figure 3a depicts an example of segmenting an audio signal. After the audio segmentation processing, only the coughing sound is retained, while noises and blank segments are cut off. The total number of data increases to more than twice the original, 4,432, with 1,091 positive samples and 3,341 negative samples. Finally, data augmentation techniques are applied to each audio segment, doubling the dataset size to a final count of 8,864 samples, which is 4.45 times the original. These processes help improve the model's generalisation ability.

Finally, non-repetitive resampling is conducted on the majority class (Negative), with the number of samples being the data quantity of the minority class (Positive). Since the negative sample is more than three times the positive sample quantity, it is possible to resample three groups of majority subsets equivalent to the minority quantity without repetition. The resampled majority subsets are combined with the minority set separately, forming three balanced datasets. Using three balanced datasets for training allowed us to produce comprehensive outcomes.

Because of diversified devices and formats, the maximum signal amplitude and sampling rates can vary significantly, especially in crowd-sourced datasets. To minimise such sample biases and other biases caused by external factors such as background noises, volume diversity, etc., the histogram of the maximum signal values is investigated. As can be seen in Fig. 3b, the maximum value of most audio is 1, but about 50% of the audio has a maximum value of less than 1. This can increase the convergence time of the model and may affect the final results. Therefore, resampling and normalisation are performed. In this study, sounds are resampled to 48,000 Hz and normalised to range from −1 to 1.

### 4.3   Feature Extraction

For each sample, multiple time-domain and frequency-domain features are extracted, including Mel spectrogram, MFCC, Chromagram, Centroid, Bandwidth, Flatness, Rolloff and Contrast.

All the features are extracted with the same frame length and hop length, thus concatenating all features as a matrix, namely a feature map. Given the human ear's sound length resolution is 20–50 ms, the frame length is set to 2,048, which is approximately 42 ms. The hop length is set to half the frame length, 1024, which allows overlap between frames to prevent the sound signal from being separated by the frames and the effective features from being extracted. To select the best set of MFCC features during feature selection, we used the Librosa Python package to extract 13, 26, and 39 coefficients as three types of features, which include MFCC features, MFCC differential features, and second-order differential features.

### 4.4   Data Analysis

First, we conduct PCA analysis on the original time signals, as shown in Fig. 4. It can be seen that the first principal component only accounts for 0.014 of the total proportion, and the boundary between positive and negative samples in the sample graph drawn with the first two components is also unclear. This

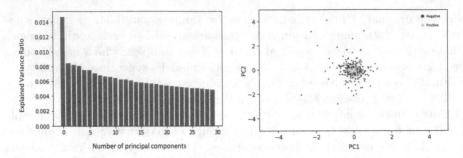

**Fig. 4.** PCA explained variance ratio (left) and first two principal components (right) of time signals in the Russian dataset. The purple points are negative samples, and the yellow points are positive samples. (Color figure online)

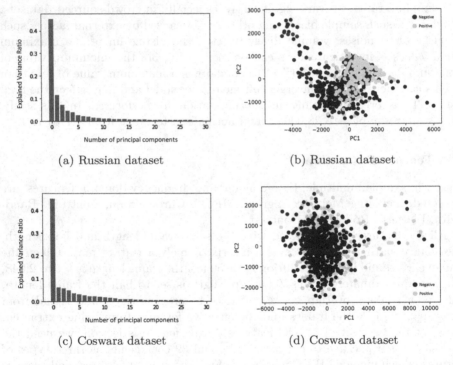

(a) Russian dataset

(b) Russian dataset

(c) Coswara dataset

(d) Coswara dataset

**Fig. 5.** PCA explained variance ratio (left) and first two principal components (right) of MFCC features. The purple and the yellow points refer to the negative and positive samples, respectively. (Color figure online)

suggests that it is difficult to obtain features suitable for classification from the time signal series or time domain features. All analyses were done using Python (version 3.8).

**Table 2.** 5-fold cross-validation accuracies with feature selection. Each entry represents the accuracy obtained using the selected features and a new feature. *Selected* means the feature is selected, and no more cross-validation is performed.

| Features | Classification accuracy | | | | |
|---|---|---|---|---|---|
| | Fold-1 | Fold-2 | Fold-3 | Fold-4 | Fold-5 |
| 39 MFCCs | 0.699 | *Selected* | *Selected* | *Selected* | *Selected* |
| 26 MFCCs | 0.683 | 0.687 | 0.698 | 0.705 | 0.716 |
| 13 MFCCs | 0.683 | 0.687 | 0.698 | 0.704 | 0.714 |
| Chromagram | 0.579 | 0.698 | 0.702 | 0.708 | 0.720 |
| Mel spectrogram | 0.601 | 0.663 | 0.673 | 0.681 | 0.792 |
| **Centroid** | - | 0.697 | 0.701 | **0.723** | *Selected* |
| **Bandwidth** | - | **0.705** | *Selected* | *Selected* | *Selected* |
| **Flatness** | - | 0.701 | **0.712** | *Selected* | *Selected* |
| Onset | - | 0.693 | 0.703 | 0.698 | 0.716 |
| ZCR | - | 0.685 | 0.702 | 0.710 | 0.718 |
| Rolloff | - | 0.700 | 0.701 | 0.709 | 0.711 |
| Contrast | 0.643 | 0.687 | 0.700 | 0.709 | 0.713 |

After the feature extraction, PCA performs quite differently. In Fig. 5 (a, b), the first two components account for over 0.5. In the sample location graph drawn using the first two components, the distribution of positive and negative samples is significantly different, with negative samples distributed more towards the left, and positive samples distributed more towards the right. This validates the effectiveness of MFCC and corroborates our research direction.

However, interestingly, the MFCC feature map extracted from the Coswara dataset does not perform the same, as shown in Fig. 5 (c, d). Although the first two components also account for over 0.5, it is difficult to see the difference in the distribution of positive and negative samples from the drawn position map. This could be due to greater noise in the Coswara dataset due to crowdsourcing. The distribution of sample points can be another reason. Despite the difficulty in differentiating, the successful COVID detection reflects the reliability of our model.

### 4.5   Feature Selection

The test accuracies in each feature selection step are recorded in Table 2. As Centroid, Bandwidth, Flatness, Onset, ZCR, and Rolloff provide less information and perform poorly as independent training features, only the remaining features were considered in the first step to optimising run time. Therefore, 39 MFCCs, centroid, bandwidth and flatness are selected as the optimal feature combination.

### 4.6   Train and Test Models

**Split Dataset.** As samples have been segmented, it is possible that different samples come from the same audio file, making samples in the training set and

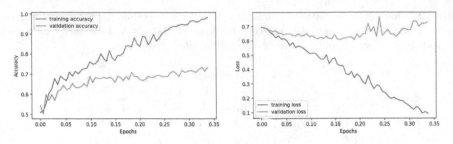

**Fig. 6.** Classification accuracy (left) and loss (right) during training.

test set similar to each other. To address this, the data are split specifically based on the source of the samples to ensure that samples from the same source are not distributed to different datasets.

**Hyperparameters.** We use Adam optimiser, whose learning rate is set at 1e-4 with $\beta$ at (0.9, 0.999). The batch size is set to 10, running 20 epochs each time to find the optimal early stopping scheme, which is usually less than 10 epochs.

**Cross Validation.** We use 5-fold cross-validation to evaluate the performance of the model. As shown in Fig. 6, this model shows serious overfitting in later epochs; we thus impose early stopping techniques to obtain the optimal model. Therefore, unlike ordinary cross-validation, this paper further divides each fold's training set into a training subset and a validation subset. The training subset is used to train the model, and the validation subset is used to decide the epoch number to stop training based on the validation result. At last, the final result of this fold is tested on the test set using the early-stopped model. Here, the position of early stopping is actually treated as a hyperparameter. Such a nested structure ensures that the choice of hyperparameters has no correlation with the test set.

### 4.7   Results

As shown in Table 3, the proposed model's test result shows an AUC-ROC of 80.0% and an accuracy of 72.3% on the Coswara dataset. As Coswara is a continuously updated crowdsourced dataset, articles using Coswara may not necessarily use the same dataset. Therefore, we compare with the results from the second DiCOVA challenge [22], which also used the Coswara dataset. The proposed model ranked 4th in all results, with a gap of 1.9% from the top AUC-ROC, indicating there is still room for improvement.

On the Russian dataset, our proposed model, **C3-PO**, achieves an AUC-ROC of 98.1%, and an accuracy of 92.7%, outperforming the Covid-Envelope model [8] by 22%.

**Table 3.** Performance comparison with other methods, including the top five teams from the DiCOVA challenge.

| Dataset | Reference | AUC-ROC | Accuracy |
|---------|-----------|---------|----------|
| Coswara | West Lab[†] | **0.819** | - |
|  | WhyNot[†] | 0.812 | - |
|  | USTCer[†] | 0.801 | - |
|  | Team SMILE[†] | 0.790 | - |
|  | ProPTIT[†] | 0.778 | - |
|  | DiCOVA Baseline [22] | 0.749 | - |
|  | **C3-PO** (Ours) | 0.800 | 0.723 |
| Russian | Covid-Envelope [8] | 0.890 | 0.683 |
|  | **C3-PO** (Ours) | **0.981** | **0.927** |

[†] Participants in the 2nd DiCOVA challenge [22]. Participants' results are available at https://competitions. codalab.org/competitions/34801#results (Track-2).

The difference in performance between the two datasets could be due to various factors. Firstly, the quality of the Coswara dataset is not as high as that of the Russian dataset; the voice information may be mixed with noise, and differences in recording devices and methods can also influence the recording results. Secondly, it is challenging to verify the annotation of crowdsourced datasets due to false information by crowdsource participants [23], which means that there is a possibility that the COVID status uploaded by users does not correspond with reality.

## 4.8 Ablation Study

Audio segmentation, data augmentation and the ensemble model are three primary methods used to handle small data volumes and data imbalance in our proposed **C3-PO** framework. To demonstrate the effectiveness of these three techniques, we perform ablation studies around them. By removing each technique one by one from the main workflow, we can see from the results the impact each technique has on model performance. The results recorded in Table 4 indicate that all three techniques contribute to a certain extent to the final performance of the model on both the Russian dataset and the Coswara dataset.

In the Coswara dataset, the model's accuracy is improved by 7.5% through these three techniques, whereas the improvement on the Russian dataset is only 2.2%. This is because the distribution difference between positive and negative samples in the Russian dataset is more significant. As we can see, the model can achieve a 90.5% accuracy rate without additional data preprocessing techniques, so the remaining room for improvement is limited. In contrast, the distribution of positive and negative sample data in the Coswara dataset is hard to distinguish linearly. More samples can help the model better identify the differences between

**Table 4.** Ablation study results. Three primary methods are added one by one in different experiments to find their significance.

| Data | Method | Exp. 1 | Exp. 2 | Exp. 3 | Exp. 4 |
|---|---|---|---|---|---|
| | Cough Segmentation | | ✓ | ✓ | ✓ |
| | Data Augmentation | | | ✓ | ✓ |
| | Ensemble Models | | | | ✓ |
| Russian | AUC-ROC | 0.959 | 0.970 | 0.975 | 0.981 |
| | Accuracy | 0.905 | 0.918 | 0.921 | 0.927 |
| Coswara | AUC-ROC | 0.731 | 0.762 | 0.781 | 0.800 |
| | Accuracy | 0.648 | 0.682 | 0.707 | 0.723 |

positive and negative samples. Notably, the most improvement was seen after the cough audio segmentation, making it the most effective method among the three.

## 5    Conclusion

This study proposes **C3-PO**, a novel framework to diagnose COVID-19 through cough sounds. Utilising data augmentation and segmentation techniques, the volume of data is increased to more than three times the original size to mitigate the impacts of a small dataset and imbalances within the dataset.

We tested the model on the crowdsourced Coswara dataset and the Russian dataset. On the Russian dataset, we achieved a state-of-the-art accuracy of 92.7% and an AUC-ROC of 98.1%, with 22% higher accuracy than the other work in the literature. On the Coswara dataset, we achieved an accuracy rate of 72.3% and an AUC-ROC of 80.0%, an AUC-ROC difference of 1.9% compared to the top-ranked models from the DiCOVA competition. Ablation studies are conducted to verify the influence of data augmentation, segmentation and ensemble models on the model's performance. The results indicate that all three methods positively contribute to the model's accuracy.

A limitation of this work is the use of crowdsourced datasets like Coswara, which run the risk of false or unverified data. A future direction for experimentation could be the application of transfer learning, transferring experience gained in other tasks, such as cough detection and speech recognition, into diagnosing COVID-19.

## References

1. Andreu-Perez, J., et al.: A generic deep learning based cough analysis system from clinically validated samples for point-of-need Covid-19 test and severity levels. IEEE Trans. Serv. Comput. **15**(3), 1220–1232 (2021). https://doi.org/10.1109/TSC.2021.3061402

2. Ashby, A.E., Meister, J.A., Soldar, G., Nguyen, K.A.: A novel cough audio segmentation framework for covid-19 detection. In: Proceedings of the Symposium on Open Data and Knowledge for a Post-Pandemic Era ODAK22, UK, pp. 1–8 (2022). https://doi.org/10.14236/ewic/ODAK22.1
3. Bhattacharya, D., et al.: Coswara: a respiratory sounds and symptoms dataset for remote screening of SARS-COV-2 infection. Sci. Data 10(1), 397 (2023)
4. Duan, H., Wei, Y., Liu, P., Yin, H.: A novel ensemble framework based on k-means and resampling for imbalanced data. Appl. Sci. 10(5), 1684 (2020)
5. Furtado, A., da Purificação, C.A.C., Badaró, R., Nascimento, E.G.S.: A light deep learning algorithm for CT diagnosis of COVID-19 Pneumonia. Diagnostics 12(7), 1527 (2022). https://doi.org/10.3390/diagnostics12071527
6. Geertsen, A., Chmelyuk, V.: Dataset of recordings of induced cough (Dec 2020). https://github.com/covid19-cough/dataset
7. Hasan, M.R., Hasan, M.M., Hossain, M.Z.: How many mel-frequency cepstral coefficients to be utilized in speech recognition? a study with the Bengali language. J. Eng. 2021(12), 817–827 (2021). https://doi.org/10.1049/tje2.12082
8. Hossain, M.Z., Uddin, M.B., Yang, Y., Ahmed, K.A.: CovidEnvelope: an automated fast approach to diagnose Covid-19 from cough signals. In: 2021 IEEE Asia-Pacific Conference on Computer Science and Data Engineering (CSDE), pp. 1–6. IEEE, Brisbane, Australia (Dec 2021). https://doi.org/10.1109/CSDE53843.2021.9718501
9. Imran, A., et al.: AI4COVID-19: AI enabled preliminary diagnosis for COVID-19 from cough samples via an app. Inform. Med. Unlocked 20, 100378 (2020). https://doi.org/10.1016/j.imu.2020.100378
10. Kassanla, S.H., Kassanib, P.H., Wooolowskic, M.J., Schmeldera, K.A., Detersa, R.: Automatic detection of coronavirus disease (Covid-19) in x-ray and CT images: a machine learning based approach. Biocybernet. Biomed. Eng. 41(3), 867–879 (2021)
11. Kranthi Kumar, L., Alphonse, P.: Covid-19 disease diagnosis with light-weight CNN using modified MFCC and enhanced GFCC from human respiratory sounds. Europ. Phys. J. Special Topics 231(18), 3329–3346 (2022)
12. Laguarta, J., Hueto, F., Subirana, B.: COVID-19 artificial intelligence diagnosis using only cough recordings. IEEE Open J. Eng. Med. Biol. 1, 275–281 (2020). https://doi.org/10.1109/OJEMB.2020.3026928
13. Liu, X., Hasan, M.R., Ahmed, K.A., Hossain, M.Z.: Machine learning to analyse omic-data for COVID-19 diagnosis and prognosis. BMC Bioinform. 24(7), 1–20 (2023). https://doi.org/10.1186/s12859-022-05127-6
14. Mahanta, S.K., Kaushik, D., Jain, S., Van Truong, H., Guha, K.: COVID-19 diagnosis from cough acoustics using convnets and data augmentation (May 2022). arXiv:2110.06123
15. Mohammed, E.A., Keyhani, M., Sanati-Nezhad, A., Hejazi, S.H., Far, B.H.: An ensemble learning approach to digital corona virus preliminary screening from cough sounds. Sci. Rep. 11(1), 15404 (2021)
16. Muguli, A., et al.: Dicova challenge: dataset, task, and baseline system for covid-19 diagnosis using acoustics. arXiv preprint arXiv:2103.09148 (2021). 10.48550/arXiv.2103.09148
17. Orlandic, L., Teijeiro, T., Atienza, D.: The coughvid crowdsourcing dataset, a corpus for the study of large-scale cough analysis algorithms. Sci. Data 8(1), 156 (2021). https://doi.org/10.1038/s41597-021-00937-4

18. Pramono, R.X.A., Imtiaz, S.A., Rodriguez-Villegas, E.: A cough-based algorithm for automatic diagnosis of pertussis. PLoS ONE **11**(9), e0162128 (2016). https://doi.org/10.1371/journal.pone.0162128

19. Richards, R.: Evidence on the accuracy of the number of reported covid-19 infections and deaths in lower-middle income countries. K4D Helpdesk Report 856 (2020). https://opendocs.ids.ac.uk/opendocs/handle/20.500.12413/15576

20. Schuller, B.W., Coppock, H., Gaskell, A.: Detecting covid-19 from breathing and coughing sounds using deep neural networks. arXiv preprint arXiv:2012.14553 (2020)

21. Sharma, N., et al.: Coswara - a database of breathing, cough, and voice sounds for Covid-19 diagnosis. In: Interspeech 2020, pp. 4811–4815 (Oct 2020). https://doi.org/10.21437/Interspeech.2020–2768

22. Sharma, N.K., Chetupalli, S.R., Bhattacharya, D., Dutta, D., Mote, P., Ganapathy, S.: The second dicova challenge: dataset and performance analysis for diagnosis of covid-19 using acoustics. In: ICASSP 2022–2022 IEEE International Conference on Acoustics, Speech and Signal Processing (ICASSP), pp. 556–560 (2022). https://doi.org/10.1109/ICASSP43922.2022.9747188

23. Sheehan, K.B.: Crowdsourcing research: data collection with amazon's mechanical turk. Commun. Monogr. **85**(1), 140–156 (2018). https://doi.org/10.1080/03637751.2017.1342043

24. Tian, S., Hu, W., Niu, L., Liu, H., Xu, H., Xiao, S.Y.: Pulmonary pathology of early-phase 2019 novel coronavirus (covid-19) pneumonia in two patients with Lung Cancer. J. Thorac. Oncol. **15**(5), 700–704 (2020). https://doi.org/10.1016/j.jtho.2020.02.010

25. World Health Organization: WHO coronavirus (COVID-19) dashboard (2023). https://covid19.who.int/

# Pseudo-label Based Unsupervised Momentum Representation Learning for Multi-domain Image Retrieval

Mingyuan Ge, Jianan Shui, Junyu Chen, and Mingyong Li[✉]

College of Computer and Information Science, Chongqing Normal University,
Chongqing, China
limingyong@cqnu.edu.cn

**Abstract.** Although many current cross-domain image retrieval researches have made good progress, most of the works is targeted at specific domains. At the same time, we also noticed that many works are based on manually annotated images. In this paper, in order to solve the above problems, we propose a new paradigm applied to multi-domain image retrieval. The specific solution is as follows: 1) The momentum contrastive mechanism is used to deeply mine the semantic features shared by images of the same category in the same domain. 2) Pseudo-labels are learned through clustering learning to provide supervised information for contrastive learning. 3) By aligning images of the same category in different domains, the aggregation of common features across multiple domains is achieved. Experiments on multi-domain datasets demonstrate the feasibility of our idea. Code will be made public in https://github.com/DannielGe/PUMR.

**Keywords:** Pseudo-label · Multi-domain · Image Retrieval

## 1 Introduction

As multi-visual domain images can be seen everywhere on the Internet, users have increasingly urgent needs for cross-domain image retrieval. It is more valuable and has more application prospects than same-domain image retrieval, because users can use any image to retrieve similar object images across visual domains. Therefore, Cross-domain image retrieval using images in one visual domain to find the same objects in another visual domain has become a hot research topic today. Specifically, cross-domain image retrieval has developed many branches, such as sketch-based image retrieval [15,30], clothing retrieval [1] and fashion image retrieval [22,27], etc.

However, many previous works have two obvious flaws. First, many works only retrieve other domains based on a certain fixed domain, which cannot be well extended to multiple different visual domains. Second, many works need to annotate existing samples with class labels or match information across domains. These limitations seriously reduce the practical application value of these works.

© The Author(s), under exclusive license to Springer Nature Switzerland AG 2024
S. Rudinac et al. (Eds.): MMM 2024, LNCS 14556, pp. 369–380, 2024.
https://doi.org/10.1007/978-3-031-53311-2_27

As far as we know, the reason why the retrieval cannot be extended to multiple domains is that the model does not effectively learn the unique characteristics of the visual domain. Therefore, we design a strategy based on momentum contrastive learning, as shown in Fig. 1. The purpose is not only to extract the feature information of the image, but also to extract the domain information due to the distinguishing feature.

In order to achieve label-free cross-domain retrieval, as shown in Fig. 2 we design a contrastive learning based on pseudo-labels. It first clusters the data in the domain to form pseudo-labels, replacing the original manual annotation category labels. Secondly, a cross-domain distance designed based on it can help achieve cross-domain alignment.

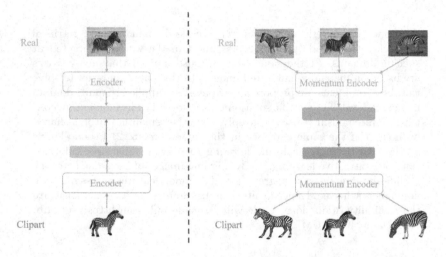

**Fig. 1.** Motivation of our work. The left side represents the traditional feature extraction method, and the encoder is generally based on CNN [8] or ViT [2]. The part on the right represents a momentum contrast feature extraction method designed by us. The momentum encoder here is built based on CNN [8]. But the core of this method is the contrastive learning based on a dynamic dictionary, which contains all samples of the same domain and the same class. The goal of this method is to learn more discriminative feature representations.

The main contributions of this paper are the following.

1. The momentum contrast mechanism is designed to preserve as much as possible the same-domain features of same class images.
2. In-domain clustering was employed to obtain pseudo-labels for each category of images. The feature representation was converted into a probabilistic representation between the samples and the pseudo-labels to create the conditions for unsupervised learning.
3. Experiments on a multi-domain dataset to demonstrate that our proposed model can perform effective retrieval not only between two domains, but also between multiple domains.

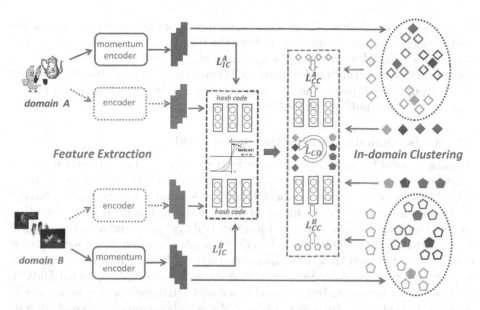

**Fig. 2.** Architecture of PUMR. The feature extraction part on the left uses a momentum contrast mechanism to learn more discriminative feature representations. The in-domain clustering on the right is used to obtain pseudo labels (i.e., the solid regular quadrilateral on the right in the figure).

## 2   Related Work

### 2.1   Domain Adaption

To bridge the domain gap, many researchers and scholars have adopted different strategies. Song et al. [21] considered fine-grained details and introduced a spatial attention module to measure the feature differences between cross-domain elements by introducing HOLEF loss between domains. However, due to the high time and labour costs of manual annotation, it is resource intensive and extremely difficult to apply it to complex environments, so the more challenging area of unsupervised cross-domain image retrieval has been given more attention. In unsupervised cross-domain image retrieval, the problem of inconsistent distribution across different domains is usually solved by Unsupervised Domain Adaptation (UDA), whose goal is to transfer knowledge learned from a source domain with fully annotated training samples to a target domain with only unannotated data. MK-MMD [17] is commonly used in early UDA approaches to measure the difference between the source and target domains. The emergence of Generative Adversarial Networks (GAN) [7] inspired several approaches [5,23]. Hoffman et al. [10] proposed to perform domain alignment in feature space, which breaks the domain barrier at the feature level or pixel level. In recent years, more researches [4,13,20] have focused more on fine-grained category-level label distribution alignment, implemented as an adversarial approach with two-domain classifiers and feature extractors. This approach differs from coarse-grained alignment by aligning each category distribution between the source and target domain data by pushing the target samples to the distribution of the source samples in each category. Xu et al. [26] proposed a weight-sharing three-branch transformation framework CDTrans (cross-domain transformer) for exact unsupervised domain adaptation, using this framework has shown better performance on a public DomainNet [19] dataset.

### 2.2   Cross-Domain Hashing Image Retrieval

Hashing is one of the most popular and effective techniques in the field of retrieval due to its fast query speed and low memory cost [24]. Hashing techniques convert high-dimensional data into compact binary codes and generate similar binary codes for similar data items. In recent years, hashing has been widely used in the field of cross-domain retrieval. Liu et al. [16] first proposed a method called Deep Sketch Hashing (DSH) for solving the complex continuous-valued distance computation problem that arises in image retrieval from hand-drawn sketches. Luo et al. [18] proposed a deep multi-task cross-domain hashing called DMCH for both cross-domain embedding and sequential attribute learning is modelled. Sequential attribute learning not only provides semantic guidance for embedding, but also generates rich attention to discriminative local details of clothing items without additional labels. Du et al. [3] proposed a novel regularised adversarial domain adaptive hashing method (R-ADAH) for cross-domain palmprint recognition based on deep hash networks. This method trains the hash network only on the source domain, and by adversarial training, the target network becomes adapted to unlabelled images and reduces the sensitivity of hyper-parameters when only domain-specific labels are provided. Zhang et al. [29] proposed a transferable hashing method called semantic guided hashing learning (SGHL), which addresses the problem of differences in data distribution between source and target domains as well as semantic differences between features and labels. While most of the above hashing methods are used between two domains, our proposed method can be used for retrieval between multiple domains, greatly improving the generalisability of hashing retrieval methods.

## 3   Proposed PUMR Method

### 3.1   Problem Setup

The essence of the cross-domain retrieval task is to retrieve an image of the same class in another domain $B$ with an image $I$ in one domain $A$. This is the same as the cross-domain retrieval task. It is assumed that data from multiple domains make up the dataset $D_{train} = \{D^*\}_{*=A,B}$ used for training. $*$ represents the domain name (Clipart, Real, etc.), we don't actually count the number of domains and their names, so we will use $A$ and $B$ to represent them. Assuming that there are $M$ and $N$ samples in domains $A$ and $B$, respectively, $D^A = \{I_i^A\}_{i=1}^M$, $D^B = \{I_j^B\}_{j=1}^N$. We need to train a mapping function $f$ to map the image $I^*$ to the same common space $O$, $O = f(I^*, \theta)$. If $I^A$ and $I^B$ belong to the same class, then $O^A$ and $O^B$ should be as identical as possible.

## 3.2   Feature Extraction with Momentum Contrast Mechanism

Both encoder and momentum encoder use the same architecture based on Resnet [9]. For the image $I$, it is fed into the momentum encoder after the enhancement operation to obtain the feature embedding $O_k$. The original image is input into the encoder to obtain the feature embedding $O_q$. For convenience of writing we have omitted the subscript $i$ and $i$ here. In unsupervised learning, there is no paired label information between $O_q$ and $O_k$. Therefore, we adopted the method of contrastive learning to extract effective features. In other words, $O_k$ is a dynamic queue container. In the initial state, it contains all pictures in this domain. When $O_k$ matches $O_q$, it is recorded as $O_{k+}$, and when the two do not match, $O_k$ is discarded out of the container. Therefore, the intra-domain contrastive learning loss can be expressed as:

$$L_{IC}^A = -log \frac{\exp\left(O_q^A O_{k+}^A / \tau\right)}{\sum_{i=1}^M \exp\left(O_q^A O_{k\,i}^A / \tau\right)} \tag{1}$$

$$L_{IC}^B = -log \frac{\exp\left(O_q^B O_{k+}^B / \tau\right)}{\sum_{j=1}^N \exp\left(O_q^B O_{k\,j}^B / \tau\right)} \tag{2}$$

where $\tau$ is a temperature hyper-parameter.

Notably, we utilize the model parameter $\theta_q$ of the encoder to update the model parameter $\theta_k$ of the momentum encoder. The main purpose is to make $\theta_k$ better smoothed. At the same time, only $\theta_q$ is involved in the backpropagation in order to reduce the number of parameters of the model.

$$\theta_k \leftarrow m\theta_k + (1 - m)\theta_q \tag{3}$$

Here $m \in [0, 1)$ is a momentum coefficient. The default setting for this is 0.999.

In order to better pull apart the mismatched $O_q$ and $O_k$ and bring closer the matched $O_q$ and $O_k$. We also utilize the mapping ability of the tanh function.

$$O_q = tanh(O_q) \tag{4}$$

## 3.3   Pseudo-label Based Contrastive Learning

We use $O_k$ as a feature for clustering. K-means [6] are applied on all image features $O_k{}_{i=1}^M$ in a single domain to obtain its $K$ cluster centers. Based on the K-means result, each sample $I_i$ is assigned with pseudo-label $y_i$. All pseudo-labels are updated after each period.

Samples from the same clusters as the query are used to form pairs of positive examples. Thus, the feature extractor is trained to bring features within the same cluster closer together while pushing different clusters apart. Specifically, the loss function for pseudo-label based Contrastive learning is given by the following equation:

$$L_{CC} = -\sum_{i \in \mathbf{I}} \frac{1}{|\mathbf{P}(i)|} \sum_{p \in \mathbf{P}(i)} log \frac{\exp\left(O_{qi}O_{kp}/\tau\right)}{\sum_{a \in \mathbf{I}} \exp\left(O_{qi}O_{ka}/\tau\right)} \qquad (5)$$

where $\mathbf{I}$ denotes the index of all samples in the same domain, and $\mathbf{P}(i)$ denotes the index of a set of samples belonging to the same cluster as $I_i$, i.e., $\mathbf{P}(i) = \{p \in \mathbf{I} : y_p = y_i\}$, $|\mathbf{P}(i)|$ is its cardinality (that is, the number of elements in that set). $O_{ka}/O_{qa}$ denote the feature representation of the $a$-th image after passing through the momentum encoder/encoder, respectively.

In addition, the clustering performed separately in their respective domains makes the class semantic features acquired by the model lack domain information, which makes the correspondence between domains ambiguous. Based on this, we design a cross-domain alignment loss function. The basic idea is to utilize the probability distribution results obtained from K-means clustering to replace the original image features to calculate the intra-domain distance and thus the cross-domain distance. The inter-domain divide is bridged by reducing the cross-domain distance through the cross-domain distance.

Given an input image $I_i$, we use the clustering center $\{C_x^A\}_{x=1}^K$, $\{C_x^B\}_{x=1}^K$ to calculate its clustering probability.

$$p_i^x = -\frac{\exp\left(O_q C_x/\gamma\right)}{\sum_{k=1}^K \exp\left(O_q C_k/\gamma\right)} \qquad (6)$$

where $\gamma$ is the temperature hyperparameter. $C_x$ denotes the centroid of $\gamma$-th cluster.

The clustering probability of $I_i$ can be expressed as $\mathbf{p}_i = [p_i^1, p_i^2, ..., p_i^K]$. $p_i^K$ represents the probability $I_i$ belong to $K$-th class. Therefore, the in-domain distance between $I_i$ and $I_j$ can be calculated using cosine distance.

$$d_{ij}^A = cosine < \mathbf{p}_i^A, \mathbf{p}_j^A > \qquad (7)$$

With order-invariant intra-domain distances, we devise a new cross-domain distance.

$$cd_{ij} = CD(d_{ij}^A, d_{ij}^B) \qquad (8)$$

Where $CD$ denotes the L2 distance between two in-domain distances.

For two well-aligned domains $A$ and $B$, $dd_{ij}$ is small because the centroids of the two domains are similar. However, when there is a large difference between the feature distributions of domains $A$ and $B$, the corresponding centroids become different. Therefore, the values of $d_{ij}^A$ and $d_{ij}^B$ will be very different, resulting in a large $dd_{ij}$. Eventually, cross domain alignment loss is written as:

$$L_{CD} = \sum_{i \in \mathbf{R}^A} \sum_{j \in \mathbf{R}^A} cd_{ij} + \sum_{i \in \mathbf{R}^B} \sum_{j \in \mathbf{R}^B} cd_{ij} \qquad (9)$$

Domain $A$ and domain $B$ are used here separately because both domains $A$ and $B$ need to be aligned. That is, aligning domains, instead of aligning between two domains, centrally aligns all the samples of a particular domain.

Finally, the objective function of the whole model is:

$$Loss = L_{IC}^A + L_{IC}^B + L_{CC}^A + L_{CC}^B + L_{CD} \tag{10}$$

## 4  Experiments

### 4.1  Dataset

DomainNet [19] contains six domains, each containing 345 categories of common objects. These domains include Clipart: a collection of clipart images; Infograph: graphical images with object-specific information; Painting: artistic depictions of objects in the form of drawings; Quickdraw: drawings for gamers around the world "Quick Draw!"; Real: photographs and real world images; Sketch: sketches of specific objects. Images from the Clipart, Infograph, Painting, Real and Sketch domains were collected by searching for category names in combination with domain names (e.g. "aircraft Painting") in different image search engines.

We measure the accuracy of the first 50/100/200 retrieved images to provide a more comprehensive evaluation of DomainNet's retrieval accuracy. Since our task is category-level cross-domain retrieval, retrieved images of the same semantic class as the query are considered correct.

### 4.2  Results Analysis

**Table 1.** Results retrieved in domains Clipart and Sketch.

| Method | Clipart → Sketch | | | Sketch → Clipart | | |
|---|---|---|---|---|---|---|
| | P@50 | P@100 | P@200 | P@50 | P@100 | P@200 |
| ID [25] | 49.46 | 46.09 | 40.44 | 54.38 | 47.12 | 37.73 |
| ProtoNCE [14] | 46.85 | 42.67 | 36.35 | 54.52 | 45.04 | 35.06 |
| CDS [12] | 45.84 | 42.37 | 37.16 | 59.13 | 48.83 | 37.4 |
| PCS [28] | 51.01 | 46.87 | 40.19 | 59.7 | 50.67 | 39.38 |
| UCDIR [11] | 56.31 | 52.74 | 47.38 | 63.07 | 57.26 | 48.17 |
| Ours | **61.28** | **57.84** | **52.57** | **71.24** | **65.28** | **55.24** |

Table 1, 2, 3, 4, 5 and 6 reports precision@k (P@k) on all cross-domain settings from cross-domain retrievals. The retrieval performance in Table 1, 2, 3, 4, 5 and 6 shows that:

1. UCDIR is the strongest baseline in our unsupervised cross-domain image retrieval task.
2. Our approach achieves the highest retrieval accuracy for almost all 12 retrieval tasks.

**Table 2.** Results retrieved in domains Infograph and Sketch.

| Method | Infograph → Sketch | | | Sketch → Infograph | | |
|---|---|---|---|---|---|---|
| | P@50 | P@100 | P@200 | P@50 | P@100 | P@200 |
| ID [25] | 30.35 | 29.04 | 26.55 | 42.20 | 34.94 | 27.52 |
| ProtoNCE [14] | 28.24 | 26.79 | 24.23 | 39.83 | 31.99 | 24.77 |
| CDS [12] | 30.55 | 29.51 | 27.00 | 46.27 | 36.11 | 27.33 |
| PCS [28] | 30.27 | 28.36 | 25.35 | 42.58 | 34.09 | 25.91 |
| UCDIR [11] | 31.29 | 29.33 | 26.54 | 43.66 | 36.14 | 28.12 |
| Ours | **32.58** | **31.51** | **29.51** | **53.11** | **43.59** | **33.79** |

3. The Quickdraw domain in DomainNet only contains some simple strokes, the Infograph domain in DomainNet contains too much character information, which leads to the largest domain gap. Our method perform less well on Painting → Quickdraw, Quickdraw → Real and Infograph → Real retrieval.
4. This method improves the retrieval of Painting → Quickdraw most significantly. Compared with the best baseline UCDIR, our P@50 is 9.45% higher.
5. Among all 12 retrieval pairs, our method performs the best for Sketch → Clipart retrieval, achieving 71.24% for P@50.
6. As shown in Fig. 3, the retrieval performance of our proposed method is higher than the UCDIR, especially in the top5 retrieval results.

**Table 3.** Results retrieved in domains Infograph and Real.

| Method | Infograph → Real | | | Real → Infograph | | |
|---|---|---|---|---|---|---|
| | P@50 | P@100 | P@200 | P@50 | P@100 | P@200 |
| ID [25] | 28.27 | 27.44 | 26.33 | 39.98 | 31.77 | 24.84 |
| ProtoNCE [14] | 28.41 | 28.53 | 28.5 | 57.01 | 41.84 | 30.33 |
| CDS [12] | 28.51 | 27.92 | 27.48 | 56.69 | 39.76 | 26.38 |
| PCS [28] | 30.56 | 30.27 | 29.68 | 55.42 | 42.13 | 30.76 |
| UCDIR [11] | 35.52 | 35.24 | 34.35 | 57.74 | 46.69 | 35.47 |
| Ours | **35.55** | **35.46** | **35.42** | **65.89** | **51.41** | **36.54** |

**Table 4.** Results retrieved in domains Clipart and Painting.

| Method | Painting → Clipart | | | Clipart → Painting | | |
|---|---|---|---|---|---|---|
| | P@50 | P@100 | P@200 | P@50 | P@100 | P@200 |
| ID [25] | 64.67 | 54.41 | 40.07 | 42.37 | 39.61 | 35.56 |
| ProtoNCE [14] | 55.44 | 43.74 | 32.59 | 39.13 | 35.87 | 32.07 |
| CDS [12] | 63.15 | 47.30 | 30.93 | 37.75 | 35.18 | 32.76 |
| PCS [28] | 63.47 | 53.21 | 41.68 | 48.83 | 46.21 | 42.1 |
| UCDIR [11] | 66.42 | 56.84 | 46.72 | 52.58 | 50.10 | 46.11 |
| Ours | **74.51** | **68.24** | **57.09** | **53.92** | **51.95** | **49.66** |

**Table 5.** Results retrieved in domains Painting and Quickdraw.

| Method | Painting → Quickdraw | | | Quickdraw → Painting | | |
|---|---|---|---|---|---|---|
| | P@50 | P@100 | P@200 | P@50 | P@100 | P@200 |
| ID [25] | 20.34 | 19.59 | 18.79 | 21.12 | 19.81 | 18.48 |
| ProtoNCE [14] | 21.63 | 21.24 | 20.56 | 23.95 | 22.84 | 21.56 |
| CDS [12] | 18.75 | 18.89 | 17.88 | 21.37 | 21.44 | 19.46 |
| PCS [28] | 25.12 | 24.65 | 23.8 | 24.03 | 23.24 | 22.13 |
| UCDIR [11] | 39.72 | 38.59 | 37.63 | 33.45 | 33.81 | 34.29 |
| Ours | **39.86** | **38.86** | **38.21** | **37.07** | **36.62** | **35.62** |

**Table 6.** Results retrieved in domains Quickdraw and Real.

| Method | Quickdraw → Real | | | Real → Quickdraw | | |
|---|---|---|---|---|---|---|
| | P@50 | P@100 | P@200 | P@50 | P@100 | P@200 |
| ID [25] | 28.27 | 27.46 | 26.32 | 123.45 | 22.79 | 22.01 |
| ProtoNCE [14] | 26.38 | 25.7 | 24.45 | 25.1 | 24.81 | 23.78 |
| CDS [12] | 19.28 | 19.14 | 18.67 | 15.36 | 15.57 | 15.82 |
| PCS [28] | 34.82 | 33.92 | 31.73 | 28.98 | 28.85 | 28.16 |
| UCDIR [11] | 42.79 | 42.75 | 42.70 | 41.90 | 42.10 | 41.59 |
| Ours | **43.19** | **43.11** | **42.71** | **45.96** | **44.81** | **42.56** |

## 4.3 Ablation Study

Model1 represents a model without a momentum comparison mechanism, Model2 represents a model without pseudo-labels as supervisory information, and Model3 represents a model without cross-domain alignment loss. We can draw the following conclusions.

1. By observing at the data of Model1 it is clear that the momentum contrast mechanism plays an important role in the whole model.
2. All three modules enhance and contribute to the overall model.

**Table 7.** The results of each module ablation experiment.

| Model | Clipart → Sketch | | | Sketch → Clipart | | |
|---|---|---|---|---|---|---|
| | P@50 | P@100 | P@200 | P@50 | P@100 | P@200 |
| Model1 | 17.96 | 17.47 | 16.84 | 17.19 | 16.79 | 16.37 |
| Model2 | 55.39 | 52.42 | 47.13 | 66.82 | 60.27 | 49.57 |
| Model3 | 60.42 | 57.45 | 52.44 | 68.86 | 63.56 | 54.61 |
| Ours | 61.28 | 57.84 | 52.57 | 71.24 | 65.28 | 55.24 |

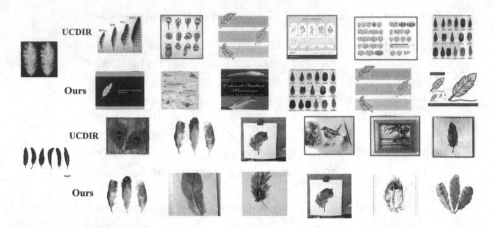

**Fig. 3.** Top 5 visualization results of PUMR and UCDIR on DomainNet. The upper part visualizes the retrieval results of the Infograph domain with image in the Real domain as the query object. The bottom part visualizes the retrieval results of the Painting domain using the Clipart domain image as the query object.

## 5   Conclusion

In this paper, we propose an unsupervised cross-domain image retrieval method that no longer uses paired cross-domain labels and has better practical value. In order to obtain image features with more domain information without losing the category features of the image, we design a momentum encoder and use the strategy of momentum contrastive learning to achieve the desired goal. In addition, we use these centroids obtained by clustering as pseudo labels for comparative learning of category features. Most importantly, we design a cross-domain distance based on these pseudo-labels to effectively measure and minimize the difference between domains to achieve alignment between domains. The ablation study verified the effectiveness of the momentum contrastive learning and pseudo-label-based contrastive learning we introduced.

**Acknowledge.** This work was supported by Humanities and social science research project of Chongqing Municipal Education Commission (22SKGH100).

## References

1. Alirezazadeh, P., Dornaika, F., Moujahid, A.: Deep learning with discriminative margin loss for cross-domain consumer-to-shop clothes retrieval. Sensors **22**(7), 2660 (2022)
2. Dosovitskiy, A., et al.: An image is worth 16x16 words: transformers for image recognition at scale. arXiv preprint arXiv:2010.11929 (2020)
3. Du, X., Zhong, D., Shao, H.: Cross-domain palmprint recognition via regularized adversarial domain adaptive hashing. IEEE Trans. Circuits Syst. Video Technol. **31**(6), 2372–2385 (2020)

4. Du, Z., Li, J., Su, H., Zhu, L., Lu, K.: Cross-domain gradient discrepancy minimization for unsupervised domain adaptation. In: Proceedings of the IEEE/CVF Conference on Computer Vision and Pattern Recognition, pp. 3937–3946 (2021)
5. Ganin, Y., et al.: Domain-adversarial training of neural networks. J. Mach. Learn. Res. **17**(1), 1–35 (2016)
6. Gao, B., Yang, Y., Gouk, H., Hospedales, T.M.: Deep clusteringwith concrete K-means. In: ICASSP 2020–2020 IEEE International Conference on Acoustics, Speech and Signal Processing (ICASSP), pp. 4252–4256. IEEE (2020)
7. Goodfellow, I., et al.: Generative adversarial networks. Commun. ACM **63**(11), 139–144 (2020)
8. He, K., Fan, H., Wu, Y., Xie, S., Girshick, R.: Momentum contrast for unsupervised visual representation learning. In: Proceedings of the IEEE/CVF Conference on Computer Vision and Pattern Recognition, pp. 9729–9738 (2020)
9. He, K., Zhang, X., Ren, S., Sun, J.: Deep residual learning for image recognition. In: Proceedings of the IEEE Conference on Computer Vision and Pattern Recognition, pp. 770–778 (2016)
10. Hoffman, J., et al.: CyCADA: cycle-consistent adversarial domain adaptation. In: International Conference on Machine Learning, pp. 1989–1998. PMLR (2018)
11. Hu, C., Lee, G.H.: Feature representation learning for unsupervised cross-domain image retrieval. In: Avidan, S., Brostow, G., Cissé, M., Farinella, G.M., Hassner, T. (eds.) Computer Vision-ECCV 2022: 17th European Conference, Tel Aviv, Israel, 23–27 October 2022, Proceedings, Part XXXVII, pp. 529–544. Springer, Cham (2022). https://doi.org/10.1007/978-3-031-19836-6_30
12. Kim, D., Saito, K., Oh, T.H., Plummer, B.A., Sclaroff, S., Saenko, K.: CDS: cross domain self-supervised pre-training. In: Proceedings of the IEEE/CVF International Conference on Computer Vision, pp. 9123–9132 (2021)
13. Li, J., Li, G., Shi, Y., Yu, Y.: Cross-domain adaptive clustering for semi-supervised domain adaptation. In: Proceedings of the IEEE/CVF Conference on Computer Vision and Pattern Recognition, pp. 2505–2514 (2021)
14. Li, J., Zhou, P., Xiong, C., Hoi, S.C.: Prototypical contrastive learning of unsupervised representations. arXiv preprint arXiv:2005.04966 (2020)
15. Lin, F., Li, M., Li, D., Hospedales, T., Song, Y.Z., Qi, Y.: Zero-shot everything sketch-based image retrieval, and in explainable style. In: Proceedings of the IEEE/CVF Conference on Computer Vision and Pattern Recognition, pp. 23349–23358 (2023)
16. Liu, L., Shen, F., Shen, Y., Liu, X., Shao, L.: Deep sketch hashing: fast free-hand sketch-based image retrieval. In: Proceedings of the IEEE Conference on Computer Vision and Pattern Recognition, pp. 2862–2871 (2017)
17. Long, M., Cao, Y., Wang, J., Jordan, M.: Learning transferable features with deep adaptation networks. In: International Conference on Machine Learning, pp. 97–105. PMLR (2015)
18. Luo, Y., Wang, Z., Huang, Z., Yang, Y., Lu, H.: Snap and find: deep discrete cross-domain garment image retrieval. arXiv preprint arXiv:1904.02887 (2019)
19. Peng, X., Bai, Q., Xia, X., Huang, Z., Saenko, K., Wang, B.: Moment matching for multi-source domain adaptation. In: Proceedings of the IEEE/CVF International Conference on Computer Vision, pp. 1406–1415 (2019)
20. Saito, K., Watanabe, K., Ushiku, Y., Harada, T.: Maximum classifier discrepancy for unsupervised domain adaptation. In: Proceedings of the IEEE Conference on Computer Vision and Pattern Recognition, pp. 3723–3732 (2018)

21. Song, J., Yu, Q., Song, Y.Z., Xiang, T., Hospedales, T.M.: Deep spatial-semantic attention for fine-grained sketch-based image retrieval. In: Proceedings of the IEEE International Conference on Computer Vision, pp. 5551–5560 (2017)

22. Tian, Y., Newsam, S., Boakye, K.: Fashion image retrieval with text feedback by additive attention compositional learning. In: Proceedings of the IEEE/CVF Winter Conference on Applications of Computer Vision, pp. 1011–1021 (2023)

23. Tzeng, E., Hoffman, J., Saenko, K., Darrell, T.: Adversarial discriminative domain adaptation. In: Proceedings of the IEEE Conference on Computer Vision and Pattern Recognition, pp. 7167–7176 (2017)

24. Wang, J., Shen, H.T., Song, J., Ji, J.: Hashing for similarity search: a survey. arXiv preprint arXiv:1408.2927 (2014)

25. Wu, Z., Xiong, Y., Yu, S.X., Lin, D.: Unsupervised feature learning via non-parametric instance discrimination. In: Proceedings of the IEEE Conference on Computer Vision and Pattern Recognition, pp. 3733–3742 (2018)

26. Xu, T., Chen, W., Wang, P., Wang, F., Li, H., Jin, R.: CDTrans: cross-domain transformer for unsupervised domain adaptation. arXiv preprint arXiv:2109.06165 (2021)

27. Yan, C., Yan, K., Zhang, Y., Wan, Y., Zhu, D.: Attribute-guided fashion image retrieval by iterative similarity learning. In: 2022 IEEE International Conference on Multimedia and Expo (ICME), pp. 1–6. IEEE (2022)

28. Yue, X., et al.: Prototypical cross-domain self-supervised learning for few-shot unsupervised domain adaptation. In: Proceedings of the IEEE/CVF Conference on Computer Vision and Pattern Recognition, pp. 13834–13844 (2021)

29. Zhang, W., Yang, X., Teng, S., Wu, N.: Semantic-guided hashing learning for domain adaptive retrieval. World Wide Web, 1–20 (2022)

30. Zhao, H., Liu, M., Li, M.: Feature fusion and metric learning network for zero-shot sketch-based image retrieval. Entropy **25**(3), 502 (2023)

# DFGait: Decomposition Fusion Representation Learning for Multimodal Gait Recognition

Jianbo Xiong, Shinan Zou, and Jin Tang[✉]

School of Automation, Central South University, Changsha, China
tjin@csu.edu.cn

**Abstract.** Multimodal gait recognition aims to utilize various gait modalities for identity recognition. Previous methods have focused on designing complex fusion techniques. However, the heterogeneity between modalities has negatively impacted recognition tasks due to distributional differences and information redundancy. Inspired by this, we have proposed a novel feature decomposition fusion (DFGait) network, combining silhouette and skeleton data. The network learns modality-shared and modality-specific feature representations for both modalities and introduces inter-modality regularization loss and intra-modality regularization loss to encourage the preservation of common and unique information between modalities, reducing modality gaps and information redundancy. Furthermore, the representations mentioned above are embedded in their own space during learning, making the fusion process challenging. Therefore, we have proposed an adversarial modality alignment learning strategy, guiding the alignment of the two modality features through the confusion of the modality discriminator to achieve maximized modality information interaction. Finally, a separable fusion module is introduced to fuse the features of the two modalities, resulting in a comprehensive gait representation. Experimental results demonstrate that our DFGait achieves state-of-the-art performance on popular gait datasets, with rank-1 accuracy of 50.30% for Gait3D and 61.42% for GREW. The source code can be obtained from https://github.com/BoyeXiong/DFGait.

**Keywords:** Biometric · Multimodal Gait recognition · Decomposition representation learning · Modality alignment · Multimodal fusion

## 1 Introduction

Gait recognition is a prominent research topic in the field of computer vision, and it is also an efficient biometric technology. It identifies individuals based on their walking patterns [24]. Due to its ability to perform recognition at long distances and its resistance to disguise, gait recognition has found wide applications in surveillance and security fields [11,19]. The raw data for gait analysis mainly consists of videos involving data from various modalities such as silhouette, skeleton, and optical flow. These rich modalities contribute to a better understanding gait recognition from a collaborative perspective.

© The Author(s), under exclusive license to Springer Nature Switzerland AG 2024
S. Rudinac et al. (Eds.): MMM 2024, LNCS 14556, pp. 381–395, 2024.
https://doi.org/10.1007/978-3-031-53311-2_28

Numerous multimodal gait recognition algorithms have been proposed to harness the abundant modal information in videos fully. For instance, mmGaitSet [31] addresses the silhouette and pose heatmap modalities and designs intra-modality and inter-modality fusion methods to achieve a more comprehensive representation. BiFusion [22] devises a bimodal fusion network specifically for merging silhouette and skeleton modality features. MMGaitFormer [3] introduces the Spatial Fusion Module (SFM) and Temporal Fusion Module (TFM) to respectively fuse features at the spatial and temporal levels of gait. Their primary focus lies in designing intricate fusion strategies to generate effective representations while disregarding the inherent heterogeneity among different modalities. Modality heterogeneity often introduces information redundancy and distributional differences, increasing the challenges in multimodal representation learning and feature fusion.

To overcome the modality heterogeneity issue, we propose a Feature Decomposition Fusion (DFGait) network based on the silhouette and skeleton modalities. The network consists of three parts: decomposition learning, modality alignment, and fusion learning. In the decomposition learning phase, we introduce deep feature separation to handle the modality heterogeneity by learning two distinct features for each modality. The first type is the modality-shared feature, aimed at capturing the shared information between different modalities. For example, both silhouette and skeleton are used to describe gait features with semantic consistency. The second type is the modality-specific feature, aimed at learning the specific information of each modality while eliminating information redundancy. For instance, the silhouette data contains distinct variations in external appearance, while the skeleton data captures the internal structural information of the body. Based on these two types of feature, we introduce inter-modal regularization loss and intra-modal regularization loss to encourage the network to retain the shared and specific information. For inter-modal regularization loss, our objective is to enhance the similarity of the shared feature distribution across different modalities while increasing the differentiation in the specific feature distribution, and this aligns with our intuition regarding the decomposition of different modality features. For intra-modal regularization loss, we enforce orthogonality between shared and specific features to facilitate feature decomposition further. Furthermore, during the process of feature decomposition, the shared features and specific features of different modalities are embedded in their respective spaces, making fusion modeling challenging. Hence, we propose an adversarial modality alignment learning strategy, where the modality discriminator is confused to align the two modalities' features and maximize the interaction of modality information. Finally, we introduce a separable fusion module to fuse the features of the two modalities, resulting in a comprehensive gait representation. The new contributions of this paper can be summarized as follows:

- We propose the Feature Decomposition Fusion (DFGait) network for multimodal gait recognition tasks. DFGait addresses the heterogeneity between modalities by separately learning modality-shared and modality-specific features. DFGait achieves a more comprehensive gait representation by incorporating modality alignment and fusion operations.

– Our proposed DFGait outperforms previous state-of-the-art methods on GREW [34] and Gait3D [33]. Comprehensive experimental results demonstrate that our approach effectively captures the commonalities and uniqueness between modalities, resulting in discriminative gait representations.

## 2    Related Work

### 2.1    Multimodal Gait Recognition

Multimodal gait recognition can be categorized into homogeneous multimodal methods and non-homogeneous multimodal approaches based on the data sources [27]. Homogeneous multimodal methods utilize data from the same original videos, resulting in explicitly semantically consistent multimodal data. For example, TransGait [14] combines silhouette and pose heatmap data to explore pedestrian gait features. Another work [1] proposes a multimodal fusion architecture that combines grayscale, optical flow, and depth images. In non-homogeneous multimodal methods, data is sourced from different sensors, and there is no explicit one-to-one correspondence between different modalities due to varying input scales. For instance, GaitCode [21] proposes using autoencoders to fuse accelerometer data and ground contact force feedback data for gait recognition. Another work [13] utilizes motion sensor data combined with video data for gait recognition, using Long Short Term Memory (LSTM) networks to model the temporal information of each modality separately, followed by evolutionary algorithms for modality fusion. GaitFi [4] proposes a multimodal recognition method using WiFi signals and videos. It employs a Lightweight Residual Convolution Network (LRCN) to extract features from each modality, and the multimodal gait representation is obtained through concatenation.

The above methods are based on aggregation fusion paradigms, using simple concatenation or designing complex fusion methods to obtain multimodal representations. However, these methods overlook the heterogeneity between different modalities. Our work addresses modality heterogeneity through decomposition representation learning and modality alignment methods, allowing us to preserve multimodal information effectively through simple fusion operations.

### 2.2    Decomposition Representation Learning

Decomposition representation learning aims to separate input data into disjoint and meaningful variables, allowing for a better understanding of the constituent parts of the variables. This approach has been widely applied in various domains, such as heterogeneous face matching [28], cross-modal retrieval based on deep mutual information estimation [8], and more. In gait recognition, GaitNet [30] proposed an autoencoder framework to separate pose and appearance features from RGB images explicitly and integrated the temporal aspect of pose features

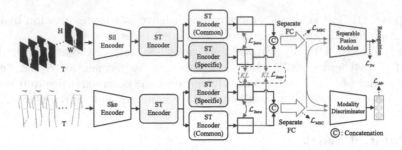

**Fig. 1.** The pipeline of our DFGait.

using LSTM. Compared to GaitNet, our method focuses on the decomposition learning of silhouette and skeleton modalities, which is advantageous for privacy preservation and robustness to different clothing/carrying conditions. Another work [15] introduced a disentangled representation learning method considering identity and covariate features for gait recognition. They first extracted identity and covariate features from the gait energy image (GEI) [9] and used an encoder to reconstruct GEIs related to and unrelated to covariates. However, this method utilizes GEI as input, which suffers from self-occlusion issues in silhouette data, and there is a risk of information loss in the temporal modeling of GEI. Moreover, relying on a single modality limits the incorporation of collaborative information from other modalities, imposing certain limitations on recognition performance.

This paper introduces the concept of multimodal decomposition representation learning to the gait recognition, analyzing the commonalities and uniqueness among multiple modalities to obtain a more comprehensive gait representation.

## 3   Method

### 3.1   Pipeline

The overall structure of DFGait is depicted in Fig. 1. We consider two primary modalities of gait: silhouette and skeleton. A batch of $B$ gait samples containing $N$ frames of two modalities are used as input to the model. The sequence corresponding to the silhouette is denoted as $sil \in \mathbb{R}^{B \times N \times C_{sil} \times H \times W}$, where $C_{sil}$ represents the number of channels for silhouette features, initially set to 1, and $H$ and $W$ denote the height and width of each silhouette input frame, respectively. The spatial structure of the skeleton sequence is represented in graph form, with the body joints as nodes and the association information between joints encoded in the adjacency matrix $A \in \mathbb{R}^{V \times V}$, where $V$ is the number of joints. Consequently, the sequence corresponding to the skeleton is $ske \in \mathbb{R}^{B \times N \times C_{ske} \times V}$, where $C_{ske}$ represents the number of channels for skeleton features, initially set to 3, representing the coordinate axes and confidence of each joint.

First, the silhouette spatial feature encoder and the skeleton spatial feature encoder are used to extract and unify the high-dimensional spatial features

$F_{sil} \in \mathbb{R}^{B \times N \times C_{sil} \times K}$ and $F_{ske} \in \mathbb{R}^{B \times N \times C_{ske} \times V}$, respectively, where K represents the number of horizontal partitions of features, and $K = V$. For these two features, the spatial-temporal feature encoder aggregates the basic spatial-temporal features and further extracts gait spatial-temporal decomposition features. Subsequently, the decomposition features of each modality are concatenated to obtain the decomposition spatial-temporal features $D_{sil} \in \mathbb{R}^{B \times C_{sil} \times 2K}$ and $D_{ske} \in \mathbb{R}^{B \times C_{ske} \times 2V}$. Finally, the decomposition modality features are aligned by the modality discriminator, and the separable fusion module fuses the aligned modality features to obtain the final feature for gait recognition, denoted as $O \in \mathbb{R}^{B \times C \times (2K+2V)}$.

## 3.2   Feature Encoder

Due to the significant structural differences between the silhouette and skeleton data in the spatial dimension, performing decomposition and fusion operations directly at the data level is challenging. To address this issue, we employ different types of spatial feature encoders to aggregate the spatial features of the two gait modalities, aiming to reduce the structural differences between the modalities in the spatial dimension. Additionally, human gait motion co-occurs in the spatial and temporal dimensions, and the gait information in these two dimensions is correlated. Therefore, we utilize a simple spatial-temporal feature encoder to aggregate the information across these two dimensions. The introductions of the three feature encoders are as follows:

**Silhouette Spatial Feature Encoder.** We have utilized the same structure as the silhouette feature extraction backbone in the GaitSet [2]. This backbone network consists of six CNN layers, with a max pooling layer inserted between every two CNN layers. The features generated by the feature extraction backbone for the silhouette data are denoted as $F'_{sil} \in \mathbb{R}^{B \times N \times C_{sil} \times H/4 \times W/4}$. Subsequently, the horizontal pooling (HP) module divides the extracted silhouette feature map into $K$ blocks in the horizontal direction. HP module applies global max pooling (GMP) and global average pooling (GAP) operations to obtain partial-level pooled features. We perform the segmentation directly along the height of the feature map, i.e., $K = W/4$. Through this process, we can obtain partial-level pooled features $F_{sil} \in \mathbb{R}^{B \times N \times C_{sil} \times K}$ for further processing in subsequent steps.

**Skeleton Spatial Feature Encoder.** As the skeleton data is a graph, we use a graph convolutional network (GCN) as the encoder. For the single-layer GCN, we used the same spatial GCN operation as in ST-GCN [29], and to ensure the capacity of the network, we stacked the GCNs with six layers. At the same time, to alleviate the over-smoothing problem caused by multi-layer stacked graph convolution, each layer of GCN is connected with residuals between them. The output feature of this encoder is $ske \in \mathbb{R}^{B \times N \times C_{ske} \times V}$.

Notably, the output feature tensors from the spatial feature encoders of the two modalities have different dimensionalities, which poses challenges for subsequent feature decomposition and fusion modeling. However, these feature tensors correspond to the structure of the human body. Using MLP to align dimensions would disrupt the original structure. Inspired by the successful application of the CLS token in Vision Transformers [5], we align the dimensionalities of the feature tensors by concatenating them with learnable parameter tensors.

**Spatial-Temporal Feature Encoder.** To model the correlation between gait features in both spatial and temporal dimensions while reducing training complexity, we employ one-dimension (1D) convolutions with a kernel size of 1 for spatial feature extraction and a kernel size of 3 for temporal feature extraction. We then integrate spatial and temporal features using the pseudo-three-dimensional (P3D) paradigm [23]. The last step involves performing max pooling operation along the temporal dimension. The input to this encoder is the output of the spatial encoders for both modalities or the output of the previous layer of the spatial-temporal encoder. The output is denoted as $ST_{sil} \in \mathbb{R}^{B \times C_{sil} \times K}$ and $ST_{ske} \in \mathbb{R}^{B \times C_{sil} \times V}$.

### 3.3  Multimodal Feature Decoupling

To address the issue of modality heterogeneity and reduce the modality gap and information redundancy, we decompose the representation into shared and specific features. Since gait is sequential, we directly employ the aforementioned spatial-temporal feature encoder as the decomposition encoder. Furthermore, we propose methods such as inter-modality regularization, intra-modality regularization, and modality semantic constraints to ensure the effectiveness of the decomposed features. We define any modality's modality-shared and modality-specific features as $ST^{sh}$ and $ST^{sp}$ respectively. The silhouette modality is represented as $ST_{sil}^{sh}, ST_{sil}^{sp} \in \mathbb{R}^{B \times C_{sil} \times K}$, and the skeleton modality is represented as $ST_{ske}^{sh}, ST_{ske}^{sp} \in \mathbb{R}^{B \times C_{ske} \times V}$.

**Inter-modality Regularization.** Based on the decomposition features of two modalities, our objective is to promote the extraction of both modality-shared and modality-specific features by increasing and decreasing the correlation between shared and private features of the two modalities. To achieve this goal, we propose a regularization loss between modalities. We utilize the Kullback-Leibler (KL) divergence to measure the correlation between features. However, the KL divergence suffers from an asymmetry issue, where different input orders of features can lead to different values. To address the bias in feature distributions caused by this asymmetry, we define the correlation between modality-shared features and modality-specific features of the two modalities as follows:

$$S_{sh} = (KL(ST_{ske}^{sh}, ST_{sil}^{sh}) + KL(ST_{sil}^{sh}, ST_{ske}^{sh}))/2 \tag{1}$$

$$S_{sp} = (KL(ST_{ske}^{sp}, ST_{sil}^{sp}) + KL(ST_{sil}^{sp}, ST_{ske}^{sp}))/2 \tag{2}$$

Subsequently, we use the correlation of modality-shared features as the numerator and the correlation of modality-specific features as the denominator, and this ensures that the optimization process simultaneously increases and decreases the correlation between modality-shared and modality-specific presentations of both modalities. The final loss function is defined as follows:

$$\mathcal{L}_{Inter} = S_{sh}/(S_{sp} + \varepsilon) \tag{3}$$

where, given that the range of KL divergence is $[0, +\infty)$, in order to avoid a denominator of zero, we set the value of $\varepsilon$ to be 0.01.

**Intra-modality Regularization.** To ensure that the modality-shared and modality-specific features effectively model different aspects within a modality and reduce information redundancy, we introduce the intra-modality regularization loss. We enforce orthogonality between the two learned features within a modality in the loss function. The specific formula is defined as follows:

$$\mathcal{L}_{Intra} = \| {ST^{sh}}^{\top} \cdot ST^{sp} \|_F^2 \tag{4}$$

where, $\| \cdot \|_F^2$ is the squared Frobenius norm.

**Modality Semantic Constraint.** As the above two regularization losses are applied during the execution, the features are continuously being decomposed. Without imposing constraints during this process, the decomposed features of the same modality would be forced to approximate orthogonal but non-representative representations. To address this, we introduce the modality semantic constraint loss. We concatenate the modality-shared and modality-specific features of a modality and pass them through a fully connected layer:

$$D = FC(Concat(ST^{sh}, ST^{sp})) \tag{5}$$

where, when the modality is the silhouette $D \in \mathbb{R}^{B \times C_{sil} \times 2K}$, skeleton $D \in \mathbb{R}^{B \times C_{ske} \times 2V}$, $K = V$. $Concat$ indicates a concatenation operation.

By fusing the shared and specific features of a modality, a more comprehensive unimodal representation is obtained, and this representation carries semantic information about a person's identity. Subsequently, we employ Triplet Loss [10] to impose semantic constraints on the decomposed features. The final semantic loss is defined as follows:

$$\mathcal{L}_{MSC} = \mathcal{L}_{Tri}(D) \tag{6}$$

## 3.4    Modality Alignment and Fusion

After obtaining the decomposed features of the two modalities, we aim to achieve a more comprehensive multimodal feature representation through fusion. However, since these features are extracted in their respective feature spaces, it

becomes challenging to perform fusion modeling directly. Therefore, we introduce a modality alignment step to maximize the interaction between the modalities before fusion. Specifically, we employ an adversarial modality alignment
strategy and a separable fusion module in this process.

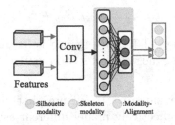

Fig. 2. Overview of the modality discriminator.

**Adversarial Modality Alignment Strategy.** We introduce adversarial learning by confusing the proposed modality discriminator to achieve modal alignment. For the confusion method, we force the two modalities to be classified
into the same category by setting the modality feature labels. At the same time,
we propose a strategy to classify both modalities into one additional category
instead of either of the two modalities, and this ensures that both modalities
belong to the same aligned potential space, thus avoiding falling into each other's
modality space. Specifically, the structure of our proposed modality discriminator is shown in Fig. 2, which first pools the dimensionality of the decomposed
features of the two modalities using a one-dimensional convolution and then
maps the features into vectors for classification using two fully connected layers.
The objective function of adversarial learning is:

$$\mathcal{L}_{Adv} = CE(m_{sil}, l) + CE(m_{ske}, l) \tag{7}$$

where, $CE(\cdot)$ denotes the cross-entropy loss, $m_{sil}$ and $m_{ske}$ represent the classification vectors for the silhouette and skeleton features, respectively, and $l$ is
the label for the confusion modality.

**Separable Fusion Module.** In the multimodal fusion approach, simple
concatenation or addition operations do not effectively facilitate the interaction between different modalities. However, complex fusion strategies such as
attention introduce many parameters. Therefore, we propose a method called
the Separable Fusion Module. Specifically, the decomposition features $D_{sil} \in
\mathbb{R}^{B \times C_{sil} \times 2K}$ and $D_{ske} \in \mathbb{R}^{B \times C_{ske} \times 2V}$ from two modalities are first concatenated
to obtain $D_{concat} \in \mathbb{R}^{B \times C \times (2K+2V)}$, where $C = C_{sil} = C_{ske}$. Then, a depthwise
separable convolution operation is applied to the concatenated features to obtain
the final output $O \in \mathbb{R}^{B \times C \times (2K+2V)}$ of the network for gait recognition.

## 3.5    Objective Optimization

We employ the separate Batch All triplet loss [10] as the loss function for gait recognition task. Combined with the aforementioned inter-modality regularization loss, intra-modality regularization loss, and modality semantic constraint loss, the final calculation of the objective function is as follows:

$$\mathcal{L}_{Total} = \mathcal{L}_{Tri} + \alpha(\mathcal{L}_{MSC}^{sil} + \mathcal{L}_{MSC}^{ske}) + \beta(\mathcal{L}_{Inter} + \mathcal{L}_{Intra}^{sil} + \mathcal{L}_{Intra}^{ske} + \mathcal{L}_{Adv}) \quad (8)$$

# 4    Experiments

## 4.1    Datasets

We conducted experiments on two standard datasets, Gait3D [33] and GREW [34], to validate the superiority of our method. We also performed further ablation experiments on the Gait3D dataset to demonstrate the positive impact of each component in our approach.

Gait3D is a 3D representation-based gait in-the-wild dataset. It contains 4000 subjects and over 25000 sequences. We keep the same training and testing setup as the official [33]. The training set consists of 3000 subjects and the testing set contains 1000 subjects. In the testing set, one sequence is registered as the probe set, while the rest are regarded as the gallery set.

GREW is one of the largest gait in-the-wild datasets. It extracted 26,345 subjects and 128,671 sequences in a large public area. We follow the same protocol in [34]. There are 20,000 subjects for training, 6,000 for testing, and 345 for validating. Two sequences are registered as the probe set in the testing set, and two become the gallery set.

## 4.2    Implementation Details

**Hyper-parameters.** For the Gait3D dataset, we set the number of channels in the convolutional layers of the silhouette spatial encoder and the skeleton spatial encoder as 32, 32, 64, 64, 128, and 128. As for the GREW dataset, due to its large number of sequences, we added two additional convolutional layers with 256 channels at the end of both encoders. In the silhouette spatial encoder, the kernel size of the first CNN layer is set to 5, while the rest are set to 3. For each layer of graph convolution in the skeleton spatial encoder, the kernel size is set to 4. In the Eq. (8), the value of $\alpha$ is set to 1, and the value of $\beta$ is set to 0.5.

**Training Details.** 1) The silhouettes of both two datasets are transformed into $64 \times 44$. The format of the skeleton is the same as the COCO dataset [18], which consists of 17 human key points. 2) Batch size is set to (32, 4) for both datasets. 3) At the training phase, randomly extract 60 frames for training. During the testing phase, all frames are used. 4) For the Gait3D dataset, the model is trained for 80K iterations with the initial Learning Rate(LR) = 1e−4, and the LR is multiplied by 0.1 at the 30K and 60K iterations. For the GREW, the model is trained for 180K iterations with the initial LR 1e−4, and the LR is multiplied by 0.1 at the 120K iterations. Adam [12] is taken as the optimizer. 5) Our model is trained and evaluated based on the OpenGait [6]. All models are implemented with PyTorch and trained in RTX 3090 GPU.

### 4.3    Comparison with State-of-the-Art Methods

**Evaluation on Gait3D.** Table 1 presents the comparison results between our proposed DFGait method and the state-of-the-art (SOTA) methods on the Gait3D dataset. It can be observed that our method achieves the highest average accuracy. Compared to the SOTA silhouette-based gait recognition method MTSGait [32], our method improves the rank-1 accuracy by 1.60%. Compared to the skeleton-based method GaitGraph [26], our method significantly improves rank-1 accuracy. Compared to the multimodal method SPMLGait [14], our method achieves a 4.00% improvement. In summary, we can conclude the following: 1) The DFGait algorithm achieves advanced results, demonstrating its robustness and advantages. 2) Our proposed DFGait based on the principles of decomposition and fusion effectively decomposes the features and reduces modality differences and information redundancy, thus obtaining a more comprehensive representation of gait.

**Table 1.** Rank-1 accuracy(%), Rank-5 accuracy (%), mAP (%) and mINP on the Gait3D dataset.

| Methods | Rank-1 | Rank-5 | mAP | mINP |
|---|---|---|---|---|
| PoseGait [16] | 0.24 | 1.08 | 0.47 | 0.34 |
| GaitGraph [26] | 6.25 | 16.23 | 5.18 | 2.42 |
| GEINet [25] | 5.40 | 14.20 | 5.06 | 3.14 |
| GaitSet [2] | 36.70 | 58.30 | 30.01 | 17.30 |
| GaitPart [7] | 28.20 | 47.60 | 21.58 | 12.36 |
| GaitGL [17] | 29.70 | 48.50 | 22.29 | 13.26 |
| MTSGait [32] | 48.70 | 67.10 | 37.63 | 21.92 |
| SMPLGait [33] | 46.30 | 64.50 | 37.16 | 22.23 |
| **Ours** | **50.30** | **69.90** | **41.05** | **25.61** |

**Table 2.** Rank-1 accuracy (%), Rank-5 accuracy (%), Rank-10 accuracy (%), and Rank-20 accuracy (%) on the GREW dataset.

| Methods | Rank-1 | Rank-5 | Rank-10 | Rank-20 |
|---|---|---|---|---|
| PoseGait [16] | 0.23 | 1.05 | 2.23 | 4.28 |
| GaitGraph [26] | 1.31 | 3.46 | 5.1 | 7.51 |
| GEINet [25] | 6.82 | 13.42 | 16.97 | 21.01 |
| GaitSet [2] | 46.28 | 63.58 | 70.26 | 76.82 |
| GaitPart [7] | 44.01 | 60.68 | 67.25 | 73.47 |
| GaitGL [17] | 47.28 | 63.56 | 69.32 | 74.18 |
| MTSGait [32] | 55.32 | 71.28 | 76.85 | 81.55 |
| TransGait [14] | 56.27 | 72.72 | 78.12 | 82.51 |
| **Ours** | **61.42** | **76.94** | **81.81** | **86.00** |

**Evaluation on the GREW.** We further evaluated the performance of DFGait on the GREW dataset, which is currently the largest publicly available in-the-wild gait dataset. Many algorithms have been optimized specifically for large-scale datasets, such as MTSGait [32] adding more convolutional layers. We consider the optimizations applied in each respective method when comparing the results. The specific comparison results are shown in Table 2, our method achieves state-of-the-art performance with an average rank-1 accuracy of 61.42%, outperforming unimodal and multimodal algorithms. This demonstrates the generalization and robustness of our method on large-scale datasets.

It is worth noting that our method outperforms TransGait [14], which is based on silhouette and pose heatmaps, by 5.15%. This highlights the effectiveness of our feature decomposition approach and the relatively low-cost fusion strategy in achieving significant performance improvement.

## 4.4  Ablation Study

**Analysis of Regularization.** We conducted experiments to evaluate the effects of different regularization methods by removing the suggested loss functions. The results are presented in Table 3. When the $\mathcal{L}_{Inter}$ is omitted, the model learns only constraints on the orthogonality as well as the semantics of the decomposed features of the two modalities, lacking constraints on the relationship between the two modalities. This can lead to the learning of redundant information between modalities, resulting in decreased performance. On the other hand, we observed that the $\mathcal{L}_{Intra}$ improves the model's performance by effectively enabling the modality-shared and modality-specific features to capture different aspects of modality information. Furthermore, the elimination of the $\mathcal{L}_{MSC}$ leads to a significant decrease in testing performance. This is because, during the execution of the two regularization losses, the features are forced to separate, potentially learning trivial and irrelevant information, negatively impacting accuracy.

**Table 3.** Analysis of Regularization.

| Model | Rank-1 | Rank-5 | mAP | mINP |
|---|---|---|---|---|
| DFGait | 50.30 | 69.90 | 41.05 | 25.61 |
| w/o $\mathcal{L}_{Inter}$ | 49.30 | 69.30 | 40.14 | 25.06 |
| w/o $\mathcal{L}_{Intra}$ | 48.90 | 69.10 | 40.22 | 25.29 |
| w/o $\mathcal{L}_{MSC}$ | 47.80 | 69.00 | 40.33 | 24.90 |

**Table 4.** Analysis of Representations.

| Methods | Rank-1 | Rank-5 | mAP | mINP |
|---|---|---|---|---|
| w/o Shared | 44.60 | 65.80 | 36.40 | 22.07 |
| w/o Specific | 45.30 | 66.50 | 37.23 | 22.79 |
| Non-Decoupled | 46.20 | 67.30 | 37.04 | 22.39 |
| Non-Aligned | 49.50 | 69.30 | 40.49 | 25.29 |
| DFGait | 50.30 | 69.90 | 41.05 | 25.61 |

**Table 5.** Different Fusion Strategies, Add denotes element-wise addition, while Cat represents concatenation.

| Methods | Rank-1 | Rank-5 | mAP | mINP | Parameters |
|---|---|---|---|---|---|
| Add fusion | 46.20 | 67.69 | 37.93 | 23.18 | 7.83463 |
| Cat fusion | 44.10 | 64.10 | 35.51 | 21.88 | 7.83463 |
| Ours | 50.30 | 69.90 | 41.05 | 25.61 | 7.90196 |

**Analysis of Representations.** To demonstrate the effectiveness of different representations, we conducted experiments where we individually removed the modality-shared and the modality-specific representation. The results are presented in Table 4. Specifically, we kept the representation learning process but only used one of the two features for fusion and prediction during testing. The results of these ablation experiments indicate that both learned representations are necessary and meaningful. Furthermore, we explored a version without any decomposition learning. The results show a significant performance gap between the model with decomposition learning and the one without, highlighting the effectiveness of decomposition learning. Additionally, we performed experiments on modality alignment before fusion, and the results demonstrate that modality alignment effectively enhances the information interaction between modalities.

**Analysis of Fusion Strategies.** We investigated the impact of different fusion strategies on the results. The results are presented in Table 5. When the separable fusion module was removed, and feature fusion was performed through addition operation, it led to inferior results. This indicates that the addition operation aggregates the linear relationship between the two modalities but fails to preserve the individual characteristics of each modality. On the other hand, simple concatenation achieved the worst performance. This is because the information from both modalities exists in parallel, allowing for the preservation of the individual modality features but without interaction between the two modalities. Our proposed fusion method, which incorporates depthwise separable convolution operation on top of concatenation, only increased the model parameters by 0.06733M while improving the rank-1 accuracy by 6.20%. This demonstrates the effectiveness of our fusion strategy, which is simple yet impactful.

**Complexity Analysis.** In Table 6, we compare the complexity of several SOTA methods with DFGait. All results are obtained from the OpenGait [6] framework and tested on the same server with a single RTX 3090 GPU. We set the batch size to 1 during the inference. The results show that our method achieves the highest rank-1 accuracy while maintaining a reasonable model complexity level.

**Table 6.** Analysis of embedding size of the feature used for inference, parameters (M) and inference time (s) on Gait3D in terms of averaged rank-1 accuracy (%).

| Methods | Embedding Size | Parameters | Inference Time | Rank-1 |
|---------|---------------|------------|----------------|--------|
| GaitSet [2] | 62 * 256 | 2.59 | 57 | 36.70 |
| GaitPart [7] | 16 * 128 | 1.2 | 162 | 28.20 |
| GaitGL [17] | 64 * 256 | 2.49 | 113 | 29.70 |
| SMPLGait [33] | 31 * 256 | 27.11 | 74 | 46.30 |
| Ours | 68 * 256 | 7.90 | 88 | 50.30 |

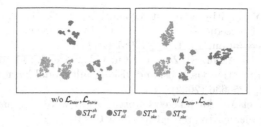

**Fig. 3.** Visualization of representations in the testing set on Gait3D. Rose and brown represent the modality-shared and modality-specific representation of the silhouette, green and cyan represent the two representations of the skeleton. (Color figure online)

**Visualization.** In Fig. 3, we utilized t-SNE [20] visualization technique to visualize the modality-shared and modality-specific representations of the two modalities on the Gait3D. From the figure, we can analyze that without using $\mathcal{L}_{Inter}$ and $\mathcal{L}_{Intra}$, there is a small gap between the two features of the skeleton modality, while there is a large gap between the shared features of the two modalities. However, when using the two regularization losses, the features of the same modality are well separated, and the gap between the shared features of the modalities is further reduced, while the gap between the modality-specific features is further expanded. The visualization results intuitively demonstrate that our method is effective in feature decomposition.

## 5 Conclusion

In this paper, we introduce DFGait, a novel network that deals with modality heterogeneity through learning shared and specific features. We apply multiple effective constraints to encourage capturing common and unique information across modalities. Additionally, we propose an adversarial modality alignment strategy for enhancing inter-modality information interaction before fusion. Ultimately, we achieve more comprehensive gait representations via separable fusion. Experimental results consistently demonstrate the superiority of our approach.

**Acknowledgement.** Project supported by Provincial Natural Science Foundation of Hunan (No. 2023JJ30697), Changsha Municipal Natural Science Foundation (No. kq2208286).

## References

1. Castro, F.M., Marin-Jimenez, M.J., Guil, N., Pérez de la Blanca, N.: Multimodal feature fusion for CNN-based gait recognition: an empirical comparison. Neural Comput. Appl. **32**, 14173–14193 (2020)
2. Chao, H., He, Y., Zhang, J., Feng, J.: GaitSet: regarding gait as a set for cross-view gait recognition. In: Proceedings of the AAAI Conference on Artificial Intelligence, vol. 33, pp. 8126–8133 (2019)

3. Cui, Y., Kang, Y.: Multi-modal gait recognition via effective spatial-temporal feature fusion. In: Proceedings of the IEEE/CVF Conference on Computer Vision and Pattern Recognition, pp. 17949–17957 (2023)
4. Deng, L., Yang, J., Yuan, S., Zou, H., Lu, C.X., Xie, L.: GaitFi: robust device-free human identification via WiFi and vision multimodal learning. IEEE Internet Things J. **10**(1), 625–636 (2022)
5. Dosovitskiy, A., et al.: An image is worth 16x16 words: transformers for image recognition at scale. arXiv preprint arXiv:2010.11929 (2020)
6. Fan, C., Liang, J., Shen, C., Hou, S., Huang, Y., Yu, S.: OpenGait: revisiting gait recognition toward better practicality (2022)
7. Fan, C., et al.: GaitPart: temporal part-based model for gait recognition. In: Proceedings of the IEEE/CVF Conference on Computer Vision and Pattern Recognition, pp. 14225–14233 (2020)
8. Guo, W., Huang, H., Kong, X., He, R.: Learning disentangled representation for cross-modal retrieval with deep mutual information estimation. In: Proceedings of the 27th ACM International Conference on Multimedia, pp. 1712–1720 (2019)
9. Han, J., Bhanu, B.: Individual recognition using gait energy image. IEEE Trans. Pattern Anal. Mach. Intell. **28**(2), 316–322 (2005)
10. Hermans, A., Beyer, L., Leibe, B.: In defense of the triplet loss for person re-identification. arXiv preprint arXiv:1703.07737 (2017)
11. Iwama, H., Muramatsu, D., Makihara, Y., Yagi, Y.: Gait verification system for criminal investigation. Inf. Media Technol. **8**(4), 1187–1199 (2013)
12. Kingma, D.P., Ba, J.: Adam: a method for stochastic optimization. arXiv preprint arXiv:1412.6980 (2014)
13. Kumar, P., Mukherjee, S., Saini, R., Kaushik, P., Roy, P.P., Dogra, D.P.: Multimodal gait recognition with inertial sensor data and video using evolutionary algorithm. IEEE Trans. Fuzzy Syst. **27**(5), 956–965 (2018)
14. Li, G., Guo, L., Zhang, R., Qian, J., Gao, S.: TransGait: multimodal-based gait recognition with set transformer. Appl. Intell. **53**(2), 1535–1547 (2023)
15. Li, X., Makihara, Y., Xu, C., Yagi, Y., Ren, M.: Gait recognition via semi-supervised disentangled representation learning to identity and covariate features. In: Proceedings of the IEEE/CVF Conference on Computer Vision and Pattern Recognition, pp. 13309–13319 (2020)
16. Liao, R., Yu, S., An, W., Huang, Y.: A model-based gait recognition method with body pose and human prior knowledge. Pattern Recogn. **98**, 107069 (2020)
17. Lin, B., Zhang, S., Yu, X.: Gait recognition via effective global-local feature representation and local temporal aggregation. In: Proceedings of the IEEE/CVF International Conference on Computer Vision, pp. 14648–14656 (2021)
18. Lin, T.-Y., et al.: Microsoft COCO: common objects in context. In: Fleet, D., Pajdla, T., Schiele, B., Tuytelaars, T. (eds.) ECCV 2014. LNCS, vol. 8693, pp. 740–755. Springer, Cham (2014). https://doi.org/10.1007/978-3-319-10602-1_48
19. Lynnerup, N., Larsen, P.K.: Gait as evidence. IET Biometrics **3**(2), 47–54 (2014)
20. Van der Maaten, L., Hinton, G.: Visualizing data using t-SNE. J. Mach. Learn. Res. **9**(11) (2008)
21. Papavasileiou, I., Qiao, Z., Zhang, C., Zhang, W., Bi, J., Han, S.: GaitCode: gait-based continuous authentication using multimodal learning and wearable sensors. Smart Health **19**, 100162 (2021)
22. Peng, Y., Ma, K., Zhang, Y., He, Z.: Learning rich features for gait recognition by integrating skeletons and silhouettes. Multimedia Tools Appl., 1–22 (2023)

23. Qiu, Z., Yao, T., Mei, T.: Learning spatio-temporal representation with pseudo-3D residual networks. In: Proceedings of the IEEE International Conference on Computer Vision, pp. 5533–5541 (2017)

24. Sepas-Moghaddam, A., Etemad, A.: Deep gait recognition: a survey. IEEE Trans. Pattern Anal. Mach. Intell. **45**(1), 264–284 (2022)

25. Shiraga, K., Makihara, Y., Muramatsu, D., Echigo, T., Yagi, Y.: GeiNet: view-invariant gait recognition using a convolutional neural network. In: 2016 International Conference on Biometrics (ICB), pp. 1–8. IEEE (2016)

26. Teepe, T., Khan, A., Gilg, J., Herzog, F., Hörmann, S., Rigoll, G.: GaitGraph: graph convolutional network for skeleton-based gait recognition. In: 2021 IEEE International Conference on Image Processing (ICIP), pp. 2314–2318. IEEE (2021)

27. Wang, Y., Huang, W., Sun, F., Xu, T., Rong, Y., Huang, J.: Deep multimodal fusion by channel exchanging. Adv. Neural. Inf. Process. Syst. **33**, 4835–4845 (2020)

28. Wu, X., Huang, H., Patel, V.M., He, R., Sun, Z.: Disentangled variational representation for heterogeneous face recognition. In: Proceedings of the AAAI Conference on Artificial Intelligence, vol. 33, pp. 9005–9012 (2019)

29. Yan, S., Xiong, Y., Lin, D.: Spatial temporal graph convolutional networks for skeleton-based action recognition. In: Proceedings of the AAAI Conference on Artificial Intelligence, vol. 32 (2018)

30. Zhang, Z., et al.: Gait recognition via disentangled representation learning. In: Proceedings of the IEEE/CVF Conference on Computer Vision and Pattern Recognition, pp. 4710–4719 (2019)

31. Zhao, L., Guo, L., Zhang, R., Xie, X., Ye, X.: mmGaitSet: multimodal based gait recognition for countering carrying and clothing changes. Appl. Intell. **52**(2), 2023–2036 (2022)

32. Zheng, J., et al.: Gait recognition in the wild with multi-hop temporal switch. In: Proceedings of the 30th ACM International Conference on Multimedia, pp. 6136–6145 (2022)

33. Zheng, J., Liu, X., Liu, W., He, L., Yan, C., Mei, T.: Gait recognition in the wild with dense 3D representations and a benchmark. In: Proceedings of the IEEE/CVF Conference on Computer Vision and Pattern Recognition, pp. 20228–20237 (2022)

34. Zhu, Z., et al.: Gait recognition in the wild: a benchmark. In: Proceedings of the IEEE/CVF International Conference on Computer Vision, pp. 14789–14799 (2021)

# MoPE: Mixture of Pooling Experts Framework for Image-Text Retrieval

Jiangfeng Li[1], Bowen Wang[1], Yongrui Qin[1], Chenxi Zhang[1],
Gang Yu[2], and Qinpei Zhao[1]([✉])

[1] School of Software Engineering, Tongji University, Shanghai 201804, China
{lijf,wangbowen,2311441,xzhang2000,qinpeizhao}@tongji.edu.cn
[2] SILC Business School, Shanghai University, Shanghai 201800, China
gyu@shu.edu.cn

**Abstract.** Image-text retrieval is a fundamental and crucial task in the field of multimodal interaction, which assists internet users in retrieving the required visual and textual information conveniently. The dominant method for image-text retrieval aims to learn a visual semantic embedding space such that related visual and textual data are close to each other. Recent research focuses on designing sophisticated pooling strategies to better aggregate visual and textual features into holistic embeddings. However, existing methods often use the same pooling operator for the whole dataset, ignoring that samples with diverse intra-modality relationships require pooling operators trained with different parameters. To tackle this issue, we propose a novel Mixture of Pooling Experts (MoPE) framework, which combines multiple pooling operators to aggregate features for different data subsets. Specifically, we introduce a novel route gating strategy in combination with an aggregation expert module to dynamically learn diverse pooling experts for samples in different data subsets. Moreover, to fully exploit the intra-modality relationships, we develop a specialized router with a self-attention gate mechanism to direct each sample to the proper pooling expert. Extensive experiments conducted on two widely used benchmark datasets, namely Flickr30K and MS-COCO, demonstrate the superiority of our method over several state-of-the-art methods.

**Keywords:** Image-Text Retrieval · Mixture of Pooling Experts · Visual Semantic Embedding · Multimodal Semantic Analysis

## 1 Introduction

With the development of the Internet, multimedia information has been rapidly spread and applied. Various online multimedia sharing platforms have been developed, such as Instagram, Twitter, etc., which enrich lives greatly. To find the required texts or images from the application quickly and accurately, we need an intelligent searching method to automatically return the answer to our query. Recent research into image-text retrieval provides approaches for image search

© The Author(s), under exclusive license to Springer Nature Switzerland AG 2024
S. Rudinac et al. (Eds.): MMM 2024, LNCS 14556, pp. 396–409, 2024.
https://doi.org/10.1007/978-3-031-53311-2_29

for given sentences or the retrieval of text descriptions from image queries. As a fundamental task in multimodal interaction that bridges the gap between vision and language, image-text retrieval has attracted extensive research attention.

Visual and non-visual semantic embedding are two paradigms in image-text retrieval methods. In visual semantic embedding (VSE) methods [5,19], image and text features are firstly extracted using separate visual and textual encoders. These features are then aggregated into the joint embedding space as fix-length vectors. Finally, similarity between these vectors is calculated. Non-visual semantic embedding (non-VSE) methods [10,24] focus on fine-grained alignment and modality interactions between image regions and sentence words. Visual semantic embedding methods don't consider cross-modal interaction so that visual features and textual features could be pre-computed [25], achieving higher efficiency in practical applications compared to the non-VSE methods.

Therefore, recent methods emphasize the focus on visual semantic embedding for efficient image-text retrieval, notably in feature aggregation [7,23]. However, existing studies [2,25] typically use the same aggregating operator trained with the same parameters to process the entire input dataset of a certain modality, ignoring data discrepancies caused by intra-modality relationship differences. Consequently, feature aggregation module learns holistic embeddings with error and noise, which impacts the performance of image-text retrieval. Concretely, both images and texts contain potential subsets, such as those for different domains or different topics. As illustrated in Fig. 1, image(a) describes some children playing games on the grass, and image(b) depicts a natural scenery about a desert. In human cognition, they are different categories of images. On the other hand, image(a) contains more entities and a more complex entity interaction relationship between 'children', 'rope', and 'adult', whereas image(b) has a simpler entity connection between 'desert', 'person', and 'sky'. For feature aggregation, image(a) and image(b) need different pooling operators due to diverse intra-modality relationship. Similarly, the semantic entity relationship of text(a) is different from text(b)'s, which indicates that they require individual feature aggregators. Additionally, according to the classical Mixture of Experts (MoE) theory [8], models should choose different parameters for each incoming example, which theoretically supports this paper's opinion.

**Text (a):** Some children are playing tug-of-war with a rope while an adult assists.

**Text(b):** A person is standing at the top of a large , rolling sand dune.

**Image (a)**                                        **Image (b)**

**Fig. 1.** The samples belonging to different subsets.

Therefore, a Mixture of Pooling Experts (MoPE) framework is proposed for tackling the issue, which uses different pooling parameters for diverse samples. Specifically, based on the principle of Mixture of Experts [8], multiple pooling operators are combined as expert sub-networks and adaptively aggregate fragment features from different data subsets. The main contributions of this paper include: (1) A route gating module is designed to adaptively select the proper pooling expert for each sample according to the intra-modality relationship, (2) An aggregation expert module is presented to aggregate fragment features from different samples into holistic global representations, and (3) An auxiliary load balancing objective is constructed to encourage a balanced load across pooling experts in the aggregation expert module.

## 2    Related Work

Image-text retrieval is a fundamental task that bridges vision and language, including two research lines, visual semantic embedding [3,5,9,20] and non-visual semantic embedding [1,4,10,24]. Due to its high practical efficiency, visual semantic embedding (VSE) has become the mainstream image-text retrieval research direction.

Wang et al. [20] propose a two-branch neural network with a maximum-margin ranking loss and novel neighborhood constraints to learn the correspondence between image and text data. And Faghri et al. [5] enhance the VSE model further by presenting a new loss based on violations incurred by relatively hard negatives, which becomes the most widely adopted objective in this domain.

Considering that the performance of VSE relies on learning a high-quality joint embedding space, this paper focuses on the improvement of feature aggregation for VSE. Existing methods usually use simple pooling aggregations, such as mean pooling [16], max pooling [12] and a combination of them [10], while some complex aggregators (e.g. self-attention [7] and intra-modal attention mechanisms [21]) consume more time and perform worse than the simple ones. Motivated by this, some studies [2,25] design an adaptive pooling strategy to learn the best combination of simple pooling methods. However, these pooling strategies provide the same aggregation operator to train the whole dataset for each modality separately, while ignoring the fact that the data includes potential subsets. As a result, the data subsets with different intra-modality relationships are trained by the same pooling parameters, resulting in final holistic embedding vectors containing errors.

Mixture of experts (MoE) is a widely used ensemble learning technique for neural network. It is first introduced by Jacobs et al. [8], which symbolizes a system composed of multiple separate networks. Each network learns to handle a subset of the complete set of training samples. Subsequently, Shazeer et al. [17] prove the effectiveness of MoE in the modern deep learning architecture through adding an MoE layer stacked between LSTM layers. Furthermore, the MoE Transformer achieves significant improvements in terms of machine translation across 100 languages [11]. Different from other MoE approaches, Switch

Transformer [6] applies a simplified routing strategy to improve computation efficiency dramatically, which only selects a single expert for each token.

## 3   Mixture of Pooling Experts Framework

The Mixture of Pooling Experts framework for image-text retrieval is illustrated in Fig. 2. We first extract features of image regions and sentence words, and then utilize the combination of the route gating module and the aggregation expert module to measure image-text similarity. In this section, the route gating module is introduced in Sect. 3.1. The aggregation expert module is described in Sect. 3.2 and the loss function module is displayed in Sect. 3.3.

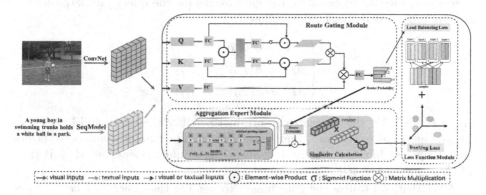

**Fig. 2.** The overall framework of our proposed method, contains three major modules: 1) Route gating module, based on the self-attention gate mechanism, calculates the routing probability and selects an expert; 2) Aggregation expert module pools features into a holistic embedding vector and calculates the cosine similarity. 3) Loss function module integrates the ranking loss and the load balancing loss to optimize image-text retrieval.

### 3.1   Route Gating Module

Formally, let $G$ be an image, $T$ be a text, and $(G, T)$ be an image-text pair. The image is represented as regions' visual features $G = \{g_i | i \in [1, n], g_i \in \mathbb{R}^d\}$, and the text is represented as textual features of words $T = \{t_j | j \in [1, s], t_j \in \mathbb{R}^d\}$, where $g_i$ is the $i$-th region's feature and $t_j$ is the $j$-th word's feature; $n$ and $s$ denote the number of regions and words, respectively; $d$ is the dimension of feature representation.

For both images and texts, we make full use of the intra-modality relationship and self-attention mechanism to route each sample to its proper pooling expert for feature aggregation. Note that here we only describe the route procedure of the visual branch, as it goes the same for the textual one. A set of region features $G = \{g_1, g_2, ..., g_n\}$ is the input of the route gating strategy, where $G \in \mathbb{R}^{n \times d}$.

Let $Q \in \mathbb{R}^{n \times d}$, $K \in \mathbb{R}^{n \times d}$, $V \in \mathbb{R}^{n \times d}$ represent query, key and value separately, where $n$ indicates the number of regions and $d$ refers to the dimension of the vectors. It can be described as:

$$\begin{cases} Q = GW^Q + b^Q \\ K = GW^K + b^K \\ V = GW^V + b^V \end{cases} \tag{1}$$

where $W^Q, W^K, W^V \in \mathbb{R}^{d \times d}$ are learnable projection matrices, and $b^Q, b^K, b^V \in \mathbb{R}^{1 \times d}$ are the bias vectors.

Given that the dot-product of the query and key may contain noise [16], we propose a gate mechanism to obtain the useful information while filtering the useless one. Firstly, we calculate the gate masks for the query and key vectors, and the process is represented as follows:

$$\begin{cases} F = Q \odot K \\ M^Q = \sigma(FW_M^Q + b_M^Q) \\ M^K = \sigma(FW_M^K + b_M^K) \end{cases} \tag{2}$$

where $F \in \mathbb{R}^{n \times d}$ is the element-wise product result of the query and key. Then, the gate masks $M^Q$, $M^K$ for $Q$ and $K$ are generated by two fully-connected layers with a sigmoid function, where $\sigma$ denotes the sigmoid operation, $W_M^Q, W_M^K \in \mathbb{R}^{d \times d}$, and $b_M^Q, b_M^K \in \mathbb{R}^{1 \times d}$.

After that, we utilize the gate masks to denoise the query and key vectors and calculate the attended value vector $attn$ by:

$$attn = Softmax(\frac{(M^Q \odot Q)(M^K \odot K)^T}{\sqrt{d}})V \tag{3}$$

where $Softmax$ function is applied to each row, $attn \in \mathbb{R}^{n \times d}$.

Then, a fully-connected layer is employed to transform the gating attended feature $attn$ to the route gating representation $Z = attnW_{attn} + b_{attn}$, $Z \in \mathbb{R}^{n \times m}$, where $m$ denotes the number of pooling experts, $W_{attn} \in \mathbb{R}^{d \times m}$, $b_{attn} \in \mathbb{R}^{1 \times m}$.

Finally, each image $G$ is routed to the pooling expert with the highest probability value, based on a softmax distribution across the $m$ pooling experts. The probability value for $G$ to expert $i$ is described as:

$$p_i(G) = \frac{e^{Z(G)_i}}{\sum_j^m e^{Z(G)_j}} \tag{4}$$

Therefore, the routing probability of the best determined pooling expert for $G$ is given by $p(G) = max\{p_i(G)\}$. With the same process on the word-level textual features $T = \{t_1, t_2, ...t_s\}$, the routing probability of the selected pooling expert for the sentence $T$ is calculated by $p(T) = max\{p_i(T)\}$.

## 3.2   Aggregation Expert Module

To measure the similarity between visual and textual modalities, we aggregate fragment features into a holistic vector by following the generalized pooling operator (GPO) [2], which aims to generate the pooling coefficients to weight the sorted feature. Therefore, the aggregation expert module comprises $m$ pooling experts, all sharing the same structure. The route gating module independently directs each sample to one of those pooling experts, and then the aggregation expert module returns the holistic embedding multiplied by the routing probability value.

Likewise, we only introduce the aggregating process of the visual branch, which is the same for the textual one. Specifically, given a set of $n$ region features $G = \{g_i\}_{i=1}^n$, the selected pooling expert aggregates it into a fix-length vector $\tilde{G} \in \mathbb{R}^d$ out from the $n$ regions. Let $max_k(\cdot)$ denote a function that extracts the top $k$-th value from a sorted list, the coefficient $\theta_k$ be the weight for the top $k$-th value among $n$ regions, and $p(G)$ represent the routing probability value for image $G$. The aggregating mechanism is formulated as:

$$\tilde{G} = p(G) \sum_{k=1}^n (\theta_k \cdot max_k(\{g_i\}_{i=1}^n)) \tag{5}$$

where $\sum_{k=1}^n \theta_k = 1$.

Therefore, each pooling expert focuses on constructing a coefficient generator for $\theta_k$, which includes: 1) Position Encoder, a positional encoding function based on trigonometric function, and 2) Position Decoder, a sequence model that generates pooling coefficients based on BiGRU [10].

*Position Encoder.* To encode every position index $k$ to a dense vector $pos_k \in \mathbb{R}^{d_k}$, a trigonometric function is used following Transformer [18], which is described as:

$$pos_k^i = \begin{cases} sin(w_j k), & when \ i = 2j, \\ cos(w_j k), & when \ i = 2j + 1, \end{cases} \tag{6}$$

where $w_j = \frac{1}{10000^{2j/d_k}}$, $d_k$ denotes the number of dimensionality for the positional encoding, and $j \in [0, \frac{d_k}{2} - 1]$

*Position Decoder.* The above dense vector $pos = \{pos_k\}_{k=1}^n$ is fed into a position decoder, which produces the sequence of pooling coefficients $\theta = \{\theta_k\}_{k=1}^n$. Specifically, the position embeddings are obtained from a BiGRU followed by a fully-connected layer, which is defined as:

$$\{h_k\}_{k=1}^n = BiGRU(\{pos_k\}_{k=1}^n) \tag{7}$$

$$\theta_k = FC(h_k) \tag{8}$$

where $h_k$ is the output of the BiGRU at the position $k$, FC denotes the fully-connected layer.

Afterwards, the holistic vector for text $T$ can be represented as $\tilde{T} \in \mathbb{R}^d$. The matching score between the visual and textual embeddings is defined as the cosine similarity, which can be calculated by:

$$Sim(\tilde{G}, \tilde{T}) = \frac{\tilde{G}^\top \tilde{T}}{||\tilde{G}|| \cdot ||\tilde{T}||} \tag{9}$$

### 3.3 Loss Function Module

**Bi-directional Triplet Ranking Loss.** The bi-directional triplet ranking objective is employed to encourage the alignment of image-text pair $(\tilde{G}, \tilde{T})$. Specially, we only consider optimizing the hardest mismatched samples which cause highest loss [5]. It can be calculated by:

$$L_R = \sum_{(\tilde{G}, \tilde{T})} [\alpha - Sim(\tilde{G}, \tilde{T}) + Sim(\tilde{G}, \tilde{T}')]_+ + [\alpha - Sim(\tilde{G}, \tilde{T}) + Sim(\tilde{G}', \tilde{T})]_+ \tag{10}$$

where $\alpha$ is a margin hyperparameter and $[x]_+ \equiv max(x, 0)$, $(\tilde{G}, \tilde{T})$ denotes the groud truth image-text pair, $(\tilde{G}, \tilde{T}')$ and $(\tilde{G}', \tilde{T})$ denote the mismatched pairs, $\tilde{T}' = argmax_{a \neq \tilde{T}} Sim(\tilde{G}, a)$ and $\tilde{G}' = argmax_{b \neq \tilde{G}} Sim(b, \tilde{T})$ are the hardest negative samples in a mini-batch.

**Load Balancing Loss.** To encourage samples routed to experts in the pooling module to be well-balanced, an auxiliary loss is implemented for load balancing [6]. For each modality, this auxiliary loss is added to the total model loss during training. Given $m$ experts and a batch $\mathcal{B}$, the loss can be described as:

$$L_B = m \sum_{i=0}^{m} (S_i \cdot P_i) \tag{11}$$

where $S_i$ denotes the proportion of samples dispatched to expert $i$, $P_i$ indicates the proportion of routing probability distributed to expert $i$. $S_i, P_i$ can be calculated by:

$$S_i = \frac{1}{\mathcal{B}} \sum_{x=0}^{\mathcal{B}} \mathbb{1}\{(argmax\ p(x)) = i\} \tag{12}$$

$$P_i = \frac{1}{\mathcal{B}} \sum_{x=0}^{\mathcal{B}} p_i(x) \tag{13}$$

where $\mathbb{1}\{condition\}$ denotes a conditional expression that the result is 1 when the condition is satisfied, 0 otherwise. Besides, $p_i(x)$ is the probability of sample $x$ routed to expert $i$ and $x$ refers to the image $G$ in the visual modality or the text $T$ in the textual modality.

As Eq. 11 demonstrates, the auxiliary loss encourages both $S_i$ and $P_i$ to have values of $\frac{1}{m}$. Moreover, the number of experts $m$ is multiplied to maintain the auxiliary loss constant as $m$ varies.

The ultimate loss is then derived by combining the triplet ranking loss and the load balancing loss, which is described as:

$$L = L_R + \beta(L_B^G + L_B^T) \tag{14}$$

where $\beta$ is a trade-off hyperparameter, $L_B^G$ and $L_B^T$ denote the load-balancing losses of the visual modality and the textual modality, respectively.

## 4  Experiment

### 4.1  Experiment Setup

**Datasets.** To evaluate effectiveness, extensive experiments are conducted on two public benchmark datasets. MS-COCO [14] dataset contains 123,287 images and each image is annotated with five sentences. Following [5], the dataset is split into 113,287 training images, 5,000 validation images and 5,000 test images. Likewise, we experiment with two kinds of evaluation setting: 1) MS-COCO 1K averaging results from 5 folds of 1K test images; and 2) MS-COCO 5K testing on the full 5K test images. Flickr30K [15] contains 31,000 images and 155,000 captions. Following the split in [10], we use 1,000 images for validation, 1,000 images for testing, and 29,000 images for training.

**Evaluation Metrics.** We choose the widely used metric Recall@K (R@K, K = 1, 5, 10), which means the percentage of true samples in the retrieved top-k lists. The evaluation tasks contain image-to-text retrieval and text-to-image retrieval. We also use RSUM, which indicates the sum of all R@K metrics, to represent the overall performance.

**Implementation Details.** Our model is implemented with the Pytorch library. We use the Adam optimizer to train the model for 35 epochs. The mini-batch size is 128 and the learning rate is set to 5e-4 with a decaying factor of 10% for every 15 epochs. The feature dimension $d$ is set to 1024. The margin hyperparameter $\alpha$ is selected as 0.2. The trade-off parameter $\beta$ is set to 0.01 and the number of experts is set to 4. We use Faster R-CNN (Resnet-101 pretrained on Visual Gnome) to extract region features. We employ BiGRU and BERT to encode each word. It enables us to compare the performance of our pooling methods with existing methods that utilize diverse text backbones.

### 4.2  Evaluation Results

We compare our proposed MoPE framework with several recent state-of-the-art methods on the two benchmarks in Table 1 and Table 2. Considering that the methods in the tables employ different textual backbones and the same visual backbone, our MoPE utilizes the corresponding textual feature extractor to compare, respectively. Table 1 presents the quantitative results on Flickr30K

**Table 1.** Comparison results of the cross-modal retrieval on the Flickr30K dataset in terms of Recall@K(R@K). The best results are highlighted in bold.

| Methods | Image-to-Text | | | Text-to-Image | | | RSUM |
|---|---|---|---|---|---|---|---|
| | R@1 | R@5 | R@10 | R@1 | R@5 | R@10 | |
| **Text Encoder:** BiGRU | | | | | | | |
| SCAN [10] | 67.4 | 90.3 | 95.8 | 48.6 | 77.7 | 85.2 | 465.0 |
| LIWE [22] | 69.6 | 90.3 | 95.6 | 51.2 | 80.4 | 87.2 | 474.3 |
| CVSE [19] | 70.5 | 88.0 | 92.7 | 54.7 | 82.2 | 88.6 | 476.7 |
| IMRAM [1] | 74.1 | 93.0 | 96.6 | 53.9 | 79.4 | 87.2 | 484.2 |
| GPO [2] | 76.5 | 94.2 | **97.7** | 56.4 | 83.4 | 89.9 | 498.1 |
| VSRN [13] | 71.3 | 90.6 | 96.0 | 54.7 | 81.8 | 88.2 | 482.6 |
| MoPE(ours) | **77.8** | **94.4** | 97.2 | **56.8** | **84.0** | **89.9** | **500.1** |
| **Text Encoder:** BERT | | | | | | | |
| CAMERA [16] | 78.0 | 95.1 | 97.9 | 60.3 | 85.9 | 91.7 | 508.9 |
| GPO [2] | 81.7 | 95.4 | 97.6 | 61.4 | 85.9 | 91.5 | 513.5 |
| VSRN++ [13] | 79.2 | 94.6 | 97.5 | 60.6 | 85.6 | 91.4 | 508.9 |
| MoPE(ours) | **82.7** | **96.0** | **98.1** | **63.6** | **87.7** | **92.7** | **520.8** |

**Table 2.** Comparison results of the cross-modal retrieval of the MS-COCO 1K setting in terms of Recall@K(R@K). The best results are highlighted in bold.

| Methods | Image-to-Text | | | Text-to-Image | | | RSUM |
|---|---|---|---|---|---|---|---|
| | R@1 | R@5 | R@10 | R@1 | R@5 | R@10 | |
| **Text Encoder:** BiGRU | | | | | | | |
| SCAN [10] | 72.7 | 94.8 | 98.4 | 58.8 | 88.4 | 94.8 | 507.9 |
| LIWE [22] | 73.2 | 95.5 | 98.2 | 57.9 | 88.3 | 94.5 | 507.6 |
| CVSE [19] | 69.2 | 93.3 | 97.5 | 55.7 | 86.9 | 93.8 | 496.4 |
| IMRAM [1] | 76.7 | 95.6 | 98.5 | 61.7 | 89.1 | 95.0 | 516.6 |
| GPO [2] | 78.5 | **96.0** | **98.7** | 61.7 | **90.3** | **95.6** | 520.8 |
| VSRN [13] | 76.2 | 94.8 | 98.2 | 62.8 | 89.7 | 95.1 | 516.8 |
| MoPE(ours) | **79.0** | 95.9 | 98.1 | **63.1** | 90.1 | 95.2 | **521.4** |
| **Text Encoder:** BERT | | | | | | | |
| CAMERA [16] | 77.5 | 96.3 | 98.8 | 63.4 | 90.9 | 95.8 | 522.7 |
| GPO [2] | 79.7 | 96.4 | 98.9 | 64.8 | 91.4 | **96.3** | 527.5 |
| VSRN++ [13] | 77.9 | 96.0 | 98.5 | 64.1 | 91.0 | 96.1 | 523.6 |
| MoPE(ours) | **80.1** | **96.4** | **98.9** | **65.1** | **91.6** | 96.1 | **528.2** |

where our proposed method outperforms recent approaches with both BiGRU and BERT. Specifically, compared with the baseline SCAN [10], our proposed MoPE obtain 7.5% relative gains on RSUM, where R@1 achieves 15.4% and 16.9% at two directions, respectively. Moreover, in comparison to the state-of-the-art approach GPO [2], we obtain relative 0.4% (498.1 → 500.1) and 1.4% (513.5 → 520.8) improvements on RSUM.

Table 2 shows the 5-fold 1K results on MS-COCO. It can be seen that our MoPE outperforms state-of-the-arts on most evaluation metrics. Specifically, based on the textual encoder BiGRU, our method obtains competitive results

**Table 3.** Comparison results of different aggregators and manually chosen pooling method evaluated with MS-COCO 5K setting in terms of Recall@K(R@K). The best results are highlighted in bold.

| Methods | Image-to-Text | | | Text-to-Image | | | RSUM |
|---|---|---|---|---|---|---|---|
| | R@1 | R@5 | R@10 | R@1 | R@5 | R@10 | |
| **Text Encoder:** BiGRU | | | | | | | |
| SelfAttn [7] | 52.3 | 80.9 | 89.5 | 35.2 | 65.8 | 77.9 | 401.6 |
| Manual | 53.6 | 82.2 | 90.3 | 37.5 | 68.0 | 79.4 | 411.0 |
| GPO [2] | 55.6 | 83.5 | 90.9 | 39.0 | 69.5 | 80.7 | 419.2 |
| ADPOOL [25] | 53.4 | 82.2 | 90.5 | 39.2 | **71.1** | **81.5** | 417.9 |
| MoPE(ours) | **55.8** | **84.0** | **91.0** | **39.3** | 69.8 | 80.9 | **420.8** |

compared with GPO [2] in terms of R@1, gaining about 2.3% improvements for text-to-image retrieval, which is the main concern in practical application. Moreover, our method is slightly lower than GPO in very few metrics, such as R@5 and R@10. We analyze that our method mainly improves the performance of precise matching as R@1 indicates, while GPO may do good at fuzzy image-text retrieval because it uses the same pooling operator for all data.

To verify the performance of our pooling strategy, we compare some advanced pooling aggregators with ours in Table 3. The experiments are evaluated under the MS-COCO 5K setting based on the textual encoder BiGRU. The manual setting indicates a carefully manually selected combination of simple pooling methods, which refers to 5-MaxPool for visual features and MeanPool for textual features. We can see that it even performs better than the complicated pooling strategy (SelfAttn [7]), as is validated in [2] and [25]. Afterwards, our pooling method achieves the best performance on most of metrics.

**Table 4.** The effect of the hyperparameters ($m$ and $\beta$) on Flickr30K dataset. The language encoder is BERT. The best results are highlighted in bold. (Default $m$ is 4 and default $\beta$ is 0.01.)

| Methods | Image-to-Text | | | Text-to-Image | | | RSUM |
|---|---|---|---|---|---|---|---|
| | R@1 | R@5 | R@10 | R@1 | R@5 | R@10 | |
| Number of Pooling Experts | | | | | | | |
| m = 2 | 81.8 | 95.9 | **98.5** | 63.2 | 86.9 | 92.2 | 518.5 |
| m = 4 | **82.7** | **96.0** | 98.1 | **63.6** | **87.7** | **92.7** | **520.8** |
| m = 8 | 81.7 | 96.0 | 98.3 | 63.1 | 87.3 | 92.2 | 518.6 |
| Trade-off Parameter for Loss | | | | | | | |
| $\beta = 0.1$ | 81.3 | **96.5** | **98.3** | 62.7 | 87.2 | 92.3 | 518.3 |
| $\beta = 0.01$ | **82.7** | 96.0 | 98.1 | **63.6** | **87.7** | **92.7** | **520.8** |
| $\beta = 0.001$ | 81.5 | 96.1 | 98.0 | 62.3 | 86.9 | 92.3 | 517.1 |

## 4.3   Ablation Analysis

In this section, we analyze the roles that different factors play in the MoPE framework. There are four aspects studied on Flickr30K with the language

**Table 5.** The effect of route gating module, aggregation expert module and load balancing loss on Flickr30K dataset. The language encoder is BERT. The best results are highlighted in bold. (Symbol "w/o X" denotes that the module X is removed.)

| Methods | Image-to-Text | | | Text-to-Image | | | RSUM |
|---|---|---|---|---|---|---|---|
| | R@1 | R@5 | R@10 | R@1 | R@5 | R@10 | |
| MoPE | **82.7** | **96.0** | **98.1** | **63.6** | **87.7** | **92.7** | **520.8** |
| SCAN$_{BERT}$ | 74.9 | 93.5 | 97.0 | 56.5 | 82.5 | 88.9 | 493.3 |
| w/o RouteGating & $L_B$ | 81.7 | 95.4 | 97.6 | 61.4 | 85.9 | 91.5 | 513.5 |
| w/o $L_B$ | 80.7 | 95.1 | 98.1 | 62.3 | 87.2 | 92.6 | 516.0 |
| w/o AggregationExpert | 75.5 | 93.6 | 96.5 | 56.6 | 83.7 | 90.2 | 496.1 |

encoder BERT for the ablation analysis: hyperparameters, route gating module, aggregation expert module and load balancing loss.

**The Effect of the Hyper Parameters.** We investigate how the number of pooling experts $m$ and the trade-off parameter $\beta$ affect the model performance. As Table 4 shows, our MoPE framework achieves better performance when $m = 4$. The less one ($m = 2$) and the larger one ($m = 8$) both have lower results on R@1, which is the most important metric for image-text retrieval. Moreover, it can be seen that a smaller trade-off parameter $\beta$ can improve the model performance, yet a too small trade-off parameter for loss ($\beta = 0.001$) limits the load balance across the pooling experts which degrades performance.

**The Effect of Route Gating Module and Load Balancing Loss.** Considering that the load balancing loss module is attached to the route gating module, we design two variants to explore the effect of these two modules: 1) **w/o RouteGating** & $L_B$ refers to the model excluding the route gating module and load balancing loss in both image and text branches; and 2) **w/o $L_B$** denotes the model removing the load balancing loss in both branches. The experiment results are shown in Table 5. Without route gating module and load balancing loss, the performance decreases obviously on all metrics compared with the original MoPE and corresponding baseline (SCAN [10] with the same text encoder BERT). Especially, RSUM reduces 1.4%, R@1 decreases 3.5% and 1.2% for image retrieval and text retrieval respectively in comparison to MoPE. In conclusion, the route gating module with load balancing loss is quite pivotal to the improvement of the aggregation for both image region features and word features, which selects a special feature aggregator for each sample. Meanwhile, when load balancing loss is removed, the result in Table 5 demonstrates that properly constraining the load balance across different feature aggregators can improve performance.

**The Effect of Aggregation Expert Module.** To justify the effectiveness of aggregation expert module, we remove it in our model and the result is denoted as **w/o AggregationExpert**. Specifically, we replace it with a carefully manually selected combination of simple pooling methods which is the same as the manually setting in Table 3 (MeanPool for textual features and 5-MaxPool for visual features). As is shown in Table 5, without the aggregation expert module, performance drops significantly in comparison to the original MoPE, particularly in terms of R@1, validating the crucial role of our aggregation expert module in the model.

### 4.4   Case Study

In order to verify the effectiveness of our MoPE, we use two representative samples for the visual modality and the textual modality, respectively, to show which pooling experts they are routed to. As illustrated in Fig. 3, Histogram 1 presents the images' probability distribution for routing to four experts and we can see that image(a) is routed to expert 1 and image(b) is allocated to expert 2. Obviously, it indicates that two images are belong to different subsets, which contain different intra-modality relationships. Besides, the relationships between semantic entities in text(a) and text(b) are different, and Histogram 2 also shows that two texts are routed to different pooing experts for feature aggregation. Compared to the probability distribution for two images, the probabilities of all experts for two texts are almost even, which indicates that the discrepancy between texts is less than the discrepancy between images.

(a)The man with pierced ears is wearing glasses and an orange hat.   (b) A black and white dog is running in a grassy garden surrounded by a white fence.

**Fig. 3.** The visualization results of our MoPE

## 5   Conclusion

In this paper, we propose a novel mixture of pooling experts framework for image-text retrieval. Concretely, we design a route gating module to adaptively allocate each sample to the appropriate pooling operator based on the intra-modality relationship. Moreover, we present an aggregation expert module that combines multiple pooling experts to aggregate fragment features for different samples. To encourage a balanced load across pooling experts, we propose an auxiliary load balancing loss. Extensive experiments are conducted to thoroughly investigate the effectiveness of our proposed method, and the results clearly demonstrate the superiority of our MoPE framework. Furthermore, our proposed framework could improve the aggregation procedure of fine-grained features, and therefore it fits a wide range of visual-textual analysis tasks for enhancing feature representation.

**Acknowledgements.** This work is supported by National Key Research and Development Program of China (2021YFC3340600), the Science and Technology Program of Shanghai, China (Grant No. 22511104300, 21ZR1423800), the Shanghai Municipal Science and Technology Major Project (2021SHZDZX0100) and the Fundamental Research Funds for the Central Universities.

**Disclaimer.** The authors have no competing interests to declare that are relevant to the content of this article.

# References

1. Chen, H., Ding, G., Liu, X., Lin, Z., Liu, J., Han, J.: IMRAM: iterative matching with recurrent attention memory for cross-modal image-text retrieval. In: Proceedings of the IEEE/CVF Conference on Computer Vision and Pattern Recognition, pp. 12655–12663 (2020)
2. Chen, J., Hu, H., Wu, H., Jiang, Y., Wang, C.: Learning the best pooling strategy for visual semantic embedding. In: Proceedings of the IEEE/CVF Conference on Computer Vision and Pattern Recognition, pp. 15789–15798 (2021)
3. Chen, T., Deng, J., Luo, J.: Adaptive offline quintuplet loss for image-text matching. In: Vedaldi, A., Bischof, H., Brox, T., Frahm, J.-M. (eds.) ECCV 2020. LNCS, vol. 12358, pp. 549–565. Springer, Cham (2020). https://doi.org/10.1007/978-3-030-58601-0_33
4. Chen, Y., et al.: More than just attention: improving cross-modal attentions with contrastive constraints for image-text matching. In: Proceedings of the IEEE/CVF Winter Conference on Applications of Computer Vision, pp. 4432–4440 (2023)
5. Faghri, F., Fleet, D.J., Kiros, J.R., Fidler, S.: VSE++: improving visual-semantic embeddings with hard negatives. arXiv preprint arXiv:1707.05612 (2017)
6. Fedus, W., Zoph, B., Shazeer, N.: Switch transformers: scaling to trillion parameter models with simple and efficient sparsity. J. Mach. Learn. Res. **23**(1), 5232–5270 (2022)
7. Han, N., Chen, J., Xiao, G., Zhang, H., Zeng, Y., Chen, H.: Fine-grained cross-modal alignment network for text-video retrieval. In: Proceedings of the 29th ACM International Conference on Multimedia, pp. 3826–3834 (2021)
8. Jacobs, R.A., Jordan, M.I., Nowlan, S.J., Hinton, G.E.: Adaptive mixtures of local experts. Neural Comput. **3**(1), 79–87 (1991)
9. Kim, S., Shim, K., Nguyen, L.T., Shim, B.: Semantic-preserving augmentation for robust image-text retrieval. In: 2023 IEEE International Conference on Acoustics, Speech and Signal Processing (ICASSP), ICASSP 2023, pp. 1–5. IEEE (2023)
10. Lee, K.H., Chen, X., Hua, G., Hu, H., He, X.: Stacked cross attention for image-text matching. In: Proceedings of the European Conference on Computer Vision (ECCV), pp. 201–216 (2018)
11. Lepikhin, D., et al.: GShard: scaling giant models with conditional computation and automatic sharding. arXiv preprint arXiv:2006.16668 (2020)
12. Li, J., Liu, L., Niu, L., Zhang, L.: Memorize, associate and match: embedding enhancement via fine-grained alignment for image-text retrieval. IEEE Trans. Image Process. **30**, 9193–9207 (2021)

13. Li, K., Zhang, Y., Li, K., Li, Y., Fu, Y.: Image-text embedding learning via visual and textual semantic reasoning. IEEE Trans. Pattern Anal. Mach. Intell. **45**(1), 641–656 (2022)

14. Lin, T.-Y., et al.: Microsoft COCO: common objects in context. In: Fleet, D., Pajdla, T., Schiele, B., Tuytelaars, T. (eds.) ECCV 2014. LNCS, vol. 8693, pp. 740–755. Springer, Cham (2014). https://doi.org/10.1007/978-3-319-10602-1_48

15. Plummer, B.A., Wang, L., Cervantes, C.M., Caicedo, J.C., Hockenmaier, J., Lazebnik, S.: Flickr30k entities: collecting region-to-phrase correspondences for richer image-to-sentence models. In: Proceedings of the IEEE International Conference on Computer Vision, pp. 2641–2649 (2015)

16. Qu, L., Liu, M., Cao, D., Nie, L., Tian, Q.: Context-aware multi-view summarization network for image-text matching. In: Proceedings of the 28th ACM International Conference on Multimedia, pp. 1047–1055 (2020)

17. Shazeer, N., et al.: Outrageously large neural networks: the sparsely-gated mixture-of-experts layer. arXiv preprint arXiv:1701.06538 (2017)

18. Vaswani, A., et al.: Attention is all you need. arXiv (2017)

19. Wang, H., Zhang, Y., Ji, Z., Pang, Y., Ma, L.: Consensus-aware visual-semantic embedding for image-text matching. In: Vedaldi, A., Bischof, H., Brox, T., Frahm, J.-M. (eds.) ECCV 2020. LNCS, vol. 12369, pp. 18–34. Springer, Cham (2020). https://doi.org/10.1007/978-3-030-58586-0_2

20. Wang, L., Li, Y., Huang, J., Lazebnik, S.: Learning two-branch neural networks for image-text matching tasks. IEEE Trans. Pattern Anal. Mach. Intell. **41**(2), 394–407 (2018)

21. Wehrmann, J., Kolling, C., Barros, R.C.: Adaptive cross-modal embeddings for image-text alignment. In: Proceedings of the AAAI Conference on Artificial Intelligence, vol. 34, pp. 12313–12320 (2020)

22. Wehrmann, J., Souza, D.M., Lopes, M.A., Barros, R.C.: Language-agnostic visual-semantic embeddings. In: Proceedings of the IEEE/CVF International Conference on Computer Vision, pp. 5804–5813 (2019)

23. Zhang, G., Wei, S., Pang, H., Zhao, Y.: Heterogeneous feature fusion and cross-modal alignment for composed image retrieval. In: Proceedings of the 29th ACM International Conference on Multimedia, pp. 5353–5362 (2021)

24. Zhang, K., Mao, Z., Wang, Q., Zhang, Y.: Negative-aware attention framework for image-text matching. In: Proceedings of the IEEE/CVF Conference on Computer Vision and Pattern Recognition, pp. 15661–15670 (2022)

25. Zhang, Z., et al.: Improving visual-semantic embedding with adaptive pooling and optimization objective. arXiv preprint arXiv:2210.02206 (2022)

# Multi-modal Video Topic Segmentation with Dual-Contrastive Domain Adaptation

Linzi Xing[1]([✉]), Quan Tran[2], Fabian Caba[2], Franck Dernoncourt[2], Seunghyun Yoon[2], Zhaowen Wang[2], Trung Bui[2], and Giuseppe Carenini[1]

[1] University of British Columbia, Vancouver, BC, Canada
{lzxing,carenini}@cs.ubc.ca
[2] Adobe Research, San Francisco, CA, USA
{qtran,caba,dernonco,syoon,zhawang,bui}@adobe.com

**Abstract.** Video topic segmentation unveils the coarse-grained semantic structure underlying videos and is essential for other video understanding tasks. Given the recent surge in multi-modal, relying solely on a single modality is arguably insufficient. On the other hand, prior solutions for similar tasks like video scene/shot segmentation cater to short videos with clear visual shifts but falter for long videos with subtle changes, such as livestreams. In this paper, we introduce a multi-modal video topic segmenter that utilizes both video transcripts and frames, bolstered by a cross-modal attention mechanism. Furthermore, we propose a dual-contrastive learning framework adhering to the unsupervised domain adaptation paradigm, enhancing our model's adaptability to longer, more semantically complex videos. Experiments on short and long video corpora demonstrate that our proposed solution, significantly surpasses baseline methods in terms of both accuracy and transferability, in both intra- and cross-domain settings.

**Keywords:** Topic Segmentation · Video Understanding · NLP

## 1 Introduction

Video Topic Segmentation aims to break stretches of videos into smaller segments consisting of video frames or clips consistently addressing a common topic. As an example given in Fig. 1, video topic segmentation does the job of segmenting a creative livestream video (e.g., from livestream platform *Behance*[1]) into a sequence of topical-coherent pieces (Seg1–Seg8), placing a boundary where a topic transition happens. This task can enhances both human-to-human and human-to-system interactions in modern social contexts, improving real-time engagement on live streaming platform [9]. In particular, the relatively coarse-grained temporal structure of the input video produced by video topic segmentation is shown to not only simplify video comprehension and helps viewers find

---

[1] https://www.behance.net/live.

---

L. Xing—Work done while the first author was an intern at Adobe Research.

© The Author(s), under exclusive license to Springer Nature Switzerland AG 2024
S. Rudinac et al. (Eds.): MMM 2024, LNCS 14556, pp. 410–424, 2024.
https://doi.org/10.1007/978-3-031-53311-2_30

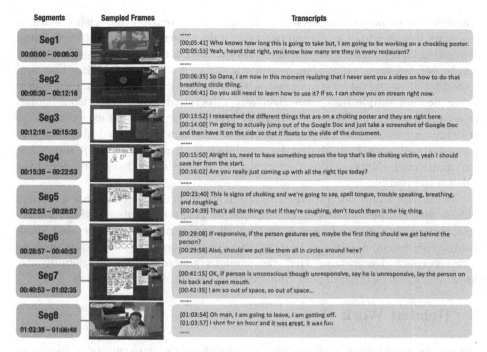

| Segments | Sampled Frames | Transcripts |
|---|---|---|
| **Seg1**<br>00:00:00 – 00:06:30 | | [00:05:41] Who knows how long this is going to take but, I am going to be working on a chocking poster.<br>[00:05:53] Yeah, heard that right, you know how many are they in every restaurant? |
| **Seg2**<br>00:06:30 – 00:12:16 | | [00:06:35] So Dana, I am now in this moment realizing that I never sent you a video on how to do that breathing circle thing.<br>[00:06:41] Do you still need to learn how to use it? If so, I can show you on stream right now. |
| **Seg3**<br>00:12:16 – 00:15:35 | | [00:13:52] I researched the different things that are on a choking poster and they are right here.<br>[00:14:00] I'm going to actually jump out of the Google Doc and just take a screenshot of Google Doc and then have it on the side so that it floats to the side of the document. |
| **Seg4**<br>00:15:35 – 00:22:53 | | [00:15:50] Alright so, need to have something across the top that's like choking victim, yeah I should save her from the start.<br>[00:16:02] Are you really just coming up with all the right tips today? |
| **Seg5**<br>00:22:53 – 00:28:57 | | [00:23:40] This is signs of choking and we're going to say, spell tongue, trouble speaking, breathing, and coughing.<br>[00:24:39] That's all the things that if they're coughing, don't touch them is the big thing. |
| **Seg6**<br>00:28:57 – 00:40:53 | | [00:29:08] If responsive, if the person gestures yes, maybe the first thing should we get behind the person?<br>[00:29:58] Also, should we put like them all in circles around here? |
| **Seg7**<br>00:40:53 – 01:02:35 | | [00:41:15] OK, if person is unconscious though unresponsive, say he is unresponsive, lay the person on his back and open mouth.<br>[00:42:35] I am so out of space, so out of space... |
| **Seg8**<br>01:02:35 – 01:06:48 | | [01:03:54] Oh man, I am going to leave, I am getting off.<br>[01:03:57] I shot for an hour and it was great, it was fun. |

**Fig. 1.** A Behance exemplar about making a choking victim poster. The left side of the figure illustrates the video's timeline after topic segmentation. The right side shows the transcript with words indicating a segment's topic **highlighted**.

content of interest easily. More importantly, it can substantially benefit other key video understanding tasks such as video summarization [29], and query-driven video localization [38].

Early computational models for video segmentation primarily targeted shot or scene detection by merely leveraging surface visual features like spatiotemporal aspects or frame colors [6,31–33]. These approaches typically measure the temporal similarity along a video's timeline to predict shot/scene boundaries. Despite the difference in definition from shot/scene segmentation (discussed in Sect. 2), the task of video topic segmentation focuses more on topic-related semantics in the video, which is not necessarily aligned with visual changes. As shown in Fig. 1, the visual background remains similar in Seg3–Seg7 for a considerable time, even though the streaming topic changes drastically. Moreover, prior video segmentation methods mostly focused on short videos with clear visual changes and simple patterns [13,16,18]. These distinctive features of short videos could be emphasized in model design or learned in supervised setups, making such models less adaptable to longer, more nuanced video content, like documentaries or instructional livestreams.

To address the aforementioned issues, in this paper, we first propose a simple yet effective multi-modal model for video topic segmentation, which can take both the aligned video transcript and visual frames as input. This ability considerably enhances the model's performance, as textual and visual features can work together to more comprehensively reflect the input video's topic-related semantics [2,17]. Similar to the formulation used for text topic segmentation [21,24,40], we treat video topic segmentation as a sequence labeling task and introduce a neural model equipped with a cross-modal attention mechanism going beyond simple fusion to effectively integrate textual and visual signals in a complementary manner. We initially conduct intra-domain experiments by training and testing our proposed segmenter on a newly collected YouTube corpus equipped with high-quality human labeling. Empirical results show that our multi-modal approach outperforms a set of baseline video segmenters by a substantial margin.

To further adapt our proposed model trained on YouTube videos to longer videos with complex visuals and semantics, we propose an unsupervised domain adaptation strategy empowered by the sliding window inference and dual-contrastive learning scheme. Further experiments on two out-of-domain long video corpora demonstrate that the model's generality can be significantly improved through applying our dual-contrastive adaptation approach.

## 2  Related Work

***Topic Segmentation*** seeks to uncover the semantic structure of a document (either monologue [40] or dialogue [39]) by dividing it, typically a sequence of sentences, into topical-coherent segments. Recently, a number of supervised neural solutions have been introduced owing to the availability of large-scale labeled corpora sampled from *Wikipedia* [21], with section marks as gold segment boundaries. Most of these neural segmentation approaches follow the same strategy to simply interpret text segmentation as a sequence labeling problem and further tackle it using a variety of hierarchical neural sequence labelers [21,23,40,41].

Inspired by above-mentioned neural text segmenters, our paper similarly frame video topic segmentation as a sequence labeling task due to the availability of textual input (video transcript) and utilize a hierarchical neural sequence labeling framework as the basic architecture of our proposal. Then we extend such framework into the multi-modal setting, with the injection of visual signals from video frames by adding a cross-modal attention network on top of it. More details will be presented in Sect. 3.

***Shot and Scene Segmentation*** are closely related to video topic segmentation but more narrowly focused. A shot is a sequence of frames from a continuous camera capture. Hence, most shot segmentation techniques mainly use the visual modality to group video frames into shots. Conversely, a scene is more semantically intricate than a shot, representing a series of related shots that depict events defined by elements like actions, places, and characters. These elements are mostly found in narrative videos such as movies. Thus, past scene segmentation techniques are primarily developed for movies exclusively, aiming to group consecutive shots into scenes based on their visual consistency, spatiotemporal features, or shot color similarity. Similar but more difficult than shot or scene segmentation, our work focuses on video topic segmentation, which can be deemed as an extension of text topic segmentation with "topic" defined as a relatively self-contained collection of semantically close information. Notably, the topic's definition relies heavily on the video's context and domain, and isn't strictly bound to the concepts of shots or scenes as previously defined. From a machine learning perspective, the more dynamic nature of shot/scene segmentation favours the visual signal, while in our task, the video topic segmentation relies more heavily on the semantic signals carried by natural language [2,17]. Therefore, our video topic segmenter is designed to extend from a topic segmentation framework for text, by integrating visual frames as the auxiliary signal.

*Contrastive Learning* algorithms aim to learn data representations by enlarging the distance between dissimilar samples and meanwhile minimizing the distance between similar samples with contrastive loss functions [27]. These techniques have been observed promising in domain adaptation for both uni-modal and cross-modal settings [5,19,20]. Recent works on using contrastive learning for video domain adaptation mainly focus on transferring models from source to target domain, where both domains consist of short videos [20]. In contrast, our work attempts to adapt the model trained on short videos to the low-resource domain containing long videos with subtler visual changes, by utilizing the semantic overlaps within a single modality or between two modalities to guide the contrastive learning process.

## 3   Neural Video Topic Segmentation

### 3.1   Problem Definition

Inspired by recent neural-based supervised approaches for text topic segmentation for text, we similarly frame video topic segmentation as a sequence labeling task, with sentences in the video transcript as the units for labeling. More precisely, given an input video with (1) the transcript as a sequence of timestamped sentences, and (2) a sequence of timestamped video frames, our model will predict a binary label for each transcript sentence to indicate whether or not the sentence indicates a topic segment boundary. Formally,

<u>Given</u>: A video $v$ with its transcript $T_v$ as a sequence of sentences $\{s_1, s_2, ..., s_n\}$ along with start time offsets $\{b_1, b_2, ..., b_n\}$/end time offsets $\{e_1, e_2, ..., e_n\}$, and video frames $X_v = \{x_1, x_2, ... , x_m\}$, as a single frame $x_i$ has timestamp $t_i$.

<u>Predict</u>: A sequence of labels $\{l_1, l_2, ..., l_{n-1}\}$ for the sequence of transcript sentences, where $l$ is a binary label, 1 means the corresponding sentence overlaps a video topic segment boundary, 0 otherwise. We do not predict the label for the last sentence $s_n$, as it is by definition equal to 1, i.e., the end of the last segment.

### 3.2   Model Architecture

Figure 2 illustrates the detailed framework of our proposed video topic segmenter, which is similar in architecture to TextSeg [21]. It comprises two hierarchically linked encoding layers: one as text encoder for contextualized encoding

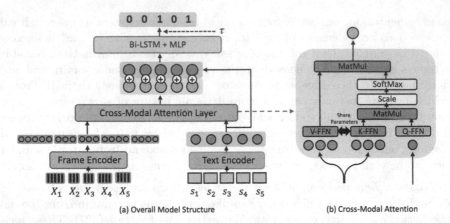

**Fig. 2. Left**: The overall architecture of our proposed multi-modal video topic segmentation model, with four sub-modules (in Sect. 3.2) coded in different colors. **Right**: The detailed illustration of cross-modal attention. (Color figure online)

within a sentence (orange in Fig. 2) and the other for contextualized encoding between sequence units (green in Fig. 2). To allow both textual and visual modalities contribute complementarily to the model's prediction, we add a frame encoder module (blue in Fig. 2) and a cross-modal attention mechanism (red in Fig. 2). This design choice extends the framework to the multi-modal setting, accepting now both the video transcript and frames as input and making prediction of topic segment boundaries based on them.

The text encoder module $E_t$ yields low-level features for sentences in the video transcript. Different from the proposal in [21] using the BiLSTM + attention as the backbone for text encoder, here we adopt the pre-trained vanilla BERT [7] in accordance to its achieved superiority on text segmentation observed in [40] and [26]. Parallel to the text encoder, a frame encoder $E_f$ is introduced to extract features for visual signals with the standard pretrained ResNet-18 [14]. Formally, given a transcript sentence $s_i$ with its time interval as $[b_i, e_i]$ and a set of video frames $X_i = \{x_1^i, ..., x_m^i\}$ associated with this sentence, where each frame (e.g., the $k^{th}$ frame) in the set has timestamp $t_k^i \in [b_i, e_i]$, we can obtain the sentence representation $tr_i = E_t(s_i)$ and its corresponding set of frame representations $FR_i = \{fr_1^i, ..., fr_m^i\}$, where $fr_k^i = E_f(x_k^i)$.

Next, we propose to use a cross-modal attention mechanism to produce a text-aware visual representation for each sentence, rather than obtaining the visual representation by naively operating mean-pooling over the set of frame representations covered by the sentence interval. This design is motivated by the observation that transcript sentences and video frames are in an one-to-many relation, and frames sharing more semantics with the text should be given more attention weights for visual representation generation. Irrelevant frames sharing no or little semantics with the text may negatively affect the quality of the fused multi-modal representation passed to the subsequent module [37]. In practice, the cross-modal attention module adopts the standard scaled dot-product attention function proposed in [36]. With the transcript sentence representation $tr_i$ and its corresponding frame representation set $FR_i = \{fr_1^i, ..., fr_m^i\}$ as query and key (value), we compute the text-aware visual representation $vr_i$ as:

$$vr_i = A_i V_i, \tag{1}$$

$$A_i = softmax(\frac{q_i K_i^T}{\sqrt{d_k}}) \tag{2}$$

where $q_i \in \mathbb{R}^{1 \times d_k}$, $K_i \in \mathbb{R}^{m \times d_k}$, and $V_i \in \mathbb{R}^{m \times d_k}$ denote the query vector, key and value matrices generated by passing the sentence representation and frame representations through three parallel feedforward layers, namely Q-FFN, K-FFN and V-FFN respectively. More formally, we have $q_i = Q\text{-}FFN(tr_i)$, $K_i = K\text{-}FFN(FR_i)$, $V_i = V\text{-}FFN(FR_i)$, where K-FFN and V-FNN share the same parameters and thus produce identical key and value matrices following [22,25].

Then all the obtained text-aware visual representations $\{vr_1, ..., vr_n\}$ are concatenated with their corresponding sentence representations $\{tr_1, ..., tr_n\}$ and fed into a BiLSTM layer which performs contextualization and returns hidden states. Next, a multilayer perceptron (MLP) followed by Softmax serves as a topic boundary predictor to make binary predictions regarding the input hidden states according to a threshold $\tau$ tuned on the validation set. More specifically, if a transcript sentence's output probability exceeds $\tau$, it's marked as 1, indicating a segment boundary. The entire model is fine-tuned using cross-entropy loss.

We train and test this model on a newly collected YouTube corpus (in Sect. 5.1) and empirically verify its in-domain effectiveness (reported in Table 2). We leverage it as the source model in the next section to help deliver the domain adaptation strategy coupled with our proposed multi-modal framework.

## 4  Long Video Adaptation

The rise of (live-)streaming platforms has increased the demand for topical segmentation of videos on these platforms. Unlike videos (e.g., on YouTube) with careful pre-editing and segment labeling provided by their creators, videos on (live-)streaming platforms (e.g., *BBC* documentaries and creative livestreams on *Behance*) are usually extremely long, with sparse visual changes, and more importantly, time-consuming to obtain segment annotations. Thus, it is impractical to learn a fully-supervised model on such videos, and the segmentation model described in Sect. 3.2 (source model) trained on short YouTube videos (source domain) might underperform when applied to these extensively long videos (target domain). To adapt the source model for the above-described target domain, we propose to equip it with two strategies, namely Sliding Window Inference (Sect. 4.1) and Dual-Contrastive Adaptation (Sect. 4.2).

### 4.1  Simple Sliding Window Inference

Due to the length discrepancy between videos from the source and target domains, directly applying the source model to the target input taking the full transcript sentence sequence as input is observed to cause extremely sparse boundary predictions [12]. Therefore, we propose to calibrate the input length by first breaking the long target input into snippets by a sliding window with the size consistent with the source input length (i.e., the average length of the YouTube videos). After applying the source model on every snippet and aggregating outputs of all snippets, we eventually make binary prediction for each sentence if its aggregated probability exceeds the threshold $\tau$ pre-tuned on the source domain. Formally, given a long video input from the target domain with length $= n$ and a fixed window size $= k$ ($n \gg k$), we can create $n - k + 1$ snippets $\{\mathbb{S}_1, ..., \mathbb{S}_{n-k+1}\}$ where each snippet consists of $k$ consecutive sentences by sliding the window over the input with the stride of 1. As a result, each transcript sentence $s_m$ can be covered by up to $k$ snippets. Once we apply the source model to all snippets, we can obtain multiple probability predictions for $s_m$. We then aggregate these probability predictions associated with $s_m$ by taking average:

$$\bar{p}_m = \frac{1}{k} \sum_{i=1}^{k} p_m^{\mathbb{S}_i} \tag{3}$$

Finally, we predict that $s_m$ falls on a segment boundary if $\bar{p}_m > \tau$.

### 4.2  Dual-Contrastive Adaptation

To transfer the pre-trained source model to the target domain while still preserving its performance on the source domain, we follow a more sophisticated unsupervised domain adaptation paradigm, which leverages both labeled source data and unlabeled target data. Specifically, we fix the frame and text encoder ($E_t$ and $E_f$) while updating the rest of the model in two steps, where the first step updates the model on unlabeled target data using two contrastive learning objectives namely *intra-modal contrastive loss* and *cross-modal contrastive loss*, while the second step updates the model on source data with the supervised training scheme described in Sect. 3.2.

As the overview of the first step shown in Fig. 3, given a collection of the paired sentence representations and their corresponding frame representation sets $\{tr_i, FR_i\}_{i=1}^{b}$ in a training batch (size $= b$) produced by the text and frame encoder from the target domain, we need two distinct projection heads to map these sentence/frame representations into a shared space. Here we use *Q-FFN* and *K-FNN* (which shares parameters with *V-FNN*) in the cross-modal attention to serve as projection heads for textual and visual modality respectively. Thus we have:

$$q_i = \text{Q-FFN}(tr_i), K_i = \text{K-FFN}(FR_i) \tag{4}$$

where $q_i$ and $K_i = \{k_1^i, ..., k_m^i\}$ denote the projected sentence embedding and the set of projected frame embeddings covered by the sentence. As frames covered by the same sentence are more likely to share similar semantics, we pull the frames attached to the same sentence closer and push the ones from different sentences far apart, by minimizing the intra-modal contrastive loss for visual modality defined as:

$$l^{intra} = -\sum_{i=1}^{b} log \frac{exp(\tilde{k}^i \cdot \tilde{k}^i)/\tau}{\sum_{j=1}^{b} exp(\tilde{k}^i \cdot \tilde{k}^j)/\tau} \tag{5}$$

where $\tilde{k}^i \in_R FR_i$ and $\tau$ is the hyper-parameter of temperature. To learn the semantic relation between modalities in the target domain, we first obtain the visual representation matched with a sentence by averaging all frames covered by the sentence, and then bring semantically close sentence-visual pairs together and push away non-related pairs by minimizing the cross-modal contrastive loss:

$$l^{cross} = -\sum_{i=1}^{b} log \frac{exp(q_i \cdot MP(K_i))/\tau}{\sum_{j=1}^{b} exp(q_i \cdot MP(K_j))/\tau} \tag{6}$$

where $MP$ denotes *Mean-Pooling*. The total dual-contrastive loss is formed as:

$$l = l^{intra} + l^{cross} \tag{7}$$

The above dual-contrastive learning phase is followed by the second step. In this step, the model is further trained on the labeled source data again to leverage the target-domain signals learned by the parameters in cross-modal attention for boundary prediction. This step ensures the model adapted to the target domain still preserves some level of effectiveness when applied to the source domain.

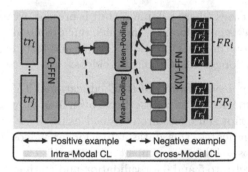

**Fig. 3.** An illustration of updating cross-modal attention module with dual-contrastive adaptation.

**Table 1.** Statistics of corpora used for training and evaluation in Sect. 5.

| Dataset | YouTube | BBC | Behance |
|---|---|---|---|
| # of Vids. | 5,422 | 11 | 575 |
| # of Tokens/Vid. | 1,411 | 2,868 | 11,554 |
| # of Sents./Vid. | 108 | 216 | 1,287 |
| # of Segs./Vid. | 6.7 | 29.1 | 5.2 |
| # of Tokens/Seg. | 209 | 99 | 2,229 |
| # of Sents./Seg. | 16 | 8 | 248 |
| # of Frames/Sec. | 10 | 4 | 4 |
| Avg. Length | 0:09:30 | 0:45:02 | 2:07:51 |

## 5   Experimental Setup

To evaluate the effectiveness and generality of our proposed video topic segmentation model (in Sect. 3) and long video adaptation strategy (in Sect. 4), we conduct experiments on two different settings, namely **Intra-Domain Segment Inference** and **Cross-Domain Segment Inference**.

## 5.1 Intra-domain Dataset – YouTube

For intra-domain segment inference, we train and test models with the data from the same domain (corpus). Due to the lack of large-scale human-annotated dataset in the field of video segmentation, we collect a novel corpus consisting of 5,422 user-generated videos randomly sampled from YouTube. During the video collecting process, we applied filtering criteria to eliminate unsuitable samples to ensure the quality of the dataset to construct. These criteria include constraints on video length (>100 s), word count (>0.5 word/second on average), chapter durations (all chapters with length >10 s), and sentence length (all sentences with length <60 tokens) inspired by [4]. Each video in this corpus is associated with a series of user-defined chapter timestamps indicating the beginning of each topic chapter contained in this video. We thereby use these available chapter beginning timestamps as ground-truth topic segment boundaries since a topic chapter is by definition a main thematic division within a video. The video transcripts are generated by YouTube ASR with token-level offsets and we further exploit a top performing punctuation restoration model [1] to boost the quality of transcript sentence segmentation. Compared with other existing video understanding corpora such as *BBC* [2] (for scene segmentation, size = 11) and *TVSum* [34] (for video summarization, size = 50), our constructed YouTube corpus is (1) larger in size; (2) covering more diverse topics; (3) with reliable segmentation which has already been specified by video creators. We split this corpus into train/dev/test portions with size: 5148/134/140.

## 5.2 Datasets for Cross-Domain Inference

To evaluate our proposal's robustness in cases where a domain-shift is present, we conduct experiments for cross-domain segment inference, in which our proposed supervised segmenter is initially trained on the YouTube video corpus, and tested on two corpora with videos significantly longer than YouTube videos:

**BBC Planet Earth** [2] consists of 11 episodes from the BBC educational TV series Planet Earth. Topic segment boundaries of the dataset have been manually annotated by human experts and transcript sentences are obtained using *Whisper*[2] [30]. As the statistics in Table 1 indicates, this dataset has longer videos with more segments than YouTube videos, while each segment covers much fewer sentences and tokens. We split this dataset into 5 and 6 for validation and testing.

**Behance Livestream** consists of 575 videos sampled from the creative livestream platform *Behance*. Livestreams, in general, are very challenging to segment into coherent sections, since they contain mixes of multi-user dialogues, with visual features that change little over the video. They are also very long, with hours of content in each video. Behance livestreams, in particular, contain another layer of challenge, since they are highly specialized in creative tasks such as animation and image edit with unknown entities and intricate visual

---

[2] https://openai.com/research/whisper.

operations. We believe finding structures in such noisy environment would be not only a good benchmark for future topic segmentation research, but also of great practical value for consumers, who would find it hard to navigate contents in a back-and-forth conversation. Thus, we collect such a corpus and annotate its videos with topic segments given the help from domain experts in animation and image edits. For each video, we pay $23.16 for human annotators to watch and create segments based on the video content. The total cost to annotate 575 videos is $13,317. We split this dataset into validation and test sets with 57 (10%), 518 (90%).

### 5.3   Baselines

We consider the following representative baselines as comparisons:

- **Random:** Given a video with its transcript consisting of $k$ sentences, we first randomly sample the segment number $b \in \{0, ..., k-1\}$ for this video, and then determine if a sentence overlaps a segment boundary with probability $\frac{b}{k}$.
- **BayesSeg** [8] originally proposed for text segmentation by predicting segment boundaries through modeling the lexical cohesion in a Bayesian context.
- **GraphSeg** [11] generates a semantic relatedness graph with sentences as nodes. Segments are then predicted as the maximal cliques in graph.
- **Cross-BERT** [26] is a supervised text topic segmenter representing candidate segment boundaries using their left and right contexts encoded with BERT. The model is trained on YouTube and applied to longer BBC and Behance videos.
- **TransNet** [35]: This model adopts stacked Dilated DCNN blocks as its basic framework and targets video segmentation on shot level. Here we use its publicly available version trained on synthesized video corpora off-the-shelf.
- **LGSS** [31] is a multi-modal movie scene segmenter pre-trained on the limited-scale *MovieScenes* corpus. We use its public version with visual frames as input.
- **X-Tiling** is our extension of *TextTiling* [15]. While TextTiling only allows textual embeddings as input, X-Tiling accepts both textual and visual embeddings and their concatenations as input to compute semantic coherence and then make segment boundary predictions for videos. For fair comparison, the input textual and visual embeddings are produced by pre-trained BERT and mean-pooling over ResNet-18 respectively.

All hyper-parameters required by the above baselines are tuned on the validation portion of the datasets included in this paper.

### 5.4   Evaluation Metrics

We apply three standard metrics in previous literature to evaluate the performances of our proposal and baselines. They are:

– $F_1$, with higher scores denoting better performance. It measures the exact match between ground truth and model's prediction.
– $P_r$ **error score** [10], which fixes the inadequacies of $P_k$ [3] and *WindowD-iff* [28], previously considered as two standard evaluation metrics for text segmentation. Concretely, $P_r$ is the mean of missing and false alarm probabilities, calculated based on the overlap between ground-truth segments and model's predictions within a certain size sliding window. Since it is a penalty metric, lower score indicates better performance.
– $mIoU$ score [42], shortened from the mean Intersection-over-Union. It is calculated by taking average over maximal IoUs of all ground-truth segments to predicted segments. Higher score indicates better performance.

## 5.5 Implementation Details

In our multi-modal video topic segmenter, we employ the [CLS] token representation from `bert-base-uncased` (dimension $d = 768$) for sentence representation and ResNet-18's avg-pooling layer output (dimension $d = 512$) for frame representation. The cross-modal attention's feedforward layers have an output dimension of 768. Our BiLSTM has 2 layers with a hidden size of 256. For YouTube training, we use the Adam optimizer with a learning rate of $1e^{-3}$ and a batch size of 16. In the long video adaptation with dual-contrastive learning, the SGD optimizer is applied with a mini-batch size of 256, learning rate of $3e^{-2}$, and softmax temperature of $1e^{-1}$. Training spans 10 epochs for both supervised learning and domain adaptation, with results averaged over 3 runs. The segmentation threshold, $\tau$, is tuned on validation sets for both intra-domain and cross-domain evaluations.

**Table 2.** Results on YouTube for intra-domain evaluation. **Bold** results indicate the best performance across all comparisons. Underlined results indicate the best performance within their own sub-section. * indicates a fully supervised setting.

| Method | Modalities | Pr ↓ | F1 ↑ | mIoU ↑ |
|---|---|---|---|---|
| Random | – | 45.84 | 48.98 | 37.76 |
| X-Tiling | Text | 39.94 | 51.56 | 50.97 |
| BayesSeg | Text | 40.70 | 50.24 | 49.69 |
| GraphSeg | Text | 38.12 | 51.41 | 51.73 |
| Cross-BERT* | Text | _32.89_ | _60.48_ | _60.00_ |
| X-Tiling | Visual | 37.53 | 52.08 | 53.24 |
| TransNet | Visual | 40.14 | 51.01 | 50.66 |
| LGSS | Visual | 39.77 | 50.87 | 51.13 |
| X-Tiling | Text + Visual | 38.78 | 52.30 | 51.45 |
| NeuralSeg* | Text | 31.91 | 63.23 | 61.36 |
| (Ours) | Visual | 50.18 | 47.59 | 16.15 |
| | Text + Visual | **30.61** | **65.29** | **63.11** |

# 6  Results and Discussion

**Intra-Domain Segment Inference:** Table 2 reports the performance of the chosen baselines and our proposal (NeuralSeg in the table) on the YouTube testing set, while NeuralSeg is trained on the YouTube training set with different input modality settings. Notably, our model significantly outperforms the best baseline, Cross-BERT, even when only using video transcripts (text) as input. But on the other hand, if we train the model by using only video frames (visual) as input, the model's performance is even considerably worse than the random baseline, possibly because that too diverse visual and topic presence in the corpus makes it difficult for the model to learn a meaningful visual input-to-prediction projection. Yet, by combining both modalities with cross-modal attention fusion, the model's performance can be further enhanced compared with only textual modality. These results confirm that the visual information itself may not be sufficient to capture a video's topic-related semantics under a supervised setting, but fusing it together with the textual information can provide a more clear picture of the video's underlying topics.

**Cross-Domain Segment Inference:** Table 3 compares the performance of the baselines and our proposed segmenter pre-trained on YouTube w/ or w/o applying our long video adaptation strategy to two challenging long video corpora. To better investigate the effectiveness of the two contrastive objectives associated with long video adaptation, Table 3 also shows the results by adding each objective (CL-Cross/CL-Intra) individually. The primary takeaway is that our long video adaptation strategy (Window + CL-Dual) significantly and consistently improves the performance achieved by the segmenter initially trained on short YouTube

**Table 3.** Results on BBC and Behance corpora for cross-domain evaluation. **Bolded** and underlined results indicate the best performance across all comparisons and within their own section. † indicates results applied contrastive domain adaptation which are significantly different ($p < 0.05$) from *Ours (Window)*.

| Method | Modalities | BBC | | | Behance | | |
|---|---|---|---|---|---|---|---|
| | | Pr ↓ | F1 ↑ | mIoU ↑ | Pr ↓ | F1 ↑ | mIoU ↑ |
| Random | – | 48.29 | 47.36 | 30.66 | 46.27 | 50.00 | 36.24 |
| X-Tiling [15] | Text | 41.59 | 50.72 | 33.27 | 45.93 | 49.95 | 43.16 |
| BayesSeg [8] | Text | 42.01 | 51.01 | 33.79 | 49.01 | 48.16 | 37.55 |
| GraphSeg [11] | Text | 45.11 | 49.28 | 30.23 | 46.76 | 50.12 | 40.38 |
| Cross-BERT [26] | Text | 44.63 | 49.70 | 31.51 | 46.15 | 49.88 | 41.02 |
| X-Tiling [15] | Visual | 44.48 | 49.89 | 33.26 | 44.56 | 50.73 | 41.81 |
| TransNet [35] | Visual | 42.54 | 49.82 | 31.12 | 46.34 | 49.88 | 40.54 |
| LGSS [31] | Visual | 42.88 | 50.02 | 32.66 | 45.67 | 50.12 | 39.90 |
| X-Tiling [15] | Text + Visual | 43.22 | 50.44 | 32.68 | 45.87 | 50.01 | 41.72 |
| Ours (Plain) | Text + Visual | 43.14 | 54.13 | 31.23 | 45.71 | 51.25 | 34.33 |
| Ours (Window) | Text + Visual | 40.66 | 54.92 | 37.50 | 43.03 | 51.83 | 47.66 |
| Ours (Window + CL-Cross) | Text + Visual | 36.45† | 54.91 | 50.92† | 42.72 | 51.68 | 48.65† |
| Ours (Window + CL-Intra) | Text + Visual | 37.14† | 55.10 | 49.22† | 42.56 | 51.84 | 48.39 |
| Ours (Window + CL-Dual) | Text + Visual | **36.01**† | **55.75** | **51.56**† | **42.25**† | **51.91** | **49.35**† |

videos (Plain). In more detail, we can observe that utilizing the sliding window inference strategy (Plain → Window) already yields noticeable performance gains on both corpora. Furthermore, adding contrastive learning objectives can further boost accuracy on two long video corpora. The greatest enhancement is seen when the model is tuned with dual-contrastive losses.

# 7  Conclusion and Future Work

We present a multi-modal video topic segmentation model accepting both video transcripts and frames as input. Further, we propose a novel unsupervised domain adaptation strategy empowered by a dual-contrastive learning framework to generalize our model pre-trained on short videos to longer videos with more complex content and subtler visual changes. Experiments on two settings (intra-domain and cross-domain segment inference) show that (1) our system achieves the SOTA performance on a newly collected YouTube corpus consisting of large-scale but short videos; (2) When we apply our long video adaption strategy, the model for short videos achieves better performance when transferred to two domains comprising long (live-)stream videos.

# References

1. Alam, T., Khan, A., Alam, F.: Punctuation restoration using transformer models for high-and low-resource languages. In: Proceedings of the Sixth Workshop on Noisy User-Generated Text (W-NUT 2020), pp. 132–142 (2020)
2. Baraldi, L., Grana, C., Cucchiara, R.: A deep siamese network for scene detection in broadcast videos. In: Proceedings of ACM MM 2015, pp. 1199–1202 (2015)
3. Beeferman, D., Berger, A., Lafferty, J.: Statistical models for text segmentation. Mach. Learn. **34**(1), 177–210 (1999). https://doi.org/10.1023/A:1007506220214
4. Cao, X., Chen, Z., Le, C., Meng, L.: Multi-modal video chapter generation. ArXiv abs/2209.12694 (2022)
5. Chen, D., Wang, D., Darrell, T., Ebrahimi, S.: Contrastive test-time adaptation. In: Proceedings of CVPR 2022, pp. 295–305 (2022)
6. Chen, S., Nie, X., Fan, D.D., Zhang, D., Bhat, V., Hamid, R.: Shot contrastive self-supervised learning for scene boundary detection. In: Proceedings of CVPR 2021, pp. 9791–9800 (2021)
7. Devlin, J., Chang, M.W., Lee, K., Toutanova, K.: BERT: pre-training of deep bidirectional transformers for language understanding. In: Proceedings of NAACL 2019, pp. 4171–4186. Association for Computational Linguistics (2019)
8. Eisenstein, J., Barzilay, R.: Bayesian unsupervised topic segmentation. In: Proceedings of EMNLP 2008, pp. 334–343 (2008)
9. Fraser, C., Kim, J., Shin, H., Brandt, J., Dontcheva, M.: Temporal segmentation of creative live streams. In: Proceedings of CHI 2020, pp. 1–12 (2020)
10. Georgescul, M., Clark, A., Armstrong, S.: An analysis of quantitative aspects in the evaluation of thematic segmentation algorithms. In: Proceedings of SIGdial 2006, pp. 144–151 (2006)
11. Glavaš, G., Nanni, F., Ponzetto, S.P.: Unsupervised text segmentation using semantic relatedness graphs. In: Proceedings of the Fifth Joint Conference on Lexical and Computational Semantics, pp. 125–130 (2016)

12. Glavas, G., Somasundaran, S.: Two-level transformer and auxiliary coherence modeling for improved text segmentation. In: Proceeding of AAAI-2020, pp. 2306–2315 (2020)
13. Gygli, M., Grabner, H., Riemenschneider, H., Van Gool, L.: Creating summaries from user videos. In: Fleet, D., Pajdla, T., Schiele, B., Tuytelaars, T. (eds.) ECCV 2014. LNCS, vol. 8695, pp. 505–520. Springer, Cham (2014). https://doi.org/10.1007/978-3-319-10584-0_33
14. He, K., Zhang, X., Ren, S., Sun, J.: Deep residual learning for image recognition. In: Proceedings of CVPR 2016, pp. 770–778 (2016)
15. Hearst, M.A.: Text tiling: segmenting text into multi-paragraph subtopic passages. Comput. Linguist. **23**(1), 33–64 (1997)
16. Jadon, S., Jasim, M.: Unsupervised video summarization framework using keyframe extraction and video skimming. In: Proceedings of ICCCA 2020, pp. 140–145 (2020)
17. James, N., Todorov, K., Hudelot, C.: Combining visual and textual modalities for multimedia ontology matching. In: Declerck, T., Granitzer, M., Grzegorzek, M., Romanelli, M., Rüger, S., Sintek, M. (eds.) SAMT 2010. LNCS, vol. 6725, pp. 95–110. Springer, Heidelberg (2011). https://doi.org/10.1007/978-3-642-23017-2_7
18. Jayaraman, D., Grauman, K.: Slow and steady feature analysis: higher order temporal coherence in video. In: Proceeding of CVPR 2016, pp. 3852–3861 (2016)
19. Kang, G., Jiang, L., Yang, Y., Hauptmann, A.G.: Contrastive adaptation network for unsupervised domain adaptation. In: Proceedings of CVPR 2019, pp. 4893–4902 (2019)
20. Kim, D., et al.: Learning cross-modal contrastive features for video domain adaptation. In: Proceedings of ICCV 2021, pp. 13598–13607 (2021)
21. Koshorek, O., Cohen, A., Mor, N., Rotman, M., Berant, J.: Text segmentation as a supervised learning task. In: Proceedings of NAACL 2018, pp. 469–473 (2018)
22. Kumar, A., Mittal, T., Manocha, D.: MCQA: multimodal co-attention based network for question answering. CoRR abs/2004.12238 (2020)
23. Li, J., Sun, A., Joty, S.: SegBot: a generic neural text segmentation model with pointer network. In: Proceedings of IJCAI-2018, pp. 4166–4172 (2018)
24. Lo, K., Jin, Y., Tan, W., Liu, M., Du, L., Buntine, W.: Transformer over pretrained transformer for neural text segmentation with enhanced topic coherence. In: EMNLP 2021 (Findings), pp. 3334–3340 (2021)
25. Lu, J., Yang, J., Batra, D., Parikh, D.: Hierarchical question-image co-attention for visual question answering. In: Proceedings of NeurIPS 2016, pp. 289–297 (2016)
26. Lukasik, M., Dadachev, B., Papineni, K., Simões, G.: Text segmentation by cross segment attention. In: Proceedings of EMNLP 2020, pp. 4707–4716 (2020)
27. van den Oord, A., Li, Y., Vinyals, O.: Representation learning with contrastive predictive coding. ArXiv abs/1807.03748 (2018)
28. Pevzner, L., Hearst, M.A.: A critique and improvement of an evaluation metric for text segmentation. Comput. Linguist. **28**(1), 19–36 (2002). https://doi.org/10.1162/089120102317341756, https://aclanthology.org/J02-1002
29. Qiu, J., et al.: Semantics-consistent cross-domain summarization via optimal transport alignment. ArXiv abs/2210.04722 (2022)
30. Radford, A., Kim, J.W., Xu, T., Brockman, G., McLeavey, C., Sutskever, I.: Robust speech recognition via large-scale weak supervision. ArXiv abs/2212.04356 (2022)
31. Rao, A., et al.: A local-to-global approach to multi-modal movie scene segmentation. In: Proceedings of CVPR 2020, pp. 10146–10155 (2020)
32. Rasheed, Z., Shah, M.: Scene detection in Hollywood movies and TV shows. In: Proceedings of CVPR 2003, p. II-343 (2003)

33. Rui, Y., Huang, T.S., Mehrotra, S.: Exploring video structure beyond the shots. In: Proceedings of the IEEE International Conference on Multimedia Computing and Systems, pp. 237–240 (1998)

34. Song, Y., Vallmitjana, J., Stent, A., Jaimes, A.: TVSum: summarizing web videos using titles. In: Proceeding of CVPR 2015, pp. 5179–5187 (2015)

35. Souček, T., Lokoč, J.: TransNet V2: an effective deep network architecture for fast shot transition detection. ArXiv abs/2008.04838 (2020)

36. Vaswani, A., et al.: Attention is all you need. In: Advances in Neural Information Processing Systems, vol. 30 (2017)

37. Wang, Z., Zhong, Y., Miao, Y., Ma, L., Specia, L.: Contrastive video-language learning with fine-grained frame sampling. In: Proceedings of AACL-IJCNLP 2022, pp. 694–705 (2022)

38. Xiao, S., et al.: Boundary proposal network for two-stage natural language video localization. In: Proceedings of AAAI-2021 (2021)

39. Xing, L., Carenini, G.: Improving unsupervised dialogue topic segmentation with utterance-pair coherence scoring. In: Proceedings of SIGdial 2021, pp. 167–177 (2021)

40. Xing, L., Hackinen, B., Carenini, G., Trebbi, F.: Improving context modeling in neural topic segmentation. In: Proceedings of AACL-IJCNLP 2020, pp. 626–636 (2020)

41. Xing, L., Huber, P., Carenini, G.: Improving topic segmentation by injecting discourse dependencies. In: Proceedings of the 3rd Workshop on Computational Approaches to Discourse, pp. 7–18 (2022)

42. Zhu, W., Pang, B., Thapliyal, A.V., Wang, W.Y., Soricut, R.: End-to-end dense video captioning as sequence generation. In: Proceedings of COLING 2022, pp. 5651–5665 (2022)

# Unsupervised Multi-collaborative Learning Network for 3D Face Reconstruction

Wenlong Lu, Suping Wu$^{(\boxtimes)}$, Xitie Zhang, and Shengjia Zhang

School of Information Engineering, Ningxia University, Yinchuan 750021, China
pswuu@nxu.edu.cn

**Abstract.** Monocular image-based 3D fine face reconstruction techniques aim to reconstruct 3D faces with rich face details from a single image. Existing methods have achieved remarkable results, but they cannot accurately extract light and perspective information, resulting in reconstructed faces with poor details and more noise. To this end, we propose a method for 3D face reconstruction using multi-collaborative learning network. Specifically, we design an illumination and view feature extraction network, which combines the ideas of FCN-style [8] point-by-point addition and UNet-style [7] channel dimension splicing and fusion. In this way, features at different scales can be better filtered and integrated, and key semantic information can be extracted, we can make full use of the effective features at different scales to obtain accurate light and view information. In addition, in order to be able to obtain a more comprehensive and realistic albedo characterisation, we propose a multi-resolution co-optimization module. Extensive experimental results on several evaluation datasets show that our method achieves significant improvements and excellent performance compared to state-of-the art methods.

**Keywords:** 3D Face Reconstruction · Multi-Collaborative · Co-Optimization · Fusion

## 1 Introduction

Three-dimensional (3D) face reconstruction has emerged as a prominent research focus within the realms of computer vision and graphics. Leveraging 2D face images for the purpose of 3D face reconstruction holds the potential to accurately recover intricate 3D face details, encompassing both the reconstruction of geometric structures and textural attributes. This technological advancement holds vast potential across diverse domains, including but not limited to face recognition [1], emotion analysis [2], face alignment [3], and the realm of virtual reality.

In recent years, the rise of deep learning has catalyzed the proliferation of supervised 3D face reconstruction methodologies. These methodologies pivot on

© The Author(s), under exclusive license to Springer Nature Switzerland AG 2024
S. Rudinac et al. (Eds.): MMM 2024, LNCS 14556, pp. 425–436, 2024.
https://doi.org/10.1007/978-3-031-53311-2_31

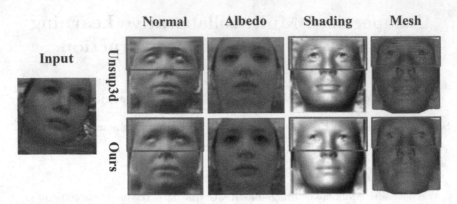

**Fig. 1.** Qualitative comparison between our approach and Unsup3d. We reconstructed the eyes and eye surrounds better, with more accurate face details and less noise, and solved the problem of reconstruction errors (eyebags shown in shading).

the utilization of images or depth information sourced from multiple perspectives, subsequently culminating in the derivation of a 3D model through processes such as matching and integration. However, the acquisition of multi-view images, depth data or large amounts of annotated data is frequently accompanied by challenges, particularly within real-world scenarios where access to sole single-view images is often the norm. In response, the domain of unsupervised single-view 3D face reconstruction algorithms has surged in importance. These algorithms proficiently reconstruct intricate 3D face models using only an image captured from a lone view, all while sidestepping the need for extensive volumes of labor-intensive labeled data. Consequently, the exploration of unsupervised single-view 3D face reconstruction algorithms has emerged as a compelling, significant avenue for research.

The current landscape of 3D face models, including the 3D Morphable Model (3DMM) [4] and Flame [5], has significantly propelled the advancement of 3D face reconstruction techniques. Nevertheless, these models encounter challenges in effectively addressing issues related to identity preservation and intricate face detail reconstruction. To enhance precision beyond the scope of 3DMM, non-parametric approaches, such as the shape-from-shading algorithm [16], have demonstrated the capacity to model 3D facial structures without relying on 3DMM assumptions. In light of the advancements in deep learning, this algorithm has been further refined by SFS-Net [17], enabling the modeling of intrinsic facial attributes. Additionally, data-driven techniques [19, 20] have been introduced to directly acquire knowledge of facial geometry under the guidance of authentic or synthetic reference data. Despite this, they still fail to accurately represent the details of the entire face and produce more noise.

Furthermore, the Unsup3d [6] methodology is founded on the principles of unsupervised face reconstruction with symmetrical constraints. Comprising multiple modules, this approach endeavors to regress various components of image formation, encompassing the 3D shape, texture, view, and light parameters of the object. The overarching objective is to ensure that the rendered image aligns with the input images, thereby maintaining visual similarity. Despite these efforts, the modules are trained using unsupervised methods on image datasets, and challenges like low image resolution and imprecise light prediction persist. The reconstruction results have problems such as abnormalities in the eyes and around the eyes, abnormal facial details, and a lot of noise. Consequently, the level of authenticity achieved in the reconstruction remains suboptimal (see Fig. 1). GAN2Shape [13] avoids weakly symmetric constraints but brings heavy per-image optimization. LiftedGAN [20] transforms the framework to a generative model but also needs optimization to address real-world images.

We propose an innovative feature extraction network that combines the UNet [7] and fully convolutional neural network (FCN) [8] frameworks. By integrating features across multiple levels, this network enhances the perception of global context within input images, thereby facilitating improved extraction of light and view features. Additionally, we propose a feature extraction module consisting of a coarse branch and a fine branch that synergizes high and low resolutions to extract albedo maps, resulting in the derivation of more comprehensive and realistic albedo representations. We achieve better results than existing methods on the BFM and CelebA datasets. In summary, this paper makes the following contributions:

1. We propose a groundbreaking feature extraction framework (FUNet). The framework coordinates FCN-style pointwise addition and UNet-style channel-dimensional splicing and fusion methods, thereby enabling feature fusion across different hierarchical layers to extract more detailed and accurate view features and light features.
2. We propose a multi-resolution co-optimization module (MRCM) feature extraction module for extracting albedo maps. This module extracts albedo maps with high and low resolutions to obtain a more comprehensive and realistic albedo representation.
3. The experimental results on the BFM dataset and the CelebA dataset show that our method has a significant improvement. Finally, we provide an ablation experiment demonstrating the effectiveness of our method.

## 2   Method

The overall goal of the network is to learn a model $\Phi$ that takes as input an image of an object instance and decouples it into albedo, depth, light, and view (see Fig. 2). Among them, light and view are predicted by a feature extraction network FUNet network proposed in this paper, where the view includes translation and rotation information. The albedo map is predicted by two encoder-decoder networks, where the input of one network is $64 \times 64$ of the original image size, and the input of the other network is the image of $128 \times 128$ size obtained by bilinear interpolation. Although the inputs of the two encoder-decoder networks are different, their output size is $64 \times 64$, so the two features can be fused directly using Eq. 2. The depth information represents the depth value of the corresponding pixel on the image, which is predicted by an encoder-decoder network, and then the normal information is calculated according to the depth information, and the shadow shading information is calculated by combining the surface normal and illumination information, which is used to represent the illumination of the corresponding point strength. Next, the shadow and albedo maps are fused to obtain the frontal view of the input image. Finally, the frontal view is combined with the predicted view to render a result similar to the input image. The entire network is trained by constraining the difference between the rendered image and the input image. The entire network in this paper does not introduce additional supervision information, and only uses the information of the input image itself for training.

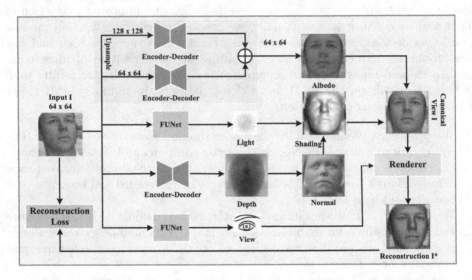

**Fig. 2. Model architecture.** This is the overall architecture of our network. Our network decomposes an input image into albedo, illumination, depth, and view. In the albedo map section, finer albedo information is extracted through our multi-resolution co-optimization module. In the light and view network, our FCN-UNet collaborative network extracts more accurate and rich light and view information. Finally, the final face model is generated based on our predicted four components, and the corresponding 2D image is drawn. The entire network operates unsupervised.

Specifically, this paper follows the framework design of the photo geometry autoencoder in Unsup3d [6], and learns four factors through the entire network: albedo, depth, light, and view. These four factors are explicitly modeled, as shown in Eq. 1.

$$\hat{I} = \Phi(d, a, v, l) = \Pi(\Lambda(d, a, l), d, v) \tag{1}$$

where $\Lambda$ is the light function, which is colored according to the albedo map $a$, the depth map $d$, and the light $l$. Then the reprojection function $\Pi$ renders and generates an image corresponding to the view $v$ and the depth map $d$.

To constrain the canonical views of $d$ and $a$ to represent full frontal faces, we adopt the following loss function based on the symmetry assumption.

$$\hat{\mathbf{I}}' = \Pi\left(\Lambda\left(a',d',l\right),d',w\right), a' = \text{flip } a, d' = \text{flip } d \qquad (2)$$

Among them, $a'$ and $d'$ are the inverted versions of $a$ and $d$, so that the two sides can be as similar as possible. Unsupervised 3D face reconstruction from images can be performed without the 3DMM assumption.

## 2.1   FCN-UNet Collaborative Network

Despite the capability of traditional unsupervised 3D face reconstruction methodologies to retrieve illumination and view particulars, their utilization of a mere 6-layer convolutional neural network for feature extraction imposes inherent limitations on the scope of captured information. This limitation can adversely affect the quality of the final reconstruction.

In order to extract more detailed and accurate view features and light features, while capturing the boundaries of the target more accurately, we designed a groundbreaking feature extraction framework called FUNet. This framework coordinates the FCN-style [8] point-by-point addition and the UNet style [7] channel dimension splicing and fusion methods, thereby achieving feature fusion across different hierarchical layers (as shown in Fig. 3).

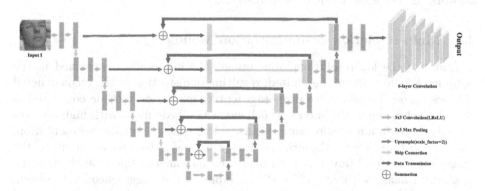

**Fig. 3.** This is our light and view extraction network, which consists of an autoencoder that follows the UNet architecture and combines FCN-style point-by-point addition and UNet-style channel dimension splicing, and an encoder composed of a 6-layer convolutional neural network.

The feature extraction network first feeds the image into an autoencoder that follows the UNet [7] architecture, where all convolution operators zero-pad the input so that the input and output have the same resolution. Thanks to the upsampling and downsampling structure of the network, multiple layers between the encoder and decoder can be fully utilized. The downsampling process helps capture global illumination information and view information in the image, while the upsampling process can restore details and local features. This combination helps generate more accurate illumination and view images. When performing cross-layer connections, the ideas of FCN-style point-by-point addition and UNet-style channel dimension splicing and fusion are combined. First, the features of the corresponding layers are added point by point, and then the results are spliced and fused in the channel dimension. Because the influence of illumination and view usually involves features of different scales, cross-layer connection helps retain low-level details, and this fusion method combining FCN and UNet, first of all, extracts more upsampling and downsampling processes directly than traditional CNN, which can obtain better details. Secondly, compared with the cross-layer connection fusion method of FCN, more channels are retained in the upsampling process, making the network spread context information to higher resolution. Finally, compared with the cross-layer connection method of UNet, the network not only fuses on the channel dimension, but also fuses point-to-point, making the cross-layer link better fused and retaining more information.connections to extract more detailed light and view information. In addition, the structure of the UNet autoencoder can accurately capture the boundaries of the object, which allows the reconstructed results to have better contours. The local receptive field of CNN limits its ability to perceive the global context information of the input image, and it does not have the other advantages mentioned above. Finally, after being processed by this structure, it is passed to an encoder composed of a 6-layer convolutional neural network to obtain the required features.

## 2.2   Multi-resolution Co-optimization Module

Considering the low-resolution input image, the information contained in the obtained albedo map will be limited, resulting in distortion of the reconstructed 3D face shape. Therefore, we propose a feature extraction module composed of thick branches and fine branches, which extracts albedo maps with high and low resolutions to obtain a more comprehensive and realistic albedo representation. The coarse branch uses the original image with size $64 \times 64$ as input to the image encoder to obtain more realistic information; the fine branch uses the upsampled image with size $128 \times 128$ as input to the image encoder to obtain more detailed and more richer information, both generate albedo features of size $64 \times 64$. Finally, fine and coarse features are fused to obtain the desired albedo map.

The high and low resolution albedo fusion formula is as follows:

$$\alpha = \lambda_1 f_1 \left( I_{64 \times 64} \right) + \lambda_2 f_2 \left( I_{128 \times 128} \right) \tag{3}$$

where $I_{64 \times 64}$ is an image with a resolution of $64 \times 64$, and $I_{128 \times 128}$ is an image with a resolution of $128 \times 128$ obtained by bilinear interpolation; $f_1()$ and $f_2()$

are the coarse branch network and the fine branch network respectively; $\lambda_1$ and $\lambda_2$ are the weight factors of the coarse branch and the fine branch respectively; $\alpha$ is the final albedo feature. Because the low-resolution image is the original image, the obtained albedo information can maintain the authenticity of the obtained features, while the high-resolution image is obtained by bilinear interpolation, which can obtain finer and richer albedo information, but the authenticity is not as good as the original image, so when the weight factors are all set to 0.5, better reconstruction results are obtained, and subsequent ablation Experiments have also verified this view.

## 3  Experiments

### 3.1  Setup

**Dataset and Metrics:** We tested our approach on two face datasets: CelebA [9]and BFM [10]. CelebA is a rich face dataset, containing 20,2599 photos of 10,177 celebrity identities, each photo is marked with features, including a face box label box, 5 face feature point coordinates and 40 attribute labels. BFM is a dataset built using the 3DMM model, collected by structural light and laser, described by 70,000 points for each model before processing, and 53,490 points for processing. We use it to evaluate the quality of a 3D reconstruction. The quality of our reconstruction is evaluated quantitatively using the ratio invariant depth error (SIDE) [11] and the mean angle deviation (MAD) between the true depth of the ground and the normal of the predicted depth calculation.

**Scale Invariant Depth Error (SIDE)** is the error between the face depth with our reconstructions and the actual face depth, which is defined as follows:

$$\text{SIDE}\,(d^*, d') = \sqrt{\frac{\sum_{u,v} \Delta_{u,v}^2}{W \times H} - \left(\frac{\sum_{u,v} \Delta_{u,v}}{W \times H}\right)^2} \tag{4}$$

In the formula, $\Delta_{u,v} = \ln(d^*) - \ln d'$, $d'$ is the depth value of the reconstructed face, and $d^*$ is the depth value of the actual face.

**Mean Angle Deviation (MAD)** is defined as the average error between the normals with our reconstructions and the actual face.

$$\text{MAD}\,(n^*, n') = \frac{\sum_{u,v} r\,(n^*, n')}{W \times H} \tag{5}$$

$r$ is the angle between two vectors starting from the same pixel, $n^*$ is the surface normal vector calculated by using the depth truth value of the data set, and $n'$ is the surface normal vector calculated by using the predicted depth value.

**Implementation Details:** We use Adam to train images with a batch size of 16. The size of our input image is $64 \times 64$, and a total of 12 epochs are trained. For ease of display, we sample the depth map etc. to $256 \times 256$.

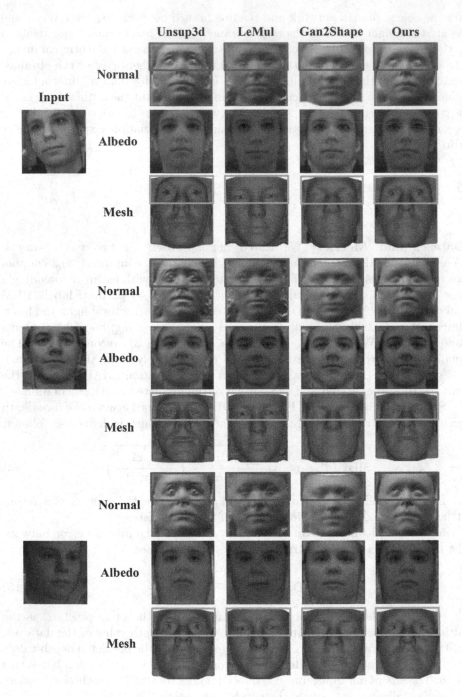

**Fig. 4.** Qualitative comparison of reconstructed albedo maps, normal maps and 3D models on the BFM dataset.

## 3.2    Comparison

**Quantitative Comparison.** We trained and tested our algorithm on the BFM dataset, the reconstruction of depth and normal is measured using the scale-invariant depth error (SIDE) and average angle deviation (MAD), the results are shown in Table 1. It can be seen that we are far ahead of the current state-of-the-art methods Unsup3d [6], LeMul [12], Gan2Shape [13], LAP [14] and L2R [15] in both metrics.

**Table 1.** Quantitative comparison with other methods on BFM dataset.

| No | Method | $SIDE(\times 10^{-2})\downarrow$ | $MAD(deg.)\downarrow$ |
|----|--------|-------------------------|------------------|
| (1) | Unsup3d  [6] | 0.793 ± 0.143 | 16.51 ± 1.56 |
| (2) | LeMul  [12] | 0.834 ± 0.169 | 15.49 ± 1.50 |
| (3) | Gan2Shape  [13] | 0.756 ± 0.156 | 15.35 ± 1.49 |
| (4) | LAP  [14] | 0.721 ± 0.128 | 15.53 ± 1.42 |
| (5) | L2R  [15] | 0.710 ± 0.139 | 14.70 ± 1.16 |
| (6) | Ours | **0.703 ± 0.136** | **14.12 ± 1.32** |

At the same time, we also compared the training time with the baseline, and the results are shown in Table 2. It can be seen that although the training time of an epoch is longer than the baseline, only 12 epochs need to be trained, while the baseline needs to be trained for 30 epochs, so the total training time is shortened by 28.9%.

**Table 2.** Compare training time to baseline.

| No | Method | Training time/epoch (s) | Number of epochs | Total time (s) |
|----|--------|------------------------|------------------|----------------|
| (1) | Unsup3d  [6] | **1238** | 30 | 37140 |
| (2) | Ours | 2202 | **12** | **26424** ↓28.9% |

**Qualitative Evaluation.** Figure 4 show the results of our method on BFM. It can be seen that the 3D model map, shape map, and albedo map reconstructed by our method are significantly better than other methods. Compared with Unsup3d [6] and LeMul [12], the eyeballs of Unsup3d [6] are abnormally protruding and there are wrong textures on the forehead, and the eyes of LeMul [12] have obvious noise points, but our method reconstructs a more natural and realistic outline. At the same time, there is no large amount of noise in LeMul [12]. Compared with Gan2Shape  [13], our method reconstructs a sharper mouth contour and better 3D face shape. At the same time, in order to better demonstrate the applicability and good effects of our method, we tested our method on the CelebA data set, as shown in Fig. 5. It can be seen that the albedo map extracted by our method has more accurate skin colors and more realistic faces, and the clarity is better than Unsup3d [6] and LeMul [12]. Among them, Gan2Shape [13] uses the trained stylegan, so the albedo map generated will be clearer, but the skin color and authenticity are still inaccurate. This demonstrates the effectiveness of our method. In this way, we can recover a finer 3D face model.

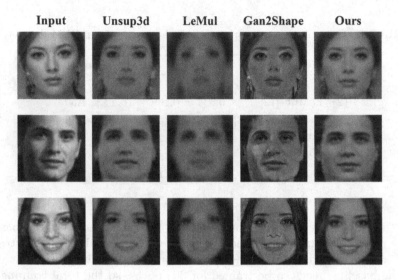

**Fig. 5.** Qualitative comparison of reconstructed albedo maps on the CelebA dataset.

## 3.3   Ablation Study

In this section, we mainly test the influence of the shape FCN-UNet collaborative network (FUNet) and multi-resolution co-optimization module (MRCM) on the performance of the model, mainly relying on the intuitive quantitative results on the BFM dataset, including scale invariant depth error (SIDE) and average angle deviation (MAD). As shown in Table 3, we show the impact of FUNet and MRCM on the model. It can be seen that FUNet has a greater impact on SIDE, and MRCM has a greater impact on MAD. The combination of the two can be obtained better results.

Finally, as shown in Table 4, we also ablated the high and low resolution feature fusion ratio in MRCM. It can be seen that the best effect can be obtained when both $\lambda_1$ and $\lambda_2$ are set to 0.5.

**Table 3.** Ablation study.

| No | Method | $SIDE(\times 10^{-2})\downarrow$ | $MAD(deg.)\downarrow$ |
|----|--------|----------------------------------|------------------------|
| (1) | w/o FUNet | $0.763 \pm 0.152$ | $14.50 \pm 1.48$ |
| (2) | w/o MRCM | $0.718 \pm 0.166$ | $14.98 \pm 1.53$ |
| (3) | Ours full | $\mathbf{0.703 \pm 0.136}$ | $\mathbf{14.12 \pm 1.43}$ |

**Table 4.** Ablation study with different fusion ratios.

| No | Method | $SIDE(\times 10^{-2})\downarrow$ | $MAD(deg.)\downarrow$ |
|----|--------|----------------------------------|------------------------|
| (1) | 0.1origin + 0.9upsampling | $0.719 \pm 0.135$ | $14.25 \pm 1.48$ |
| (2) | 0.3origin + 0.7upsampling | $0.711 \pm 0.142$ | $14.14 \pm 1.37$ |
| (3) | 0.7origin + 0.3upsampling | $0.708 \pm 0.129$ | $14.36 \pm 1.46$ |
| (4) | 0.9origin + 0.1upsampling | $0.712 \pm 0.134$ | $14.56 \pm 1.54$ |
| (5) | Ours(0.5origin + 0.5upsampling) | $\mathbf{0.703 \pm 0.136}$ | $\mathbf{14.12 \pm 1.43}$ |

# 4    Conclusion

This paper presents an innovative feature extraction network that combines the UNet and fully convolutional neural network (FCN) frameworks. By integrating features across multiple levels, this network enhances the perception of global context within input images, thereby facilitating improved extraction of light and view features. Additionally, we propose a feature extraction module that synergizes high and low resolutions to extract albedo maps, resulting in the derivation of more comprehensive and realistic albedo representations. Extensive quantitative and qualitative experiments demonstrate that our approach outperforms state-of-the-art methods. In terms of future work, our forthcoming objectives could encompass further refinement of the outcomes pertaining to the reconstruction of intricate face textures.

**Acknowledgements.** This work was supported by National Natural Science Foundation of China under Grant 62062056, and in part by the Ningxia Graduate Education and Teaching Reform Research and Practice Project 2021.

# References

1. Bahroun, S., Abed, R., Zagrouba, E.: Deep 3D-LBP: CNN-based fusion of shape modeling and texture descriptors for accurate face recognition. Vis. Comput. **39**, 239–254 (2021)
2. Jin, H., Wang, X., Lian, Y., Hua, J.: Emotion information visualization through learning of 3D morphable face model. Vis. Comput. **35**, 535–548 (2019)
3. Guo, J., Zhu, X., Yang, Y., Yang, F., Lei, Z., Li, S.Z.: Towards fast, accurate and stable 3D dense face alignment. In: Vedaldi, A., Bischof, H., Brox, T., Frahm, J.-M. (eds.) ECCV 2020. LNCS, vol. 12364, pp. 152–168. Springer, Cham (2020). https://doi.org/10.1007/978-3-030-58529-7_10

4. Blanz, V., Vetter, T.: A morphable model for the synthesis of 3D faces. In: Proceedings of the 26th Annual Conference on Computer Graphics and Interactive Techniques (1999)

5. Li, T., Bolkart, T., Black, M.J., Li, H., Romero, J.: Learning a model of facial shape and expression from 4D scans. ACM Trans. Graph. (TOG) **36**, 1–17 (2017)

6. Wu, S., Rupprecht, C., Vedaldi, A.: Unsupervised learning of probably symmetric deformable 3D objects from images in the wild. In: 2020 IEEE/CVF Conference on Computer Vision and Pattern Recognition (CVPR), pp. 1–10 (2020)

7. Ronneberger, O., Fischer, P., Brox, T.: U-Net: convolutional networks for biomedical image segmentation. In: Navab, N., Hornegger, J., Wells, W.M., Frangi, A.F. (eds.) MICCAI 2015. LNCS, vol. 9351, pp. 234–241. Springer, Cham (2015). https://doi.org/10.1007/978-3-319-24574-4_28

8. Shelhamer, E., Long, J., Darrell, T.: Fully convolutional networks for semantic segmentation. In: 2015 IEEE Conference on Computer Vision and Pattern Recognition (CVPR), pp. 3431–3440 (2014)

9. Liu, Z., Luo, P., Wang, X., Tang, X.: Deep learning face attributes in the wild. IEEE In: 2015 International Conference on Computer Vision (ICCV), pp. 3730–3738 (2014)

10. Paysan, P., Knothe, R., Amberg, B., Romdhani, S., Vetter, T.: A 3D face model for pose and illumination invariant face recognition. In: 2009 Sixth IEEE International Conference on Advanced Video and Signal Based Surveillance, pp. 296–301 (2009)

11. Eigen, D., Puhrsch, C., Fergus, R.: Depth map prediction from a single image using a multi-scale deep network. In: NIPS (2014)

12. Ho, L., Tran, A., Phung, Q., Hoai, M.: Toward realistic single-view 3D object reconstruction with unsupervised learning from multiple images. In: 2021 IEEE/CVF International Conference on Computer Vision (ICCV), pp. 12580–12590 (2021)

13. Pan, X., Dai, B., Liu, Z., Loy, C.C., Luo, P.: Do 2D GANs Know 3D Shape? Unsupervised 3D shape reconstruction from 2D Image GANs. arXiv: 2011.00844 (2020)

14. Zhang, Z., et al.: Learning to aggregate and personalize 3d face from in-the-wild photo collection. In: 2021 IEEE/CVF Conference on Computer Vision and Pattern Recognition (CVPR), pp. 14209–14219 (2021)

15. Zhang, Z., et al.: Learning to restore 3D face from in-the-wild degraded images. In: 2022 IEEE/CVF Conference on Computer Vision and Pattern Recognition (CVPR), pp. 4227–4237 (2022)

16. Zhang, R., Tsai, P., Cryer, J.E., Shah, M.: Shape from shading: a survey. IEEE Trans. Pattern Anal. Mach. Intell. **21**, 690–706 (1999)

17. Sengupta, S., Kanazawa, A., Castillo, C.D., Jacobs, D.W.: SfSNet: learning shape, reflectance and illuminance of faces in the wild. In: 2018 IEEE/CVF Conference on Computer Vision and Pattern Recognition, pp. 6296–6305 (2017)

18. Abrevaya, V.F., Boukhayma, A., Torr, P.H., Boyer, E.: Cross-modal deep face normals with deactivable skip connections. In: 2020 IEEE/CVF Conference on Computer Vision and Pattern Recognition (CVPR), pp. 4978–4988 (2020)

19. Zeng, X., Peng, X., Qiao, Y.: DF2Net: a dense-fine-finer network for detailed 3D face reconstruction. In: 2019 IEEE/CVF International Conference on Computer Vision (ICCV), pp. 2315–2324 (2019)

20. Shi, Y., Aggarwal, D., Jain, A.K.: Lifting 2D StyleGAN for 3D-aware face generation. In: 2021 IEEE/CVF Conference on Computer Vision and Pattern Recognition (CVPR), pp. 6254–6262 (2020)

# A Region Based Non-overlapping Reference Speech Estimation Method for Speaker Extraction

Yiru Zhang[1], Zeke Li[2,3], Bijing Liu[3,4], Haiwei Fan[2], Yong Yang[3,4], and Qun Yang[1(✉)]

[1] Nanjing University of Aeronautics and Astronautics, Nanjing, China
qun.yang@nuaa.edu.cn
[2] State Grid Fujian Electric Power Dispatching and Control Center, Fuzhou, China
[3] NARI Technology Co., Ltd., Nanjing, China
[4] Beijing Kedong Electric Power Control System Co., Ltd., Beijing, China

**Abstract.** Speaker extraction is a technique that separates the target speech from multi-talker mixtures using a priori information about the target speaker, such as pre-enrolled reference speech. However, in real-world scenarios, the mixture speech is partially overlapped continuous long speech and obtaining a priori information is often challenging. Hence, we propose a framework to estimate the reference speech of participating speakers from non-overlapping input regions and extract target speech. To accurately estimate the regions, we adopt the idea of region proposal to generate multiple speech segment proposals of non-overlapping regions from the input speech mixtures. And then, we cluster these proposed segments into clusters to obtain the best reference speech of each speaker. We conduct experiment on simulated meeting-style test set with different overlap ratio based on LibriSpeech. The experimental results show that the region proposal method can achieve the best performance in speech extraction compared to other reference speech estimation methods.

**Keywords:** Speaker extraction · Continuous speech separation · Region proposal

## 1 Introduction

In multi-talker communications, the speech signals often overlap with interference speech of another speaker. The overlapped speech may affect the performance of downstream tasks. In recent years, some speaker extraction models has been proposed to extracting the target speaker's speech using pre-enrolled signals about the target speaker [1–3]. Speaker extraction can track a certain speaker and focus on speaker identity differentiation. However, applying speaker extraction models to real-world scenarios remains a pressing problem.

---

Y. Zhang and Z. Li—Contributed equally to this research.

ⓒ The Author(s), under exclusive license to Springer Nature Switzerland AG 2024
S. Rudinac et al. (Eds.): MMM 2024, LNCS 14556, pp. 437–447, 2024.
https://doi.org/10.1007/978-3-031-53311-2_32

Currently, most speaker extraction models are based on two assumptions, which are often invalid in real-world scenarios. First, the input speech is highly overlapped, whereas in realistic, speech is a sparsely overlapped continuous long audio stream. For example, in natural meeting, the overlap ratio is usually below 20% [4] and can be hours long. Therefore, the model has to ensure both the separation accuracy for the overlapping segments and the distortion lessness for the overlap-free segments [5]. In addition, the target speaker extraction has to estimate the silence when the target speaker is inactive in a segment [6], which would otherwise generate a large number of insertion errors in speech recognition. Second, the model requires a priori information about the target speaker in advance, such as external speaker enrollments or visual information, etc. However, in practice, this information is difficult to obtain when the users are not a fixed population.

For the first problem, some researchers have proposed the task of continuous speech separation (CSS) [5] and universal speaker extraction [6]. Some researchers [7,8] introduce speaker activity information and visual cue for the sparsely overlapped conditions. For the second problem, estimating reference speech from input speech is an attractive research direction. For example, [9,10] estimate the speaker embeddings from input mixture speech to guide the separation of target speakers. Some approaches [11] propose to obtain speaker identity estimates from preliminary separation results. Nevertheless, these methods requiring the number of speakers to be fixed in advance, which limits their application.

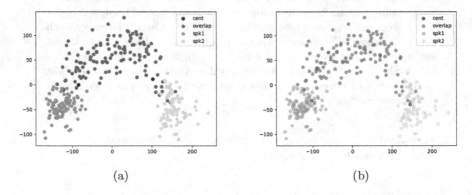

(a)                                         (b)

**Fig. 1.** The PCA plots of speaker embeddings obtained from a sparsely overlapped 2-talker mixture, which contain the single speech of two speakers, the overlapped speech and the estimated cluster centroids. (a) Clustering of all segments. (b) Clustering of estimated non-overlapping segments.

Previous studies have typically addressed these two problems separately. However, in real-world scenarios, both problems coexist. Although sparsely overlapped speech poses a challenge, we can leverage it to obtain participants' reference speech from a large number of single speech segments. CSSUSI [12] has proposed to use clustering to construct an inventory of speaker embeddings from the mixed signals. However, this method apply clustering to all segments without distinguishing whether they overlap or not, which may create some additional clusters representing overlapping regions and shift the clustering centroids towards overlapping regions. As Fig. 1(a) illustrates, when the clustering does not distinguish whether the signals are overlapping or not, the obtained clustering centroids will be biased towards the other speaker. While when we distinguish whether the speech overlaps, the clustering centroids falls in the correct speakers, as Fig. 1(b) illustrates. To mitigate this performance degradation, CSSUSI [12] chose the speech separation using speaker inventory (SSUSI) [13] model, which can employ permutation invariant training (PIT) [14] to compensate for the biased information and wrong selection. However, this method deprives the model of the advantage of being able to track a particular speaker, and makes it difficult to match speaker identities to recognized speech in some applications, such as audio subtitle generation.

Therefore, we propose a speaker extraction framework that does not require external speaker enrollment. We estimate each involved speaker in non-overlapping regions of the input speech, and generates corresponding individual speech embedding for each speaker. This requires the estimated reference audio to be as accurate as possible. (1) We adopt the idea of anchors generation in region proposal network (RPN) [15] and propose a speaker reference selection method. It can generate high confidence non-overlapping region proposals based on overlap detection model. (2) We use clustering methods to obtain the clusters of speakers from non-overlapping regions and estimate the embedding of each speaker. (3) We use an universal speaker extraction model [6] to keep generating the extracted speech for each speaker based on the estimated speaker embeddings. (4) We conduct our experiments on meeting-like long recordings generated by LibriSpeech [16]. The experimental results show that our method outperforms other models in all four scenarios.[1]

## 2   Methodology

### 2.1   Overall Network Structure

Let $y$ denote a continuously mixture speech signal including $N$ speakers and $s_i$ denote the individual target signal of $i-th$ speaker. Since speakers take turns, $y$ contains mostly non-overlapping single speech and some overlapping speech.

$$y = \sum_{i=1}^{N} s_i \tag{1}$$

---

[1] Our code are available online at https://github.com/Ease-3600/TSE-with-ref-selection.

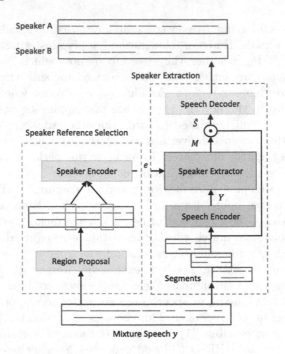

**Fig. 2.** The network architecture of our framework.

Our goal is to extract the speech for each speaker without introducing external enrollment information. Thus, we input only the mixture speech $y$ into the framework, which usually lasts for a long time. We estimate the identity information of each speaker involved from non-overlapping region of y, and extract the speech accordingly. As Fig. 2 shows, the overall structure of our framework contains two branches: a speaker reference selection branch and a speaker extraction branch.

The speaker reference selection branch includes two parts: the region proposal module and the speaker encoder. The region proposal module generates region proposals of different speakers, which are a series of speech segments. The speaker encoder extracts a fixed-dimensional speaker embedding from the region proposals.

For the speaker extraction branch, it includes three components: speech encoder, speaker extractor, and speech decoder. The speech encoder encodes the waveform into acoustic representation. The speaker extractor estimates a mask of target speaker from the mixture speech representation to filter out the speech of other speakers. And the speech decoder can reconstruct the speech signal from the masked speech representation.

In the following sections, we will first introduce the details of the region proposal module for speaker reference selection branch. And then, we will introduce the process of speaker extraction branch.

## 2.2    Region Proposal for Speaker Reference Selection

We use the speaker reference selection branch to estimate the speaker identity information. Given a mixture speech $y$, we first generate non-overlapping speech regions that best represent the participating speakers using the region proposal module. Next, the speaker encoder extracts a fixed dimensional embedding e from the proposed regions as the identity information of the target speaker.

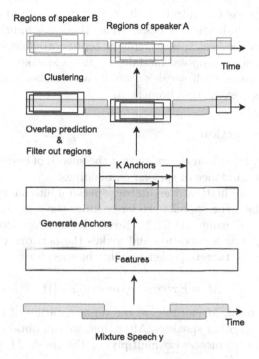

**Fig. 3.** The process of region proposal module.

The key to an estimated speaker embedding that accurately represents speaker identity lies in the region proposal module. It was proposed in Faster R-CNN [15] for object detection and applied by R-CRNN [17] and RPNSD [18] for audio event detection and speaker diarization. The RPN for object detection generates 2-d region proposals while the RPN for audio processing generates 1-d speech segment proposals. In this paper, to obtain accurate speaker embeddings, it is necessary to ensure that the proposed regions are non-overlapping speech with high confidence. Thus, we adopt the idea of region proposal and simplify the process for non-overlapping region estimation as shown in Fig. 3.

Given a mixture speech $y$, we first extract the features of the speech. And then, we generate a set of fixed length anchors centered around each time step. In this paper, we use 20 anchors to cover the speech segments from about 1 s to 20 s.

After that, we feed the generated anchors into a pre-trained overlap detection network to filter out segments with low confidence of non-overlap prediction. Consequently, we can obtain a series of single speaker segments with high confidence. Traditional overlap detection is usually difficult to accurately predict the boundaries. While we use the idea of region proposal to predict speech segments with different lengths and obtain accurate non-overlapping segments.

Finally, we extract the speaker embedding of the filtered proposal segments and cluster them into several clusters. we choose the region closest to the centroids of the clusters as the output of the region proposal module. By this way, we can naturally exclude the few cases where there are multiple speakers in a segment and obtain the best speech segment representing each speaker.

Following the region proposal module, we use a speaker encoder to encode the proposed regions of each speaker into speaker embedding, which we then pass into the speaker extraction branch.

### 2.3   Speaker Extraction

In the speaker extraction branch, we extract the speech of corresponding speaker iteratively with the guidance of speaker embeddings.

As Fig. 2 shows, we first cut the mixture speech $y$ into short segments with a sliding window. Then, the speech encoder encodes the segments into spectrum-like representation $Y$ using 1D-CNN. Next, the speaker extractor takes the speaker embedding $e$ as a condition and makes the network estimate the mask $M$ which only lets the target speaker pass in the speech representation $Y$:

$$M = Extractor(concat([Y, e]))  \qquad (2)$$

The estimated mask $M$ can reserve the representation of the target speaker and filter out those of other speakers. After that, we can obtain the target speech representation $\hat{S}$ by element-wise multiplying the mask $M$ with the mixture speech representation $Y$.

$$\hat{S} = Y \odot M  \qquad (3)$$

Finally, the speech decoder reconstructs the time-domain target signal $\hat{s}$ from the estimated speech representation $\hat{S}$.

## 3   Experiments

### 3.1   Dataset

In our experiments, we use the "train-clean-100" subset of LibriMix [19] as our training set. It contains 100 h of speech from 251 speakers with a sampling rate of 16kHz. Libri2Mix generate 13,900 utterances of mixtures by selecting random utterances from two speakers. For speaker extraction task, we choose each speaker in the mixtures as the target speaker by turns.

We evaluate our framework on simulated 100 meetings based on Lib-riSpeech [16], and each meetings is about 180 s long with 20%, 30% and 40% overlap ratio.

## 3.2 Experimental Settings

The speaker extraction model in our paper is based on the backbone of SPEX+ [20]. The speech encoder and speech decoder are 1D-CNNs of multi-scales. And the speaker extractor contains 4 temporal convolutional network (TCN) stacks. In our experiment, we randomly cut 4 s long segments from the utterances as the input of speaker extraction model.

The experiments for the speaker reference selection are based on the pyannote toolkit [21, 22]. Due to the lengthy input speech, instead of generating anchors for each timestep, we employ a pre-trained overlap detection model to determine the approximate point in time when anchors need to be generated. The model is also used later to filter out the region with low confidence. Then, we employ ECAPA-TDNN [23] to obtain the speaker embeddings of proposed regions and apply agglomerative hierarchical clustering (AHC) to group the speaker embeddings into different clusters.

## 3.3 Baselines

We compare our proposed method with different methods that can select reference speech from non-overlapping regions [24, 25].

(1) oracle: We choose the longest non-overlapping regions of each speaker using the ground-truth segmentation information. It represents the condition that the selected reference speech is correct, which is the topline of our framework.
(2) Speaker Diarization: We use speaker diarization and choose the longest segment of estimated non-overlapping regions.
(3) Cluster all segments: Similar to CSSUSI [12], the method splits the mixture recording into 5 s segments and directly extracts the embeddings of each segment. And then, it applies clustering and obtains the cluster centroids.
(4) Overlap Detection: We perform overlap detection on the mixture speech. Then, we split the non-overlapping regions using a sliding window and obtain the cluster centroids. Due to the length of non-overlapping segments is uncertain, the window size is difficult to determine. In our experiment, we set the window size and step to 5 s and 0.5 s to ensure the accuracy of non-overlapping speech within a window.

**Table 1.** SI-SDR and accuracy of speaker reference selection on test set with different overlap ratio.

| Method | System | SI-SDR(dB) | | | Acc.(%) | | |
|---|---|---|---|---|---|---|---|
| | | 20% | 30% | 40% | 20% | 30% | 40% |
| 1 | Oracle (topline) | 10.78 | 10.44 | 10.28 | 100 | 100 | 100 |
| 2 | Speaker Diarization | 8.69 | 8.77 | 8.35 | 89.0 | 86.0 | 81.0 |
| 3 | Cluster all segments | 9.23 | 8.75 | 9.14 | 82.8 | 79.5 | 83.6 |
| 4 | Overlap Detection | 9.20 | 8.89 | 8.74 | **99.0** | **99.5** | **97.0** |
| 5 | Ours | **9.49** | **9.11** | **9.22** | 89.6 | 89.2 | 94.0 |
| 6 | Ours+WavLM [26] | 10.12 | 9.72 | 9.74 | 89.6 | 89.2 | 94.0 |

## 3.4 Evaluation Metrics

For all of the experiments, we use the scale-invariant signal-to-noise ratio(SI-SDR) [27] to evaluate the quality of speech extracted from overlapped speech. Since we focus on comparing different reference speech selection methods, we also use the accuracy of speaker reference selection as an evaluation metric. We consider segments containing only the target speaker as correct and calculate the proportion of correct segments.

## 3.5 Experimental Results

In this section, we first compare our proposed reference selection method with baselines. In Table 1, we show the speaker extraction performance under a variety of conditions. For conditions with different overlap ratio, our method outperforms other methods and is closest to method 1(topline) in terms of SI-SDR.

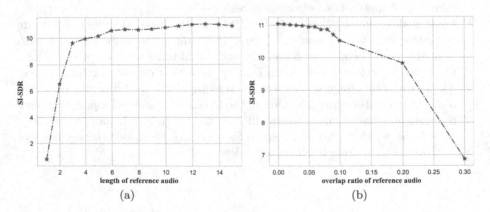

(a)                                    (b)

**Fig. 4.** The plot of speaker extraction performance (SI-SDR) for different length and overlap ratio of reference speech.

It demonstrates the effectiveness of our proposed method for speaker extraction without external speaker enrollment.

Nevertheless, we notice that our method is less accurate for reference speech selection than method 4, but obtains better speaker extraction performance. It suggests that the performance of speaker extraction is not completely affected by the reference speech selection accuracy. Therefore, we explored the effect of the length of the reference speech and the overlap with other speakers on the speaker extraction performance. As shown in Fig. 4(a), we plot the SI-SDR of the speaker extraction model using different lengths of reference speech. The experimental result shows that the longer reference speech helps the speaker extraction. And the performance drops off sharply when the length of reference speech is less than 5 s. In Fig. 4(b), we add different ratios of interfering speech to the reference speech and compare the model performance. We observe that the speaker extraction model can tolerate the reference speech with a small amount of overlap (about 5%), which is consistent with the observation of speaker embedding extraction that the less overlap there is, the closer it is to the main speaker. Therefore, compared to the the method 4 where the length of the reference speech is fixed at 5 s, our proposed method can select speech regions with flexible lengths that can better represent the identity of the speaker.

Although our proposed method obtains high-quality reference audio from the input speech, it is not perfect. There is still a small amount of interfering speech in the estimated reference speech, which causes our model's performance to deteriorate in comparison to the topline. Because of this, we further use the robust WavLM [26] model in place of our speaker encoder. This is due to WavLM's ability to produce consistent speaker embedding despite the presence of interfering speakers in the reference speech. The experimental results of Method 6 demonstrate that the combined use of our system and WavLM considerably reduces the performance difference with topline.

## 4   Conclusion

In this paper, we proposed a speaker extraction framework with two branches for real-world continuous mixture speech that does not require external speaker enrollment. The speaker reference selection branch estimates the speaker embedding from non-overlapping regions of the input speech. In order to obtain non-overlapping regions accurately, we proposed a module based on the idea of region proposal. For the speaker extraction branch, we extracted the speech of each speaker with the guidance of speaker embedding. We trained a universal speaker extraction model to handle sparsely overlapped speech. The experiments on simulated continuous speech showed that the reference speech estimated by our region proposal module enabled the model achieve the closest performance to the topline. As technology advances, more models for real-world speaker extraction will be proposed. Our approach can further liberate these models from the limitation of difficult access to priori speaker information.

# References

1. Xu, C., Rao, W., Chng, E.S., Li, H.: Spex: multi-scale time domain speaker extraction network. IEEE/ACM Trans. Audio Speech Lang. Process. **28**, 1370–1384 (2020). https://doi.org/10.1109/TASLP.2020.2987429
2. Wang, Q., et al.: VoiceFilter: targeted voice separation by speaker-conditioned spectrogram masking. In: Proceedings of Interspeech 2019, pp. 2728–2732 (2019). https://doi.org/10.21437/Interspeech. 2019-1101
3. ŽmolíkovŽmolíková, K., et al.: Speakerbeam: speaker aware neural network for target speaker extraction in speech mixtures. IEEE J. Selected Topics Signal Process. **13**(4), 800–814 (2019). https://doi.org/10.1109/JSTSP.2019.2922820
4. Özgür Çetin, Shriberg, E.: Analysis of overlaps in meetings by dialog factors, hot spots, speakers, and collection site: insights for automatic speech recognition. In: Proceedings of Interspeech 2006, pp. paper 1915-Mon2A2O.6 (2006). https://doi.org/10.21437/Interspeech. 2006-91
5. Chen, Z., et al.: Continuous speech separation: Dataset and analysis. In: ICASSP 2020–2020 IEEE International Conference on Acoustics, Speech and Signal Processing (ICASSP), pp. 7284–7288 (2020). https://doi.org/10.1109/ICASSP40776.2020.9053426
6. Borsdorf, M., Xu, C., Li, H., Schultz, T.: Universal speaker extraction in the presence and absence of target speakers for speech of one and two talkers. In: Proceedings of Interspeech 2021, pp. 1469–1473 (2021). https://doi.org/10.21437/Interspeech. 2021-1939
7. Delcroix, M., Zmolíková, K., Ochiai, T., Kinoshita, K., Nakatani, T.: Speaker activity driven neural speech extraction. In: IEEE International Conference on Acoustics, Speech and Signal Processing, ICASSP 2021, Toronto, ON, Canada, 6–11 June 2021, pp. 6099–6103. IEEE (2021). https://doi.org/10.1109/ICASSP39728.2021.9414998
8. Pan, Z., Ge, M., Li, H.: Usev: universal speaker extraction with visual cue. IEEE/ACM Trans. Audio Speech Lang. Process. **30**, 3032–3045 (2022). https://doi.org/10.1109/TASLP.2022.3205759
9. Zeghidour, N., Grangier, D.: Wavesplit: end-to-end speech separation by speaker clustering. IEEE/ACM Trans. Audio Speech Lang. Process. **29**, 2840–2849 (2021). https://doi.org/10.1109/TASLP.2021.3099291
10. Han, H., Chung, S.W., Kang, H.G.: MIRNet: learning multiple identities representations in overlapped speech. In: Proceedings of Interspeech 2020, pp. 4303–4307 (2020). https://doi.org/10.21437/Interspeech. 2020-2076
11. Byun, J., Shin, J.W.: Monaural speech separation using speaker embedding from preliminary separation. IEEE/ACM Trans. Audio Speech Lang. Process. **29**, 2753–2763 (2021). https://doi.org/10.1109/TASLP.2021.3101617
12. Han, C., et al.: Continuous speech separation using speaker inventory for long recording. In: Proceedings of Interspeech 2021, pp. 3036–3040 (2021). https://doi.org/10.21437/Interspeech. 2021-338
13. Wang, P., et al.: Speech separation using speaker inventory. In: 2019 IEEE Automatic Speech Recognition and Understanding Workshop (ASRU), pp. 230–236 (2019). https://doi.org/10.1109/ASRU46091.2019.9003884
14. Yu, D., Kolbæk, M., Tan, Z.H., Jensen, J.: Permutation invariant training of deep models for speaker-independent multi-talker speech separation. In: 2017 IEEE International Conference on Acoustics, Speech and Signal Processing (ICASSP), pp. 241–245 (2017). https://doi.org/10.1109/ICASSP.2017.7952154

15. Ren, S., He, K., Girshick, R., Sun, J.: Faster r-cnn: towards real-time object detection with region proposal networks. IEEE Trans. Pattern Anal. Mach. Intell. **39**(6), 1137–1149 (2017). https://doi.org/10.1109/TPAMI.2016.2577031
16. Panayotov, V., Chen, G., Povey, D., Khudanpur, S.: Librispeech: an asr corpus based on public domain audio books. In: 2015 IEEE International Conference on Acoustics, Speech and Signal Processing (ICASSP), pp. 5206–5210 (2015). https://doi.org/10.1109/ICASSP.2015.7178964
17. Kao, C.C., Wang, W., Sun, M., Wang, C.: R-CRNN: region-based convolutional recurrent neural network for audio event detection. In: Proc. Interspeech 2018, pp. 1358–1362 (2018). https://doi.org/10.21437/Interspeech. 2018–2323
18. Huang, Z., Watanabe, S., Fujita, Y., García, P., Shao, Y., Povey, D., Khudanpur, S.: Speaker diarization with region proposal network. In: ICASSP 2020–2020 IEEE International Conference on Acoustics, Speech and Signal Processing (ICASSP), pp. 6514–6518 (2020). https://doi.org/10.1109/ICASSP40776.2020.9053760
19. Cosentino, J., Pariente, M., Cornell, S., Deleforge, A., Vincent, E.: Librimix: an open-source dataset for generalizable speech separation (2020)
20. Ge, M., Xu, C., Wang, L., Chng, E.S., Dang, J., Li, H.: SpEx+: a Complete Time Domain Speaker Extraction Network. In: Proceedings of Interspeech 2020, pp. 1406–1410 (2020). https://doi.org/10.21437/Interspeech. 2020–1397
21. Bredin, H., et al.: pyannote.audio: neural building blocks for speaker diarization. In: ICASSP 2020, IEEE International Conference on Acoustics, Speech, and Signal Processing (2020)
22. Bredin, H., Laurent, A.: End-to-end speaker segmentation for overlap-aware resegmentation. In: Proceedings of Interspeech 2021 (2021)
23. Desplanques, B., Thienpondt, J., Demuynck, K.: ECAPA-TDNN: emphasized channel attention, propagation and aggregation in tdnn based speaker verification. In: Proceedings of Interspeech 2020, pp. 3830–3834 (2020). https://doi.org/10.21437/Interspeech. 2020–2650
24. Kanda, N., Horiguchi, S., Fujita, Y., Xue, Y., Nagamatsu, K., Watanabe, S.: Simultaneous speech recognition and speaker diarization for monaural dialogue recordings with target-speaker acoustic models. In: 2019 IEEE Automatic Speech Recognition and Understanding Workshop (ASRU), pp. 31–38 (2019). https://doi.org/10.1109/ASRU46091.2019.9004009
25. Manohar, V., Chen, S.J., Wang, Z., Fujita, Y., Watanabe, S., Khudanpur, S.: Acoustic modeling for overlapping speech recognition: Jhu chime-5 challenge system. In: ICASSP 2019–2019 IEEE International Conference on Acoustics, Speech and Signal Processing (ICASSP), pp. 6665–6669 (2019). https://doi.org/10.1109/ICASSP.2019.8682556
26. Chen, S., et al.: Wavlm: large-scale self-supervised pre-training for full stack speech processing. IEEE J. Selected Topics Signal Process. **16**(6), 1505–1518 (2022). https://doi.org/10.1109/JSTSP.2022.3188113
27. Roux, J.L., Wisdom, S., Erdogan, H., Hershey, J.R.: Sdr - half-baked or well done? In: ICASSP 2019–2019 IEEE International Conference on Acoustics, Speech and Signal Processing (ICASSP), pp. 626–630 (2019). https://doi.org/10.1109/ICASSP.2019.8683855

# Self-supervised Edge Structure Learning for Multi-view Stereo and Parallel Optimization

Pan Li, Suping Wu[✉], Xitie Zhang, Yuxin Peng, Boyang Zhang,
and Bin Wang

Ningxia University, Ningxia 750000, China
pswuu@nxu.edu.cn

**Abstract.** Recent studies have witnessed that many self-supervised methods obtain clear progress on the multi-view stereo (MVS). However, existing methods ignore the edge structure information of the reconstructed target, which includes the outer silhouette and the edge information of the internal structure. This may lead to less satisfactory edges and completeness of the reconstruction result. To solve this problem, we propose an extractor for extracting edge structure maps, and we innovatively design an edge structure Loss to constrain the network to pay more attention to edge structure features of the reference view to improve the texture details of the reconstruction results. Specially, we utilize the idea of constructing cost volume in multi-view stereo and warp the edge structure map of the source view to the reference view to provide reliable self-supervision. In addition, we design a masking mechanism that combines local and global properties, which ensures robustness and improves the reconstruction completeness of the model for complex samples. Furthermore, we adopt an effective parallel acceleration approach to improve the training speed and reconstruction efficiency. Extensive experiments on the DTU and Tanks&Temples benchmarks demonstrate that our method improves both accuracy and completeness in comparison with other unsupervised work. In addition, our parallel method improves efficiency while ensuring accuracy. The code will be published.

**Keywords:** Multi-view stereo · 3D reconstruction · Edge structure · Parallel acceleration

## 1 Introduction

Multi-View Stereo (MVS) aims to reconstruct a dense 3D model from a series of scene images and corresponding camera parameters. The learning-based method [1–5] adopt CNNs to extract features and build a cost volume using differentiable homography. The cost volume is regularized to predict the depth map, then fuse the depth map into the point cloud model, which is the most common pipeline for MVS. Whereas most of the well-performing results are based on the availability

© The Author(s), under exclusive license to Springer Nature Switzerland AG 2024
S. Rudinac et al. (Eds.): MMM 2024, LNCS 14556, pp. 448–461, 2024.
https://doi.org/10.1007/978-3-031-53311-2_33

of large-scale 3D ground truth, which relies heavily on the supervision of real labels, that are not easily obtainable [6]. Thus our work is based on unsupervised learning to implement MVS.

MVSNet [7] is a classical work in the field of multi-view stereo, and many variants based on MVSNet [7] were proposed to improve its performance. CVP-MVSNet [8] is a multi-view deep estimation network architecture based on a coarse-to-fine strategy, where 3D ConvNet for cost volume regularization works on different scales of features and supports training with low-resolution images and testing with high-resolution images. To reduce memory consumption, R-MVSNet [2] utilizes recurrent neural network in place of 3D CNN regularization by processing the cost volume along the depth dimension, which reduces the memory requirements but increases the running time. Fast-MVSNet [5] utilizes a sparse cost volume and a Gaussian Newton layer, and it aims primarily to increase the speed of MVSNet [7]. UCSNet [9] narrows the range of cost volume and depth, thus making the model as small as possible. The first self-supervised MVS work was proposed in [1], which solves the issue of multiple views requiring a large number of real labels by self-supervised signals. The network proposed in [10] is symmetric and predicts the depth maps of all views simultaneously, which enforces cross-view consistency of the multi-view depth maps during the training and testing stages. [11] proposes a new multi-metric loss function by merging the pixel loss function and feature loss function to learn the intrinsic constraints corresponding to matching from different perspectives, which enhances the robustness and completeness of point cloud reconstruction. [12] employs an intra-view adaptive aggregation module and an inter-view adaptive aggregation module to aggregate context-aware features and the cost volumes of different views. [13] uses a view adaptive aggregation module to aggregate the context aware features of multiple scales and the cost of different views to achieve better accuracy. And in realistic scenarios, the existing edge extraction methods cannot handle various disturbances [14], such as the reflections of the scene, image position, etc.

These methods ignore edge structure information or extract cluttered edge structure information and are disturbed by information in challenging scenes that leads to a lack of performance. To solve the aforementioned problem, we design an edge structure information extractor with better disturbance resistance, aiming to add prior knowledge of the edge structure to the original MVS pipeline to provide effective edge structure information. In addition, we design an edge structure loss to constrain self-supervised signal called the edge structure consistency loss. Finally, we design a masking mechanism to narrow the mask and extend it to the whole image chunking area. This enhances the robustness while considering both local and global potential disturbances. Experiments show that the framework is very effective, our contributions can be summarized as:

- We propose a self-supervised framework with joint edge structure learning and masking mechanism abbreviation as ESM, which is simple and effective.
- We propose a novel self-supervised loss based on edge structure information, which can ensure more effective edge structure information learning.

- The experimental results show that our proposed method can simultaneously improve the accuracy and completeness compared with unsupervised
methods.

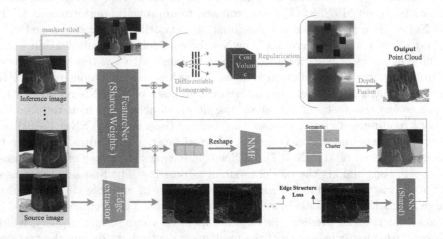

**Fig. 1.** An overview of our framework for joint Edge Structure and Masking mechanism
(ESM). The first row is the masking mechanism, while the second row is based on
MVSNet and the masking mechanism together to achieve depth estimation. The NMF
in the third row generates the semantics, and the edge extractor in the last row brings
up the edge structure information.

## 2   Method

Given multiple view images of a target, the self-supervised edge structure learning network generates corresponding edge structure maps with 2D geometric
contours and 2D geometric edges.

The training process of the whole reconstruction network is shown in Fig. 1
and consists of three parts. 1) The depth estimation module of the depth estimation network estimates the depth map of the reconstructed target, and the
semantic segmentation module extracts semantic features to provide semantic
consistency. (Sect. 2.1) 2) An edge structure information extractor in the edge
structure learning network, which is used to extract the edge structure map of the
input image to provide more detailed edge structure features for depth estimation
(including silhouette shape and internal edges, Sect. 2.2). 3) The masking mechanism network, we propose an uniform non-repetitive random masking mechanism, which the input image is input to the depth estimation module to obtain
the masked depth map, providing mask-enhanced consistency and improving the
anti-interference capability and robustness of the model (Sect. 2.3).

## 2.1   Depth Estimation Network

We adopt the backbone network of MVSNet [7] and the semantic branch in JDACS [15] as the depth estimation module and the semantic module in the depth estimation network.

For $N$ input images $I_{i=1}^{N}$ in the MVSNet [7], the $I_1$ denotes the reference view and the remaining images $I_{i=2}^{N}$ as source views. First, the depth estimation module extracts features from $N$ input images with CNNs, then constructs variance-based cost volume by differentiable homography warping, and regularizes cost volume by 3D U-Net [16]. The reference view and the source views have correlated rotation and translation parameters $(R, T)$, and we utilize the differentiable homography to compute the corresponding pixel $P_{i,j}$ by warping the source feature $F_{i=2}^{N}$:

$$D_{1,j} = K_1^{-1} \cdot P \cdot d_j \; , \; P_{i,j} = K_i \cdot (R_{1,i} \cdot D_{1,j} + T_{1,i}) \qquad (1)$$

where $d_j$ denotes the depth hypothesis, $K$ denotes the intrinsic matrix. Given $P_{i,j}$ and the source feature map $F_i$, we reconstruct the warped source feature map via the differentiable bilinear interpolation. Second, the semantic module exploit Non-negative Matrix Factorization (NMF) [17] mines the implicit common segments from the multi-view images. By decomposing the feature maps of all input images $I_i$, the corresponding semantic feature maps $P$ are available and approximated with the constraint of Frobenius Norm [18]:

$$\| A - P \cdot Q \|_F, P \ge 0, Q \ge 0 \qquad (2)$$

where $A$ represents the feature map, $Q$ represents the semantic clusters. NMF [17] activates pre-trained CNN layers to discover the corresponding semantics between images, so as to provide semantic consistency between multiple views.

The loss function for the depth estimation network is defined as follows:

$$L_{DESS} = \lambda_3 L_{PC} + \lambda_4 L_{SC} \qquad (3)$$

where $\lambda_3 = 0.067$, $\lambda_4 = 0.1$. MVSNet [7] utilize photometric consistency loss $L_{PC}$, JDACS [15] use the semantic consistency loss $L_{SC}$ to achieve constraints.

## 2.2   Edge Structure Learning Network

In previous unsupervised methods, the outer silhouette and inner edge of the reconstruction results are highly ambiguous or sparse, and some approaches even fail to reconstruct the edges. Supervised multi-view stereo utilizes hand-crafted semantic annotations to provide extra supervision that improves performance. However, due to the huge variety of scenarios and the expensive cost of manual annotations in MVS task, we choose to mine the edge structure information from multi-view images via self-supervised edge structure learning network.

Our edge structure learning network extracts the edge structure map of the reconstructed target using an edge structure extractor, and subsequently employs the same network structure and shared parameters as the feature extraction network in the depth estimation network. As shown in Fig. 1, our edge structure extractor extracts boundary structure information [19], and we prove its effectiveness in this paper. In the Fig. 2, the edge structure extractor first transforms the processed image into 3-channel image data using the *mean* and *std* parameters of the dataloader after inverse normalization and inverse tensor. We smooth the image data using Gaussian filtering, which is done to remove noise. The gradient amplitude $G$ and the gradient direction $\theta$ of the image are calculated for the smoothed image using the Sobel operator, which is calculated as follows:

$$G = \sqrt{G_x^2 + G_y^2}, \theta = \tan^{-1}(\frac{G_y}{G_x}) \tag{4}$$

where $G_x$ and $G_y$ denote the amplitude of the gradient in the x and y directions, respectively. Then a non-maximal constraint is performed to eliminate a large portion of non-edge points. In the last step, We experimentally analyze the edge structure map to choose a weak threshold $t_{min} = 100$ and a strong threshold $t_{max} = 150$ as the double thresholds for the output edge structure map. The output edge structure map is fed to our feature CNN network, which empirically extracts edge structure features by combining the feature network of MVSNet [7] branch and the pretrained VGG-19 [20] network of semantic information extraction branch. So we use the shared CNN network to effectively ensure that the edge structure features are consistent with the feature dimension of the original branch.

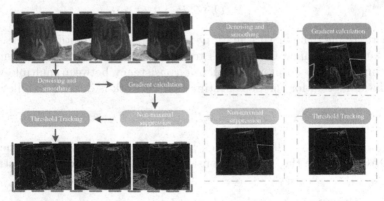

**Fig. 2.** An overview of our edge structure extractor.

As shown in Fig. 2, we add the edge structure features extracted by the shared feature network to the depth and semantic features. After combining the edge structure features, the depth estimation network can obtain more complete edge structure information of the construction target, and the semantic extraction branch gets accurate feature information to ensure a better NMF solution. The adaptive nature of this shared feature network means that our extractor is highly friendly to any feature network. With the edge structure maps $E$ extracted from the extractor, we can design a self-supervised constraint based on the consistency of the target edge structure. Similar to the photometric consistency in Sect. 2.1, which primary idea to warp the edge structure map $E_{i,j}$ of all other source views to the reference edge structure map $E'_{i,j}$. Specifically, the edge structure loss $L_{ES}$ is as follows:

$$L_{ES} = -\sum_{i=2}^{N} \frac{\sum_{j=1}^{HW} [E_{i,j} log(E'_{i,j})] M_{i,j}}{\|M_{i,j}\|_1} \tag{5}$$

As above Eq. 3, we use cross-entropy loss to measure the variability between $E_{i,j}$ and $E'_{i,j}$, where $j$ denotes the j-th pixel, $M_{i,j}$ are the valid pixels after warping.

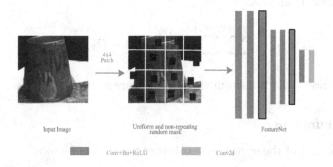

**Fig. 3.** An overview of our masking mechanism.

## 2.3   Masking Mechanism

The benefits of data augmentation in self-supervised learning have been demonstrated by backbone networks and some recent works [21,22]. For the current information density of image data, the spatial redundancy of the image is large, and a pixel or part of the image can recover the reconstruction by its neighboring pixels.

In order to reduce this difference in information density and to make the network more focused on reconstructing information such as the completeness and edges of the target, we propose a masking mechanism network, which randomly masks each local block of the image to reduce redundancy. Furthermore, this masking mechanism also presents a challenging self-supervised task that requires a more comprehensive understanding of the image than merely a low-level single mask. The intuition is that the masking mechanism offers challenging samples which bring a negative impact on the unsupervised loss and hence provides robustness to variations and disturbances for the proposed method. Briefly, the random vectors $\theta_x$ and $\theta_y$ are defined to parameterize an arbitrary augmentation $I \rightarrow I_{\theta_x,\theta_y}$ on image I. $\theta_x$ denotes the mask parameter in the x-direction and $\theta_y$ denotes the mask parameter in the y-direction. As shown in Fig. 3, we partition the input image equally into local blocks, and determine the size and position of the mask on each local block according to the random parameters $\theta_x$ and $\theta_y$. In this manner, a uniform non-repetitive random mask over the whole image is achieved. By studying the dimensionality of the dataset images, we use a 4 × 4 local block to segment the image for better performance.

After the image is masked, we use masking augmentation loss to reduce the depth map difference between the masked image and the original image, thus ensuring that the effective regions of the masked image can execute self-supervised data augmentation consistency. The prediction of a regular forward pass for original images I in depth estimation branch is denoted as D. Accordingly, the prediction of masked images $I_\theta$ is denoted as $D_\theta$. In a contrastive manner, the mask-augmentation consistency is ensured by minimizing the difference between D and $D_\theta$:

$$L_{MM} = \sum^{4 \times 4 \, (\theta_x, \theta_y)} \sum_\theta \frac{1}{\|M_\theta\|_1} \|(D - D_\theta) \cdot M_\theta\|_2 \tag{6}$$

where $M_\theta$ denotes the transformed unmasked areas.

### 2.4 Loss Function

The loss function of the overall network is expressed as follows:

$$L_{Total} = L_{DESS} + \lambda_1 L_{ES} + \lambda_2 L_{MM} \tag{7}$$

where $\lambda$ is the hyperparameter, $\lambda_1 = 0.15$, $\lambda_2 = 0.1$. $L_{DESS}$ is a loss function for the depth estimation network, $L_{ES}$ is a loss function for the edge structure learning network, and $L_{MM}$ is a loss function for the masking mechanism network.

### 2.5 Parallel Optimization

The problem of inefficient training has always existed with multi-view 3D reconstruction, large datasets, multiple views as input and construct cost volume all seriously affect the training speed. This section proposes a multi-view object reconstruction based on parallel acceleration algorithm, which include distributed data parallelism and pipelined model parallelism.

We use two GPUs in this paper to utilize distributed data parallelism for automatic distribution to multiple processes running on the GPU. At the API level, Pytorch provides the *torch.distributed.launcher*. During execution, we use *launcher* to obtain the sequence number (*rank*) of the current process, and the execution file uses rank to determine the process that input data. Finally, the model is packed using distributed DataParallel, as in the global averaging process and *Allreduce* is executed to aggregate the gradients computed by the two GPUs and synchronize the results. Once the gradient computation process is ready, the reconstructed data and model can be loaded directly into the GPU used by the current process and backpropagated normally. Based on distributed data parallelism, the model is further dividing the original data into batches and sending them to the GPU for training. The data before partitioning is the mini-batch, and partitioning on the mini-batch is the micro-batch $M$. The first 0 in $F(0,0)$ denotes the GPU number and the second 0 denotes the micro-batch number. The overall representation is that we put the 0th micro-batch of input data into the 0th GPU, and so on, resulting in multiple micro-batches of data stored per GPU to train the model in parallel. The time complexity under pipeline parallelism is shown in Eq. 8. Where K denotes the number of GPUs. It is experimentally demonstrated that when $M \geq 4K$, the resulting idle time share has a negligible impact on the final training duration.

$$O(\frac{K-1}{K+M-1}) \tag{8}$$

pipelined model parallelism is applied to multi-view object reconstruction, combining the feature network and the costome network into $GPU0$, and combining the feature network and the regularisation network into $GPU1$. The costom network on $GPU0$ construct the costom and the input to the $GPU1$. The features obtained from the feature network on $GPU1$ are input to the costom network on $GPU0$ to obtain the costom, which is then input to $GPU1$ for subsequent operations. This constitutes a parallel way of pipelining the model, thus reducing the idle time generated by the GPU during the training of previous models, which wastes computational resources.

## 3    Experiments

In this section, we conduct comprehensive experiments to evaluate the proposed method. The hardware environment for this section is the proposed method trained on an Nvidia RTX2080 GPU (12194 MB) with an Intel(R) Xeon(R) CPU E5-2603 v3 @ 1.60 GHz and an operating system of Ubuntu 16.1. The proposed method uses Python version 3.6, implemented in the Pytorch 1.2.0-GPU framework.

### 3.1    Implementation Datails

**Datasets**
The DTU dataset [23] is a large-scale indoor MVS dataset, which contains 128 different scenes with 49 views under different lighting conditions. All scenes are collected under laboratory environments with the same camera trajectory. We follow the configuration in MVSNet and apply the DTU dataset [23] to train and evaluate the network. The Tanks&Temples dataset [24] is a large-scale outdoor MVS dataset, which mainly applies to 3D reconstruction. The evaluation data of the dataset is divided into intermediate and advanced sets, including a total of 14 sceneries.

**Training**
We implement the proposed approach based on the most concise MVSNet [7], we only use DTU training set to train the model without any ground truth depth map. In default, the hyperparameters of the training phase follow the

same settings of the unsupervised MVSNet. The batch size is set to 1, and the memory consumed is less than 10 G in GPU, the optimizer selects Adam. During the training on DTU dataset, we set the number of input images to $N = 7$ and the width and height to 640 and 512, respectively. We set the number of depth hypothesis for depth prediction on MVSNet to 48, and train 10 epochs with learning rate of $1 \times 10^{-3}$, which decrease by half times for every two epochs.

## Evaluation

We evaluate the proposed method on the DTU evaluation set and both intermediate and advanced sets of the Tanks&Temples dataset [24]. To achieve the generalization performance of the method, we evaluate the intermediate and advanced sets of Tanks&Temples using the model trained on DTU without fine-tuning. In the DTU benchmark, *Accuracy* (Acc.) is measured as the quality of the reconstruction, *Completeness* (Comp.) is measured as how many surfaces were captured, *Overall* is the average of Acc. and Comp., and these are composite error metrics. The evaluation metrics for parallelism are *runtime* and *speedupratio*, where *runtime* is the average runtime per epoch and batch, and *speedupratio* is the ratio of serial training time to parallel training time of the network to measure the effectiveness of the parallel speedup algorithm.

**Table 1.** Geo. represents traditional geometric methods. UnSup. represents unsupervised methods.

| Methods | Geo. | | | | UnSup. | | |
|---------|------|------|------|------|--------|-------|------|
| | Furu [25] | Tola [26] | Camp [27] | Gipuma [28] | MVSNet | JDACS | **Ours** |
| Acc.↓ | 0.613 | 0.342 | 0.835 | 0.283 | 0.881 | 0.571 | **0.544** |
| Comp.↓ | 0.941 | 1.190 | 0.554 | 0.873 | 1.073 | 0.515 | **0.495** |
| Overall.↓ | 0.777 | 0.766 | 0.694 | 0.578 | 0.977 | 0.543 | **0.519** |

## Quantitative Results

We evaluated our method on the official metrics of the DTU dataset [23] and compared it with other unsupervised methods [2,15,29–31]. The quantitative results are shown in Table 1 and Table 2. We can conclude that our method

**Table 2.** Comparison with other unsupervised methods. Our (cvp) shows the feature extraction with classical pyramid network [8] replacing the original feature network.

| Methods | UnSup. | | | | | |
|---------|--------|-----------|-------|----------|-------------|-----------|
| | MVS² | M³ VSNet | JDACS | **Ours** | End-to-End MV | **Ours (cvp)** |
| Acc.↓ | 0.760 | 0.636 | 0.571 | 0.544 | **0.391** | 0.399 |
| Comp.↓ | 0.515 | 0.531 | 0.515 | 0.495 | 0.429 | **0.352** |
| Overall.↓ | 0.637 | 0.583 | 0.543 | 0.519 | 0.411 | **0.376** |

reconstructs a more complete point cloud than the traditional methods, and outperforms the previous unsupervised methods in terms of all official metrics.

## Qualitative Results

The Fig. 4 shows a more qualitative comparison of our method on the DTU dataset [23] with the self-supervised method [15]. In Fig. 4, blue boxes indicate that our method provides more accurate boundary and texture details, and red boxes indicate better performance in terms of completeness of the 3D point cloud. In addition, the partial point cloud reconstruction results of the Tank&Temples datasets [24] are shown in Fig. 5. As seen in Fig. 5, the reconstruction results of the generalization experiments are of high quality even for challenging sequences. We also present the qualitative results from multiple views of the tanks dataset in Fig. 6.

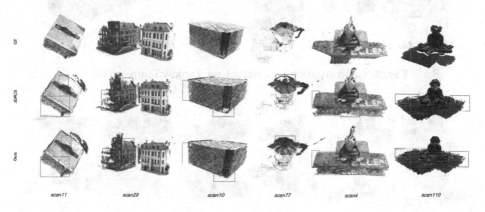

**Fig. 4.** Qualitative comparison in 3D reconstruction between Ours and JDACS on DTU dataset.

## 3.2 Ablation Study

In this section, we provide ablation experiments to analyze the impact of each part of the proposed framework (ESM) on the reconstruction. The Overall metric of 0.532 with just the masking mechanism can be seen in Table 3, which demonstrates the benefit of the masking mechanism in self-supervised learning and the improved masking enhancement consistency by comparing the output of the

**Table 3.** We trained the network with different combinations of these modules (w/o indicates without).

| Model | Acc.↓ | Comp.↓ | Overall↓ |
|---|---|---|---|
| Ours w/o Mask-Augmentation | 0.564 | 0.503 | 0.533 |
| Ours w/o Edge-Structure | 0.561 | 0.502 | 0.532 |
| Ours | **0.544** | **0.495** | **0.519** |

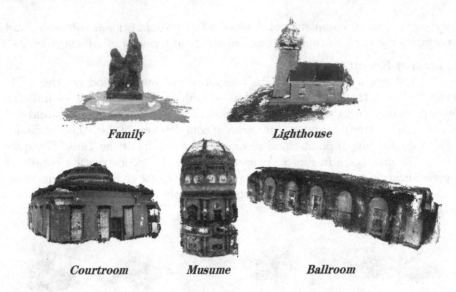

*Family*                    *Lighthouse*

*Courtroom*        *Musume*        *Ballroom*

**Fig. 5.** Generalization experiment on Tanks&Temples [24].

**Fig. 6.** Qualitative results from multiple perspectives on Tanks&Temples dataset.

original data and the masked images. The Overall metric of 0.533 using only the edge structure reaffirms the effectiveness of the edge structure for reconstructing the point cloud.

### 3.3   Parallel Results

In this section, we compare not only the running times and speed-up ratios of the methods before and after parallel acceleration, but also the evaluation metrics after acceleration on the DTU dataset. As shown in Table 4, the parallel acceleration optimisation outperforms the previous one in terms of training time per cycle, training time per batch and speed-up ratio, which indicates that the training speed of multi-view reconstruction is improved after the parallel acceleration. Through a comprehensive analysis of the evaluation metrics, the overall performance of the parallel optimisation is great, and the corresponding 3D point cloud can be effectively reconstructed while speeding up the training time and improving efficiency.

As shown in Table 5, we compare the training time and the speed-up ratio at batch size of 1–5 respectively, and the results show that the parallel optimisation method has the best parallel speed-up at batch size of 4.

**Table 4.** We compared the methods after parallel optimisation and before optimisation in terms of evaluation metrics, training time and speed-up ratios.

| Methods | Acc.↓ | Comp.↓ | Overall.↓ | Epoch Runtime/s↓ | Speedup Ratio↑ |
|---|---|---|---|---|---|
| Before Parallel Optimization | 0.544 | 0.495 | 0.519 | 51482.4 | 1 |
| After parallel optimization | **0.541** | **0.495** | **0.518** | **34685.4** | **1.48** |

**Table 5.** We demonstrate that the parallel optimization performs best with a batch size of 4 by comparing training times and speedup ratios for batch sizes of 1–5.

| Runtime | Batchsize = 1 | Batchsize = 2 | Batchsize = 3 | Batchsize = 4 | Batchsize = 5 |
|---|---|---|---|---|---|
| Epoch Runtime/s | 34685.44 | 28319.5 | 26288.94 | 24596.88 | 22926.6 |
| Batch Runtime/s | 1.28 | 2.09 | 2.91 | 3.63 | 4.23 |
| Speedup ratio | 1.48 | 1.42 | 1.31 | **1.49** | 1.44 |

To analyse the effectiveness of parallelism algorithm, train them on the base-line network, respectively. The results in the ablation experiments in Table 6 confirm that distributed data parallelism also makes significant improvements in parallelism and achieves better speedup ratios by further disrupting the order of data on individual GPUs. The parallelism of pipeline models improves the memory occupied by large models on GPUs, solving the problem of large models that can only be trained with multiple small memory GPUs.

**Table 6.** We train the network with different parallel optimizations (w/o indicates without).

| Model Architecture | Epoch Runtime/s↓ | Batch Runtime/s↓ | Speedup Ratio↑ |
|---|---|---|---|
| Before Parallel Optimization | 51482.4 | 1.9 | 1 |
| w/o Distributed data parallelism | 41185.92 | 1.52 | 1.25 |
| w/o Pipeline model parallelism | 37934.4 | 1.4 | 1.35 |
| After Parallel Optimization | **34685.44** | **1.28** | **1.48** |

## 4    Conclusion

In this work, we present a novel self-supervised method including an edge structure learning network and masking mechanism network. Compared with other unsupervised MVS methods, our method achieves richer boundaries and the overall structure becomes denser. In order to improve the efficiency of the multi-view stereo reconstruction, we use a combination of distributed data parallelism and pipeline model parallelism to parallelize and optimize the proposed method. Experimental results show better results in terms of both accuracy and completeness, demonstrating the effectiveness of our proposed self-supervised method. The reduction in runtime and the speed-up ratio also demonstrate that parallel optimization improves the speed and efficiency of multi-view stereo training.

**Acknowledgements.** This work was supported by National Natural Science Foundation of China under Grant 62062056, and in part by the Ningxia Graduate Education and Teaching Reform Research and Practice Project 2021.

# References

1. Khot, T., Agrawal, S., Tulsiani, S., Mertz, C., Lucey, S., Hebert, M.: Learning unsupervised multi-view stereopsis via robust photometric consistency. arXiv:abs/1905.02706 (2019)
2. Yao, Y., Luo, Z., Li, S., Shen, T., Fang, T., Quan, L.: Recurrent mvsnet for high-resolution multi-view stereo depth inference, pp. 5520–5529 (2019)
3. Ji, M., Gall, J., Zheng, H., Liu, Y., Fang, L.: Surfacenet: an end-to-end 3D neural network for multiview stereopsis. In: IEEE International Conference on Computer Vision (ICCV), pp. 2326–2334 (2017)
4. Xue, Y., et al.: MVSCRF: learning multi-view stereo with conditional random fields. In: IEEE/CVF International Conference on Computer Vision (ICCV), pp. 4311–4320 (2019)
5. Yu, Z., Gao, S.: Fast-mvsnet: sparse-to-dense multi-view stereo with learned propagation and gauss-newton refinement. In: 2020 IEEE/CVF Conference on Computer Vision and Pattern Recognition (CVPR), pp. 1946–1955 (2020)
6. Zhong, Y., Li, H., Dai, Y.: Open-world stereo video matching with deep RNN. In: Ferrari, V., Hebert, M., Sminchisescu, C., Weiss, Y. (eds.) ECCV 2018. LNCS, vol. 11206, pp. 104–119. Springer, Cham (2018). https://doi.org/10.1007/978-3-030-01216-8_7
7. Yao, Y., Luo, Z., Li, S., Fang, T., Quan, L.: MVSNet: depth inference for unstructured multi-view stereo. In: Ferrari, V., Hebert, M., Sminchisescu, C., Weiss, Y. (eds.) ECCV 2018. LNCS, vol. 11212, pp. 785–801. Springer, Cham (2018). https://doi.org/10.1007/978-3-030-01237-3_47
8. Yang, J., Mao, W., Alvarez, J.M., Liu, M.: Cost volume pyramid based depth inference for multi-view stereo. In: Proceedings of the IEEE/CVF Conference on Computer Vision and Pattern Recognition, pp. 4877–4886 (2020)
9. Cheng, S., et al.: Deep stereo using adaptive thin volume representation with uncertainty awareness. In: Proceedings of the IEEE/CVF Conference on Computer Vision and Pattern Recognition, pp. 2524–2534 (2020)
10. Kolmogorov, V., Zabih, R.: Computing visual correspondence with occlusions using graph cuts. In: Proceedings Eighth IEEE International Conference on Computer Vision. ICCV, volume 2, pp. 508–515. IEEE (2001)
11. Guo, X., Yang, K., Yang, W., Wang, X., Li, H.: Group-wise correlation stereo network. In: Proceedings of the IEEE/CVF Conference on Computer Vision and Pattern Recognition, pp. 3273–3282 (2019)

12. Hirschmüller, H., Innocent, P.R., Garibaldi, J.: Real-time correlation-based stereo vision with reduced border errors. Int. J. Comput. Vis. **47**, 229–246 (2002)
13. Min, C., Chen, Y., Wei, Z., Zhu, Q., Wang, G.: Aa-rmvsnet: adaptive aggregation recurrent multi-view stereo network. In: 2021 IEEE/CVF International Conference on Computer Vision (ICCV), pp. 6167–6176 (2021)
14. Lin, K., Li, L., Zhang, J., Zheng, X., Wu, S.: High-resolution multi-view stereo with dynamic depth edge flow. In: IEEE International Conference on Multimedia and Expo (ICME), pp. 1–6 (2021)
15. Zhou, Z., Qiao, Y., Kang, W., Wu, Q., Xu, H.: Self-supervised multi-view stereo via effective co-segmentation and data-augmentation. In: Proceedings of the AAAI Conference on Artificial Intelligence, vol. 35, no. 4, pp. 3030–3038 (2021)
16. Çiçek, Ö., Abdulkadir, A., Lienkamp, S.S., Brox, T., Ronneberger, O.: 3D U-net: learning dense volumetric segmentation from sparse annotation. In: Ourselin, S., Joskowicz, L., Sabuncu, M.R., Unal, G., Wells, W. (eds.) MICCAI 2016. LNCS, vol. 9901, pp. 424–432. Springer, Cham (2016). https://doi.org/10.1007/978-3-319-46723-8_49
17. Seung, H.S., Lee, D.: Algorithms for non-negative matrix factorization (2000)
18. Ding, X., He, C., Simon, H.D.: On the equivalence of nonnegative matrix factorization and spectral clustering. In: Proceedings of the 2005 SIAM International Conference on Data Mining, pp. 606–610 (2005)
19. Canny, J.: A computational approach to edge detection. In: Fischler, M.A., Firschein, O. (eds.) Readings in Computer Vision, pp. 184–203. Morgan Kaufmann, San Francisco (CA) (1987)
20. Simonyan, K., Zisserman, A.: Very deep convolutional networks for large-scale image recognition. arXiv preprint arXiv:1409.1556 (2014)
21. Hovy, Z., Luong, E., Xie, M.-T., Dai, Q., Le, Q.V.: Unsupervised data augmentation for consistency training. arXiv (2019)
22. Norouzi, S., Chen, M., Kornblith, T., Hinton, G.: A simple framework for contrastive learning of visual representations. arXiv (2020)
23. Vogiatzis, R.R., Tola, G., Aanæs, E., Jensen, H., Dahl, A.B.: Large-scale data for multiple-view stereopsis. Int. J. Comput. Vis. **120**, 153–168 (2016)
24. Zhou, J., Knapitsch, Q.-Y., Park, A., Koltun, V.: Tanks and temples: benchmarking large-scale scene reconstruction. ACM **36**, 1–13 (2017)
25. Furukawa, Y., Ponce, J.: Accurate, dense, and robust multiview stereopsis (2009)
26. Tola, E., Strecha, C., Fua, P.: Efficient large scale multi-view stereo for ultra high resolution image sets (2011)
27. Campbell, N.D.F., Vogiatzis, G., Hernández, C., Cipolla, R.: Using multiple hypotheses to improve depth-maps for multi-view stereo. In: Forsyth, D., Torr, P., Zisserman, A. (eds.) ECCV 2008. LNCS, vol. 5302, pp. 766–779. Springer, Heidelberg (2008). https://doi.org/10.1007/978-3-540-88682-2_58
28. Galliani, S., Lasinger, K., Schindler, K.: Massively parallel multiview stereopsis by surface normal diffusion. In: IEEE International Conference on Computer Vision (2015)
29. Rao, Z., Dai, Y., Zhu, Z., Li, B.: Mvs2: deep unsupervised multi-view stereo with multi-view symmetry. arXiv:abs/2203.14237:1–8 (2019)
30. Huang, C., He,Y., Liu, J., Huang, B., Yi, H., Liu, X.: M3vsnet: unsupervised multi-metric multi-view stereo network. In: IEEE International Conference on Image Processing (ICIP), pp. 3163–3167 (2021)
31. Chen, Q., Poullis, C.: End-to-end multi-view structure-from-motion with hyper-correlation volumes. arXiv preprint arXiv:2209.06926 (2022)

# Prototype-Enhanced Hypergraph Learning for Heterogeneous Information Networks

Shuai Wang[1]([✉]), Jiayi Shen[1], Athanasios Efthymiou[1], Stevan Rudinac[1], Monika Kackovic[1], Nachoem Wijnberg[1,2], and Marcel Worring[1]

[1] University of Amsterdam, Amsterdam, The Netherlands
s.wang3@uva.nl
[2] College of Business and Economics, University of Johannesburg, Johannesburg, South Africa

**Abstract.** The variety and complexity of relations in multimedia data lead to Heterogeneous Information Networks (HINs). Capturing the semantics from such networks requires approaches capable of utilizing the full richness of the HINs. Existing methods for modeling HINs employ techniques originally designed for graph neural networks, and HINs decomposition analysis, like using manually predefined metapaths. In this paper, we introduce a novel prototype-enhanced hypergraph learning approach for node classification in HINs. Using hypergraphs instead of graphs, our method captures higher-order relationships among nodes and extracts semantic information without relying on metapaths. Our method leverages the power of prototypes to improve the robustness of the hypergraph learning process and creates the potential to provide human-interpretable insights into the underlying network structure. Extensive experiments on three real-world HINs demonstrate the effectiveness of our method.

**Keywords:** Heterogeneous Information Network · Hypergraph · Prototype · Multimodal Learning · Multimedia Modelling

## 1 Introduction

Many multimedia collections can be effectively formulated as HINs, where different types of nodes and edges embody multiple types of entities and relations. For example, as shown in Fig. 1, the visual arts network WikiArt has several types of nodes: Painting, Artist, and Time, as well as different types of relations, each associated with different semantics, such as Artist $\xrightarrow{\text{paints}}$ Painting, Painting $\xrightarrow{\text{belongs to}}$ Time. These relations can be aggregated to give rise to higher-order semantic associations. For instance, the triadic (ternary) relationship Painting-Artist-Painting represents a co-creation relationship, while Painting-Time-Painting conveys a contemporary connection. Modeling the relational and semantic richness of HINs requires the development of specialized models for their effective analysis and interpretation.

© The Author(s), under exclusive license to Springer Nature Switzerland AG 2024
S. Rudinac et al. (Eds.): MMM 2024, LNCS 14556, pp. 462–476, 2024.
https://doi.org/10.1007/978-3-031-53311-2_34

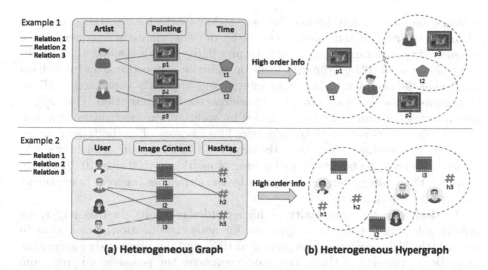

**Fig. 1.** Comparison between conventional heterogeneous graphs (e.g., an art network or a content recommendation network) and their corresponding heterogeneous hypergraph. In a conventional heterogeneous graph or network, different nodes are connected by different pairwise links and cannot explicitly capture the high-order complex relation among those nodes. For example, in the art network, the interactions are not only among artists creating paintings and paintings belonging to a time period but also high-order information, e.g., *an artist created different paintings at different times*. Or, in the recommendation network, *contents with the same tag are recommended to different users*.

Recent years have brought rapid development of Graph Neural Networks (GNNs) in pursuit of performance improvement in graph representation learning [26,27]. GNNs are primarily designed for homogeneous graphs associated with a single type of nodes and edges, and follow a neighborhood aggregation scheme to capture the structural information of a graph [13,21]. Thus, most GNNs are not well-equipped to deal with HINs, which also have rich semantic information induced by different types of nodes, as well as by varied structural information [20].

Various Heterogeneous Graph Neural Networks (HGNNs) have been introduced as effective tools for the extraction and incorporation of semantic knowledge, yielding remarkable performance in representation learning for HINs [22, 29]. With regard to their approach to relation handling, these techniques can be broadly grouped into two categories: *Metapath-based methods* and *Metapath-free methods*. Metapath-based methodologies leverage metapath-sequential arrangements of node types and edge types. Due to the semantic expressiveness of metapaths, many techniques initially extract various substructures from the original HINs, each possessing distinct semantic characteristics. This extraction process is guided by a set of predefined metapaths, which subsequently serve as the topology for representation learning on these substructures [3,7,23,31,33]. Although Metapath-based methods have achieved state-of-the-art performance on plenty of tasks, they are usually limited in that (1) Metapaths have to be specified in advance, requiring domain-specific knowledge or even exhaustive enumeration of schemes, strategies often associated with prohibitively high manual and computational costs. (2) They primarily focus on pairwise connections, and it is hard to capture the complex higher-order interactions implicitly contained in HINs. Metapath-free methods are proposed to address the first limitation. They aggregate information from neighboring nodes by an attention mechanism or a usually relation-dependent graph transformer. This category of methods operates by using one-hop relations as input to the layers of a GNN and subsequently stacking multiple layers to facilitate the learning of multi-hop relations [10,16,17,32]. However, this strategy can be challenged by the intricacies inherent in capturing higher-order relations.

To deal with the complexity of higher-order relations, in this paper, we present a hypergraph learning approach for node classification, which aims to preserve the high-order relations present in HINs and simultaneously capture the semantic information in them. Our model leverages the power of a hypergraph representation, a structure that generalizes graphs by allowing edges to connect more than two nodes. By representing higher-order relationships more explicitly, hypergraphs provide a natural framework for capturing complex dependencies and group information. Traditional hypergraph approaches aimed to simplify a hypergraph to a regular graph, e.g., by applying star or clique expansion [5,28]. They facilitate learning hypergraph representations using spectral theory, but that inherently leads to loss of information. Recently, there have been works on applying deep neural network message passing to propagate vertex-hyperedge-vertex information through the hypergraph. This allows direct learning from the hypergraph topology [1,8,34]. Doing so avoids reducing the high-order relations into pairwise ones, hence no information loss, and provides a new way to model semantic information without relying on metapaths. In line with previous studies on HINs [16,23], we specifically focus on utilizing symbolic relations. This means we do not incorporate visual or image content to define network structure. While our method can connect nodes based on similarity, we have chosen to prioritize symbolic relationships in this study for practical reasons.

Modeling symbolic relations with current hypergraph models is known to be sensitive to noisy information in the nodes and hyperedges [1]. Hence, our method utilizes prototypes, representative nodes that summarize groups or similar entities in the data structure, in two ways to regularize the hypergraph learning process. First, we design a prototype-based hyperedge regularization, where each hyperedge serves as a prototype and forces its nodes not to be too far from it in the embedding space to stabilize the optimization. Second, to further improve the robustness of our model for node classification against noise or small changes in the initial samples, we utilize learnable prototypes for creating classifiers and learn multiple prototypes to represent different classes.

Our contributions are three-fold:

1. We introduce a novel approach that uses the power of hypergraphs and prototypes for node classification in HINs. The hypergraphs allow for higher-order

relationships among nodes to capture complex semantic dependencies and group information in the data.

2. We show how to use prototypes, being representative landmarks, to enhance the robustness and interpretability of hypergraph model learning for HINs.
3. We demonstrate the effectiveness of the proposed method through experiments conducted on multiple real-world HINs benchmarks.

## 2   Related Work

In this section, we reflect on related work about heterogeneous information networks in Sect. 2.1 and hypergraph learning in Sect. 2.2.

### 2.1   Heterogeneous Information Networks

Different from homogeneous networks, heterogeneous networks consist of different types of nodes and edges. Most recent methods for analyzing heterogeneous graphs concentrate on decomposing them into homogeneous sub-graphs and deploying GNNs. Metapath-based methods first extract substructures according to hand-crafted metapaths and then learn node representations based on these sub-structures. For instance, HAN, MAGNN, SeHGNN., are representative methods that applie hierarchical attention to aggregate information from metapath-based neighbors [7,23,31]. In contrast, Metapath free methods adhere to a different paradigm, aggregating messages from neighbors within the immediate one-hop vicinity akin to traditional GNNs, disregarding the specific node types involved. However, they augment this process with additional modules, such as attention mechanisms, to encode semantic information like node types and edge types into the propagated messages, thereby enriching the data representation. GTN [32] can discover valuable meta-paths automatically with the intuition that a meta-path neighbor graph can be obtained by multiplying the adjacency matrices of several sub-graphs. However, GTN consumes a gigantic amount of time, and memory, e.g., 400 × time and 120 × the memory of Graph Attention Network (GAT) [16]. HGT [10] builds on the transformer to handle large academic heterogeneous graphs, focusing on handling web-scale graphs via graph sampling strategy. HGB [16] instead uses a GAT network as the backbone, with the help of learnable edge-type attention, L2 normalization, and residual attention for node representation generation. However, all of the above methods mostly focus on the pairwise relations in the network and ignore high-order relations in HINs, leading to semantic information loss.

### 2.2   Hypergraph Learning

Hypergraph learning is related to graph learning since a hypergraph is a graph generalization that allows edges to connect 2 to $n$ nodes, where $n$ is the number of nodes/vertices. Hypergraph learning was first introduced in [35] and can be seen as a propagation process along the hypergraph structure in analyzing categorical

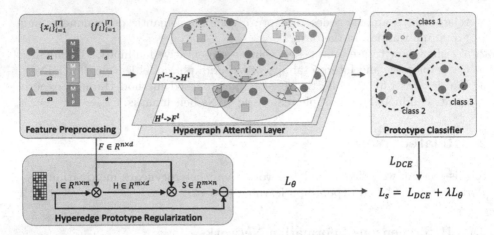

**Fig. 2. An illustration of prototype-enhanced hypergraph learning model.** First, linear layers map heterogeneous nodes with varying embedding lengths into a shared space. Then, high-order message passing occurs among different nodes based on the topology of the hyperedges. Hyperedge prototype regularization constrains node embeddings based on their proximity to their respective hyperedges. Finally, nodes are classified by learnable prototype-based classifiers according to their representations.

data with complex relationships. It conducts transductive learning and aims to minimize the label difference among vertices having stronger connections in the hypergraph. Inspired by the immense success of deep learning, recently effective approaches to deep learning on hypergraphs have been proposed [9]. Hypergraph neural networks design a vertex-hyperedge-vertex information propagating pattern to iteratively learn the data representation [1,2,5,8,11,34]. In recent years, there has been a surge in research focusing on the utilization of hypergraph learning to capture high-order interactions in multimedia analysis. Hypergraphs can be constructed by including global and local features, or tag information to learn the relevance of images in tasks of tag-based image retrieval [24]. Hypergraphs can also be applied adaptively to capture the relations in Multi-Label Image Classification [25] and 360-degree Image Quality Assessment [6]. Recently, there has been growing interest in utilizing hypergraphs to model structured data in HINs [14,15]. However, these approaches necessitate predefined metapaths as a foundation for constructing hyperedges. This reliance on metapaths introduces several challenges, including the need for manual specification, making them less suitable for diverse datasets. Additionally, capturing high-order information directly from the data can be challenging within this metapath-centric paradigm.

## 3    Methodology

Here, we present our methodology closely following Fig. 2. We start by providing some general notation.

**Notations** A heterogeneous hypergraph is represented as $\mathcal{G} = \{\mathcal{V}, \mathcal{E}, \mathcal{T}\}$, where $\mathcal{V} = \{v_1, v_2, ..., v_n\}$ is the node set, $\mathcal{E} = \{e_1, e_2, ..., e_m\}$ represents the set of hyperedges, and $\mathcal{T}$ is the set of node types. Each hyperedge has 2 or more nodes. When $|\mathcal{T}| \geq 2$, the hypergraph is heterogeneous. The relationship between nodes and hyperedges can be represented by an incidence matrix $I \in \mathbb{R}^{|\mathcal{V}| \times |\mathcal{E}|}$, with entries defined as:

$$I(v, e) = \begin{cases} 1, & \text{if } v \in e \\ 0, & \text{otherwise} \end{cases}$$

Let $D_e \in \mathbb{R}^{|\mathcal{E}| \times |\mathcal{E}|}$ denotes the diagonal matrices containing hyperedge degrees, where $D_e(i, i) = \sum_{u \in \mathcal{V}} I(u, i)$. The hyperedge degree is a valuable parameter for normalization purposes.

**Problem Statement.** In the context of HIN, we capture and retain their implicit high-order relations, effectively forming a heterogeneous hypergraph denoted as $\mathcal{G}$. Then we aim at learning the low dimensional representations $\mathbf{f} \in \mathbb{R}^d$ for nodes in $\mathcal{G}$ with $d \ll |v|$ while fully considering both the high-order relations and heterogeneity implied in $\mathcal{G}$. This representation can be used for downstream predictive applications such as node classification.

## 3.1 Feature Preprocessing

Due to the heterogeneity of nodes, different types of nodes are originally represented in different semantic/feature spaces associated with their specific probability distributions. Therefore, for each node type $t_i$, we design the learnable type-specific transformation matrix $M_{t_i}$ to project the heterogeneous nodes with varying embedding lengths into the same dimension. This allows for messages to be passed among them in a common space. The projection process can be represented as follows:

$$\mathbf{f}_i = M_{t_i} \cdot \mathbf{x}_i \tag{1}$$

where $\mathbf{f}_i$ and $\mathbf{x}_i$ are the projected and original features of node $v_i$, respectively. With the learnable type-specific projection operation, we address the network heterogeneity induced by the node type. Following this transformation, projected features of all nodes are unified to the same dimension, facilitating the subsequent aggregation process in the next model component.

## 3.2 Hypergraph Attention Layer

To capture the heterogeneous high-order context information on hypergraph, we employ node and hyperedge attention mechanisms. For layer $l_0$, we define node representation $F^0 = \{\mathbf{f}_1^{l_0}, \mathbf{f}_2^{l_0}, ..., \mathbf{f}_n^{l_0}\}$ and incidence matrix $I \in \mathbb{R}^{n \times m}$. The target of the heterogeneous hypergraph layer $l$ is to update node representations through hypergraph message passing by calculating hypergraph attention.

**Node-level Attention.** In the first step, we calculate hyperedge representation $H^l = \{\mathbf{h}_1^l, \mathbf{h}_2^l, ..., \mathbf{h}_m^l\}$ given node embedding $F^{l-1} \in \mathbb{R}^{n \times d}$

$$\mathbf{h}_j^l = \sigma \left( \sum_{v_k \in e_j} \alpha_{jk} W_h \mathbf{f}_k^{l-1} \right), \tag{2}$$

where $\sigma$ is a nonlinearity such as LeakyReLU, and $W_h$ is a trainable matrix, $\alpha_{jk}$ is a coefficient to control how much information is contributed from node $v_k$ to hyperedge $e_j$, and it can be computed by:

$$\alpha_{jk} = \frac{\exp\left(\mathbf{a}_1^T \mathbf{u}_k\right)}{\sum_{v_p \in e_j} \exp\left(\mathbf{a}_1^T \mathbf{u}_p\right)}, \tag{3}$$

$$\mathbf{u}_k = \text{LeakyReLU}\left(W_h \mathbf{f}_k^{l-1}\right),$$

where $\mathbf{a}_1^T$ is a trainable weight vector.

**Hyperedge-level Attention.** With all hyperedge representation $\{\mathbf{h}_j^l \mid \forall e_j \in f_i\}$, where $f_i$ is the set of associated hyperedges given vertex $v_i$. Then we update node representation $F^l = \{\mathbf{f}_1^l, \mathbf{f}_2^l, ..., \mathbf{f}_n^l\}$ based on updated hyperedge representations $H^l$.

$$\mathbf{f}_i^l = \sigma \left( \sum_{e_k \in f_i} \beta_{ik} W_e \mathbf{h}_k^l \right),$$

$$\beta_{ik} = \frac{\exp\left(\mathbf{a}_2^T \mathbf{v}_k\right)}{\sum_{v_q \in e_i} \exp\left(\mathbf{a}_2^T \mathbf{v}_q\right)}, \tag{4}$$

$$\mathbf{v}_k = \text{LeakyReLU}\left(\left[W_h \mathbf{f}_k^{l-1} \| W_e \mathbf{h}_i^l\right]\right),$$

where $f_i^l$ is the update representation of node $v_i$ and $W_e$ is a weight matrix. $\beta_{ik}$ denotes the attention coefficient of hyperedge $\mathbf{h}_k$ to node $\mathbf{f}_i$. $\mathbf{a}_2^T$ is another weight vector measuring the importance of the hyperedges. $\|$ here is the concatenation operation.

We extend hypergraph attention (HAT) into multi-head hypergraph attention(MH-HAT) by concatenating multiple HATs together to expand the model's representation ability.

$$\text{MH-HAT}(F, I) = \overset{K}{\underset{i=1}{\|}} \sigma(\text{HAT}_i(F, I)). \tag{5}$$

By harnessing the MH-HAT structure, we effectively capture the high-order relationship and semantic information present in HINs data.

## 3.3    Learnable Prototype Classifier

To enhance the robustness of hypergraph message passing, we completely replace the softmax layer used in conventional neural networks. Instead, we utilize a node classification approach rooted in learnable prototypes. Here, we assume each class has an equal number of $k$ prototypes, and this assumption can be effortlessly relaxed in other use cases. The prototypes are denoted as $\mathbf{m}_{ij}$ where $i \in \{1, 2, ..., c\}$ represents the index of the category and $j \in \{1, 2, ..., k\}$ represents the index of the prototypes in each category.

In the prediction phase, utilizing an input pattern $\mathbf{f}$ generated by the heterogeneous message passing module, we firstly employ a linear layer to integrate it, represented as $g(\mathbf{f}^l; \theta)$. Then, we measure the distance of an input pattern to all prototypes and classify it into the category associated with the nearest prototype.

$$\hat{y} = \arg \min_{i \in c, j \in k} \|\mathbf{m}_{ij} - \mathbf{z}\|,$$
$$\mathbf{z} = g(\mathbf{f}; \theta). \tag{6}$$

We use distance-based cross-entropy loss (DCE) [30] to measure the similarity between the samples and the prototypes. Thus, for a sample characterized by feature $\mathbf{x}$ and category $y$, the probability of $(\mathbf{x}, y)$ belonging to the prototype $\mathbf{m}_{ij}$ can be measured by the distance between them:

$$p(y|\mathbf{z}) - p(\mathbf{z} \in \mathbf{m}_{ij} \mid \mathbf{z}) = -\|\mathbf{z} - \mathbf{m}_{ij}\|_2^2. \tag{7}$$

Based on the probability of $\mathbf{x}$, we can define the cross entropy (CE) in our framework as:

$$L_{DCE} = l((\mathbf{z}, y); \theta, M) = -\log p(y|\mathbf{z}) \tag{8}$$

## 3.4    Hyperedge Prototype Regularization

Hypergraph modeling has been observed to exhibit heightened sensitivity to noise [1], and the presence of heterogeneity further amplifies this sensitivity. To mitigate the destabilizing impact of noisy nodes, we introduce a novel hyperedge-based regularization technique, tailored to enhance training stability. In this regularization scheme, each hyperedge assumes the role of a prototype, imposing constraints that compel its associated nodes to maintain a defined proximity (could be incidence matrix $I$ or learned attention map) in the embedding space. This strategic approach aims to curtail the influence of noise on the message-passing dynamics along the hypergraph topology.

As mentioned in Notations, we use incidence matrix $I \in \mathbb{R}^{n \times m}$ to denote the presence of nodes in different hyperedges. Hence, $I$ indicates the relations between nodes and hyperedges. Before applying hypergraph message passing, we can get the hyperedges representation $H \in \mathbb{R}^{m \times d}$ by node representation $F \in \mathbb{R}^{n \times d}$ and incidence matrix $I$, respectively. Then, we define our regularization normalized by diagonal matrices of hyperedge degree $D_e \in \mathbb{R}^{m \times m}$ as:

$$L_\theta = I - FH^T D_e^{-1}.$$
$$= I - FF^T I D_e^{-1}$$
$$(9)$$

Then, the total loss function is defined as:

$$L_s = L_{DCE} + \lambda L_\theta, \tag{10}$$

where $\lambda$ denotes the weight to balance the above two tasks. Through the introduced regularization, we mitigate the influence of noise and further enhance the model's robustness throughout the training process.

## 4    Experimental Setup

In this section, we introduce the datasets and evaluation metrics, followed by a detailed explanation of method implementations.

### 4.1    Datasets

We perform experimental evaluations on three well-established multimedia art and heterogeneous academic structure datasets presented in Table 1. The first dataset, WikiART_Artists from [4], represents a diverse compilation of artworks with 17k paintings from 23 most famous artists and 240 creation time periods. The other two datasets, DBLP and ACM, are obtained from the Heterogeneous Graph Benchmark (HGB) [16], which are heterogeneous but do not contain a visual component. We adopt them there to facilitate comparison with a broader range of existing methods. We strategically connect target nodes (with labels to be classified) using pertinent relationships to construct hyperedges within the datasets. For instance, in the WikiART dataset, we link each artwork to its corresponding artist and creation time. In the ACM dataset, the target nodes are research papers, and we form hyperedges by linking each paper to its respective authors, references and venue. Similarly, within the DBLP dataset, authors are targets and connected to their associated papers, the venue, and their terms.

**Table 1.** Statistics of HIN datasets

| Dataset | Node | Node Type | Hyperedges | Target | Class |
|---------|--------|-----------|------------|---------|-------|
| WikiArt | 18,048 | 3 | 17,820 | artwork | 12 |
| ACM | 10,942 | 3 | 3025 | paper | 3 |
| DBLP | 26,128 | 4 | 4057 | author | 4 |

## 4.2   Implementation Details

Node classification in our experiments follows a transductive setting using the train/val/test split from the ArtSAGENet [4] and HGB benchmark [16]. We implement our method in PyTorch and optimize it using the Adam optimizer [12] with an initial learning rate of 0.001. Hyperparameter settings for all baselines are consistent with those reported in their original papers. Early stopping is applied with a patience rate of 60 and all reported scores are averages from 5 different random seeds. We use one prototype per class, hidden dimension = 512, layer = 2 to WikiArt dataset, hidden dimension = 64, layer = 3, to ACM and DBLP datasets. For $\lambda$ value, we use $10^{-3}$ for DBLP, and $10^{-6}$ for WikiArt and ACM. Our experiments reveal that the optimal $\lambda$ value is contingent on the dataset's heterogeneity. Specifically, datasets with a greater variety of node types tend to benefit from larger $\lambda$ values.

# 5   Experimental Results

In this section, several experiments and their results are discussed to answer the following research questions:

1. Is deep learning on heterogeneous hypergraphs effective in node classification for HINs?
2. What are the benefits of the prototype classifier and prototype-based hyper-edge regularization in node classification for HINs?
3. Can prototypes facilitate the interpretation of HINs?

## 5.1   Heterogeneous Hypergraph Modelling

To answer whether heterogeneous hypergraphs are effective in node classification for HIN data, we conduct a transductive node classification experiment. Table 2 and Table 3 show the results of our methods on three datasets compared with the results of baselines, including metapath-based methods (RGCN [19], Het-GNN [33], HAN [23], MAGNN [7]) and metapath free-based methods (GTN [32], HetSANN [36], HGT [10], He-GCN [13], He-GAT [21]). Our approach is the overall best performer. A nuanced picture emerges when we delve into the differences between Micro-F1 and Macro-F1 scores. While it undeniably outperforms alternative methods in terms of Micro-F1 scores, the advantages of our approach concerning the Macro-F1 measure are not as readily apparent. This observation implies potential challenges in effectively handling imbalanced data. Varying number of nodes in imbalanced datasets lead to imbalanced hyperedge density, which results in data sparsity issues, challenging learning meaningful patterns for the underrepresented class nodes.

**Table 2. Semi-supervised node classification task results on WikiArt dataset.** Bold means best performance, and underline means second best.

| | | GTN | HGT | He-GCN | He-GAT | Ours |
|---|---|---|---|---|---|---|
| WikiArt | Micro-F1 | 91.25±0.48 | 91.17±0.54 | 90.81 ± 0.51 | 89.22 ±0.27 | **92.10 ± 0.64** |
| | Macro-F1 | 78.12±0.84 | 77.56±0.91 | **78.73 ± 0.87** | 74.9 ±2.06 | 78.58±0.95 |

## 5.2 Learnable Prototype Classifier and Prototype Regularization for Hypergraph Modeling

To understand the contribution of prototype regularization and the prototype classifier to the model's overall performance, we performed an ablation study, where we trained our methods with and without the prototype classifier and prototype regularization. Results in Table 4 show that training a heterogeneous hypergraph model with prototype classifier and regularization can improve both the Macro-F1 and Micro-F1 performance of HIN modeling. The result also shows that a learnable prototype classifier contributes more to the performance than prototype regularization. For a more intuitive comparison, we learn the node embedding space on the proposed model and project the learned embedding into a 2-dimensional space by UMAP [18]. Figure 3 shows that there is a clearer inter-class decision boundary for classes with learnable prototype classifiers and regularization, and intra-class clusters are also more compact.

**Table 3. Semi-supervised node classification task results on ACM and DBLP datasets.** Bold font denotes the best-performing results. Each model was run five runs, and we report the mean ± standard deviation. Baseline performance metrics are extracted from the HGB benchmark [16], with 'Mph' representing the metapath.

| | | ACM | | DBLP | |
|---|---|---|---|---|---|
| | | Micro-F1 | Macro-F1 | Micro-F1 | Macro-F1 |
| w/ Mph | RGCN | 91.41±0.75 | 91.55±0.74 | 92.07±0.50 | 91.52±0.50 |
| | HetGNN | 86.05±0.25 | 85.91±0.25 | 92.33±0.41 | 91.76±0.43 |
| | HAN | 90.79±0.43 | 90.89±0.43 | 92.05±0.62 | 91.67±0.49 |
| | MAGNN | 90.77±0.65 | 90.88±0.64 | 93.76±0.45 | 93.28±0.51 |
| w/o Mph | GTN | 91.20±0.71 | 91.31±0.70 | 93.97±0.54 | 93.52±0.55 |
| | RSHN | 90.32±1.54 | 90.50±1.51 | 93.81±0.55 | 93.34±0.58 |
| | HetSANN | 89.91±0.37 | 90.02±0.35 | 80.56±1.50 | 78.55±2.42 |
| | HGT | 91.00±0.76 | 91.12±0.76 | 93.49±0.25 | 93.01±0.23 |
| | He-GCN | 92.12±0.23 | 92.17±0.24 | 91.47±0.34 | 90.84±0.32 |
| | He-GAT | 92.19±0.93 | 92.26±0.94 | 93.39±0.30 | **93.83 ± 0.27** |
| | Ours | **93.30 ± 0.59** | **93.49 ± 0.56** | **94.12 ± 0.65** | 93.09±0.44 |

**Table 4. Performance comparison between the model with/without Prototype Regularization and Prototype Classifier.** The $\sqrt{}$ indicates the model with our proposed modules. The $\times$ indicates the model with standard softmax classifier and no regualization.

| Module | | WikiArt | | ACM | |
|---|---|---|---|---|---|
| P-Regularization | P-Classifier | Micro-F1 | Macro-F1 | Micro-F1 | Macro-F1 |
| $\times$ | $\times$ | 88.83± 2.55 | 74.25±2.31 | 90.42±1.36 | 90.61±1.30 |
| $\sqrt{}$ | $\times$ | 89.06± 1.51 | 75.21±0.81 | 91.04±0.67 | 91.21±0.68 |
| $\times$ | $\sqrt{}$ | 91.23±0.64 | 77.53±1.04 | 93.05±0.22 | 93.15±0.23 |
| $\sqrt{}$ | $\sqrt{}$ | **92.10 ± 0.64** | **78.58 ± 0.95** | **93.30 ± 0.59** | **93.49 ± 0.56** |

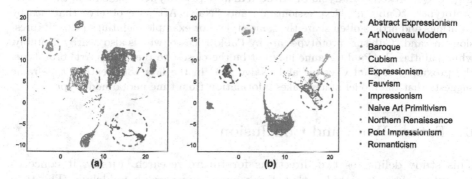

- Abstract Expressionism
- Art Nouveau Modern
- Baroque
- Cubism
- Expressionism
- Fauvism
- Impressionism
- Naive Art Primitivism
- Northern Renaissance
- Post Impressionism
- Romanticism

**Fig. 3. UMAP representation vectors of nodes on WikiArt dataset.** Subfigures (a) and (b) indicate painting representations learned without and with prototypes. The color represents artistic style. We observe that when using prototype-enhanced learning, the distribution of the node representations is more clustered for easily classified styles like Baroque and Romanticism. For hard to determine classes like Post Impressionism, the paintings are grouped together in (b) compared with the separation observed in (a), and they also become more distant from other classes.

## 5.3 Prototype for Interpreting HINs

To investigate the interpretative potential of prototype components for the HINs, we utilized three distinct nearby style prototypes based on the embedding space depicted in Fig. 3. Figure 4 unveils each style prototype's representative paintings and artists. Notably, the prototypes also unveil overlapping artists in their closest associations, reaffirming the interconnected nature of these artistic styles.

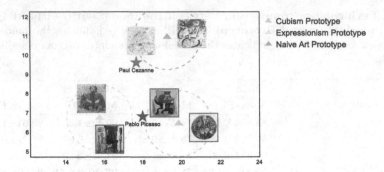

**Fig. 4. Qualitative analysis of three learned prototypes**. Examining three style prototypes, "Cubism," "Expressionism", and "Naive Art", we observe that nearby paintings and artists often share the same style. For example, paintings near "Expressionism" and "Cubism" prototypes are by Pablo Picasso, who is also nearby. Notably, while paintings by Paul Cezanne may not be the closest to the 'Naive Art' prototype, the proximity of Paul Cezanne and 'Naive Art' in the late 18th-century timeframe suggests that the model indeed takes information from time period node type.

## 6    Future Work and Conclusion

This study delineates two directions for future research. Firstly, it concerns the intricacies associated with heterogeneous hypergraph modeling. The two-step attention mechanism (Node/Hyperedge-level Attention) deployed in our framework may pose limitations in effectively modeling large-scale Heterogeneous Information Networks (HINs). In future work, we will explore techniques like sampling or partitioning to offer a solution. The second direction aims to advance the construction of a hyperedge. Our current methodology does not require building metapaths, and we have demonstrated the ease of building hyperedge based on target nodes. However, to broaden the applicability of our hypergraph modeling approach, further exploration into alternative hyperedge construction methods is warranted.

In this paper, we have introduced prototype-enhanced hypergraph learning, a novel framework for modeling heregeneous information networks without the need to build metapaths. Our framework enables effective analysis and information propagation across diverse entities within hypergraph structures. To demonstrate the effectiveness of our method, we have conducted numerical and qualitative experiments on several heterogeneous and multimedia datasets, showcasing its capabilities in capturing the rich relationships present in complex networks.

## References

1. Arya, D., Gupta, D.K., Rudinac, S., Worring, M.: Adaptive neural message passing for inductive learning on hypergraphs. arXiv preprint arXiv:2109.10683 (2021)

2. Ding, K., Wang, J., Li, J., Li, D., Liu, H.: Be more with less: hypergraph attention networks for inductive text classification. In: Proceedings of the 2020 Conference on Empirical Methods in Natural Language Processing (2020)

3. Dong, Y., Chawla, N.V., Swami, A.: Metapath2vec: scalable representation learning for heterogeneous networks. In: Proceedings of the 23rd ACM SIGKDD International Conference on Knowledge Discovery and Data Mining (2017)

4. Efthymiou, A., Rudinac, S., Kackovic, M., Worring, M., Wijnberg, N.: Graph neural networks for knowledge enhanced visual representation of paintings. In: Proceedings of the 29th ACM International Conference on Multimedia (2021)

5. Feng, Y., You, H., Zhang, Z., Ji, R., Gao, Y.: Hypergraph neural networks (2019)

6. Fu, J., Hou, C., Zhou, W., Xu, J., Chen, Z.: Adaptive hypergraph convolutional network for no-reference 360-degree image quality assessment. In: Proceedings of the 30th ACM International Conference on Multimedia (2022)

7. Fu, X., Zhang, J., Meng, Z., King, I.: MAGNN: metapath aggregated graph neural network for heterogeneous graph embedding. In: Proceedings of The Web Conference 2020 (2020)

8. Gao, Y., Feng, Y., Ji, S., Ji, R.: HGNN+: general hypergraph neural networks. IEEE Trans. Pattern Anal. Mach. Intell. 35, 3181–3199 (2023)

9. Gao, Y., Zhang, Z., Lin, H., Zhao, X., Du, S., Zou, C.: Hypergraph learning: methods and practices. IEEE Trans. Pattern Anal. Mach. Intell. 25, 2548–2566 (2022)

10. Hu, Z., Dong, Y., Wang, K., Sun, Y.: Heterogeneous graph transformer. In: Proceedings of The Web Conference (2020)

11. Huang, J., Yang, J.: UniGNN: a unified framework for graph and hypergraph neural networks. arXiv preprint arXiv:2105.00956 (2021)

12. Kingma, D.P., Ba, J.: Adam: a method for stochastic optimization. arXiv preprint arXiv:1412.6980 (2014)

13. Kipf, T.N., Welling, M.: Semi-supervised classification with graph convolutional networks. arXiv preprint arXiv:1609.02907 (2016)

14. Li, M., Zhang, Y., Li, X., Zhang, Y., Yin, B.: Hypergraph transformer neural networks. ACM Trans. Knowl. Discov. Data 17, 1–22 (2023)

15. Liu, J., Song, L., Wang, G., Shang, X.: Meta-HGT: metapath-aware hypergraph transformer for heterogeneous information network embedding. Neural Networks (2023)

16. Lv, Q., et al.: Are we really making much progress? Revisiting, benchmarking and refining heterogeneous graph neural networks. In: Proceedings of the 27th ACM SIGKDD Conference on Knowledge Discovery & Data Mining (2021)

17. Mao, Q., Liu, Z., Liu, C., Sun, J.: Hinormer: Representation learning on heterogeneous information networks with graph transformer. In: Proceedings of the ACM Web Conference 2023, pp. 599–610 (2023)

18. McInnes, L., Healy, J., Melville, J.: Umap: Uniform manifold approximation and projection for dimension reduction. arXiv preprint arXiv:1802.03426 (2018)

19. Schlichtkrull, M., Kipf, T.N., Bloem, P., Van Den Berg, R., Titov, I., Welling, M.: Modeling relational data with graph convolutional networks. In: The Semantic Web: 15th International Conference (2018)

20. Shi, C., Li, Y., Zhang, J., Sun, Y., Yu, P.S.: A survey of heterogeneous information network analysis. IEEE Transactions on Knowledge and Data Engineering (2017)

21. Veličković, P., Cucurull, G., Casanova, A., Romero, A., Liò, P., Bengio, Y.: Graph attention networks. In: International Conference on Learning Representations (2018)

22. Wang, X., Bo, D., Shi, C., Fan, S., Ye, Y., Yu, P.S.: A survey on heterogeneous graph embedding: methods, techniques, applications and sources. IEEE Trans. Big Data **9**, 415–436 (2023)
23. Wang, X., et al.: Heterogeneous graph attention network. In: The World Wide Web Conference (2019)
24. Wang, Y., Zhu, L., Qian, X., Han, J.: Joint hypergraph learning for tag-based image retrieval. IEEE Trans. Image Process. **27**, 4437–4451 (2018)
25. Wu, X., Chen, Q., Li, W., Xiao, Y., Hu, B.: AdahGNN: adaptive hypergraph neural networks for multi-label image classification. In: Proceedings of the 28th ACM International Conference on Multimedia (2020)
26. Wu, Z., Pan, S., Chen, F., Long, G., Zhang, C., Yu, P.S.: A comprehensive survey on graph neural networks. IEEE Trans. Neural Netw. Learn. Syst. **32**, 4–24 (2021)
27. Xie, Y., Xu, Z., Zhang, J., Wang, Z., Ji, S.: Self-supervised learning of graph neural networks: a unified review. IEEE Trans. Pattern Anal. Mach. Intell. **45**, 2412–2419 (2022)
28. Yadati, N., Nimishakavi, M., Yadav, P., Nitin, V., Louis, A., Talukdar, P.: Hyper-GCN: a new method for training graph convolutional networks on hypergraphs. In: Advances in Neural Information Processing Systems, vol. 32 (2019)
29. Yang, C., Xiao, Y., Zhang, Y., Sun, Y., Han, J.: Heterogeneous network representation learning: A unified framework with survey and benchmark. IEEE Trans. Knowl. Data Eng. **34**, 4854–4873 (2022)
30. Yang, H.M., Zhang, X.Y., Yin, F., Liu, C.L.: Robust classification with convolutional prototype learning. In: Proceedings of the IEEE Conference on Computer Vision and Pattern Recognition (2018)
31. Yang, X., Yan, M., Pan, S., Ye, X., Fan, D.: Simple and efficient heterogeneous graph neural network. In: Proceedings of the AAAI Conference on Artificial Intelligence, vol. 37, pp. 10816–10824 (2023)
32. Yun, S., Jeong, M., Kim, R., Kang, J., Kim, H.J.: Graph transformer networks. In: Advances in Neural Information Processing Systems (2019)
33. Zhang, C., Song, D., Huang, C., Swami, A., Chawla, N.V.: Heterogeneous graph neural network. In: Proceedings of the 25th ACM SIGKDD International Conference on Knowledge Discovery & Data Mining (2019)
34. Zhang, H., Liu, X., Zhang, J.: HEGEL: hypergraph transformer for long document summarization. In: Proceedings of the 2022 Conference on Empirical Methods in Natural Language Processing (2022)
35. Zhou, D., Huang, J., Schölkopf, B.: Learning with hypergraphs: clustering, classification, and embedding. In: Advances in Neural Information Processing Systems (2006)
36. Zhu, S., Zhou, C., Pan, S., Zhu, X., Wang, B.: Relation structure-aware heterogeneous graph neural network. In: IEEE International Conference on Data Mining (2019)

# A Language-Based Solution to Enable Metaverse Retrieval

Ali Abdari[1,2]($\boxtimes$) (iD), Alex Falcon[1] (iD), and Giuseppe Serra[1] (iD)

[1] University of Udine, Udine, Italy
abdari.ali@spes.uniud.it
[2] University of Naples Federico II, Naples, Italy

**Abstract.** Recently, the Metaverse is becoming increasingly attractive, with millions of users accessing the many available virtual worlds. However, how do users find the one Metaverse which best fits their current interests? So far, the search process is mostly done by word of mouth, or by advertisement on technology-oriented websites. However, the lack of search engines similar to those available for other multimedia formats (e.g., YouTube for videos) is showing its limitations, since it is often cumbersome to find a Metaverse based on some specific interests using the available methods, while also making it difficult to discover user-created ones which lack strong advertisement. To address this limitation, we propose to use language to naturally describe the desired contents of the Metaverse a user wishes to find. Second, we highlight that, differently from more conventional 3D scenes, Metaverse scenarios represent a more complex data format since they often contain one or more types of multimedia which influence the relevance of the scenario itself to a user query. Therefore, in this work, we create a novel task, called Text-to-Metaverse retrieval, which aims at modeling these aspects while also taking the cross-modal relations with the textual data into account. Since we are the first ones to tackle this problem, we also collect a dataset of 33000 Metaverses, each of which consists of a 3D scene enriched with multimedia content. Finally, we design and implement a deep learning framework based on contrastive learning, resulting in a thorough experimental setup.

**Keywords:** Multimedia · Text-to-Multimedia Retrieval · Cross-modal understanding · Metaverse · Contrastive Learning

## 1 Introduction

Nowadays, the Metaverse is becoming an increasingly popular pastime where the users relax using a multitude of different applications, mostly revolving around entertainment, including social-oriented games (e.g., Decentraland[1] and Roblox[2]), fitness activity (e.g., "The thrill of the fight" and "Archer VR", two

---

[1] https://decentraland.org/.
[2] https://www.roblox.com/.

© The Author(s), under exclusive license to Springer Nature Switzerland AG 2024
S. Rudinac et al. (Eds.): MMM 2024, LNCS 14556, pp. 477–488, 2024.
https://doi.org/10.1007/978-3-031-53311-2_35

**Fig. 1.** An illustration of the Text-to-Metaverse retrieval task, which requires the model to rank the available Metaverses by estimating their relevance to a user-defined query.

virtual reality applications for boxing and archery, respectively), and much more. Notably, the popularity of the Metaverse is ever-increasing and so far has reached a total of around 400 million users logging in monthly [18]. As an example, the amount of users logging in daily into the Roblox has grown from around 10 million at the end of 2018 to a staggering 70 million during the first part of 2023 [11]. Nonetheless, the Metaverse is not all about entertainment: in fact, many Metaverses are being created for more professional use cases, for instance for industrial training [4] or predictive maintenance applications [3,23].

The process of discovering a Metaverse that best matches one's interests is unfortunately not easy: the current means for discovering Metaverses are word-of-mouth (e.g., on social media, dedicated forums, etc.) and advertisement in technology-oriented websites. Therefore, similarly to what happened with other formats of digital content, there is a need for the user to express their interests through a natural language query. For instance, this happened with Google Images for visual data, Instagram or TikTok for videos, and YouTube for audio. To address the lack of research interest in this topic, in this paper, we propose a novel task, called Text-to-Metaverse retrieval, whose objective is to rank a list of Metaverses based on their relevance to a user-defined query, as illustrated in Fig. 1. In this way, we empower the users and give them more freedom in choosing how they want to express their interests.

A second point to note is that Metaverses have greater diversity and dynamism than conventional 3D scenes. In fact, Metaverses contain one or more types of multimedia data. For instance, a user could travel millions of years back in time within a Metaverse and interact with digital reconstructions of extinct animals, whereas a virtual shopping outlet could be hosted in a Metaverse so users can try on different clothes. Interestingly, multimedia content plays a fundamental role in defining the relevance of the Metaverse to the user query within the search engine. Moreover, their multimedia content can also change over time. For instance, the exhibition hosted at a Metaverse museum could be temporary and therefore it could significantly change from time to time. Unfortunately, there are two major issues with the public datasets available for 3D scenes: first, they do not contain multiple instances of the same Metaverse at different times; second, their scenes do not present any multimedia content. Therefore, to benchmark the progress on the Text-to-Metaverse retrieval task, we collect a dataset composed of around 33000 Metaverses, each of which contains a painting, hence focusing our work on art-related Metaverses.

Finally, we design and implement several solutions based on recent advancements in deep learning and cross-modal retrieval. Our results show that a solution to the Text-to-Metaverse retrieval is feasible and may provide a great tool for users to filter irrelevant Metaverses, further motivating the research on this very important and emergent topic.

The main contributions of our work can be summarized as follows:

- We introduce and motivate the Text-to-Metaverse retrieval task, which is becoming an increasingly important task for Metaverse understanding and exploration.
- We collect a dataset of around 33000 art-related Metaverses, and annotate each of them with a textual description of its contents.
- We design and implement a framework comprising several solutions based on deep learning, which experimentally motivates the feasibility of solutions for this important topic.

The rest of the paper is organized as follows. In Sect. 2, we review the literature on related topics, highlighting the lack of datasets on Metaverses and the need for further work on this topic. In Sect. 3, we describe in detail our methodology, which is based on recent advancements in 3D scene understanding and cross-modal retrieval. Section 4 reports the experimental results observed on our dataset. Finally, in Sect. 5, we highlight future work and draw conclusions on our research study.

## 2   Related Work

In this Section, we explore the literature related to the proposed task, Text-to-Metaverse retrieval. First, in Sect. 2.1 we review the literature on Metaverse-related technologies, to contextualize the work and highlight the lack of research directions in text-guided retrieval for Metaverses. Second, the area on cross-modal retrieval has greatly grown over the past few years and gives us access to a vast knowledge related to how to perform text-guided retrieval (Sect. 2.2).

### 2.1   Background on Metaverse-Related Research

The term Metaverse was first mentioned in the sci-fi novel "Snow Crash" by Neal Stephenson, who envisioned it as the successor of the Internet and as a form of multiplayer online game with user-controlled digital avatars. More recently, Lee et al. defined it as a virtual world where synthetic elements are blended into the real world [15]. Nowadays, extended reality technologies are used to enable users to access the Metaverse, and new devices are being presented frequently as the technology for it becomes more accessible to customers. Along the same track, new use cases are being developed at an increasing rate, resulting in many applications ranging from entertainment, such as digital museums [6,14] and concerts

[12], passing through smart healthcare and assistance [13,16], to industry-ready applications, such digital shopping and virtual try-on [7,24] to predictive maintenance on industrial equipment [3,23]. Notably, these applications are often supported by foundational research being performed on related problems, such as human pose estimation [27,33] and content generation [26].

However, the task of searching for a Metaverse based on user interests has not been addressed in the literature so far, highlighting the importance of this new research direction.

## 2.2    Cross-Modal Understanding and Retrieval Applications

The process of finding query-relevant content in a large dataset is commonly regarded as a query-guided retrieval problem. During the past few years, multimedia retrieval has become increasingly popular due to the large increase in user-generated content on social media and the need to filter irrelevant content for a specific user. Specifically, considerable advancements were obtained in text-guided retrieval for videos [10,17], images [5,21], and audio [29,30], mostly thanks to the surge of deep representations. A common way to learn neural models for cross-modal retrieval consists in the use of contrastive loss functions to automatically learn how to address the domain gap [19,22]. In fact, contrastive loss functions aim at maximizing the similarity of paired samples in the dataset, e.g., an image and its caption, allowing for the subsequent use of the learned functions to map the queries into the same embedding space and perform retrieval.

Interestingly, the 3D scene retrieval problem, which could be seen as greatly related to the proposed Text-to-Metaverse retrieval, has not been addressed in the literature so far. In fact, the research on this topic is often focused on specific aspects of the scene, such as identifying the text [25,28] or the objects present in the scenes [20], or retrieving 3D scenes using 2d images [1,2] or sketches as the input queries [32]. In particular, differently from the SHREC challenge [1,2], the Text-to-Metaverse retrieval task requires free-form text queries and the Metaverses contain multimedia content which is not available in the 3D scenes used in SHREC. Therefore, the datasets available for the aforementioned tasks are not usable for our task, raising the necessity to collect a dataset for it.

## 3    Proposed Methodology

In this Section, we describe the three main components of our methodology. First, in Sect. 3.1, we provide a thorough description of the network architecture used in all our experiments. Second, Sect. 3.2 describes the steps we performed to collect the dataset used in the experiments.

### 3.1    Network Architecture

Figure 2 presents an overview of the proposed architecture. As can be seen, it is composed of three main components: MUM, the Metaverse Understanding Module; TUM, the Textual Understanding Module; and, finally, the Text-Metaverse

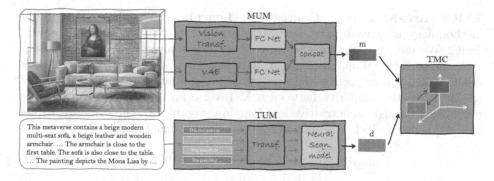

**Fig. 2.** Overview of the network architecture used in the experiments. A thorough discussion is present in Sect. 3.1.

Contrastive learning framework (TMC). All of these components are described in detail in the rest of this section.

**MUM: Metaverse Understanding Module.** As the name suggests, the MUM is used to model the Metaverse scenarios under analysis. It is assumed that the paintings can be extracted from the room. Then, the paintings are initially processed separately from the scene. Specifically, the representation of the scene $\rho_s$ is obtained by using a recent Variational Autoencoder [31], whereas a Vision Transformer trained with CLIP is used for the painting representation $\rho_p$ [21]. Then, a deep network, FCNet, is used to learn two separate representations of the inputs, using a similarly structured architecture:

$$r_1 = \text{ReLU}(W_1 \rho_\star + b_1) \tag{1}$$
$$r_2 = \text{ReLU}(W_2 \, \text{BatchNorm}(\delta_1(r_1))) + b_2 \tag{2}$$
$$r_\star = W_3 \, \text{BatchNorm}(\delta_2(r_2))) + b_3 \tag{3}$$

where $\rho_\star$ is either $\rho_s$ or $\rho_p$, and $\delta$ is the dropout operator. The output of this FCNet is $r_s \in R^{1 \times \frac{H}{2}}$ when the input is $\rho_s$, and $r_p \in R^{1 \times \frac{H}{2}}$ when the input is $\rho_p$. Finally, the output of MUM is a vectorial representation $m$ of size $1 \times H$ obtained by concatenating $r_s$ and $r_p$.

**TUM: Textual Understanding Module.** To obtain the representation of the textual data, TUM consisting of the following steps is employed. First, the Metaverse descriptions are separated into sentences by using the period as a splitting term. Then, a representation of each sentence is obtained by using the same CLIP module used in the previous section. This process is used since the descriptions are very long, hence difficult to model with standard natural language processing techniques (e.g., BERT has a context of 512 tokens [8]). After obtaining the sentence-level representations, a contextual representation for the full description is obtained through a neural sequence model (more details in Sect. 4). Ultimately, the final representation $d \in R^{1 \times H}$ for the description is given by the last hidden state of the sequence model.

**TMC: Text-Metaverse Contrastive Learning.** The final step of our methodology is given by the Text-Metaverse contrastive learning framework, TMC. We use this approach because of its great success in recent years in cross-modal retrieval applications [21]. Given a batch of B Metaverses and their descriptions, the respective representations are obtained through MUM and TUM. Then, the similarity between a Metaverse $m_i$ and its description $d_i$ is maximized through a contrastive loss function. Specifically, the standard triplet loss [22] is used, which is described by the following equations:

$$\mathcal{L}_{mt,i,j} = relu(0, \Delta + sim(s_i, d_j) - sim(s_i, d_i)) \tag{4}$$

$$\mathcal{L}_{tm,i,j} = relu(0, \Delta + sim(s_j, d_i) - sim(s_i, d_i)) \tag{5}$$

$$\mathcal{L} = \frac{1}{B} \sum_i \sum_{j \neq i} \mathcal{L}_{tm,i,j} + \mathcal{L}_{mt,i,j} \tag{6}$$

where $\Delta$ is a fixed margin, and $\mathcal{L}$ is the final loss used to perform the training.

### 3.2   Dataset Collection

As mentioned in the previous sections, Metaverses can be seen as 3D scenes in which there is additional multimedia content playing a fundamental role in defining the purpose of the Metaverse itself. However, as mentioned in Sect. 2.2, there are no suitable public datasets which contain multimedia-enriched 3D scenes. Therefore, in the following, we describe the process followed to collect the dataset.

**Collection of Suitable Metaverse Scenarios.** To create the Metaverses suitable for our task, we started from 3384 indoor scenarios designed by professionals [9] to provide high-quality 3D scenes. These scenarios feature several types of furniture, including lamps, beds, and wardrobes, among others. Then, inspired by virtual Metaverses covering museums and similar art exhibitions, we decided to make a selection of famous paintings and put those into the 3D scenes to form the final Metaverses. Specifically, we selected ten paintings and then paired each of them to each scenario, resulting in a total number of 33840 Metaverse scenarios.

**Textual Descriptions.** To obtain the descriptions for our Metaverses, we devised an automatic procedure consisting of two main phases.

First, given a scenario, the description is obtained by putting together the automatically created sentences, which are separately created for each furniture piece. These are obtained by gathering a set of tags from the 3d-front dataset, and then by putting them together via a template. The tags are divided into categories (e.g., "Wardrobe", "Bed"), style (e.g., "Chinese", "Industrial"), theme (e.g., "Wrought Iron", "Texture Mark"), and material tags (e.g., "Marble", "Cloth"). The template for the first furniture is the following: "This room contains <N> <Category> with <Style> style, <Theme> theme, and <Material>.", where N identifies the quantity of repetitions for a given instance, and the other is the tags for that furniture piece. Then, for the other pieces, it slightly changes to: "Additionally, it also contains <N> <Category> with <Style> style, <Theme> theme, and <Material>.". After that, a similar procedure is followed to describe the position of the furniture with respect to other furniture pieces. These positional sentences use a template, "The <N> <Category1> with <Style1> style, <Theme1> theme, and <Material1> is <Distance> from <N> <Category2> with <Style2> style, <Theme2> theme, and <Material2>.", to provide intuition on the distance separating two objects in the scenario. The words we use for the distance ("so close", "close", "far", and "so far") are determined by the standard deviation of the distribution of distances.

The second step involves the description of the paintings. After describing the scenario, the description of the painting is created using the following template: "Also, in this room there is a painting called <Name> by <Author>, which <Description>.", where the description is obtained by asking ChatGPT with the following prompt: "provide two sentences for this painting: <Name> by <Author>" and manually adapted.

Finally, given a scenario and a painting, we put together their descriptions to obtain the final one. On average, the descriptions are 626.91 tokens long (ranging from 162 to 2014), with an average of 23.33 sentences (ranging from 7 to 79).

## 4    Experimental Results

In this section, several experiments are performed to analyze the behavior of the model when architectural changes are applied to MUM (Sect. 4.1) or TUM (Sect. 4.2), followed by a discussion of the results and the limitations of our methodology. To perform the evaluation on the Text-to-Metaverse retrieval task, standard metrics are used: the recall rates at K (R@K, with K = 1, 5, 10, 50, 100), the median rank (MedR), and the mean rank (MR). All the code and data are available at https://github.com/aliabdari/NLP_to_rank_artistic_Metaverses.

### 4.1    How to Model the Metaverses?

To model the Metaverses under analysis, we consider an alternative to the proposed MUM. The proposed model can be identified as a late fusion approach (LF), since the representation for the scene and the painting are first treated separately, and then fused. Therefore, here we explore an early fusion approach (EF). Specifically, the CLIP representation of the painting and that of the scene are early fused through concatenation, after learning a linear transformation from 512 CLIP features to D features (same as the scene representation). Then, their concatenation is passed through a FCNet (Sect. 3) to learn a joint representation. The results of the comparison are shown in Table 1. Two major observations can be made. First, both the solutions achieve very high R@100

**Table 1.** Comparison across different alternatives for our Metaverse understanding module. Details in Sect. 4.1.

| Method | R@1 | R@5 | R@10 | R@50 | R@100 | MedR | MR |
|--------|-----|-----|------|------|-------|------|----|
| EF | 1.8 | 5.1 | 17.9 | 55.6 | 74.7 | 42.0 | 82.4 |
| LF | **15.3** | **41.1** | **55.6** | **84.4** | **92.1** | **8.0** | **37.7** |

**Table 2.** Comparison across different alternatives for our textual understanding module. Details in Sect. 4.2.

| Method | R@1 | R@5 | R@10 | R@50 | R@100 | MedR | MR |
|--------|-----|-----|------|------|-------|------|----|
| Mean | 7.5 | 24.8 | 37.2 | 74.5 | 86.9 | 19.0 | 52.8 |
| GRU | 10.8 | 35.9 | 51.8 | 83.2 | 90.9 | 10.0 | 40.6 |
| BiGRU | **15.3** | **41.1** | **55.6** | **84.4** | **92.1** | **8.0** | **37.7** |
| LSTM | 4.1 | 16.3 | 28.6 | 65.7 | 79.5 | 27.0 | 96.0 |
| BiLSTM | 13.2 | 37.1 | 52.2 | 82.4 | 90.7 | 10.0 | 43.6 |

(74.7% and 92.1%). However, it does not mean that the problem is solved since retrieving the most suitable Metaverse among the top 100 Metaverses is far from perfect from the user's perspective. On the other hand, a R@1 of 1.8% and 15.3% shows that there is a lot of room for improvement on both the early and late fusion approaches. Second, the late fusion approach achieves far better performance in our experimental setup. This may be due to the great domain gap between the two modalities (scene and painting) which makes it difficult to obtain a meaningful representation by early concatenating the two separate representations.

## 4.2   How to Model the Descriptions?

Similar to the previous analysis, here we aim to explore different variations for the neural sequence model used in TUM. Specifically, we explore standard methods: temporal average pooling ("Mean"), LSTM, Bidirectional LSTM ("BiLSTM"), GRU, and Bidirectional GRU ("BiGRU"). The results are reported in Table 2. Interestingly, the results show that pooling the sentence representations achieves better performance than using a standard LSTM (e.g., Mean achieves 24.8% R@5, whereas the LSTM leads to 16.3%). On the other hand, the GRU-based methods perform better than their LSTM counterpart and, in particular, the BiGRU achieves the best overall performance (15.3% R@1 compared to 13.2% obtained by BiLSTM). These results may be explained by the fact that the LSTM has more parameters than the GRU, which may require more diverse data to outperform the simpler GRU architecture. Second, in both cases, bidirectionality is fundamental to achieve a better contextual understanding.

### 4.3   Limitations

In this section, the limitations of our work are highlighted.

**About the Metaverse Modeling.** Our Metaverse Understanding Module uses the recent advancements in scene understanding achieved by Yang et al. [31] to model the scenario. A different approach is based on large pretrained vision-language models, e.g., CLIP [21]. Different from the scene understanding technique, it would allow for improved spatial reasoning, hence enabling a better understanding of the relations between objects and furniture. Moreover, the language-supported training also reduces the domain gap between the two modalities, possibly leading to a smoother training behavior. This possibility is a promising future direction for our work.

**About the Multimedia Content.** In our work, the focus is on painting-related Metaverses. However, there are other forms of art, such as mosaics, videos, and 3D artifacts (statues, jewels, armors, etc.). Moreover, Metaverses could also contain multimedia content which is not artistic per se, such as game-like or digital traveling experiences. Therefore, future work on this topic should also strive to include and model other media formats.

### 4.4   Implementation Details

In our experiments, we used $B = 64$, $D = 200$, $H = 256$. Specifically, the value for $D$ was given by the previous work on scene understanding which was used to get the scene representation [31]. To optimize the parameters, we used Adam with an initial learning rate of 0.008, which was reduced by 25% after 17 epochs. The available data is split into three sets with a 70/15/15 ratio, and we ensure that each scene is not present in multiple sets at the same time. The training lasts 30 epochs, and $\Delta = 0.25$ in our loss function. The performance on the testing set is done using the best model on the validation set. To perform the experiments, we used PyTorch 1.12.1 on a machine equipped with an RTX A5000 GPU, an Intel Xeon E5-1620, and 16 GB of RAM.

## 5   Conclusions

The Metaverse is becoming increasingly popular among the users for daily use. However, filtering irrelevant Metaverses based on a user query is a difficult task which has not been addressed by research so far. Therefore, in this paper, we introduced the Text-to-Metaverse retrieval task, inspired by the recent advancements in multimedia retrieval. Since the public datasets do not have multimedia-rich 3D scenes, we collected a dataset of 33840 Metaverses, each paired with a detailed textual description. We compared the proposed methodology to several alternatives both for the Metaverse and the textual modeling modules. Finally, we highlighted the limitations of our work which may serve as inspiration for promising research directions.

**Acknowledgments.** This work was supported by the Department Strategic Plan (PSD) of the University of Udine-Interdepartmental Project on Artificial Intelligence (2020-25), MUR Progetti di Ricerca di Rilevante Interesse Nazionale (PRIN) 2022 (project code 2022YTE579), and by TechStar Srl, Italy. Also, we thank Beatrice Portelli for helping with the illustrations and for the useful feedback during the preparation of this work.

# References

1. Abdul-Rashid, H., et al.: Shrec'18 track: 2D image-based 3D scene retrieval. Training **700**, 70 (2018)
2. Abdul-Rashid, H., et al.: Shrec'19 track: extended 2D scene image-based 3D scene retrieval. Training (per class) **700**, 70 (2019)
3. Agnusdei, G.P., Elia, V., Gnoni, M.G.: A classification proposal of digital twin applications in the safety domain. Comput. Ind. Eng. **154**, 107137 (2021)
4. Almeida, L.G.G., de Vasconcelos, N.V., Winkler, I., Catapan, M.F.: Innovating industrial training with immersive metaverses: a method for developing cross-platform virtual reality environments. Appl. Sci. **13**, 8915 (2023)
5. Cheng, Y., Zhu, X., Qian, J., Wen, F., Liu, P.: Cross-modal graph matching network for image-text retrieval. ACM Trans. Multimedia Comput. Commun. Appl. **18**(4), 1–23 (2022)
6. Choi, H.S., Kim, S.H.: A content service deployment plan for metaverse museum exhibitions-centering on the combination of beacons and HMDs. Int. J. Inf. Manag. **37**(1), 1519–1527 (2017)
7. Dawson, A., et al.: Data-driven consumer engagement, virtual immersive shopping experiences, and blockchain-based digital assets in the retail metaverse. J. Self-Gov. Manag. Econ. **10**(2), 52–66 (2022)
8. Devlin, J., Chang, M.W., Lee, K.: Google, KT, language, AI: bert: pre-training of deep bidirectional transformers for language understanding. In: Proceedings of NAACL-HLT, pp. 4171–4186 (2019)
9. Fu, H., et al.: 3D-front: 3D furnished rooms with layouts and semantics. In: Proceedings of the IEEE/CVF International Conference on Computer Vision, pp. 10933–10942 (2021)
10. Ge, Y., et al.: Bridging video-text retrieval with multiple choice questions. In: Proceedings of the IEEE/CVF Conference on Computer Vision and Pattern Recognition, pp. 16167–16176 (2022)
11. J., C.: Daily active users (dau) of roblox games worldwide from 4th quarter 2018 to 2nd quarter 2023. Technical report, Statista (2023). https://www.statista.com/statistics/1192573/daily-active-users-global-roblox/
12. Jin, C., Wu, F., Wang, J., Liu, Y., Guan, Z., Han, Z.: Metamgc: a music generation framework for concerts in metaverse. J. Audio Speech Music Proc. 31 (2022)
13. Laaki, H., Miche, Y., Tammi, K.: Prototyping a digital twin for real time remote control over mobile networks: Application of remote surgery. IEEE Access **7**, 20325–20336 (2019)

14. Lee, H.K., Park, S., Lee, Y.: A proposal of virtual museum metaverse content for the mz generation. Dig. Creat. **33**(2), 79–95 (2022)
15. Lee, L.H., et al.: All one needs to know about metaverse: a complete survey on technological singularity, virtual ecosystem, and research agenda. arXiv preprint arXiv:2110.05352 (2021)
16. Liu, Y., et al.: A novel cloud-based framework for the elderly healthcare services using digital twin. IEEE Access **7**, 49088–49101 (2019)
17. Luo, H., et al.: Clip4clip: an empirical study of clip for end to end video clip retrieval and captioning. Neurocomputing **508**, 293–304 (2022)
18. Metaversed: The metaverse reaches 400m monthly active users. Technical report, Metaversed Consulting (2023). https://www.metaversed.consulting/blog/the-metaverse-reaches-400m-active-users
19. Miech, A., Alayrac, J.B., Smaira, L., Laptev, I., Sivic, J., Zisserman, A.: End-to-end learning of visual representations from uncurated instructional videos. In: Proceedings of IEEE/CVF CVPR, pp. 9879–9889 (2020)
20. Nguyen, T., Gopalan, N., Patel, R., Corsaro, M., Pavlick, E., Tellex, S.: Robot object retrieval with contextual natural language queries. arXiv preprint arXiv:2006.13253 (2020)
21. Radford, A., et al.: Learning transferable visual models from natural language supervision. In: International Conference on Machine Learning, pp. 8748–8763. PMLR (2021)
22. Schroff, F., Kalenichenko, D., Philbin, J.: FaceNet: a unified embedding for face recognition and clustering. In: Proceedings of the IEEE Conference on Computer Vision and Pattern Recognition, pp. 815–823 (2015)
23. Siyaev, A., Jo, G.S.: Towards aircraft maintenance metaverse using speech interactions with virtual objects in mixed reality. Sensors **21**(6), 2066 (2021)
24. Song, W., Gong, Y., Wang, Y.: VTONShoes: Virtual try-on of shoes in augmented reality on a mobile device. In: IEEE ISMAR, pp. 234–242 (2022)
25. Wang, H., Bai, X., Yang, M., Zhu, S., Wang, J., Liu, W.: Scene text retrieval via joint text detection and similarity learning. In: Proceedings of the IEEE/CVF Conference on Computer Vision and Pattern Recognition, pp. 4558–4567 (2021)
26. Wang, J., Chen, S., Liu, Y., Lau, R.: Intelligent metaverse scene content construction. IEEE Access **11**, 76222–76241 (2023). https://doi.org/10.1109/ACCESS.2023.3297873
27. Wang, X., Wang, Y., Shi, Y., Zhang, W., Zheng, Q.: AvatarMeeting: an augmented reality remote interaction system with personalized avatars. In: Proceedings of the 28th ACMMM, pp. 4533–4535 (2020)
28. Wen, L., Wang, Y., Zhang, D., Chen, G.: Visual matching is enough for scene text retrieval. In: Proceedings of the Sixteenth ACM International Conference on Web Search and Data Mining, pp. 447–455 (2023)
29. Wu, Y., Chen, K., Zhang, T., Hui, Y., Berg-Kirkpatrick, T., Dubnov, S.: Large-scale contrastive language-audio pretraining with feature fusion and keyword-to-caption augmentation. In: ICASSP 2023–2023 IEEE International Conference on Acoustics, Speech and Signal Processing (ICASSP), pp. 1–5. IEEE (2023)
30. Xin, Y., Yang, D., Zou, Y.: Improving text-audio retrieval by text-aware attention pooling and prior matrix revised loss. In: ICASSP 2023–2023 IEEE International Conference on Acoustics, Speech and Signal Processing (ICASSP), pp. 1–5. IEEE (2023)
31. Yang, H., et al.: Scene synthesis via uncertainty-driven attribute synchronization. In: Proceedings of the IEEE/CVF ICCV, pp. 5630–5640 (2021)

32. Yuan, J., Abdul-Rashid, H., Li, B., Lu, Y.: Sketch/image-based 3d scene retrieval: Benchmark, algorithm, evaluation. In: 2019 IEEE Conference on Multimedia Information Processing and Retrieval (MIPR), pp. 264–269. IEEE (2019)
33. Zhou, Y., Huang, H., Yuan, S., Zou, H., Xie, L., Yang, J.: Metafi++: Wifi-enabled transformer-based human pose estimation for metaverse avatar simulation. IEEE Internet Things J. **10**(16), 14128–14136 (2023). https://doi.org/10.1109/JIOT. 2023.3262940

# Part-Aware Prompt Tuning for Weakly Supervised Referring Expression Grounding

Chenlin Zhao[1,2,3], Jiabo Ye[3,5], Yaguang Song[4], Ming Yan[3],
Xiaoshan Yang[1,2,4], and Changsheng Xu[1,2,4(✉)]

[1] State Key Laboratory of Multimodal Artificial Intelligence Systems(MAIS),
Institute of Automation, Chinese Academy of Sciences (CASIA), Beijing, China
zhaochenlin2021@ia.ac.cn, xiaoshan.yang@nlpr.ia.ac.cn
[2] School of Artificial Intelligence, University of Chinese Academy of Science (UCAS),
Beijing, China
[3] Damo Academy, Alibaba Group, Beijing, China
jiabo.ye@stu.encu.edu.cn, ym119608@alibaba-inc.com
[4] Peng Cheng Laboratory, Shenzhen, China
songyg01@pcl.ac.cn, csxu@nlpr.ia.ac.cn
[5] East China Normal University, Shanghai, China

**Abstract.** Referring expression grounding represents a complex multi-modal task that merits meticulous investigation. To alleviate the conventional methods' reliance on fine-grained supervised data, there is a pressing need to explore visual grounding techniques under the weakly-supervised setting, encompassing only image-text pairs. Weakly supervised method with pretrained multimodal model has achieved impressive results; however, during the inference phase, it fails to generate a comprehensive attention map for entities, consequently leading to a reduction in inference accuracy. In this study, we introduce Part-aware Prompt Tuning (PPT), an innovative weakly supervised method. By dividing the entities extracted by the detector into different parts to optimize the part-aware prompt during the training phase, these prompt can guide the attention of pretrained multimodal model during the inference phase to obtain a more comprehensive focus on the whole entity, thereby enhancing inference accuracy. Empirical validation on two benchmark datasets, RefCOCO and RefCOCO+, underscores the remarkable superiority of our proposed method over prior referring expression grounding methods.

**Keywords:** Referring Expression Grounding · Prompt Tuning · Weakly Supervised

## 1 Introduction

Referring expression grounding plays a key role in many domains of human-machine interaction, including but not limited to robot navigation [21,28] and

---

C. Zhao and J. Ye—Equal contribution.

© The Author(s), under exclusive license to Springer Nature Switzerland AG 2024
S. Rudinac et al. (Eds.): MMM 2024, LNCS 14556, pp. 489–502, 2024.
https://doi.org/10.1007/978-3-031-53311-2_36

older lady     *the head of* older lady     *the hands of* older lady     *the legs of* older lady     Blend Result

(a)                                    (b)                                    (c)

**Fig. 1.** (a) The attention map of the model under the original text. (b) The attention map of the model under the text with different prompts. (c) The blend result of all three attention maps under the text with different prompts, which demonstrates by conducting multiple inferences based on different templates and overlaying the obtained outcomes. The highlighted white regions represent the model's attention.

visual object disambiguation [7,16]. Conventional supervised frameworks [15,23–25] manage solving this problem predominantly lean on the utilization of sizable, costly datasets with bounding box annotations, which is extremely costly. Hence, the exploration of methods under weakly-supervised conditions, where there is no bounding box annotation information and only pairs of images and texts are available, becomes highly significant.

Existing weakly supervised methods can be divided into two categories based on whether or not using pretrained multi-modal models as the base model. Early studies [2,12,13,22] employ detectors [8,20] to provide the spatial locations of entities within images and subsequently rank these entities. While such methods require a relatively smaller volume of training data and are cost-effective, their precision falls short of the ideal standard. Recently, weakly supervised methods based on pretrained multimodal models, e.g. [10,26], demonstrate remarkable performance. During inference, the model leverages the cross-modal encoder's alignment results to generate attention map, which is then used to mark the regions of text-image alignment on the image. However, in practice, the areas where text and image can align effectively are relatively limited. The attention mechanism often fails to encompass the entirety of an object, focusing primarily on its most discriminative regions while disregarding other facets of the object, as shown in Fig. 1.

To solve this issue, the key lies in redirecting the model's attention to the entirety of the entity. By framing entity components as prompts and appending them to the textual input, we can steer the model's focus towards different segments of the image, as illustrated in Fig. 1. The original text, *older lady*, elicits a notably limited attention span on the image. By separately introducing three prompts, *the head of*, *the hands of*, and *the legs of*, preceding the text, we effectively prompt the model to attend to various aspects of the entity within the image. Nonetheless, template-based prompt approaches restrict the model's attention to specific object categories, rendering it challenging to discover readily available templates for other object classes that could serve as prompts. Thus, there arises a need to render prompts amenable to learning, enabling the model to obtain appropriate prompts for different objects through optimization. In light of this, we propose a method based on prompt tuning, called Part-aware Prompt Tuning (PPT), which extends the model's attention span to encompass the entire object, consequently selecting detection boxes that attend comprehensively to the object. Employing three distinct learnable prompts to guide the model's focus on the upper, middle, and lower segments of the object facilitates a more comprehensive coverage of the object, as opposed to fixating solely on a fragment of the object. Consequently, during the inference process, we achieve greater efficacy in selecting detection boxes that encapsulate the complete object, thereby enhancing the model's performance. The contributions of this work are as follows:

- We proposed a novel framework, Part-aware Prompt Tuning (PPT), to address the challenge of weakly supervised referring expression grounding, addressing the issue of the model's attention being concentrated on only a small part of the entity during the inference stage, which had resulted in erroneous bounding box selection.
- We innovatively partition the entities extracted by the detector into different parts to optimize the part-aware prompt during the training phase, which enables us to utilize the part-aware prompt during the inference phase to guide the pretrained multimodal model in locating the corresponding parts of the entities. Subsequently, we combine the results obtained using different part-aware prompts to infer the complete regions of entity.
- We conducted experiments on two widely adopted datasets, RefCOCO and RefCOCO+, the results of which demonstrate that our model has achieved competitive performance.

## 2    Related Work

### 2.1    Referring Expression Grounding (REG)

Fully supervised referring expression grounding typically employs two methods: two-stage methods [15, 24, 25] and one-stage methods [4, 9, 11]. Two-stage methods divide the visual grounding process into two steps: generating candidate regions and ranking these regions based on the similarity. In contrast, one-stage methods directly locate the inference objects from the image using a multimodal matching network.

Yu et al. [25] extract visual features directly from objects in close proximity to each other and have demonstrated that paying attention to the relationships between objects enhances model performance. TransVG [4] combines the DETR [1] encoder to extract visual features and found that directly regressing bounding box coordinates results in better localization accuracy. These methods have proven to be highly effective but often rely on the availability of a substantial amount of pixel-level fine-grained annotation data, which can be costly.

## 2.2 Weakly Supervised Referring Expression Grounding (WSREG)

To alleviate the need for expensive fine-grained annotations, the weakly supervised setting for referring expression grounding (REG) has been proposed, which involve training data that consists solely of pairs of images and corresponding descriptions or queries, without providing fine-grained annotations in the form of bounding boxes. The absence of fine-grained annotation makes the commonly used bounding box regression loss ineffective in weakly supervised scenarios.

However, some methods [3,17,22] still have achieved promising results under weakly supervised conditions. GroundR [17] reconstructs phrases by applying attention mechanisms to visual features to establish alignment between entities in the image and entities in the language. Similarly, KAC-Net [2] employs a method similar to GroundR but leverages visual consistency and knowledge from object categories to improve grounding accuracy. Nevertheless, these methods are limited by the lack of fine-grained supervision data, making it challenging to align the visual and linguistic modals in feature space, often resulting in lower inference accuracy.

To better align the visual and linguistic modals, some methods [10,12,19, 26,27] fine-tune visual-language pre-trained (VLP) models on weakly supervised visual grounding datasets. Due to modality alignment capabilities already present in VLP models during their pre-training phase, the absence of fine-grained supervision data does not significantly impact the model's inference accuracy. ALBEF [10], for instance, employs contrastive learning with paired image-text data and utilizes a momentum distillation approach to effectively align the two modalities during training phrase. X-VLM [26] builds upon ALBEF by introducing multi-grained inputs, enhancing the model's performance.

During the inference phase, these methods often convert the image-text matching results output by the cross-modal encoder into an attention map and utilize a detector to provide bounding boxes for objects in the image. They then use the attention map to filter and select the final prediction results from these boxes. However, attention maps tend to focus primarily on the most distinctive portions of the target entity, neglecting a significant portion of the entity's surrounding edge regions. This limitation, to some extent, restricts the effectiveness of these methods. Therefore, in the inference phase, effectively guiding the model's attention towards the entire region of the entity and forming a comprehensive attention map will become the key to this issue.

## 3 Method

### 3.1 Problem Setting and Overview

The aim of weakly supervised referring expression grounding is to locate an entity mentioned in a noun phrase in an image that pairs with it without fine-grained bounding box annotation. Given a picture $I$ and a query $T$ that describes an object in that picture, our task is to output the position $P$ of the object mentioned in $T$ in the image. The schematic diagram of our method is illustrated in

Fig. 2. This paper addresses the issue of potential inaccurate localization during the inference stage of pre-trained multi-modal models due to the incomplete coverage of the entity's region by the attention map. To mitigate this problem, we introduce part-aware prompts to direct the model's attention simultaneously to both part and core of entities.

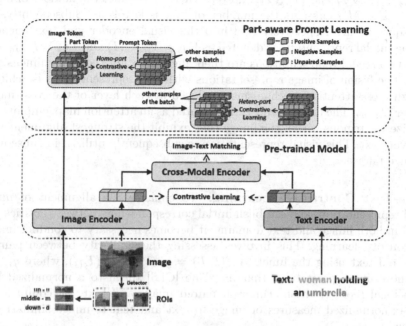

**Fig. 2.** Illustration of PPT. The alignment and fusion of the two modalities' information are accomplished within the pre-trained model. The attention map generated by image-text matching is used for subsequent inference. Visual features are concatenated with part features, while text is prepended with part-aware prompts. Both sets of concatenated features are input into the part-aware prompt learning module to optimize part-aware prompts.

## 3.2   Pre-trained Multi-modal Model Pipeline

In this section, we demonstrate the processes involving the encoding of visual and linguistic modal using the pretrained multimodal model, the alignment of visual and linguistic features, as well as the procedures for cross-modal fusion and the acquisition of cross-modal attention maps.

**Modal Encoding and Fusion.** We employed the same visual encoder $V_{trans}$, text encoder $L_{trans}$ and cross-modal encoder $C_{trans}$ as those present in X-VLM [26]. Given a paired input of image and text denoted as $I$ and $T$ respectively, the visual encoder subdivides the image into equal-sized patches and encodes them into a sequential format. For a 224×224 input image $I$ and a patch size of 32×32, this process yields a sequence of length $N = 49$. The text encoder, on the other hand, directly encodes the input text $T$ into a token sequence. Additionally, a special token denoted as [CLS] is added to each sequence to represent modal representations. Consequently, we obtain sequences in the form of $\{V_{cls}, I_1, ..., I_N\}$ and $\{T_{cls}, T_1, ..., T_M\}$, where $N$ signifies the number of image patches, and $M$ represents the number of linguistic tokens. Subsequently, the two sequences are respectively feed into the visual encoder and text encoder, yielding modal representations denoted as $V_{cls} = V_{trans}(I)$ and $T_{cls} = L_{trans}(T)$. These two modal representations are then aligned within the same dimensional space. The fusion of image representations with text representations is achieved through cross-attention mechanisms employed at each layer of the cross-modal encoder $C_{trans}$. The fusion result is represented as an attention map, employed to visualize the regions within the image that exhibit the strongest alignment with the given text. This visual representation is subsequently utilized in subsequent inference phrase.

**Image-Text Contrastive Learning.** To enhance the alignment of multimodal representations and establish initial correspondences between entities presented in both image and text domains, it becomes necessary to conduct image-text contrast learning. This involves assessing the similarity between pairs of image and text using the function $S(I, T) = g_V(V_{cls})^\top g_T(T_{cls})$, where $g_V$ and $g_T$ denote linear projections that map the [CLS] token to a normalized low-dimensional representation. For each paired image and text, we compute the softmax-normalized measures of image-to-text and text-to-image similarity as follows:

$$p_j^{i2t}(I) = \frac{exp(s(I, T_j)/\tau)}{\sum_{j=1}^{J} exp(s(I, T_j)/\tau)}, \quad p_j^{t2i}(T) = \frac{exp(s(T, I_j)/\tau)}{\sum_{j=1}^{J} exp(s(T, I_j)/\tau)} \quad (1)$$

where $\tau$ is a learnable temperature parameter and J is the batch size.

Let $q^{i2t}(f_V^u)$ and $q^{t2i}(f_T^u)$ denote the ground-truth one-hot similarity, where negative pairs have a probability of 0 and the positive pair has a probability of 1. The loss function of image-text contrastive (ITC) can be defined as:

$$L_{itc} = \frac{1}{2} E_{(I,T) \sim D}[H(q^{i2t}(I), p^{i2t}(I)) + H(q^{t2i}(T), p^{t2i}(T))] \quad (2)$$

where $H$ denotes the cross-entropy between $p$ and $q$.

**Image-Text Matching.** In order to enhance the model's capability for aligning images and text, we can introduce Image-Text Matching (ITM). The alignment between images and text can be measured through the output of a multimodal encoder. ITM allows the features of images and text to undergo evaluation for matching after passing through multimodal layers. The matching results calculate feature similarity within a batch, identifying the text with the highest similarity to an image, excluding its own corresponding text, as a negative sample. These negative samples are then used to construct a batch of hard negatives, enhancing the training difficulty for the binary classification task. The expression for ITM loss is as follows:

$$L_{itm} = E_{(I,T)\sim D} H(p^{itm}, q^{itm}(I,T)) \tag{3}$$

where $p^{itm}$ is a 2-dimensional one-hot vector representing the ground-truth label. The model computes the GradCAM [18] map on the cross-attention maps of the third layer of the cross-modal encoder, with gradients obtained from minimizing the text-image similarity $S_{itm}$. The resulting GradCAM map, denoted as $G$, is subsequently employed during the inference stage to filter the bounding boxes provided by detector.

### 3.3 Part-Aware Prompt Tuning

The GradCAM map obtained through image-text matching often focuses only on a small, highly relevant region of the entity, while neglecting the other parts of the entity. This can lead to inaccurate results when selecting bounding boxes during the inference phase. To avoid this, we introduce a learnable prompt placed before the text to guide the model's attention. We employ part-based prompt and devised two categories of contrastive losses to optimize it.

**Part-Aware Prompt Construct.** Template-based prompts are effective only for certain categories capable of specifying specific part names. For instance, when dealing with categories like *bowl*, it becomes challenging to precisely define the components of the entity, thus making it difficult to formulate appropriate templates. The Part-aware prompt represents an advanced iteration of template-based prompts, as during training, it aggregates abstract part information from entities of different categories, allowing it to concurrently comprehend diverse categories of entities requiring inference.

Since the object in the image is generally top-down, for example, the person from top to bottom is three parts including head, body and legs. We divided the entity into three parts: *upper*, *middle* and *lower*, and defined three learnable part-aware prompts $\{T_u, T_m, T_d\}$, corresponding to these parts. Here we use $T_u$ as an example, and the other prompts follow a similar pattern. The prompt is matrices with dimensions of $\mathbb{R}^{k \times d}$, where $k$ represents the number of prompt tokens, and $d$ represents the embedding dimension. We added a placeholder of length $k$ in front of the input text to obtain the concatenated text $T'$, and then embedded it into a token sequence $\{T_{cls}, T_{p_1}, ..., T_{p_k}, T_1, T_2, ..., T_N\}$, where the dimension of $T_{p_k}$ is $\mathbb{R}^{1 \times d}$. Next, we replaced the $\{T_{p_1}, ..., T_{p_k}\}$ part with $T_u$, resulting in the sequence $\{T_{cls}, T_u, T_1, T_2, ..., T_N\}$. This was done to ensure that the sequence's positional encoding remained unchanged before and after adding the prompts. After passing through the encoder, we obtained text features that include the prompts: $\{f_T^u, f_T^m, f_T^d\}$.

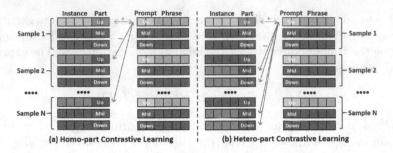

(a) Homo-part Contrastive Learning    (b) Hetero-part Contrastive Learning

**Fig. 3.** (a) Illustration of homo-part contrastive learning. Negative samples consist of samples with the same prompt and the same part in a similar position. (b) Illustration of hetero-part contrastive learning. Negative samples consist of samples with different prompts and parts.

**Homo-Part Contrastive Learning.** We need to make the model be able to distinguish the entity from which this part comes when selecting a single kind of prompt, so we set the homo-part contrastive learning, as shown in Fig. 3. We use a pre-trained detector to detect the entity $\{i_1, ..., i_n\}$ in the input image $I$, and select the optimal entity $i_0$ according to the similarity between the label output by the detector and the origin text. Then the entity is divided into three equally sized parts from top to bottom, producing three entity parts, respectively, $i_u$, $i_m$, $i_d$. The representations of each part are spliced with the representation of the whole entity, and we can obtain a set of visual features $\{f_V^u, f_V^m, f_V^d\}$. This very whole entity can provide a reference for the selected part representations, indicating which part of the entity has been selected for splicing when training.

Here we use *upper* prompt as an example, and the other prompts follow a similar pattern. For a pair of visual and linguistic features with the same part-aware prompt $(f_V^u, f_T^u)$, when the model learns the concept of parts, we should divide all of them with the same parts in a mini-batch into groups, that is, divide into three groups respectively. Then, in each group, we regard the paired splicing representations as positive samples, and the other non-paired representations in the group as negative samples, and calculate the similarity of vision to text and text to vision respectively as follows:

$$p_j^{i2t}(f_V^u) = \frac{exp(s(f_V^u, f_{Tj}^u)/\tau)}{\sum_{j=1}^{J} exp(s(f_V^u, f_{Tj}^u)/\tau)} \tag{4}$$

$$p_j^{t2i}(f_T^u) = \frac{exp(s(f_T^u, f_{Vj}^u)/\tau)}{\sum_{j=1}^{J} exp(s(f_T^u, f_{Vj}^u)/\tau)} \tag{5}$$

where $\tau$ is a learnable temperature parameter and J is the batch size.

Let $q^{i2t}(f_V^u)$ and $q^{t2i}(f_T^u)$ denote the ground-truth one-hot similarity, where negative pairs have a probability of 0 and the positive pair has a probability of 1. The loss function of our homo-part contrast learning can be defined as

$$L_{homo}^u = \frac{1}{2} E_{(f_V^u, f_T^u) \sim D}[H(q^{i2t}(f_V^u), p^{i2t}(f_V^u)) + H(q^{t2i}(f_T^u), p^{t2i}(f_T^u))] \tag{6}$$

$$L_{homo} = \frac{1}{3}(L_{homo}^u + L_{homo}^m + L_{homo}^d) \qquad (7)$$

where $H$ denotes the cross-entropy between $p$ and $q$, $L_{homo}^m$ and $L_{homo}^d$ denotes the homo-part contrastive losses obtained when the prompt is *middle* and *lower* respectively. We also need to have the prompt refer to different positions separately, so we introduce contrastive learning, as shown in Fig. 3. Our positive samples are still paired splicing representations, but the negative samples become representations that using different prompt. Here we also use *upper* prompt as an example, and the other prompts follow a similar pattern. The paired concatenated visual feature and linguistic feature in $(f_T^u, f_V^u)$ are a set of positive samples, while $(f_T^u, (f_V^m, f_V^d))$ are a set of negative samples, because the prompt in the text is *upper* and the prompt in the image is *middle* and *lower*. The similarity of vision to text and text to vision respectively as follows:

$$p_j^{i2t}(f_v^u) = \frac{exp(s(f_v^u, (f_{Tj}^m, f_{Tj}^d))/\tau)}{\sum_{j=1}^{J} exp(s(f_v^u, (f_{Tj}^m, f_{Tj}^d))/\tau)} \qquad (8)$$

$$p_j^{t2i}(f_T^u) = \frac{exp(s(f_T^u, (f_{Vj}^m, f_{Vj}^d))/\tau)}{\sum_{j=1}^{J} exp(s(f_T^u, (f_{Vj}^m, f_{Vj}^d))/\tau)} \qquad (9)$$

where $\tau$ is a learnable temperature parameter and J is the batch size.

Similarly, let $q^{i2t}(f_V^u)$ and $q^{t2i}(f_T^u)$ denote the ground-truth one-hot similarity, where negative pairs have a probability of 0 and the positive pair has a probability of 1. The loss function of our hetero-part contrast learning can be defined as:

$$L_{hetero}^u = \frac{1}{2}E_{(f_V^u, f_T^u) \sim D}[H(q^{i2t}(f_V^u), p^{i2t}(f_V^u)) + H(q^{t2i}(f_T^u), p^{t2i}(f_T^u))] \quad (10)$$

$$L_{hetero} = \frac{1}{3}(L_{hetero}^u + L_{hetero}^m + L_{hetero}^d) \qquad (11)$$

where $H$ denotes the cross-entropy between $p$ and $q$, $L_{hetero}^m$ and $L_{hetero}^d$ denotes the hetero-part contrastive losses obtained when the prompt is *middle* and *lower* respectively.

In Sect. 3.3, we endow the prompt with the capability to perceive various categories of entities through homo-part contrastive learning, and sensitize it to distinguish between the core part and peripheral parts of entities via hetero-part contrastive learning. As a result, upon the addition of the corresponding prompt in front of the text, the model can infer the spatial location of the corresponding entity part.

## 3.4   Train and Inference

**Training Objective.** The training objective of the whole model can be expressed as:

$$L = L_{itc} + L_{itm} + L_{homo} + L_{hetero} \qquad (12)$$

**Inference.** In the inference phase, we concatenate each of the three optimized prompts with text in the same way as in Sect. 3.3, thereby yielding three modified text with part-aware prompts. Subsequently, the image, the modified text, and the original text are separately feed into the model to obtain image-text matching results, computing the attention maps in the forms of GradCAM [18] map, denoted as $\{G_0, G_u, G_m, G_d\}$. Here, $G_0$ signifies the outcome obtained from inputting original text, while $G_u$, $G_m$, and $G_d$ correspond to the results gained by introducing the *upper*, *middle*, and *lower* prompts, respectively. These intermediate GradCAM map outputs, $G_u$, $G_m$, and $G_d$, are subjected to a weighted averaging procedure in conjunction with $G_0$. The optimal weights are found through parameter search. These outputs are then merged as the final GradCAM map representation, denoted as $G$. Subsequently, we leverage a pre-trained detector to generate an array of bounding boxes for the original image. The bounding box exhibiting the highest degree of overlap with the attention region in the Grad-CAM map $G$ is then selected as the prediction outcome, based on the calculated overlap metric.

## 4  Experiment

### 4.1  Dataset

We conducted our experiments on two datasets, RefCOCO [25] and RefCOCO+ [25]. RefCOCO and RefCOCO+ contain 19,994 and 19,992 images, 50,000 and 49,856 annotated objects, and 142,209 and 141,564 annotated texts, respectively, from the COCO dataset. Reasoning with RefCOCO+ is a little more difficult than with RefCOCO because there is no relative positional information to aid positioning. We adopted UNC split, which includes four parts: train, val, testA and testB.

### 4.2  Evaluation Metric

As in previous work [10,12,19,26,27], we use standard rules for the evaluation of the model. Under this rule, top-1 accuracy is counted. If the intersection ratio between the output region and ground-truth is greater than 0.5, we believe that this positioning result is accurate.

### 4.3  Implementation Details

The image encoder for our model is a ViT [6], initialized with Swin Transformer [14]. The text encoder and cross-modal encoder are initialized using the first and last six layers of BERT [5], respectively. During the training phase, the batch size was set to 10 and AdamW was used as the optimizer, whose weight decay was set to $1e^{-3}$. We conducted our experiment on 8 NVIDIA V100 GPUs with the learning rate and prompt learning rate of $2e^{-5}$ and $2e^{-7}$. The length $k$ of part-aware prompt was set to 10.

**Table 1.** Comparison with previous methods [10,12,19,26,27] on two benchmark datasets, RefCOCO [25] and RefCOCO+ [25].

| Method | Venue | RefCOCO | | | RefCOCO+ | | |
|--------|-------|---------|--------|--------|----------|--------|--------|
| | | val | test A | test B | val | test A | test B |
| ARN [12] | ICCV'19 | 32.17 | 35.35 | 30.28 | 32.78 | 34.35 | 32.13 |
| CCL [27] | NeurIPS'20 | 34.78 | 37.64 | 32.59 | 34.29 | 36.91 | 33.56 |
| ReCLIP [19] | ACL'22 | 54.04 | 58.60 | 49.54 | 55.07 | 60.47 | 47.41 |
| ALBEF [10] | NeurIPS'21 | 55.92 | 64.57 | 45.95 | 58.46 | 65.89 | 46.25 |
| X-VLM [26] | ICML'22 | 67.78 | 76.81 | 56.93 | 67.46 | 76.53 | 56.29 |
| PPT(Ours) | – | **68.16** | **77.53** | **58.22** | **68.62** | **77.32** | **57.24** |

## 4.4 Results

Table 1 presents the results on two benchmark datasets RefCOCO [25] and Ref-COCO+ [25], where our method outperforms previous approaches. As depicted in Table 1, compared to the baseline method X-VLM [26], we achieve significant improvements of 1.3% and 1.3% on the RefCOCO and RefCOCO+ respectively.

From the experimental results, it can be observed that in RefCOCO, testB exhibits the best performance among the three test types. This is primarily attributed to the fact that in testB, each image requiring inference may contain multiple objects and involve complex scenarios. Our designed approach, aided by homo-part contrastive learning, effectively establishes the connection between prompts and the entities to be inferred, thereby enhancing the generalization capability of prompts across diverse entity categories.

Furthermore, in comparison between the two datasets, our method demonstrates a more pronounced average improvement on RefCOCO. This is possibly due to the relative spatial information contained in the textual descriptions of RefCOCO. In contrast to baseline models, our approach segments entities into multiple parts with positional relationships, thereby complementing each other. This complementary information aids in learning part-aware prompts and enhances the accuracy of inference. In contrast, RefCOCO+ lacks relative positional information, restricting the perceptual capability of prompts for part information, which results in less noticeable improvements in accuracy on Ref-COCO+.

## 4.5 Ablation Study

To assess the effectiveness of two key components in our proposed method, we present the results of our ablation experiments in Table 2. As shown, homo-part contrastive exhibits slightly superior performance, enhancing the adaptability of the part-aware prompt to entities in the text by treating different entities as negative samples. Without this process, the prompt would be unable to reference specific parts of entities. Meanwhile, hetero-part contrastive continues to demonstrate effectiveness, as it enhances the sensitivity of the part-aware prompt to positional information by treating samples with different part-aware prompts as negative samples.

**Table 2.** Efficiency of homo-part contrastive (Homo) and hetero-part contrastive (Hetero) on RefCOCO+ [25].

| Methods | Homo | Hetero | val | testA | testB |
|---|---|---|---|---|---|
| X-VLM [26] | | | 67.46 | 76.53 | 56.29 |
| | √ | | 68.40 | 76.98 | 57.15 |
| | | √ | 68.01 | 76.86 | 56.53 |
| Ours | √ | √ | **68.62** | **77.32** | **57.24** |

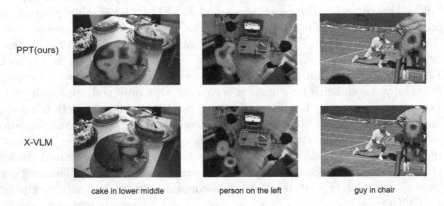

PPT(ours)

X-VLM

cake in lower middle          person on the left          guy in chair

**Fig. 4.** Inference results of X-VLM and ours, based on GradCAM visualizations on the cross-attention maps of the cross-modal encoder

### 4.6   Qualitative Analysis

Figure 4 displays several inference result generated by the proposed method and X-VLM [26]. In contrast to X-VLM, these results indicate that our approach has successfully broadened the model's attention regions during inference, thereby mitigating the grounding bias caused by overly limited attention to the target entity.

# 5 Conclusion

In this study, we have identified the issue of model attention being overly concentrated on small portions of entities, which has been overlooked by weakly supervised referring expression grounding methods which use pre-trained multimodal models. To address this, we have proposed a learnable prompt, known as the Part-aware Prompt, to guide model attention during the inference stage. Experimental results on two widely used datasets demonstrate the effectiveness of our approach.

# References

1. Carion, N., Massa, F., Synnaeve, G., Usunier, N., Kirillov, A., Zagoruyko, S.: End-to-end object detection with transformers. In: Vedaldi, A., Bischof, H., Brox, T., Frahm, J.-M. (eds.) ECCV 2020. LNCS, vol. 12346, pp. 213–229. Springer, Cham (2020). https://doi.org/10.1007/978-3-030-58452-8_13
2. Chen, K., Gao, J., Nevatia, R.: Knowledge aided consistency for weakly supervised phrase grounding. In: Proceedings of the IEEE Conference on Computer Vision and Pattern Recognition, pp. 4042–4050 (2018)
3. Datta, S., Sikka, K., Roy, A., Ahuja, K., Parikh, D., Divakaran, A.: Align2ground: weakly supervised phrase grounding guided by image-caption alignment. In: Proceedings of the IEEE/CVF ICCV, pp. 2601–2610 (2019)
4. Deng, J., Yang, Z., Chen, T., Zhou, W., Li, H.: Transvg: end-to-end visual grounding with transformers. In: Proceedings of the IEEE/CVF ICCV, pp. 1769–1779 (2021)
5. Devlin, J., Chang, M.W., Lee, K., Toutanova, K.: Bert: pre-training of deep bidirectional transformers for language understanding. arXiv preprint arXiv:1810.04805 (2018)
6. Dosovitskiy, A., et al.: An image is worth 16x16 words: transformers for image recognition at scale. arXiv preprint arXiv:2010.11929 (2020)
7. Gella, S., Lapata, M., Keller, F.: Unsupervised visual sense disambiguation for verbs using multimodal embeddings. arXiv preprint arXiv:1603.09188 (2016)
8. Girshick, R., Donahue, J., Darrell, T., Malik, J.: Rich feature hierarchies for accurate object detection and semantic segmentation. In: Proceedings of the IEEE Conference on Computer Vision and Pattern Recognition, pp. 580–587 (2014)
9. Kamath, A., Singh, M., LeCun, Y., Synnaeve, G., Misra, I., Carion, N.: Mdetr-modulated detection for end-to-end multi-modal understanding. In: Proceedings of the IEEE/CVF ICCV, pp. 1780–1790 (2021)
10. Li, J., Selvaraju, R., Gotmare, A., Joty, S., Xiong, C., Hoi, S.C.H.: Align before fuse: vision and language representation learning with momentum distillation. Adv. Neural. Inf. Process. Syst. **34**, 9694–9705 (2021)
11. Liao, Y., et al.: A real-time cross-modality correlation filtering method for referring expression comprehension. In: Proceedings of the IEEE/CVF Conference on Computer Vision and Pattern Recognition, pp. 10880–10889 (2020)
12. Liu, X., Li, L., Wang, S., Zha, Z.J., Meng, D., Huang, Q.: Adaptive reconstruction network for weakly supervised referring expression grounding. In: Proceedings of the IEEE/CVF ICCV, pp. 2611–2620 (2019)

13. Liu, Y., Wan, B., Ma, L., He, X.: Relation-aware instance refinement for weakly supervised visual grounding. In: Proceedings of the IEEE/CVF Conference on Computer Vision and Pattern Recognition, pp. 5612–5621 (2021)

14. Liu, Z., et al.: Swin transformer: hierarchical vision transformer using shifted windows. In: Proceedings of the IEEE/CVF ICCV, pp. 10012–10022 (2021)

15. Mao, J., Huang, J., Toshev, A., Camburu, O., Murphy, K.: Generation and comprehension of unambiguous object descriptions. In: 2016 IEEE Conference on Computer Vision and Pattern Recognition (CVPR) (2016)

16. Raganato, A., Calixto, I., Ushio, A., Camacho-Collados, J., Pilehvar, M.T.: Semeval-2023 task 1: visual word sense disambiguation. In: Proceedings of the The 17th International Workshop on Semantic Evaluation (SemEval-2023), pp. 2227–2234 (2023)

17. Rohrbach, A., Rohrbach, M., Hu, R., Darrell, T., Schiele, B.: Grounding of textual phrases in images by reconstruction. In: Leibe, B., Matas, J., Sebe, N., Welling, M. (eds.) ECCV 2016. LNCS, vol. 9905, pp. 817–834. Springer, Cham (2016). https://doi.org/10.1007/978-3-319-46448-0_49

18. Selvaraju, R.R., Cogswell, M., Das, A., Vedantam, R., Parikh, D., Batra, D.: Gradcam: visual explanations from deep networks via gradient-based localization. In: Proceedings of the IEEE ICCV, pp. 618–626 (2017)

19. Subramanian, S., Merrill, W., Darrell, T., Gardner, M., Singh, S., Rohrbach, A.: Reclip: a strong zero-shot baseline for referring expression comprehension. arXiv preprint arXiv:2204.05991 (2022)

20. Viola, P., Jones, M.: Rapid object detection using a boosted cascade of simple features. In: Proceedings of the 2001 IEEE Computer Society Conference on Computer Vision and Pattern Recognition, CVPR 2001, vol. 1, pp. I-I. IEEE (2001)

21. Wang, X., et al.: Reinforced cross-modal matching and self-supervised imitation learning for vision-language navigation. In: Proceedings of the IEEE/CVF Conference on Computer Vision and Pattern Recognition, pp. 6629–6638 (2019)

22. Xiao, F., Sigal, L., Jae Lee, Y.: Weakly-supervised visual grounding of phrases with linguistic structures. In: Proceedings of the IEEE Conference on Computer Vision and Pattern Recognition, pp. 5945–5954 (2017)

23. Yang, L., Xu, Y., Yuan, C., Liu, W., Li, B., Hu, W.: Improving visual grounding with visual-linguistic verification and iterative reasoning. In: Proceedings of the IEEE/CVF Conference on Computer Vision and Pattern Recognition, pp. 9499–9508 (2022)

24. Yin, W., Liwei, H., Jing, L.: Svetlana: learning two-branch neural networks for image-text matching tasks. IEEE Trans. Pattern Anal. Mach. Intell. (2019)

25. Yu, L., Poirson, P., Yang, S., Berg, A.C., Berg, T.L.: Modeling context in referring expressions. In: Leibe, B., Matas, J., Sebe, N., Welling, M. (eds.) ECCV 2016. LNCS, vol. 9906, pp. 69–85. Springer, Cham (2016). https://doi.org/10.1007/978-3-319-46475-6_5

26. Zeng, Y., Zhang, X., Li, H.: Multi-grained vision language pre-training: aligning texts with visual concepts. arXiv preprint arXiv:2111.08276 (2021)

27. Zhang, Z., Zhao, Z., Lin, Z., He, X., et al.: Counterfactual contrastive learning for weakly-supervised vision-language grounding. Adv. Neural. Inf. Process. Syst. **33**, 18123–18134 (2020)

28. Zhu, F., Zhu, Y., Chang, X., Liang, X.: Vision-language navigation with self-supervised auxiliary reasoning tasks. In: Proceedings of the IEEE/CVF Conference on Computer Vision and Pattern Recognition, pp. 10012–10022 (2020)

# Adversarially Robust Deepfake Detection via Adversarial Feature Similarity Learning

Sarwar Khan[1,2,3]($\boxtimes$), Jun-Cheng Chen[1,2], Wen-Hung Liao[2,3], and Chu-Song Chen[4]

[1] Research Center for Information Technology Innovation, Academia Sinica, Taipei, Taiwan
say2sarwar@gmail.com, pullpull@citi.sinica.edu.tw
[2] Social Networks Human-Centered Computing, TIGP, Academia Sinica, Taipei, Taiwan
whliao@cs.nccu.edu.tw
[3] Computer Science, National Chengchi University, Taipei, Taiwan
[4] Computer Science and Information Engineering, National Taiwan University, Taipei, Taiwan
chusong@csie.ntu.edu.tw

**Abstract.** Deepfake technology has raised concerns about the authenticity of digital content, necessitating the development of effective detection methods. However, the widespread availability of deepfakes has given rise to a new challenge in the form of adversarial attacks. Adversaries can manipulate deepfake videos with small, imperceptible perturbations that can deceive the detection models into producing incorrect outputs. To tackle this critical issue, we introduce Adversarial Feature Similarity Learning (AFSL), which integrates three fundamental deep feature learning paradigms. By optimizing the similarity between samples and weight vectors, our approach aims to distinguish between real and fake instances. Additionally, we aim to maximize the similarity between both adversarially perturbed examples and unperturbed examples, regardless of their real or fake nature. Moreover, we introduce a regularization technique that maximizes the dissimilarity between real and fake samples, ensuring a clear separation between these two categories. With extensive experiments on popular deepfake datasets, including FaceForensics++, FaceShifter, and DeeperForensics, the proposed method outperforms other standard adversarial training-based defense methods significantly. This further demonstrates the effectiveness of our approach to protecting deepfake detectors from adversarial attacks.

**Keywords:** Adversarial attack · Adversarial training · Deepfake video detection · Forgery detector

## 1 Introduction

Deepfakes are synthetic videos in which a person's face is altered to resemble a different individual, resulting in the production of highly realistic footage

© The Author(s), under exclusive license to Springer Nature Switzerland AG 2024
S. Rudinac et al. (Eds.): MMM 2024, LNCS 14556, pp. 503–516, 2024.
https://doi.org/10.1007/978-3-031-53311-2_37

depicting events that never actually took place [7]. Deepfake technology has captivated and alarmed society by offering the ability to create remarkably convincing and misleading media, which in turn threatens the authenticity of digital content. In response to these concerns, researchers have diligently worked to develop effective deepfake detection methods [1, 8, 15, 16, 35, 36, 45].

Deepfake detectors have shown promising performance under normal conditions, accurately identifying manipulated videos. However, a new challenge has emerged in the form of adversarial attacks, where small and imperceptible perturbations can deceive the detection models into producing incorrect outputs [10, 14, 17, 18, 26, 32]. An adversarial example is a manipulated input intentionally designed to deceive a classification model [28]. Adversarial deepfakes [18] leverage the pre-softmax layer to compute the loss and iteratively calculate gradients, allowing for the creation of adversarial fakes that can successfully evade detection. Statistical consistency attack (StatAttack) robust [17] looks into statistical consistency between real and fake and uses degradation techniques to create transferable deepfake adversarial attacks. This poses a significant threat to the reliability and effectiveness of deepfake detection systems, as it undermines their ability to distinguish between genuine and manipulated content.

Adversarial training [28] tackles adversarial attacks through a *min-max* optimization but often at the cost of reduced performance on normal inputs. To address this, TRADES [44] is a surrogate loss for adversarial training utilizing cross-entropy loss supervising and the distance loss between the features of clean and adversarial examples as the regularization. However, training deepfake detectors for robustness remains challenging due to the inclusion of fake images. Our study is motivated by the recognition that adversaries can exploit misclassification, aiming to develop effective strategies for adversarially robust deepfake detection.

In pursuit of this objective, we develop an Adversarial Feature Similarity Learning (AFSL) objective function that optimizes similarity across three fundamental paradigms of deep feature learning. First, we optimize the similarity between samples and weight vectors, specifically focusing on differentiating between real and fake instances. Secondly, our objective is to maximize the similarity between samples, considering both adversarially perturbed examples and unperturbed examples, where the perturbed instances can be either real or fake. Finally, we introduce a regularization approach that aims to maximize the dissimilarity between real and fake samples, ensuring a clear separation and distinct representation of these two categories. This comprehensive approach enables effective deepfake detection by enhancing discrimination between real and fake content and mitigating the impact of adversarial perturbations. We conduct extensive experiments on the FaceForensics++ [35], FaceShifter [23], and DeeperForensics [19] datasets, evaluating the performance of our proposed method. Impressively, our method outperforms widely used adversarial training-based defense methods by a significant margin. This demonstrates the effectiveness of our approach to help deepfake detectors fight against various adversarial attacks.

# 2    Related Work

## 2.1    Deepfake Creation and Detection

The development of Generative Adversarial Networks (GANs) and their diverse variants has yielded remarkable outcomes in image generation and manipulation, consequently facilitating the emergence of deepfake technology. By leveraging GANs, deepfake has enabled the creation of fabricated images or videos across various categories. The current deepfake generation techniques include various approaches, such as complete face synthesis [21], face identity swap [11], and face manipulation [12]. The utilization of these generation methods by malicious applications can greatly jeopardize public information security. Nonetheless, it is crucial to recognize that the misuse of deepfake technology raises additional concerns regarding security and privacy, extending to sensitive areas such as politics, religion, and pornography [38,42]. Meanwhile study conducted in Thailand [37] explores Thai perspectives on deepfake AI images, emphasizing both creative interests and potential risks. It suggests the Thai government take a proactive role in regulating and raising awareness to harness the technology's creative potential while addressing concerns related to data protection and image copyright.

To mitigate the potential misuse of deepfake technologies, various Deep Neural Network (DNN) methods have been proposed for detecting deepfake inputs. Deepfake detection primarily involves the binary classification of distinguishing between fake and real inputs. In the realm of deepfake detection methods, some methods focus on extracting spatial information [8,14–16,26,45], whereas others delve into analyzing the differences in frequency information [9] between fake and real inputs. LipForensics [16] employs a lips extraction technique from facial images and leverages a combination of a pretrained feature extractor and a temporal convolutional network to train an effective deepfake detector. FTCN [45] proposed exceptional generalization across various manipulation scenarios by enforcing a uniform spatial convolutional kernel size of one. RealForensics [15] aims to enhance forgery detection performance and improve generalization across different datasets by leveraging real talking faces through self-supervision using spatiotemporal features. These methods achieve remarkable detection results within their respective experimental configurations by harnessing the formidable feature extraction capabilities of DNNs.

## 2.2    Adversarial Examples

Adversarial examples exploit the vulnerability of deep learning models by intentionally designing inputs that cause the models to make mistakes or misclassify data [3]. Gradient-based adversarial attacks are extremely effective against deep learning models in image [3,24,25,28], video [30,41], and audio [5,33,34] domain. FGSM [13], PGD [28], CW [3], and StatAttack [17] represent potent adversarial attack techniques that employ distinct optimization strategies to generate perturbations capable of deceiving classification models. Similar to other deep learning models, deepfake detectors are vulnerable to adversarial attacks, making them a significant threat within the realm of deepfakes.

Adversarial deepfakes [18] present a robust white-box attack (RWB) setting, where the attacker possesses full access to the model, as well as a robust black-box (RBB) attack setting, where the attacker lacks access to the model. These settings are achieved by utilizing the pre-softmax layer and employing diverse transformations. Likewise, Neekhara et al. [32] investigate the transferability of adversarial attacks in forgery detectors and propose a universal attack approach, demonstrating the effectiveness of adversarial examples across different models and architectures. In addition, Statistical Consistency Attack (StatAttack) [17] introduces a transferable approach that leverages adversarial statistical consistency through the minimization of a distribution-aware loss, enabling it to circumvent deepfake detectors effectively.

A multitude of defense strategies have been proposed to combat adversarial examples in both image and video domains, adversarial training has exhibited commendable performance in mitigating the impact of such attacks [20,27,28, 43,44]. Although adversarial training has shown effectiveness against various forms of adversarial attacks, it often comes at the cost of reduced performance on unperturbed data. To address this, Deep image prior (DIP) [10] utilizes a GAN network to remove perturbations from the input, making the model robust against adversarial attacks, and also incorporates regularization techniques as a defense strategy in deepfake detection. To handle another growing concern of deepfake technology exacerbated by the use of face masks during the pandemic, Alnaim et al. [2] propose a novel deepfake face mask dataset and detection model with identifying face-mask-related deepfakes. Previous research has demonstrated the effectiveness of defending against various attack methods by utilizing projected gradient descent (PGD) as a defense mechanism, validating the proposal by Madry et al., [28]. This further supports the notion that robust defenses, specifically tailored to counter PGD attacks, can provide efficient protection against different first-order attack methods.

## 3    Adversarial Feature Similarity Learning

**Notation:** We consider $f_\theta(\cdot)$ as the feature encoder from the deepfake detector and $\theta$ representing its learnable model parameters. The variable $x \in \mathcal{X}$ denotes an input frame or a video clip from a video depending on if $f_\theta(\cdot)$ is a frame-based or a video-based deepfake detector method, where $x$ can be either real or fake samples. In addition, $x_{\text{adv}}$ represents the adversarial example generated using $x \in \{x^{real}, x^{fake}\}$ where $x^{real} \in \mathcal{X}$ and $x^{fake} \in \mathcal{X}$ represent real and fake samples, which can be either an individual frame or a video clip depending on whether $f_\theta(\cdot)$ is a frame-based or video-based detector. $y \in \{0,1\}$ denotes the label, where 0 indicates a fake class and 1 real class. Adversarial Feature Similarity Learning.

## 3.1  Overview

Our objective is to develop an adversarially robust deepfake detector that effectively mitigates the impact of adversarial attacks while preserving the performance on unperturbed data. We address this problem by discerning the features of real and fake samples and their corresponding adversarial counterparts. We hypothesize that an adversarial attack will shift the features in the opposite direction, irrespective of whether the input is real or fake. However, deepfake detectors trained solely with adversarial training will not effectively learn the desirable features to distinguish between real and fake samples in the presence of adversarial attacks. To overcome this limitation, we proposed a novel loss function to effectively separate the two classes i.e. (Real and Fake) under most conditions. Figure 1 provides a comprehensive illustration of our framework. We adopt a three-step approach. Separating unperturbed (Real and Fake) in Sect. 3.2, using adversarial similarity loss to make the detector robust in Sect. 3.3, and finally similarity regularization loss to preserve the unperturbed performance in Sect. 3.4. Additionally, we formulate the final loss function in Sect. 3.5.

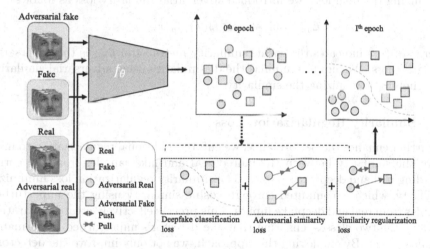

**Fig. 1.** Framework for adversarial feature similarity learning. First, we select a pair of real and deepfake samples and create adversarial perturbation for the corresponding inputs. Then, we generate the features of real, fake, and their adversarial samples. Finally, through the proposed loss function, the model can learn a better representation to separate real samples from fake ones along with their adversarial counterparts, where the backbone $f_\theta$ is from the deepfake detector.

## 3.2  Deepfake Classification Loss

The deepfake classification loss is realized using a supervised loss that utilizes a logit-adjusted variant of binary cross-entropy (LBCE) [29], denoted as $\mathcal{L}_{LBCE}(f_\theta(x), y)$ to tackle the potential issues of class imbalance. The supervised deepfake classification loss function aims to maximize the dissimilarity between real and fake samples, while simultaneously addressing the class imbalance in the dataset. This approach helps to refine the model's discrimination abilities, enabling it to better differentiate between real and fake videos. The deepfake classification loss, $\mathcal{L}_{\text{dcl}}$, is denoted as follows:

$$\mathcal{L}_{\text{dcl}} = \mathcal{L}_{LBCE}\left(f_\theta(x), y\right), \tag{1}$$

### 3.3 Adversarial Similarity Loss

Let $f_\theta(x)$ and $f_\theta(x_{\text{adv}})$ respectively represent the mapping from an input sample $x \in \{x^{\text{real}}, x^{\text{fake}}\}$ and from an adversarial input sample $x_{\text{adv}} \in \{x^{\text{real}}_{\text{adv}}, x^{\text{fake}}_{\text{adv}}\}$ to their corresponding embedding spaces. To perform adversarial training, we generate adversarial examples $x_{\text{adv}}$ from $x \in \{x^{\text{real}}, x^{\text{fake}}\}$ by employing the PGD adversarial attack method. The objective is to maximize the cosine similarity $sim\left(f_\theta(x), f_\theta(x_{\text{adv}})\right)$ between $x$ and $x_{\text{adv}}$ to bring them closer as they represent the same class and to avoid adversarial examples from being misclassified to the other class (i.e., real to fake and fake to real). To achieve this objective while minimizing the final loss, we introduce adversarial similarity loss as follows

$$\mathcal{L}_{asl} = (1 - sim\left(f_\theta(x), f_\theta(x_{\text{adv}})\right)), \tag{2}$$

where $sim(\cdot, \cdot)$ indicates the cosine similarity metric, and $\mathcal{L}_{asl}$ is the adversarial similarity loss. We aim to minimize the dissimilarity using adversarial similarity loss, thereby maximizing the similarity.

### 3.4 Similarity Regularization Loss

To further enhance the performance, we enforce additional regularization to minimize the similarity between the paired real and fake samples from the corresponding real and deepfake samples. The similarity regularization loss minimizes similarity, which is computed through cosine similarity using the unperturbed samples from real and deepfake inputs. This process effectively creates separation between the two classes, ensuring that the detector's unperturbed performance remains intact. By employing this approach, we not only improve the detector's robustness but also preserve its unperturbed performance. Similarity regularized loss is calculated as follows:

$$\mathcal{L}_{\text{srl}} = sim\left(f_\theta(x^{\text{real}}), f_\theta(x^{\text{fake}})\right), \tag{3}$$

where $\mathcal{L}_{\text{srl}}$ is the similarity regularization loss, $\mathcal{L}$ is cosine similarity, and respectively $x^{\text{real}}, x^{\text{fake}} \in \mathcal{X}$ are real and fake pair input.

### 3.5 Final Loss Function

In this section, we formulate the final loss function for minimization based on the three previously discussed components. The objective function is presented in Eq. 4.

$$\mathcal{L}_{\text{afsl}} = \mathcal{L}_{\text{dcl}} + \beta_1 \mathcal{L}_{asl} + \beta_2 \mathcal{L}_{\text{srl}}, \tag{4}$$

where $\mathcal{L}_{dcl}$ denotes the deepfake classification loss, while $\mathcal{L}_{asl}$ and $\mathcal{L}_{srl}$ correspond to the adversarial similarity loss and similarity regularization loss respectively. To control the influence of the regularization terms, we set the scaling factors $\beta_1$ and $\beta_2$ to 1 and 0.1, respectively. By incorporating these components into the loss function $\mathcal{L}_{afsl}$, we aim to enhance the robustness of the detectors while preserving their unperturbed performance.

Unlike TRADES, our approach utilizes similarity regularization, capturing finer differences between real and fake videos and thereby enabling the extraction of more accurate features.

**Table 1.** Video level AUC (%) for deepfake detectors when testing on each deepfake type of FF++ after training on the remaining three types. "No Attack" denotes that an adversarial attack is not applied while "PGD10" denotes that the Projected Gradient descent (PGD) attack is applied to the input. The types of deepfakes are DeepFakes (DF), FaceSwap (FS), Face2Face (F2F), and NeuralTextures (NT). **All the numbers of the baseline methods shown in this Table are reproduced based on the default settings of the officially released implementations, and the performance discrepancies from their papers may be due to the released versions or hyperparameters being different from the ones used in the experiments of the papers.**

| Methods | No Attack | | | | PGD10 | | | |
|---|---|---|---|---|---|---|---|---|
| | DF | FS | F2F | NT | DF | FS | F2F | NT |
| RealForensics [15] | 91.5 | 89.6 | 92.3 | 92.6 | 0.7 | 0.8 | 1.2 | 1.6 |
| LipForensics [16] | 90.8 | 87.1 | 92.0 | 91.8 | 2.8 | 2.4 | 1.0 | 2.6 |
| FTCN [45] | 91.6 | 90.0 | 91.6 | 92.8 | 1.0 | 1.2 | 1.7 | 2.4 |
| Patch-based [4] | 85.7 | 57.4 | 84.6 | 80.3 | 3.8 | 2.6 | 1.8 | 4.2 |
| Xception [35] | 83.1 | 50.6 | 81.5 | 76.9 | 1.0 | 4.7 | 0.8 | 0.2 |

## 4 Experimental Description

### 4.1 Implementation Details

The faces are extracted through the utilization of face detection and alignment techniques and video clips comprising 25 frames. For the training process, the video clips are randomly cropped to dimensions $140 \times 140$ and subsequently resized to $112 \times 112$. Horizontal flipping and grayscale transformation with a probability of 0.5 are applied along with random masking. For sequence-based deepfake detection, we employ the Channel-Separated Convolutional Network (CSN) [39]. We refer interested readers to [15,39] for more details about CSN. The optimization process utilizes the Adam optimizer with a learning rate of $3 \times 10^{-4}$. The model is trained for 150 epochs. We used RealForensics self-supervised pretrained on Lip Reading in the Wild (LRW) dataset[1] This pretrained model

---

[1] Pretrained model: https://github.com/ahaliassos/RealForensics.

serves as the starting point and provides the initial push to effectively capture the relevant features, thereby enhancing the model's generalization capabilities.

While for frame-based detection, we utilize XceptionNet [6] and MesoNet [1][2] We optimize the frame-based model using a Stochastic gradient descent (SGD) optimizer with a learning rate of $2 \times 10^{-3}$ and the model is trained for 150 epochs with a batch size of 16. For further details about the frame-based model, interested readers are directed to [1,31]. Normalization ($L_2$-norm) is applied to all features in both sequence-based and frame-based detectors.

**Datasets:** FaceForensics++ (FF++), is comprised of 1,000 authentic videos and 4,000 deepfake videos. Unless specified otherwise, the mildly compressed version of the dataset (c23) was utilized. Other datasets used in the experiments are FaceShifter [23] and DeeperForensics [19], featuring different face-swapping techniques applied to FF++ real videos.

**Evaluation metrics:** We utilize accuracy and area under the receiver operating characteristic curve (AUC) metrics. For video-level assessment, we uniformly sample non-overlapping clips from a single video and average their predictions.

**Table 2.** AUC (%) scores for video-level detection on the FF++ dataset, containing four deepfake methods. Models train on three methods and test on the remaining method. We employ PGD5 for adversarial training and PGD10 for testing purposes. $L_\infty$ is allowed distortion for adversarial attacks. $L_\infty = 0$ Adversarial Feature Similarity Learning means no adversarial attack is applied.

| Method | $L_\infty$ | DF | FS | F2F | NT | Avg |
|---|---|---|---|---|---|---|
| RealForensics [15] | 0 | 91.5 | 89.6 | 92.3 | 92.6 | 91.5 |
| RealForensics | 8/255 | 0.7 | 0.8 | 1.2 | 1.6 | 1.0 |
| RealForensics + AT [28] | 8/255 | 76.3 | 74.1 | 73.7 | 70.1 | 73.5 |
| RealForensics + TRADES [44] | 8/255 | 78.2 | 75.4 | 80.7 | 72.5 | 76.7 |
| RealForensics + AFSL (Ours) | 0 | 89.4 | 87.6 | 90.4 | 91.7 | 89.7 |
| AFSL (Ours) | 8/255 | 79.0 | 77.2 | 82.6 | 75.6 | 78.6 |
| RealForensics + AFSL (Ours) | 8/255 | 81.5 | 79.7 | 84.1 | 78.1 | 80.8 |

## 4.2    Victim Models: Deepfake Detectors

In our work, we assess the vulnerability of top-performing deepfake detectors to adversarial attacks. We employ video-based detectors, namely RealForensics [15], LipForensics [16], and FTCN [45]. In addition, we incorporate frame-by-frame based detectors, such as Patch-based [4] and Xception [35]. All the models are tested on each of the four methods using FF++ after training on the remaining three. Table 1 presents the AUC score of all detectors under two conditions: "No Attack," where no adversarial attack is applied to the input, and "PGD10," where an adversarial attack is applied to the input video.

---

[2] Pretrained model: https://github.com/paarthneekhara/AdversarialDeepFakes.

As observed from the results, all the deepfake detectors demonstrate vulnerability to adversarial attacks. Our proposed loss function offers the advantage of seamless integration with most deepfake detectors. For the evaluation of our method, we select RealForensics from sequence-based detectors, along with XceptionNet and MesoNet from frame-by-frame detectors. As we cannot evaluate every detector, we choose the top-performing detector from each category.

**Table 3.** Average video-level AUC (%) for adversarial attacks is computed by training the model on three methods and testing it on a fourth method. The reported values represent the average scores across the entire test dataset, encompassing all four methods, under both white-box and black-box adversarial attacks.

| Methods | No Attack | PGD10 | RWA [18] | CW2 | SA [17] | UI [32] | RBB [18] |
|---|---|---|---|---|---|---|---|
| RealForensics [15] | **91.5** | 1.1 | 1.4 | 0.0 | 0.0 | 0.0 | 36.8 |
| RealForensics + AT [28] | 78.4 | 73.5 | 79.5 | 78.3 | 63.6 | 66.9 | 80.5 |
| RealForensics + TRADES [44] | 84.0 | 76.7 | 76.1 | 78.1 | 67.2 | 69.3 | 82.4 |
| AFSL (Ours) | 87.3 | 78.3 | 79.1 | 79.7 | 68.5 | 68.7 | 86.1 |
| RealForensics + AFSL (Ours) | 89.8 | **80.7** | **81.3** | **82.8** | **73.9** | **74.7** | **87.5** |

**Table 4.** Video level AUC(%) for unseen datasets: DeeperForensics and FaceShifter under different adversarial attacks.

| Methods | No Attack | PGD10 | RWA [18] | CW2 | SA [17] | UI [32] | RBB [18] |
|---|---|---|---|---|---|---|---|
| **Deeper Forensics** | | | | | | | |
| RealForensics [15] | **93.6** | 1.0 | 0.0 | 0.0 | 0.0 | 0.0 | 42.2 |
| RealForensics + AT [28] | 84.5 | 78.2 | 76.4 | 75.0 | 65.4 | 69.7 | 81.3 |
| RealForensics + TRADES [44] | 88.1 | 78.0 | 79.3 | 77.9 | 69.7 | **85.6** | 85.8 |
| AFSL (Ours) | 90.2 | 80.3 | 79.9 | 80.4 | 69.4 | 83.5 | 87.3 |
| RealForensics + AFSL (Ours) | 92.9 | **83.6** | **81.2** | **83.5** | **72.8** | 85.6 | **89.1** |
| **FaceShifter** | | | | | | | |
| RealForensics [15] | **91.7** | 1.0 | 1.0 | 1.0 | 0.0 | 1.0 | 37.7 |
| RealForensics + AT [28] | 83.6 | 76.2 | 74.0 | 75.8 | 60.7 | 73.1 | 79.9 |
| RealForensics + TRADES [44] | 87.1 | 79.3 | 77.9 | 78.2 | **67.3** | 72.8 | 84.6 |
| AFSL (Ours) | 88.2 | 79.6 | 78.1 | 81.4 | 65.1 | 74.6 | 84.3 |
| RealForensics + AFSL (Ours) | 89.4 | **81.7** | **80.7** | **84.6** | 67.1 | **78.1** | **86.5** |

We employ pretrained weights from RealForensics(see footnote 1) and fine-tune LipForensics, FTCN[3], Patch-based[4], and Xception. We follow the exact instructions for pre-processing provided in their official code to replicate the results. While we do observe a decline in performance compared to the reported results, this could potentially be attributed to the absence of supplementary data. We solely report the reproduced results as we aim to improve model robustness against adversarial attacks.

---

[3] https://github.com/yinglinzheng/FTCN.
[4] https://github.com/chail/patch-forensics.

## 4.3   Robust Cross-Manipulation Generalization

Most deepfake detectors typically conduct generalization experiments to assess their performance. These experiments involve training the detectors on three methods and testing them on the remaining techniques using the FF++ dataset [15,36]. In this study, we follow the same protocol and introduce adversarial attacks during both the training and testing stages, utilizing a sequence-based model. To make the comparison fair, we utilized self-supervised pretrained weights of RealForensics for all methods using the CSN model. The results in Table 2 demonstrate that our proposed method achieves adversarially robust generalization to unseen adversarial deepfakes. While RealForensics achieves the best result with unperturbed input, it fails to withstand adversarial attacks. On the other hand, our proposed method performs well under both adversarially perturbed and unperturbed data compared with adversarial training (AT) and TRADES. AT is the baseline while TRADES is the state-of-the-art method in terms of both clean and robust performance. Table 3 presents the average AUC score on FF++ using white-box and black-box attacks. RBB is a black-box attack generated using ResNet3D [40], while PGD, CW, StatAttack, Universal, and RWA are white-box attacks. Our proposed method AFSL outperforms all previous defense techniques across all types of adversarial attacks.

**Table 5.** Frame-level Accuracy (%) of deepfake detector on FF++ dataset under different adversarial attacks using XceptionNet and MesoNet models.

| XceptionNet | | | | | |
|---|---|---|---|---|---|
| Methods | PGD10 | CW2 | RWA [18] | SA [17] | RBB [18] |
| Adversarial Deepfakes [18] | 7.3 | 8.7 | 1.9 | 0.0 | 48.3 |
| Vanilla AT [28] | 76.0 | 72.7 | 73.1 | 58.9 | 80.1 |
| TRADES [44] | 80.8 | 79.4 | 77.2 | 76.8 | 84.7 |
| AFSL (Ours) | **81.7** | **82.9** | **80.6** | **77.5** | **85.7** |
| MesoNet | | | | | |
| Adversarial Deepfakes [18] | 6.2 | 7.3 | 0.0 | 0.0 | 44.6 |
| Vanilla AT [28] | 74.3 | 70.7 | 71.9 | 55.6 | 67.4 |
| TRADES [44] | 78.2 | **77.1** | 72.9 | **73.7** | 79.3 |
| AFSL (Ours) | **79.5** | 76.9 | **74.1** | 72.6 | **81.3** |

We also evaluate the robust generalization across datasets by training the model on FF++ using all manipulation methods and testing it on two datasets, DeeperForensics and Faceshifter. Table 4 presents the AUC results for both datasets. We compare the proposed method with state-of-the-art under stronger white-box attacks and black-box attacks. The robust AUC confirms that the proposed method performs well compared to the baseline method and other defenses when exposed to various types of adversarial attacks.

## 4.4   Evaluation on Frame-Based Detectors

To showcase the effectiveness of our proposed approach, we incorporated two frame-by-frame based deepfake detectors, namely XceptionNet [6] and MesoNet [1]. These detectors are CNN-based classification models that independently classify each frame as either real or fake. Table 5 presents the performance of both models against state-of-the-art white-box adversarial attacks. For robust black-box (RBB) attacks, we generated perturbations from pre-trained clean models without accessing their parameters. This allows us to evaluate the robustness of the detectors in scenarios where they are not aware of each other internal architecture or parameters.

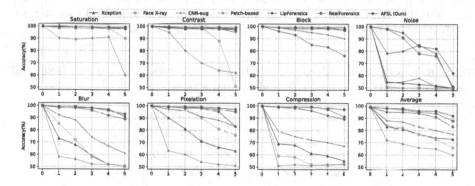

**Fig. 2.** Robustness to unseen distortions: Video level AUC scores (%) varying with the severity level of different distortions. Average is the mean value at each severity level.

## 4.5   Robustness to Common Distortions

In addition to robust generalization across different manipulations and resistance against adversarial attacks, deepfake detectors must also withstand common distortions that videos may encounter online. To assess the robustness of the detector against unfamiliar distortions, we follow the settings of [15,16]. During training on the FF++ dataset, we limit the augmentation techniques to horizontal flipping and random cropping for grayscale inputs. This approach helps prevent any potential interactions between the training distortions and those used during testing. Following [19], we use seven different types of distortions with five levels of severity. Figure 2 presents the results of each distortion with five levels of severity using the proposed method and state-of-the-art methods. The proposed method outperforms both frame-based and sequence-based methods. The inclusion of an adversarial training term in the loss function acts as regularization to increase model robustness against common distortions, which is in line with the findings of Kireev et al. [22]. In comparison to previous methods, the proposed approach performs well under most conditions, highlighting its effectiveness in tackling distortions commonly encountered in real-world scenarios.

## 5    Ablation Study

In this section, we analyze various components of our proposed method to comprehend the factors contributing to its performance. We conduct ablation experiments under PGD adversarial attack to inspect its robust generalization. The training is performed on FaceForensics++, and we evaluate the model on FaceShifter and DeeperForensics datasets, reporting the AUC score. Table 6 displays the results of the ablation study. We use deepfake classification loss as the first term in the loss function as S1 to train the model without any adversarial training. Unfortunately, this model proves to be inadequate in defending against adversarial attacks. Next, by training the model with the S2 setting without the similarity regularization, we can significantly improve the AUC scores against adversarial attacks. Finally, with all three loss components in the S3 setting, we can further improve the robustness by about 2% in AUC score.

**Table 6.** Impact of various components. Robust AUC (%) on FaceShifter (FSh) and DeeperForensics (DFo) using PGD10 adversarial attack.

| Settings | Losses | | | AUC (%) | |
|---|---|---|---|---|---|
| | $\mathcal{L}_{dcl}$ | $\mathcal{L}_{asl}$ | $\mathcal{L}_{srl}$ | DFo | FSh |
| S1 | ✓ | ✗ | ✗ | 1.0 | 1.0 |
| S2 | ✓ | ✓ | ✗ | 81.3 | 79.1 |
| S3 | ✓ | ✓ | ✓ | **83.6** | **81.7** |

## 6    Conclusion and Future Work

This paper introduces a novel approach Adversarial Feature Similarity Learning (AFSL) for enhancing the robustness of a deepfake detector against adversarial attacks. We propose an adversarially robust loss function, specifically designed to detect fake videos even when subjected to deliberate adversarial perturbations. Our experimental results demonstrate the effectiveness of the proposed method under unperturbed input but also against common distortions. The future work will consider self-supervised learning using the proposed loss function, such as pairing real and fake samples in self-supervised adversarial defense.

**Acknowledgment.** This research is supported by National Science and Technology Council, Taiwan (R.O.C), under the grant number of NSTC-111-2634-F-002-022, 110-2221-E-001-009-MY2, 112-2634-F-001-001-MBK, and Academia Sinica under the grant number of AS-CDA-112-M09. In addition, we would like to express our gratitude for the valuable contributions and guidance from these organizations, which have been instrumental in achieving the goals of this research.

# References

1. Afchar, D., Nozick, V., Yamagishi, J., Echizen, I.: MesoNet: a compact facial video forgery detection network. In: WIFS, pp. 1–7 (2018)
2. Alnaim, N.M., Almutairi, Z.M., Alsuwat, M.S., Alalawi, H.H., Alshobaili, A., Alenezi, F.S.: DFFMD: a deepfake face mask dataset for infectious disease era with deepfake detection algorithms. IEEE Access, 16711–16722 (2023)
3. Carlini, N., Wagner, D.: Adversarial examples are not easily detected: Bypassing ten detection methods. In: AIS, pp. 3–14 (2017)
4. Chai, L., Bau, D., Lim, S.N., Isola, P.: What makes fake images detectable? understanding properties that generalize. In: ECCV, pp. 103–120 (2020)
5. Chen, G., et al.: Towards understanding and mitigating audio adversarial examples for speaker recognition. TDSC (2022)
6. Chollet, F.: Xception: deep learning with depthwise separable convolutions. In: CVPR, pp. 1251–1258 (2017)
7. Deepfakes: faceswap. In: GitHub (2017). Accessed 14 Jun 2023. https://github.com/deepfakes/faceswap
8. Dong, S., Wang, J., Ji, R., Liang, J., Fan, H., Ge, Z.: Implicit identity leakage: the stumbling block to improving deepfake detection generalization. In: CVPR, pp. 3994–4004 (2023)
9. Frank, J., Eisenhofer, T., Schönherr, L., Fischer, A., Kolossa, D., Holz, T.: Leveraging frequency analysis for deep fake image recognition. In: ICML, pp. 3247–3258 (2020)
10. Gandhi, A., Jain, S.: Adversarial perturbations fool deepfake detectors. In: IJCNN, pp. 1–8 (2020)
11. Gao, G., Huang, H., Fu, C., Li, Z., He, R.: Information bottleneck disentanglement for identity swapping. In: CVPR, pp. 3404–3413 (2021)
12. Gao, Y., et al.: High-fidelity and arbitrary face editing. In: CVPR, pp. 16115–16124 (2021)
13. Goodfellow, I.J., Shlens, J., Szegedy, C.: Explaining and harnessing adversarial examples. In: ICLR (2015)
14. Guan, J., et al.: Delving into sequential patches for deepfake detection. arXiv preprint arXiv:2207.02803 (2022)
15. Haliassos, A., Mira, R., Petridis, S., Pantic, M.: Leveraging real talking faces via self-supervision for robust forgery detection. In: CVPR, pp. 14950–14962 (2022)
16. Haliassos, A., Vougioukas, K., Petridis, S., Pantic, M.: Lips don't lie: a generalisable and robust approach to face forgery detection. In: CVPR, pp. 5039–5049 (2021)
17. Hou, Y., Guo, Q., Huang, Y., Xie, X., Ma, L., Zhao, J.: Evading deepfake detectors via adversarial statistical consistency. In: CVPR, pp. 12271–12280 (2023)
18. Hussain, S., Neekhara, P., Jere, M., Koushanfar, F., McAuley, J.: Adversarial deepfakes: evaluating vulnerability of deepfake detectors to adversarial examples. In: WACV, pp. 3348–3357 (2021)
19. Jiang, L., Li, R., Wu, W., Qian, C., Loy, C.C.: DeeperForensics-1.0: a large-scale dataset for real-world face forgery detection. In: CVPR, pp. 2889–2898 (2020)
20. Jiang, Z., Chen, T., Chen, T., Wang, Z.: Robust pre-training by adversarial contrastive learning. In: NIPS, pp. 16199–16210 (2020)
21. Karras, T., Laine, S., Aila, T.: A style-based generator architecture for generative adversarial networks. In: CVPR, pp. 4401–4410 (2019)
22. Kireev, K., Andriushchenko, M., Flammarion, N.: On the effectiveness of adversarial training against common corruptions. In: UAI, pp. 1012–1021 (2022)

23. Li, L., Bao, J., Yang, H., Chen, D., Wen, F.: Advancing high fidelity identity swapping for forgery detection. In: CVPR, pp. 5074–5083 (2020)
24. Li, Z., et al.: Sibling-attack: rethinking transferable adversarial attacks against face recognition. In: CVPR, pp. 24626–24637 (2023)
25. Liang, K., Xiao, B.: StyLess: boosting the transferability of adversarial examples. In: CVPR, pp. 8163–8172 (2023)
26. Liu, B., Liu, B., Ding, M., Zhu, T., Yu, X.: TI2Net: temporal identity inconsistency network for deepfake detection. In: WACV, pp. 4691–4700 (2023)
27. Lo, S.Y., Patel, V.M.: Defending against multiple and unforeseen adversarial videos. In: TIP, pp. 962–973 (2021)
28. Madry, A., Makelov, A., Schmidt, L., Tsipras, D., Vladu, A.: Towards deep learning models resistant to adversarial attacks. In: ICLR (2018)
29. Menon, A.K., Jayasumana, S., Rawat, A.S., Jain, H., Veit, A., Kumar, S.: Long-tail learning via logit adjustment. In: ICLR (2021)
30. Mumcu, F., Doshi, K., Yilmaz, Y.: Adversarial machine learning attacks against video anomaly detection systems. In: CVPR, pp. 206–213 (2022)
31. Neekhara, P.: Adversarial deepfake. In: GitHub (2019). Accessed 14 Jun 2023. https://github.com/paarthneekhara/AdversarialDeepFakes
32. Neekhara, P., Dolhansky, B., Bitton, J., Ferrer, C.C.: Adversarial threats to deepfake detection: a practical perspective. In: CVPR, pp. 923–932 (2021)
33. Neekhara, P., Hussain, S., Pandey, P., Dubnov, S., McAuley, J., Koushanfar, F.: Universal adversarial perturbations for speech recognition systems. arXiv preprint arXiv:1905.03828 (2019)
34. Qin, Y., Carlini, N., Cottrell, G., Goodfellow, I., Raffel, C.: Imperceptible, robust, and targeted adversarial examples for automatic speech recognition. In: ICML, pp. 5231–5240 (2019)
35. Rossler, A., Cozzolino, D., Verdoliva, L., Riess, C., Thies, J., Nießner, M.: Face-Forensics++: learning to detect manipulated facial images. In: CVPR, pp. 1–11 (2019)
36. Shahzad, S.A., Hashmi, A., Khan, S., Peng, Y.T., Tsao, Y., Wang, H.M.: Lip sync matters: a novel multimodal forgery detector. In: APSIPA, pp. 1885–1892 (2022)
37. Songja, R., Promboot, I., Haetanurak, B., Kerdvibulvech, C.: Deepfake AI images: should deepfakes be banned in Thailand? AI and Ethics, pp. 1–13 (2023)
38. Spivak, R.: deepfakes: the newest way to commit one of the oldest crimes. HeinOnline, p. 339 (2018)
39. Tran, D., Wang, H., Torresani, L., Feiszli, M.: Video classification with channel-separated convolutional networks. In: ICCV, pp. 5552–5561 (2019)
40. Tran, D., Wang, H., Torresani, L., Ray, J., LeCun, Y., Paluri, M.: A closer look at spatiotemporal convolutions for action recognition. In: CVPR, pp. 6450–6459 (2018)
41. Wang, H., et al.: Understanding the robustness of skeleton-based action recognition under adversarial attack. In: CVPR, pp. 14656–14665 (2021)
42. Yadlin-Segal, A., Oppenheim, Y.: Whose dystopia is it anyway? deepfakes and social media regulation. In: Convergence, pp. 36–51 (2021)
43. Yang, C., Ding, L., Chen, Y., Li, H.: Defending against GAN-based deepfake attacks via transformation-aware adversarial faces. In: IJCNN, pp. 1–8 (2021)
44. Zhang, H., Yu, Y., Jiao, J., Xing, E., El Ghaoui, L., Jordan, M.: Theoretically principled trade-off between robustness and accuracy. In: ICML, pp. 7472–7482 (2019)
45. Zheng, Y., Bao, J., Chen, D., Zeng, M., Wen, F.: Exploring temporal coherence for more general video face forgery detection. In: ICCV, pp. 15044–15054 (2021)

# A Multidimensional Taxonomy Model for Music Tangible User Interfaces

Adriano Baratè[ID] and Luca Andrea Ludovico[✉][ID]

LIM – University of Milan, via G. Celoria 18, Milan, Italy
{adriano.barate,luca.ludovico}@unimi.it

**Abstract.** This paper focuses on music tangible user interfaces, namely human-computer interfaces in which a user interacts with digital information through the physical environment with music intention. Music tangible user interfaces can be employed in many scenarios, including music education for beginners, gamification approaches, musical expression for impaired users, and physical and cognitive rehabilitation. After analyzing the state of the art, we will propose a model of multidimensional taxonomy that presents multiple axes along which music tangible user interfaces can be placed. Such a model is expected to help educators, therapists, and interested users choose the best-fitting solution for their specific needs.

**Keywords:** Music · Multidimensional modeling · Tangible User Interfaces · Education · Rehabilitation

## 1 Introduction

Human-computer interaction (HCI) is a research field dealing with the design and use of computer technology and focusing on the interfaces between users and computers. In this context, tangible user interfaces (TUIs) can be employed as a replacement for graphical user interfaces (GUIs), more common in computing systems. With respect to GUIs, TUIs are based on real physical objects rather than graphical representations on a screen. The key idea of TUIs is to give digital information a physical form and let it serve as a representation and control for digital information. A TUI allows users to manipulate digital information with their hands and perceive it with their senses.

One of the pioneers in tangible user interfaces is Hiroshi Ishii, a professor who heads the *Tangible Media Group* at the MIT Media Lab. His particular vision for TUIs, called *Tangible Bits*, is to give physical form to digital information, making bits directly manipulable and perceptible [11]. According to Kim and Maher [13], there are five basic properties that characterize TUIs: 1) space-multiplex both input and output; 2) concurrent access and manipulation of interface components; 3) strong specific devices; 4) spatially-aware computational devices; 5) spatial reconfigurability of devices.

© The Author(s), under exclusive license to Springer Nature Switzerland AG 2024
S. Rudinac et al. (Eds.): MMM 2024, LNCS 14556, pp. 517–531, 2024.
https://doi.org/10.1007/978-3-031-53311-2_38

A clear advantage of TUIs is the user experience, which implies a physical interaction between the user and the interface. Consequent positive aspects are usability and accessibility since the user intuitively knows how to use the interface after understanding the function of the physical object. For this reason, TUIs are often used as enabling technologies in rehabilitation for elderly people and physically or cognitively impaired users.

Currently, there are different research areas and applications related to TUIs. For instance, in tangible tabletop interaction, physical objects are moved upon a multi-touch surface; moreover, physical objects can be used as ambient displays or integrated inside embodied user interfaces; as a final example, tangible augmented reality implies that virtual objects are "attached" to physically manipulated objects.

Some important concepts have to be defined before starting our discussion. A *tangible* is a physical part of the system with which the user can interact. Examples may include wooden cubes, 3D-fabricated objects, building bricks, etc. A *physical icon*, or simply *phicon*, is the tangible computing equivalent of an icon in a traditional GUI. It holds a reference to some digital object and thereby conveys meaning. Finally, a *fiducial marker*, or simply *fiducial*, is an object placed in the field of view of an imaging system used as a point of reference or a measure.

In this paper, we will focus on a specific category of TUIs, namely those oriented to musical applications, known as *music TUIs*. The final goal is to provide a multidimensional taxonomy to explore the applicability of music TUIs in different scenarios and guide users (e.g., educators, therapists, etc.) in the best-fitting choice.

## 2    Music Tangible User Interfaces

As mentioned in Sect. 1, TUIs aim to give a physical form to digital information and computation, thus facilitating direct manipulation of bits. The advantages of this approach are multiple, but, in our opinion, they have been effectively summarized by the following sentence: empowering collaboration, learning, and decision-making through digital technology while taking advantage of our human ability to grasp and manipulate physical objects and materials [11].

In the music field, physical interaction is very common in a performing context. Suffice it to say that musicians are used to interacting with their instruments through different parts of the body. This is particularly true for "traditional" instruments, such as the piano, violin, or trumpet; however, tangible interfaces are also common for controlling virtual, digital, and augmented instruments. Conversely, other music-related activities, such as music theory or composition, usually imply reading, understanding, and writing, but they are rarely based on or reinforced by tactile aspects.

In the digital domain, music TUIs apply the concept of tangible interfaces to the music field. TUIs have been used considerably and profitably in musical performance, education, and music therapy. A tangible interface, by implying something "real" and "concrete", offers a physical way to interact with music and sound parameters. The main advantages include: offering alternative interfaces to make music; lowering the barriers towards the comprehension of music parameters (e.g., pitch, rhythm, and timbre); providing an intuitive way to control music parameters in a creative or performative context; improving peer-to-peer interaction in a highly collaborative music-generation environment; introducing an engaging approach for the achievement of music-related education goals (theory, instrumental practice, original composition, etc.); employing music as a means to increase users' motivation towards activities with extra-musical purposes, such as rehabilitation.

Music TUIs also foster so-called soft skills, such as social (teamwork, communication, positive attitude, etc.), thinking (creativity, problem-solving, decision making, etc.), and negotiating (coping with time, stress, emotions) skills [1].

Clearly, the specific technology chosen to implement a given music TUI plays a fundamental role. For example, the extent of peer cooperation could be limited by the sensing capabilities of the system (e.g., the number of sensitive points on a touch screen, the number of blocks that a webcam can simultaneously detect, etc.). As another example, a simplified interface could be a perfect fit for beginners who need a few parameters to be controlled in an intuitive way, but it could not meet the requirements of a skilled musician who is visually impaired.

Theoretical approaches and available products are continuously changing, so the list provided below should be constantly updated. For this reason, our discussion does not claim to be exhaustive or comprehensive. Nevertheless, it is worth remarking that such an effort aims to be functional to the design and validation of the taxonomy (see Sect. 3) by providing a number of meaningful and heterogeneous examples.

In our presentation of the state of the art, first, we will mention some publicly available and/or commercial products, and then we will review relevant scientific literature. Please note that this subdivision is not always clear-cut.

Concerning technology-enhanced TUIs for music, available devices are very heterogeneous. In fact, music TUIs can play a number of different roles: synthesizers, sequencers, remote controllers for music and sound parameters, interfaces for musical expression, music-related games, and so on.

Probably, the most famous example of commercially available music TUI is the *Reactable*. This device is an electronic musical instrument with a tabletop tangible user interface. The system is composed of a round translucent table, typically used in a darkened environment, that acts as a backlit display, as shown in Fig. 1. By placing blocks called *tangibles* on the table, and interfacing with the visual display via the tangibles or fingertips, a virtual synthesizer is operated, thus creating music or sound effects. Even if born in an academic context [12], developed in 2003 by a research team at the Pompeu Fabra University in Barcelona, it has been used by renowned artists such as Björk in their live performances. Currently, there are two versions of the *Reactable*: *Reactable Live!* and the *Reactable experience*. The former is a smaller, more portable version designed for professional musicians. The latter is more similar to the original *Reactable* and suited for installations in public spaces.

**Fig. 1.** The *Reactable*.          **Fig. 2.** The *Kibo*'s body and tangibles.

Another example of a commercial product based on a tangible interface is *Kibo* by Kodaly S.r.l. [3]. The *Kibo* is a wooden board that presents eight distinct and easy-to-recognize tangibles (see Fig. 2). These geometric shapes can be inserted into and removed from the corresponding slots, thus triggering events encoded in the form of MIDI messages. The device is also sensitive to pressure variations on single tangibles and individual dynamic responses can be communicated via MIDI messages. As an additional controller, there is a knob that can be rotated clockwise and counterclockwise. Finally, the device is equipped with a gyroscope. The *Kibo* ecosystem, made of one or more *Kibo* units and an app that receives MIDI messages and synthesizes sound, supports up to 7 devices simultaneously, without perceivable latency. This aspect is particularly interesting for collaborative experiences [5] and in rehabilitation contexts [4]. Being a fully compatible controller, the *Kibo* can also be integrated into any MIDI setup without the intervention of the app as a mediator.

*BeatBearing* is not a commercial product, but rather a do-it-yourself (DIY) project. This tangible rhythm sequencer is made of a computer interface overlaid with the grid pattern of metal washers and ball bearings. The system is controlled by an Arduino microcontroller.

A special category of music TUIs is the one relying on the use of unmarked building bricks to represent music concepts. The position, color, orientation, and shape of bricks are examples of available physical parameters that can be mapped onto musical parameters according to different strategies. Due to their easy-to-use and familiar interface, these music TUIs are often used in music-oriented educational activities, especially for beginners and/or preschool. For example, blocks can be used to create melodic and rhythmical patterns, where pitches depend on the vertical position over the board, sound attacks on the horizontal position, and note lengths on the bricks' widths. A purely rhythmical application was explored in the scientific literature [17]. More articulated educational activities are supported by products such as *LEGOTECHNO* and *Music Blocks* experiences where bricks are tracked using a camera mounted underneath a transparent baseplate or multiple cameras above it [16]. Finally, it is worth citing *LEGO Music*, a multitouch tabletop application that uses LEGO bricks to illustrate composition principles and musical operations, such as transposition, inversion, and retrograde [20]. Most experiences based on building blocks have the main goal of teaching music parameters through a playful approach. In this sense, an original approach is discussed in [6] and [15], where the ultimate objective is to convey computational thinking skills due to the combined use of music and bricks.

Music TUIs can also be implemented in the form of smartphone apps, thus taking advantage of the sensors intrinsically provided by this kind of device. For the sake of brevity, we will not consider the huge number of apps implementing digital musical instruments where the tangible interaction is limited to the smartphone's touchscreen. For instance, *PhonHarp* is a hybrid digital/physical musical instrument for mobile phones that exploits the vocal tract [25]. The tangible characterization of *PhonHarp* embraces both the tactile interactions of the musicians with the screen and the involvement of their phonatory apparatus in the actual shaping of sound. Similarly, *Pocket Talkbox* is an Android and iPhone app that presents an 8-element grid for musical notes and implements a talk box effect by bringing the smartphone loudspeaker near the mouth.

Concerning the scientific literature, the most relevant examples should be described in papers presenting the strings "music(al) tangible user interface(s)" or "music TUI(s)" in their title. To this end, we first queried Google Scholar, a well-known search engine for scientific publications. This search returned the following results. Costanza *et al.* presented *Audio d-touch*, a system that uses a consumer-grade web camera and customizable block objects to provide an interactive tangible interface for a variety of time-based musical tasks (e.g., sequencing, drum editing, and collaborative composition) [8].

Graham and Hull designed a tangible user interface called *iCandy* that tries to restore the sensation of physical albums for the electronic music stored in iTunes, also providing a method for easy access to recorded media [9]. A very recent proposal that somehow recalls this concept is *Discover*, a user interface for Spotify aiming to couple digital-music streaming and the tangible experience of putting a record on.

Hu *et al.* described a tangible user interface that uses taps on a hard surface as the mode of interaction between the user and the information device [10]. The surface is constructed by a matrix of small acoustic sensing modules that work together to track both the location of taps on the surface and the rhythm of tap sequences. A set of LEDs provides visual feedback and lighting effects. Even if the system can be seen as a generic input/output tool for the interface with digital devices, the paper presents a prototype of a digital audio player, called *Music Wall*, as an embodiment of the proposal.

Waranusast *et al.* introduced *muSurface*, an interactive surface with the ability to interact with tangible musical symbols. Thanks to this prototype, children can explore musical symbols by placing phicons on the virtual musical staffs displayed on the playing surface and listening to the melody produced by the system [28].

Villafuerte *et al.* authored an original contribution to the acquisition of social abilities through music TUIs by children with autism spectrum condition [27]. Even if their contribution assessed the potential of the already mentioned *reactable*, the research focus is different from other papers that simply describe its functionalities. For this reason, we choose to mention such a paper in the current section.

We decided to further extend the field of investigation by using the keywords "tangible" or"TUI" in conjunction with "music" or "audio". This research returned a higher number of scientific works dealing with specific music TUIs.

In 2001, Paradiso *et al.* reviewed some initiatives based on magnetic tags, including *Musical Trinkets*, an installation based on tagged objects publicly exhibited first at SIGGRAPH 2000 and, several months later, at SMAU in Milan [21].

In 2003, Newton-Dunn *et al.* described a way to control a dynamic polyrhythmic sequencer using physical artifacts [19]. Specifically, *Block Jam* is a music TUI that controls a dynamic polyrhythmic sequencer using 26 physical artifacts called blocks as the input. The functional and topological statuses of the blocks are tightly coupled to an ad hoc sequencer, which interprets the user's arrangement of the blocks as meaningful musical phrases and structures. Many *Block Jam* systems can be linked to create a network that further extends the principle of collaboration. Both novice and musically trained users can experience music-making thanks to tangible and visual language. The idea of a tangible sequencer addressing the preparation and improvisation of electronic music is also the foundation of a platform called *mixiTUI*. [22]. Modular sound synthesis is addressed by the *Spyractable* described in [24]. This platform reconfigures the functionality of the *Reactable* and redesigns most features, adjusting them to the needs of a synthesizer. The *MusicCube*, described in [2] and dating back to 2005, is a wireless cube-like object that lets users physically interact with music collections using gestures to shuffle music and a rotary dial with a button for song navigation and volume control. Speech and non-speech feedback communicate the current mode and song title. Similarly, in 2008 *AudioCubes*, a distributed cube interface based on the interaction range for sound design [26], was released.

*Algo.Rhythm* is a tangible toolkit whose main goal is to transmit computational concepts through tangible and audio feedback, more specifically provided by drum beats [23]. Various drum bots can be arranged and physically connected. Their functions include recording beat patterns from the external world or other upstream bots, replaying patterns in selectable ways, and passing down the rhythm to neighbors in the 3D space. The educational goal is to teach users computational concepts, such as sequential execution, iteration, or forking, through music. The prototype, made of plywood shaped with a laser cutter, was tested at the first Pittsburgh Mini Maker Faire in 2012.

A more recent framework documented in the literature is *Note Code*, a music programming puzzle game designed as a tangible device coupled with a graphical user interface [14]. The actions of tapping patterns and placing boxes in proximity allow the user to program these entities to store sets of notes, play them back, and activate different subcomponents or neighboring boxes. The GUI adds a dimension of viewing the created programs and interacting with a set of puzzles that help discover the various computational concepts in the pursuit of creating target tunes and optimizing the program made. Both *Algo.Rhythm* and *Note Code* fall into the category of music TUIs mainly intended to develop computational thinking through music and physical interaction.

## 3  A Multidimensional Taxonomy for TUIs

The taxonomy we propose to classify TUIs is hierarchically articulated in three levels called *domains*, *dimensions*, and *subdimensions* respectively. For the sake of clarity, we will typographically denote them through different conventions: SMALL CAPS + BOLD for domains, SMALL CAPS ONLY for dimensions, and *italics* for subdimensions.

It is worth remarking that our proposal is a **model** of taxonomy rather than a taxonomy itself and, as such, it does not claim to be exhaustive or definitive. Its content can be revised depending on the user's needs (e.g., extended or reduced as it regards the number of hierarchical levels, modified in the node values at each level, etc.). Consequently, our main purpose is to provide and demonstrate a methodological approach to the classification of TUIs.

This multilevel parent-child structure recalls that of the hierarchical trees, so far. Each TUI can be classified according to the taxonomy by being the leaf node for one or more subdimensions. In this way, the structure is now more the one of a tree, but rather a directed acyclic graph.

0.  METADATA — This domain addresses a set of data describing music TUIs at a high level of abstraction through names, numbers, dates, etc. about the products. The dimensions belonging to this domain should be self-explanatory. With respect to other domains and dimensions, in this case the subdimensions cannot be predetermined on the base of theoretical speculations; only the analysis of the products to be classified can return node values. For this reason, the lists of subdimensions provided below give mere examples. In particular, we will take into consideration the *Kibo* and the *Reactable*, marked through [K] and [R] only for the sake of clarity.

0.0.  DEVELOPER NAMES – subdimensions are textual data such as: 0.0.0. *Marcos Alonso* [R]; 0.0.1. *Mattia Davide Amico* [K]; 0.0.2. *Günther Geiger* [R]; 0.0.3. *Sergi Jordà* [R]; 0.0.4. *Martin Kaltenbrunner* [R]; ...

0.1.  MANUFACTURING COMPANY – subdimensions are textual data such as: 0.1.0. *Kodaly S.r.l.* [K]; 0.1.1. *Reactable Systems* [R]; ...

0.2.  INVOLVED INSTITUTIONS – subdimensions are textual data such as: 0.2.0. *Music Technology Group, Universitat Pompeu Fabra, Barcelona (Spain)* [R]; 0.2.1. *Polytechnic University of Milan, Milan (Italy)* [K]; 0.2.2. *University of Milan, Milan (Italy)* [K]; ...

0.3.  RELEASE YEAR – subdimensions are timestamps that should present continuity in a given range such as: ...; *2007* [R]; *2008*; *2009*; *2010* [R, mobile version]; ...; *2019* [K]; ... (please note that the granularity of date ranges can be reconfigured, if necessary);

By way of example, other dimensions can involve sales data (the PRICE RANGE, the NUMBER OF PRODUCTS distributed on the market, the NUMBER

OF ESTIMATED OR EXPECTED ACTIVE USERS, etc.), focus on new versions and editions, and so on.

1. **TECHNOLOGY** — This domain addresses the technical and technological aspects of the TUI itself and the requirements to make it work.

  1.0. INVOLVED OBJECTS – This dimension focuses on the characteristics of the physical objects with which the user has to interact. Subdimensions fall into these categories:

  1.0.0. *Common objects* – This is the case of objects used in everyday life in a technologically augmented environment. An example is the tracking of kitchen tools or colored balls via cameras;

  1.0.1. *Expressly-designed objects* – This subdimension embraces ad hoc physical objects designed for a specific TUI experience and provided with the system. This kind of object is often called tangibles or fiducials. Examples include 3D-printed shapes and connectable units;

  1.0.2. *Reinterpretations of common objects* – This scenario stands in the middle between the previous ones, involving objects whose usage mode changes in the context of a TUI experience. A typical example is provided by standard building blocks for a construction game turned into tangibles;

  1.0.3. *User-buildable objects* – This subdimension addresses those objects that are not physically provided with the TUI system, but can be produced (and often customized) by users themselves, e.g. drawn on paper or fabricated with a 3D printer;

  1.0.4. *Dematerialized objects* – This scenario involves 3D objects that are not tangible, but are represented in a virtual physical space (e.g., shapes drawn on a touch screen or projected as holograms that can be manipulated via special gloves).

  1.1. SENSING APPROACHES – This dimension addresses the technological approaches employed to detect the interactions of users and tangible interfaces. Subdimensions include:

  1.1.0. *Stand-alone systems* – In this scenario, the TUI embeds all the sensors needed to detect user interaction;

  1.1.1. *Computer-based systems* – In this case, standard PC equipment is required to acquire data, process them, and eventually pass them to a sound-generation system. In other words, the TUI system includes a computer.

  1.1.2. *Wearable devices* – Sensors to detect user movements and interactions with physical objects are directly applied to the user's body.

  1.2. SENSING DEVICES – This dimension focuses on the devices used to detect movements and interaction. Subdimensions could be represented by specific devices, such as a webcam or a linear membrane potentiometer, but, in our opinion, it is more effective to stay at a higher level of abstraction and focus on sensor categories. Clearly, a given TUI could simultaneously depend on more than one sensor category. Examples: 1.2.0. *Position sensors*; 1.2.1. *Pressure sensors*; 1.2.2. *Temperature sensors*; 1.2.3. *Force sensors*; 1.2.4. *Vibration sensors*; 1.2.5. *Piezo sensors*; 1.2.6. *Fluid property sensors*; 1.2.7. *Humidity sensors*; 1.2.8. *Strain gauges*; 1.2.9. *Photo optic/image sensors*; 1.2.10. *Flow and level switches*.

2. MUSICAL SCOPE — This domain is intended to catch the intentions and goals of TUIs from a musical point of view. In general, a single solution is expected to adhere to multiple subdimensions even along the same dimension.

   2.0. GOALS – This dimension aims to identify the main objectives of the TUI under exam. In fact, a TUI does not necessarily have a primary music goal; it could employ music as a facilitating means or a reinforcement aspect. Subdimensions fall into these categories: 2.0.0. *Music creation/creativity*; 2.0.1. *Music education*; 2.0.2. *Development of soft skills* (peer-to-peer cooperation, problem-solving, abstraction skills, etc.); 2.0.3. *Cognitive and/or motor rehabilitation*.

   2.1. ARTISTIC PROCESSES — Starting from the idea that a TUI finally produces music, this dimension takes into account the artistic activities defined in [18].

   2.1.0. *Creating* – Activities that range from music composition to improvisation and music programming;

   2.1.1. *Performing* – TUIs that help students in learning instrumental techniques, singing, and conducting;

   2.1.2. *Responding* – Activities such as listening and describing, analyzing, and evaluating [7].

   2.2. FUNCTIONS — This dimension takes into account the multiple roles that a TUI can play in the context of a musical performance. Most products and solutions are expected to assume multiple roles in this dimension. The list of subdimensions includes: 2.2.0. *Controller*; 2.2.1. *Sequencer*; 2.2.2. *Synthesizer*; 2.2.3. *Music parameter editor*.

3. TARGET AUDIENCE — This domain aims to characterize the potential range of users. Once again, a single solution is expected to adhere to multiple subdimensions even along the same dimension.

   3.0. USER AGE – This dimension groups potential users into age categories. We have adopted a well-established and commonly accepted taxonomy, but age ranges can be fine-tuned depending on the user's needs. The groups we propose are: 3.0.0. *Children* (0–14 years); 3.0.1. *Youth* (15–24 years); 3.0.2. *Adults* (25–64 years); 3.0.3. *Seniors* (65 years and over).

   3.1. NUMBER OF SIMULTANEOUS USERS – This dimension takes into account the potential of the TUI in terms of user cooperation. subdimensions are: 3.1.0. *Single user*; 3.1.1. *Pair* (2 users); 3.1.2. *Small group* (3–10 users); 3.1.3. *Big group* (more than 10 users).

   3.2. EDUCATIONAL CONTEXT – This dimension explores the different contexts in which the TUI can be used for educational purposes. These environments include: 3.2.0. *Pre-school*; 3.2.1. *Primary school*; 3.2.2. *Lower secondary school*; 3.2.3. *Higher secondary school*; 3.2.4. *University and conservatories*; 3.2.5. *Lifelong learning*; 3.2.6. *None/not defined*.

The model of taxonomy presented here aims to be a paradigmatic example of the multiple aspects that characterize music TUIs and their use. We reaffirm that the specific definitions of domains, dimensions, and subdimensions could be revised depending on users' needs.

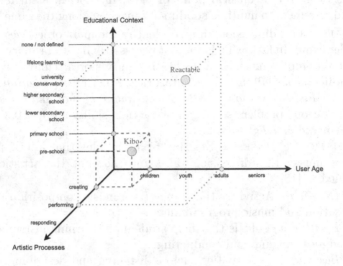

**Fig. 3.** The position of the *Reactable* and the *Kibo* in a 3D space.

## 4   The Taxonomy in Practice

In our vision, the taxonomy can be seen both as a theoretical instrument aiming to highlight the multiple dimensions for describing TUIs and as a practical tool that lets users identify the most suitable approaches described in the scientific literature or products on the market.

The first step includes both a thorough recognition of available TUIs and a careful review of the scientific literature. The solutions retrieved have to be tagged in accordance with the proposed multidimensional taxonomy. In some cases, this operation is quite straightforward, since the products under exam can be clearly and uniquely classified; in other cases, the tags of a TUI can be multiple for a single domain/dimension, or they have to be inferred on the basis of a merely theoretical description.

After this step, we can explore the taxonomy in a hierarchical way. To this end, it would be sufficient to represent information in a text-based format such as XML or JSON. The navigation of the tree starts from the top level (**DOMAIN**), crosses the second one (DIMENSION), goes down the third one (*subdimension*), and finally brings to the TUIs tagged accordingly. This kind of description, which presents the structure of the taxonomy as a tree, is very clear and easy to navigate, but not particularly informative. For example, it would not aggregate products across multiple subdimensions nor allow the selection of TUIs simultaneously belonging to multiple subdimensions of the same dimension, since, in a tree representation, a TUI instance can be the child for only one parent.

If, on the one hand, there are well-established human- and machine-readable textual representations for a tree, on the other hand, it is hard to find a similar level of standardization and readability in the description of a directed acyclic graph. Options include YAML nested lists with anchors and the DOT language from Graphviz. Another possibility is to enter information into a database. These solutions provide the user with the highest flexibility and power.

Examples of queries on the taxonomy are:

- "Investigate a single property" – For example, a selection by *small group* (subdimension) would extract all TUIs addressing a NUMBER OF SIMULTANEOUS USERS (dimension) in the range 3–10 as the TARGET AUDIENCE (domain). In this context, the METADATA domain can also be handy. For instance, a selection by *Laboratory of Music Informatics, University of Milan* (subdimension) as it concerns INVOLVED INSTITUTIONS would return all TUIs contributed by this research group;
- "Focus on many children of the same parent", i.e. siblings – For instance, a query by *pre-school* and *primary school* (subdimensions) would return all the TUIs addressing both EDUCATIONAL CONTEXTS (dimension) as it regards the TARGET AUDIENCE (domain);
- "Extract multiple children that are not siblings" – For example, a query combining *years 2000–2009*, *image sensors*, *dematerialized objects*, and *lower secondary school* would return all TUIs released in the first decade of the century, addressing lower secondary school students, and adopting webcams (or similar technologies) to detect custom-made tangibles.

Finally, it is worth mentioning the possibility of providing graphical representations of the axes of the taxonomy and locating the position of TUIs in multidimensional space. For the sake of clarity, Fig. 3 narrows the potential of the graphical comparison to 3 axes, which are, specifically, 3 dimensions under the SCOPE domain.

The various ways to explore the taxonomy (e.g., database queries, XPath expressions, graphical representations, etc.) are expected to let users easily find the applications and approaches that best suit their needs.

To show the potential of the taxonomy, we will discuss two cases – namely the *Reactable* and the *Kibo* – adopting a bottom-up approach. We will intentionally ignore the METADATA domain since specific examples of use have already been provided in Sect. 3.

The first TUI that we are going to analyze is the *Reactable*, a commercially-available TUI described in Sect. 2. Concerning the domain of TECHNOLOGY and, specifically, the dimension of INVOLVED OBJECTS, the *Reactable* employs *expressly-designed objects*. There are similar initiatives relying on *user-buildable objects* and using *reacTIVision*, an open source, cross-platform computer vision framework for the fast and robust tracking of fiducial markers attached to physical objects. Concerning the dimension of SENSING APPROACHES, the original *Reactable* is a *stand-alone system*. As regards the dimension of SENSING DEVICES, it uses *photo optic/image sensors*. Now, let us focus on the **MUSICAL SCOPE** domain. The GOALS are mainly *music creation* and *creativity*; nevertheless, the *Reactable* can be profitably adopted also in *music education* and *rehabilitation* contexts. The ARTISTIC PROCESSES potentially involved are both *creating* and *performing*. The FUNCTIONS assumed by the system include those of a *controller*, a *sequencer*, and a *synthesizer*. Finally, let us analyze the **TARGET AUDIENCE** domain. RegardingUSER AGE, the *Reactable* was probably conceived for *adults* and, to a lesser extent, the *youth*; *seniors* and *children* can take part in the experience as well. The NUMBER OF SIMULTANEOUS USERS includes *single user*, *pair*, and *small group*. Finally, concerning EDUCATIONAL CONTEXT, the most likely choice is *none*, since the *Reactable* mainly has an artistic intention.

The second TUI under exam is the *Kibo*, another commercially-available product described in Sect. 2 whose characteristics are significantly different from those of the *Reactable*. Thus, it will be interesting to notice if and how the taxonomy is able to remark these differences. Concerning the domain of **TECHNOLOGY** and, specifically, the dimension of INVOLVED OBJECTS, it employs *expressly-designed objects*, too. Concerning the dimension of SENSING APPROACHES, it is closer to a *computer-based system*, since the wooden board is basically a mere musical controller. As regards the dimension of SENSING DEVICES, it uses both *position* and *pressure* sensors. Now, let us focus on the **MUSICAL SCOPE** domain. The GOALS embrace *music creation/creativity*, *music education*, and *cognitive and/or motor rehabilitation*. The main ARTISTIC PROCESS involved is *creating*. Even if the whole system embeds a synthesizer, implemented by the accompanying mobile app, the FUNCTION assumed by the board is that of a *controller*. Finally, let us analyze the **TARGET AUDIENCE** domain. Regarding USER AGE, the *Kibo* was mainly designed for *children* and *youth*. The NUMBER OF SIMULTANEOUS USERS is limited to a *single user* or, at most, a *pair*. The fact that multiple devices can be connected together can clearly extend such a number but is not an intrinsic characteristic of the single device. Finally, concerning the EDUCATIONAL CONTEXT, the primary choices are *pre-school* and *primary school*.

## 5   Conclusions and Future Work

In this paper, we addressed music TUIs as tools that can be profitably used in many activities, including musical expression, music education, music-supported rehabilitation, and gamification. As demonstrated by the analysis of the state of the art, the types of devices that fall within the definition of TUI are extremely heterogeneous in their purposes, technical characteristics, expected users, etc. For this reason, it is necessary to design a theoretical framework and implement tools capable of marking different levels of analysis and interpretation.

In our opinion, a solution is offered by a multidimensional taxonomy made of three hierarchical levels, called domains, dimensions, and subdimensions. Since our proposal does not claim to be comprehensive, what we proposed is a methodological approach rather than a specific taxonomy.

With regard to future work, one goal will be to carry out the study of both unconsidered and brand-new solutions to determine further levels of analysis to be embedded into the model. Another evolution will be the design and implementation of a web platform to gather TUI data, organize them according to our multidimensional model, and equip the user with an intuitive query tool to navigate the taxonomy.

# References

1. Alex, K.: Soft skills. S. Chand Publishing (2009)
2. Alonso, M.B., Keyson, D.V.: MusicCube: making digital music tangible. In: CHI '05 Extended Abstracts on Human Factors in Computing Systems, pp. 1176–1179. CHI EA '05, ACM, New York, NY, USA (2005). https://doi.org/10.1145/1056808. 1056870
3. Amico, M.D., Ludovico, L.A.: Kibo: a MIDI controller with a tangible user interface for music education. In: International Conference on Computer Supported Education, pp. 613–619. SCITEPRESS (2020)
4. Baratè, A., Korsten, H., Ludovico, L.A.: A music tangible user interface for the cognitive and motor rehabilitation of elderly people. In: Proceedings of the 6th International Conference on Computer-Human Interaction Research and Applications (CHIRA 2022), pp. 121–128. SCITEPRESS - Science and Technology Publications, Lda (2022)
5. Baratè, A., Korsten, H., Ludovico, L.A., Oriolo, E.: Music tangible user interfaces and vulnerable users: state of the art and experimentation. In: 5th International Conference on CHIRA 2021 and 6th International Conference on CHIRA 2022, Communications in Computer and Information Science, vol. 1882, pp. 1–25. Springer Science and Business Media Deutschland GmbH (2023). https://doi.org/ 10.1007/978-3-031-41962-1_1
6. Baratè, A., Ludovico, L.A., Malchiodi, D.: Fostering computational thinking in primary school through a LEGO®-based music notation. In: Procedia Computer Science. Knowledge-Based and Intelligent Information & Engineering Systems: Proceedings of the 21st International Conference, KES 2017, 6–8 September 2017, Marseille, France, vol. 112, pp. 1334–1344 (2017). https://doi.org/10.1016/j.procs. 2017.08.018
7. Bauer, W., Harris, J., Hofer, M.: Music learning activity types (2012). http:// activitytypes.wm.edu/MusicLearningATs-June2012.pdf. Accessed 30 Sept 2022
8. Costanza, E., Shelley, S.B., Robinson, J.: Introducing audio D-touch: a tangible user interface for music composition and performance. In: Proceedings of the 2003 International Conference on Digital Audio EffectsDAFx03 (01/01/03) (2003), https://eprints.soton.ac.uk/270957/
9. Graham, J., Hull, J.J.: iCandy: a tangible user interface for iTunes. In: CHI '08 Extended Abstracts on Human Factors in Computing Systems, pp. 2343–2348. CHI EA '08, ACM, New York, NY, USA (2008). https://doi.org/10.1145/1358628. 1358681
10. Hu, C., Tung, K., Lau, L.: Music Wall: a tangible user interface using tapping as an interactive technique. In: Lee, S., Choo, H., Ha, S., Shin, I.C. (eds.) APCHI 2008. LNCS, vol. 5068, pp. 284–291. Springer, Heidelberg (2008). https://doi.org/ 10.1007/978-3-540-70585-7_32

11. Ishii, H.: The tangible user interface and its evolution. Commun. ACM **51**(6), 32–36 (2008). https://doi.org/10.1145/1349026.1349034
12. Jordà, S., Geiger, G., Alonso, M., Kaltenbrunner, M.: The reacTable: exploring the synergy between live music performance and tabletop tangible interfaces. In: Proceedings of 1st International Conference on Tangible and Embedded Interaction, pp. 139–146 (2007)
13. Kim, M.J., Maher, M.L.: The impact of tangible user interfaces on designers' spatial cognition. Hum.-Comput. Interact. **23**(2), 101–137 (2008)
14. Kumar, V., Dargan, T., Dwivedi, U., Vijay, P.: Note Code: a tangible music programming puzzle tool. In: Proceedings of the 9th International Conference on Tangible, Embedded, and Embodied Interaction, pp. 625–629 (2015)
15. Ludovico, L.A., Malchiodi, D., Zecca, L.: A multimodal LEGO®-based learning activity mixing musical notation and computer programming. In: MIE 2017: Proceedings of the 1st ACM SIGCHI International Workshop on Multimodal Interaction for Education, pp. 44–48. ACM, New York, NY (2017). https://doi.org/10.1145/3139513.3139519
16. Miotti, B., Bassani, L., Cauteruccio, E., Morandi, M.: MusicBlocks: an innovative tool for learning the foundations of music. In: CSEDU (1), pp. 475–484 (2022)
17. Mullett, S.: Lego beats music manipulatives (2017). https://www.letsplaykidsmusic.com/lego-beats-music-manipulatives/. Accessed 30 Sept 2022
18. National Coalition for Core Arts Standards Archives: what are the national core arts standards? https://www.nationalartsstandards.org/content/national-core-arts-standards. Accessed 10 May 2021
19. Newton-Dunn, H., Nakano, H., Gibson, J.: Block Jam: a tangible interface for interactive music. J. New Music Res. **32**(4), 383–393 (2003)
20. Oestermeier, U., Mock, P., Edelmann, J., Gerjets, P.: LEGO Music: learning composition with bricks. In: Proceedings of the 14th International Conference on Interaction Design and Children, pp. 283–286. IDC '15, ACM, New York, NY, USA (2015). https://doi.org/10.1145/2771839.2771897
21. Paradiso, J.A., Hsiao, K.y., Benbasat, A.: Tangible music interfaces using passive magnetic tags. In: Proceedings of the 2001 Conference on New Interfaces for Musical Expression, pp. 1–4. NIME '01, National University of Singapore, SGP (2001)
22. Pedersen, E.W., Hornbæk, K.: mixiTUI: a tangible sequencer for electronic live performances. In: Proceedings of the 3rd International Conference on Tangible and Embedded Interaction, pp. 223–230 (2009)
23. Peng, H.: Algorithm Rhythm: computational thinking through tangible music device. In: Proceedings of the 6th International Conference on Tangible, Embedded and Embodied Interaction, pp. 401–402. TEI '12, ACM, New York, NY, USA (2012). https://doi.org/10.1145/2148131.2148234
24. Potidis, S., Spyrou, T.: Spyractable: a tangible user interface modular synthesizer. In: Kurosu, M. (ed.) HCI 2014. LNCS, vol. 8511, pp. 600–611. Springer, Cham (2014). https://doi.org/10.1007/978-3-319-07230-2_57
25. Presti, G., Adriano, D., Avanzini, F., Baratè, A., Ludovico, L.A.: PhonHarp: a hybrid digital-physical musical instrument for mobile phones exploiting the vocal tract. In: AM'21: Proceedings of the 16th International Audio Mostly Conference: Sonic experiences in the era of the Internet of Sounds, pp. 276–279. ACM International Conference Proceeding Series, ACM (2021). https://doi.org/10.1145/3478384.3478413

26. Schiettecatte, B., Vanderdonckt, J.: AudioCubes: a distributed cube tangible interface based on interaction range for sound design. In: Proceedings 2nd International Conference on Tangible and Embedded Interaction, pp. 3–10. TEI '08, ACM, New York, NY, USA (2008). https://doi.org/10.1145/1347390.1347394
27. Villafuerte, L., Markova, M., Jorda, S.: Acquisition of social abilities through musical tangible user interface: children with autism spectrum condition and the reactable. In: CHI '12 Extended Abstracts on Human Factors in Computing Systems, pp. 745–760. CHI EA '12, ACM, New York, NY, USA (2012). https://doi.org/10.1145/2212776.2212847
28. Waranusast, R., Bang-ngoen, A., Thipakorn, J.: Interactive tangible user interface for music learning. In: 2013 28th International Conference on Image and Vision Computing New Zealand (IVCNZ 2013), pp. 400–405 (2013). https://doi.org/10.1109/IVCNZ.2013.6727048

# Author Index

© The Editor(s) (if applicable) and The Author(s), under exclusive license
to Springer Nature Switzerland AG 2024
S. Rudinac et al. (Eds.): MMM 2024, LNCS 14556, pp. 533–535, 2024.
https://doi.org/10.1007/978-3-031-53311-2